PRINCIPLES OF
FUEL CELLS

PRINCIPLES OF
FUEL CELLS

Xianguo Li

Taylor & Francis
Taylor & Francis Group
New York London

Published in 2006 by
Taylor & Francis Group
270 Madison Avenue
New York, NY 10016

Published in Great Britain by
Taylor & Francis Group
4 Park Square
Milton Park, Abingdon
Oxon OX14 4RN

International Standard Book Number-10: 1-59169-022-6 (Hardcover)
International Standard Book Number-13: 978-1-59169-022-1 (Hardcover)
Library of Congress Card Number 2005021228

Library of Congress Cataloging-in-Publication Data

Li, Xianguo, 1962-
    Principles of fuel cells / by Xianguo Li.
        p. cm.
    ISBN 1-59169-022-6 (alk. paper)
    1. Fuel cells. I. Title.

TK2931.L5 2005
621.31'2429--dc22   2005021228

Taylor & Francis Group
is the Academic Division of T&F Informa plc.

**Visit the Taylor & Francis Web site at
http://www.taylorandfrancis.com**

Contents

Preface

Consumption of energy (or useful energy, exergy, to be thermodynamically correct) has become a daily necessity in the modern world, dramatically increasing along with improvements to the quality of life, the industrialization of developing nations, and the increase of world population. At present, the majority of energy required worldwide is met by the combustion of fossil fuels. These resources have become an essential and integral part of modern life, increasingly relied upon since the Industrial Revolution. It has long been recognized that this excessive energy consumption has a significant adverse impact on the environment, resulting in increased health risks to all life forms and the threat of global climate change. The diminishing fossil fuel reserve is also intensifying international tensions, transforming national security, and contributing to high inflation. Fuel cell technology has the potential to meet the extraordinary energy needs of our modern civilization and lessen the undesirable effects of energy consumption.

Fuel cells are being developed at an increasingly fast pace with many academic researchers entering the field. The study of fuel cells is growing in popularity among students in mechanical, chemical, and electrical engineering, environmental studies and engineering, as well as materials science and engineering, and fuel cell courses are offered at many institutions. The need for an introductory text on fuel cells has motivated the writing of this book, and my experience in offering such a course at the University of Victoria and University of Waterloo over the past decade has formed the basis of the book.

Although the primary audience is intended to be undergraduate senior-level and first-year graduate students in mechanical engineering, students in chemical and electrical engineering, environmental studies and engineering as well as materials science and engineering may find the text useful and suitable. Many examples and problems are provided to help students understand the materials presented; the book also will be a useful reference to practicing engineers and other professionals.

The text is organized into 10 chapters. The first four chapters are considered essential and fundamental for a first course on fuel cells. The first chapter is devoted to the basics of fuel cells, including how a fuel cell works, the classification (or naming convention) of fuel cells, from fuel cell components to systems, as well as a brief description of fuel cell history. The thermodynamic aspect of fuel cells is provided in chapter 2, covering the basic thermodynamic performance of a fuel cell called the reversible cell potential (difference), as well as the influence of the operating conditions such as the cell temperature, pressure, reactant concentrations and utilization. The chemical to electrical efficiency of the energy conversion in fuel cells is analyzed in detail as well. Chapter 3 provides a brief overview of the electrochemistry occurring in fuel cells, covering such topics as kinetics and the related cell voltage (energy) loss mechanism called activation overpotential (or polarization). Also touched upon are the electric double layers near the electrode surface, the reaction mechanism in fuel cell electrodes, and electrocatalysis. Chapter 4 is concerned with the transport phenomena of mass, momentum, and energy. The transport of charged species (ions) through the electrolyte is also addressed. The two other mechanisms for cell voltage loss, namely, the concentration and ohmic overpotential (or polarization), are introduced based on the mass transfer processes in different media.

Following the coverage of the fundamentals in the first four chapters, chapters 5 through 10 are focused on each of the six major types of fuel cells: the alkaline fuel cells (AFCs), phosphoric acid fuel cells (PAFCs), proton exchange membrane fuel cells (PEMFCs), molten carbonate fuel cells (MCFCs), solid oxide fuel cells (SOFCs), and direct methanol fuel cells (DMFCs). For each type of fuel cell, the following is described: major advantages and disadvantages as well as the major technical barriers; the operational principle of the fuel cell; typical cell components and configurations with the typical design of stacks and systems; materials used and the construction methods; and the impact on the performance for various operating and design parameters. Finally, a brief discussion is outlined for the future direction of the R&D for each type of fuel cell. These six chapters are more related to the practice and the state-of-the-art development of fuel cells. Therefore, it is hoped that practicing engineers can also benefit from the materials presented, while students can gain a glimpse of practical fuel cell technology.

I would like to acknowledge with appreciation all my students, especially my graduate students and research associates, who have in no small way contributed to my own learning and appreciation of the subject and writing of the text. The typing of the text and drawing of the illustrations with various kinds and makes of software over the long hours of late nights, weekends, holidays, and vacations has been a great fun and learning experience. So often success nears when frustration mounts! My wife often laughs over the years that I have labored for so long for this text, while she had already labored a couple of times for my son, Randy, and daughter, Jennifer. So above all, my heartfelt gratitude is due to my wife and my children for their continued patience, understanding, encouragement, and support throughout the writing of this text, and for their sharing of my happiness, my disappointment, my sorrows, and my joys over the years. What a life!

Xianguo Li
Waterloo, Ontario, Canada

INTRODUCTION

1.1 INTRODUCTION

Fuel cells are environmentally friendly devices for energy conversion and power generation, and are one of the most promising candidates as a zero-emission power source. Hence, they are often regarded as one of the advanced energy technologies of the future. In reality, fuel cells are one of the oldest energy conversion devices known to humankind, although their development and deployment for practical applications lag far behind other competitive technologies, mainly heat engines such as steam turbine and the internal combustion engine. Presently, heat engines employing combustion of fossil fuels have become an essential and integral part of modern civilization, being increasingly relied upon since the Industrial Revolution.

This almost exclusive reliance on combustion of fossil fuels has resulted in severe local air pollution due to the pollutant emissions, including SO_x, NO_x, CO, and particulates, which pose a severe threat to the health of millions of people living in many of the world's urban areas. It continues to contribute significantly to the increase in atmospheric carbon dioxide concentrations, thus intensifying the prospect of global warming and threatening the very existence of our civilization and humankind on planet Earth. The specter of climate change is becoming so ominous that it is the substantial reduction of greenhouse gas emissions that is driving research and development into cleaner, more efficient energy technology and alternative energy sources and carriers (primary and secondary fuels). In addition to the health and environmental concerns, a steady depletion of the world's limited fossil fuel reserves calls for new energy technology for energy conversion and power generation, which is more energy efficient than the conventional heat engine with minimal or no pollutant emissions and also compatible with renewable energy sources or carries for sustainable development. Fuel cells have been identified as one of the most promising and potential energy technologies which meet all of these requirements for energy security, economic growth, and environmental sustainability.

Therefore, this chapter discusses the motivation for the study, research, development, and deployment of fuel cells and then describes some basic concepts and ideas of fuel cells by raising a series of questions. Some of the fundamental questions are: What is a fuel cell? How does it work? Why is it regarded as an advanced energy technology of the future? By addressing these questions, we begin to appreciate the physical and electrochemical mechanisms that underlie fuel cell operation and the relevance and significance of fuel cells in our efforts to address industrial and environmental problems and develop an environmentally benign energy system sustainable in the future. We next illustrate a typical fuel cell system for power generation with various components described. Then we provide the classification and overview of fuel cells, present a brief account of historical development, and illustrate fuel cells as an emerging energy technology of the future. Finally, the scope and outline of the book is given along with the objective for the book.

1.2   MOTIVATION FOR FUEL CELLS

As discussed earlier, conventional power generation by using heat engines based on combustion of fossil fuels produces a significant amount of pollutant emissions that increasingly contribute to the degradation of the environment on which humans and other life forms dearly depend. A variety of alternative approaches has been proposed and implemented for power generation, such as hydroelectric power, wind, wave, solar, bioenergy, geothermal energy, and so on. These renewable energy sources can be used with relative ease to generate electricity for utility applications, but for example, they are subject to seasonal and irregular fluctuations in terms of the amount of energy available and are limited in the quantity of harvestable energy. Further, these renewable energy sources and associated power generation technologies remain virtually impossible for direct utilization in transportation applications, which presently account for a significant amount of pollutant emissions.

The intermittent nature of power generation from the renewable sources favors the use of energy storage for power provision. Although batteries, super-capacitors, and flywheels, can be used as energy storage devices, they have not been shown to be an ideal option, especially for transportation applications.[1−5] It has been proposed that the most promising option for environmentally benign operation is to use hydrogen as the energy carrier, produced from the renewable energy sources such as solar energy and hydroelectric power, and to adopt fuel cells as the clean and efficient means of energy conversion and power generation for mobile applications. Figure 1.1 shows the relative amount of carbon dioxide produced by various fuels for one unit of energy consumed. It is seen that hydrogen, as expected, does not produce carbon dioxide at all and that hydrogen is most efficiently used in fuel cells for power generation as opposed to combustion in heat engines. The beauty of the hydrogen option lies in the fact that its production from water produces only oxygen as a byproduct, and its recombination with oxygen from the air forms water once again when power is needed

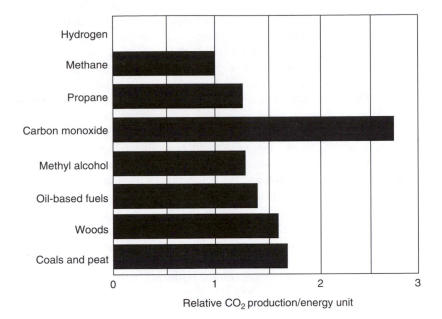

Figure 1.1 Comparison of the relative amount of carbon dioxide production from various fuels for one unit of energy. Note wood is commonly regarded as CO_2-neutral fuel for the entire cycle since CO_2 is absorbed during the wood growth stage.

and produced. Figure 1.1 also shows that methane yields the lowest carbon dioxide production among the hydrocarbon family of fuels due to the highest hydrogen-to-carbon ratio of its molecular structure. It might also be pointed out that methyl alcohol or methanol produces slightly more carbon dioxide emissions than propane in this comparison, with their carbon dioxide emissions above only those from methane. Both methane and methanol can be directly or indirectly used in fuel cells for electricity generation and both can be manufactured from the fossil fuels or biomass resources. Renewable sources are the only possible and plausible path because fossil fuels have been dismissed as a viable option for a sustainable future. Therefore, a sustainable future energy system will include fuel cells, with hydrogen, methane and methanol derived from renewable sources, to produce electricity in areas where and in times when power is needed or electricity is in demand.

Fuel cells can also be employed for power generation from fossil fuels and biomass-derived materials with reduced pollutant emissions due to higher practical efficiencies. The option of fuel cells with fossil fuels can be regarded as an interim method of introducing fuel cells into the market place without significant barriers arising from the lack of fuel distribution infrastructure needed for a future sustainable energy system. Biomass-derived fuels include methane from municipal solid wastes, sewage sludge, forestry residues, landfill sites, and oil-field flare gases and can be used efficiently by fuel cells for power generation, instead of their direct release into the atmosphere. Hundreds of thousands of landfill sites can be found in the world that could be economically exploited and the deployment of small unattended power plants of fuel cells running on these gases could make them highly profitable sources of energy. Biofuels from agricultural and animal waste are another attractive source of energy for fuel cell implementation. Biofuel options can be implemented in an overall carbon dioxide neutral process if appropriate methods are used.

Furthermore, a significant portion of the energy need is in the demand for electricity, which is, at present, almost exclusively produced by large power plants burning fossil fuels and then distributed through networks of high-voltage transmission wires over long distances. Even though these large power plants have optimal energy efficiencies because of their size, electric energy losses occur during long distance transmission, amounting to 7%–8% of electricity generated in Europe and as much as 10% in North America. As such, overall energy efficiency is reduced when accounting is done at the end-user site. In addition, these networks of transmission wires may not function properly all the time, especially when they are needed most, as evidenced by what happened during the ice storm in Quebec in January 1998. Logically, electricity generation should be decentralized and located at or near where electricity is needed. Because of their unique characteristics for electricity generation, fuel cells can be used for cogeneration, on-site, and distributed power generation without the need of complex, long-distance transmission networks. Hydrogen fuel cells thus can make the energy system more decentralized, and contribute to a more secure energy supply system.

On the other hand, the fossil fuel reserve is quite limited and is unevenly distributed throughout the world. The latter also leads to severe geopolitics and regional conflicts that threaten world peace. The limited fossil fuel reserve is being rapidly depleted and will not be able to support world energy demand in a "business as usual" manner for long. It has been estimated[6] that natural gas and petroleum oils may last another half-century, while coals and nuclear energy may be sufficient for only another few hundred years. This limited supply and huge demand situation drives up the cost of fossil fuels rapidly, as it is being encountered in many parts of the world today. The end of the cheap oil era is rapidly approaching.[7] In comparison, renewable energy sources such as solar, wind, geothermal, biomass, and tidal waves may be regarded as lasting forever. Therefore, future energy technology for power generation must be compatible with the renewable energy sources, and fuel cells meet this requirement nicely.

In summary, fuel cells are an energy conversion technology, with operation on hydrogen to produce electric power on demand. Therefore, they are not a primary energy source and they do not compete with electricity generation from the renewable energy sources. Rather they are complementary to each other and fuel cells are compatible with the renewable energy sources and carriers for sustainable development. In due time with adequate development, fuel cells will become an integral part of a future sustainable energy chain.

1.3   FUEL CELL BASICS

This section describes the fundamental concepts, ideas, and operational principles of fuel cells, their pros and cons, and practical applications. It also points out the multidisciplinary knowledge needed for fuel cell study, research and development with an outline of typical fuel cell analysis and simulation. So here comes the very first question:

1.3.1 What Is a Fuel Cell?

A simple, yet general, definition provides sufficient response to this question: A fuel cell is an electrochemical device that converts chemical energy of reactants (both fuel and oxidant) directly into electrical energy. In short, we may define a fuel cell simply as an energy conversion device for power generation, similar to batteries and heat engines.

Comparison with Battery   A battery, for the time being, means primary battery because the secondary battery functions in essentially the same way as the primary battery except that some of its reactants can be regenerated (or rechargeable) by using an external electric power. Hence a battery has the reactants built into itself as an integral part. In another words, a battery stores reactants (both fuel and oxidant) inside itself in a manner that is often referred to as onboard storage, in addition to being an energy conversion device. This interpretation of a battery is illustrated in Figure 1.2.

Therefore, we summarize the characteristics of a battery as follows: A battery is really an energy storage device; the maximum amount of useful energy available from a battery is determined by the amount of chemical reactants stored within the battery itself in the first place (i.e., during manufacturing); and the battery will cease to produce electric energy when the chemical reactants are all consumed. Therefore,

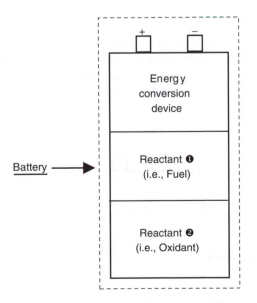

Figure 1.2 Schematic of a battery as represented by the components within the dashed box.

in principle, the lifetime of a battery is quite limited, often depending on the amount of both reactants stored onboard. However, because both the fuel and the oxidant are stored together inside the battery at the beginning, a very slow electrochemical reaction does take place, accompanied with battery component corrosions, even without connection to an external load. This slow "leakage" of useful energy in the absence of a load also affects operational lifetime as well as the energy storage time of a battery, hence limiting the usefulness of a battery as an energy storage device. In addition, the electrode of a battery not only participates in the electrochemical reactions producing electric energy, but also is consumed in the process. Therefore, the electrode in a battery is not stable during the energy conversion process. As a matter of fact, the lifetime of a battery really depends on the lifetime of its electrode.

In comparison, a fuel cell is only an energy conversion device to which reactants are supplied continuously, as shown schematically in Figure 1.3. Clearly, the fuel and the oxidant are stored outside the fuel cell and during the operation the electrochemical reaction products are rejected from the fuel cell. Therefore, we summarize the characteristics of a fuel cell in the following statements: A fuel cell can produce electric energy output for as long as the fuel and oxidant are supplied. Because the reactants are stored externally, the amount of useful energy derivable from a fuel cell is inexhaustible as long as the reactants are available. Consequently the lifetime of a fuel cell is unlimited in principle, as long as the reactants are supplied and products are removed continuously. In reality, however, degradation or malfunction of components limits the practical operational life of fuel cells. Further, there is no "leakage" of useful energy from a fuel cell, hence no corrosion of cell components during the offload period because the reactants are stopped from delivery to the fuel cell from their external storages. In addition, electrodes in a fuel cell are stable during the energy conversion process; they facilitate the electrochemical reactions but keep their physical and chemical characteristics without being consumed at the same time.

However, battery and fuel cells have a number of similarities. They are both electrochemical devices (or more often referred to as cells); they both produce electric energy *directly* from electrochemical reaction of fuel and oxidant; they both have similar components called **electrodes** and **electrolytes**. It might be mentioned that an electrochemical cell produces electric power output through electrochemical reactions

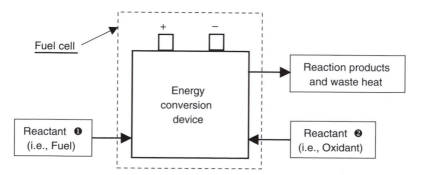

Figure 1.3 Schematic of a fuel cell as represented by the components within the dashed box.

which occur inside and consume reacting chemical substances. This is a "reverse" process of an electrolytic cell. Whereas a battery, in a strict sense, refers to a group of individual electrochemical cells connected together, although it is sometimes misused to stand for a single cell. The word "*battery*" is attributed to Benjamin Franklin, who first used it to refer to a group of electric capacitors, also called Leyden jars.

Comparison with Heat Engine   On the other hand, a heat engine-driven generator, as shown in Figure 1.4, can also convert the chemical energy of reactants into electric energy. However, chemical energy is converted into heat (thermal energy) first through the exothermic chemical reaction often referred to as combustion; thermal energy is then converted into mechanical energy (kinetic energy) by the heat engine, and finally mechanical energy is transformed into electric energy by an electric generator driven by the heat engine. Clearly this is a multistep energy-conversion process involving several intermediate energy forms and conversion processes in several different devices, as shown in Figure 1.5. Further, a part of the conversion processes is based upon a heat engine, which operates between a low temperature (T_L) and a high-temperature (T_H) thermal energy reservoirs. Hence, its maximum possible efficiency is limited by the Carnot efficiency given as

$$\eta_{Carnot} = 1 - \frac{T_L}{T_H} \tag{1.1}$$

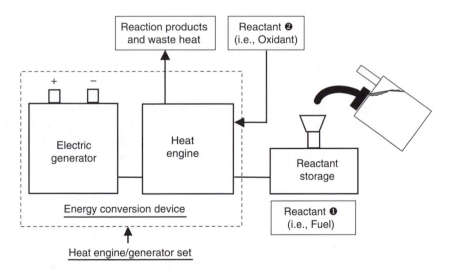

Figure 1.4 Schematic of a heat engine-driven generator as an energy conversion device, represented by the components within the dashed box.

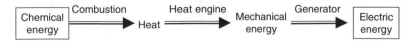

Figure 1.5 The multistep nature of electric power generation involving heat engines as shown in Figure 1.4.

as a direct result of the Second Law of Thermodynamics. Because in heat engines, high temperature T_H is achieved through the combustion of fuel with oxygen in the air, T_H is significantly limited due to the dilution of nitrogen present in the air and endothermic gas dissociation reactions. The latter also leads to harmful pollutant formation, causing environmental degradation. In practice, T_H is further drastically reduced to a sufficiently low value in order for the materials for heat engine construction to have sufficient mechanical strengths during heat engine operation. As a result, a heat engine has a quite low energy conversion efficiency, as explained in detail in the next chapter. Finally, the process in the heat engine-driven generator set also involves moving parts in the heat engine and electric generator, thus frictional losses and wear occur there, further reducing the overall energy conversion efficiency from the chemical energy to the electric energy. In addition, regular maintenance of moving components is required for the proper operation of the mechanical components involved.

Although both fuel cells and heat engines are energy conversion devices with reactants stored externally, fuel cells operate isothermally at a given temperature and electric energy is generated directly in one step through electrochemical reactions of the reactants without any intermediate forms of energy involved. Therefore, it might be expected, as it is the case, that fuel cells will have higher electrical energy efficiency than heat engines. Since they are free of moving mechanical parts during operation, fuel cells can work reliably in an unattended mode without vibration and noise, another added advantage of fuel cells. This will result in a low maintenance cost and is particularly advantageous in practical applications such as space exploration and remote generation.

1.3.2 How Does a Fuel Cell Work?

A fuel cell is composed of three active components: a fuel electrode (anode), an oxidant electrode (cathode), and an electrolyte sandwiched inbetween. Figure 1.6 illustrates the basic operational principle of fuel cells with a typical acid electrolyte fuel cell. It is seen that molecular hydrogen is delivered from a gas flow stream to one of the electrodes, often named anode (or fuel electrode), then the hydrogen reacts electrochemically in the anode as follows:

$$\text{Anode half-cell reaction:} \quad H_2 \Longrightarrow 2H^+ + 2e^- \tag{1.2}$$

It is often said that hydrogen (fuel) is oxidized at the anode/electrolyte interface into hydrogen ion or proton H^+ and gives up electron e^-. The protons migrate through the (acid) electrolyte, while the electrons are forced to transfer through an external circuit, both arriving at another electrode that is often referred to as cathode (or oxidant electrode). At the cathode, the protons and electrons react with the oxygen supplied from an external gas flow stream, forming water:

$$\text{Cathode half-cell reaction:} \quad \frac{1}{2}O_2 + 2H^+ + 2e^- \Longrightarrow H_2O \tag{1.3}$$

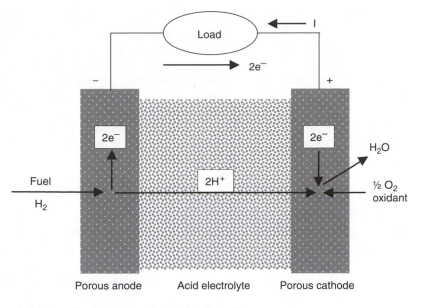

Figure 1.6 Schematic of a typical acid electrolyte fuel cell.

Thus, oxygen is being reduced into water at the cathode by combining with H^+ and e^-. Now both the electric current and mass transfer form a complete circuit. The electrons going through the external electric circuit do work on the electric load, constituting the useful electric energy output from the fuel cell. At the same time, waste heat is also generated due to electrochemical reactions occurring at the anode and the cathode, as well as due to protons migrating through the electrolyte and electrons transporting in the solid portion of the electrodes and the external circuit. As a result, the overall cell reaction can be obtained by summing the above two half-cell reactions to yield:

$$\text{Overall cell reaction:} \quad H_2 + \frac{1}{2}O_2 \Longrightarrow H_2O + W + \text{Waste Heat} \qquad (1.4)$$

where "W" stands for the useful electric energy output from the fuel cell. Although the half-cell reactions may be quite different in different types of fuel cells (which are described later), the overall cell reaction remains exactly the same as Equation (1.4).

Therefore, the byproduct of electrochemical reactions described here is water and waste heat. They should be removed continuously from the cell in order to maintain its continuous isothermal operation for electric power generation. This need for continuous removal of water and heat results in the so-called **water and heat (thermal) management**, which may become two critical issues for the design and operation of some types of fuel cells. In general, they are not easy tasks to be accomplished.

Important Physical and Chemical Phenomena   The aforementioned description of the processes occurring in a fuel cell is a gross underestimate of what really happens. It has been identified that many physical (i.e., transport of mass, momentum, and energy) and chemical processes are involved in the overall electrochemical reactions in the porous fuel cell electrodes that influence the performance of fuel cells. The transport processes involving the mass transfer of reactants and products play a prominent role in the performance of porous electrodes in fuel cells and those involving heat transfer and thermal management are important in fuel cell systems. Some of the important physical and chemical processes occurring in porous fuel cell electrodes during electrochemical reactions are for liquid-electrolyte fuel cells[8]:

1.   First, the reactant stream consists of multicomponent gas mixture; for example, the fuel stream typically contains hydrogen and water vapor, as well as carbon dioxide and even some carbon monoxide; whereas the oxidant stream usually has oxygen, nitrogen, water vapor, carbon dioxide, etc. The molecular reactant (such as H_2 or O_2) is transferred to the porous electrode surface from the reactant stream through the mechanism of convection; and then transported through the porous electrode, primarily by diffusion, to reach the gas/electrolyte interface.
2.   The reactant dissolves into the liquid electrolyte at the two-phase interface.
3.   The dissolved reactant then diffuses through the liquid electrolyte to arrive at the electrode surface.
4.   Some pre-electrochemical homogeneous or heterogeneous chemical reactions may occur, such as electrode corrosion reaction, or impurities in the reactant stream may react with the electrolyte.
5.   Electroactive species (which could be reactant themselves as well as impurities in the reactant stream) are adsorbed onto the solid electrode surface.
6.   Adsorbed species may migrate on the solid electrode surface, principally by the mechanism of diffusion.
7.   Electrochemical reactions then occur on the electrode surface wetted by the electrolyte — the so-called **three phase boundary**, giving rise to electrically charged species (or ions and electrons).
8.   Electrically charged species and other neutral reaction products such as water, still adsorbed on the electrode surface, may migrate along the surface due to diffusion in what has been referred to as post-electrochemical surface migration.
9.   The adsorbed reaction products become desorbed.
10. Some post-electrochemical homogeneous or heterogeneous chemical reactions may occur.
11. Electrochemical reaction products (neutral species, ions and electrons) are transported away from the electrode surface, mainly by diffusion but also for the ions, influenced by the electric field set up between the anode and cathode. The electron motion is dominated by the electric field effect.
12. Neutral reaction products diffuse through the electrolyte to reach the reactant gas/electrolyte interface.
13. Finally, the products will be transported out of the electrode and the cell in gas form.

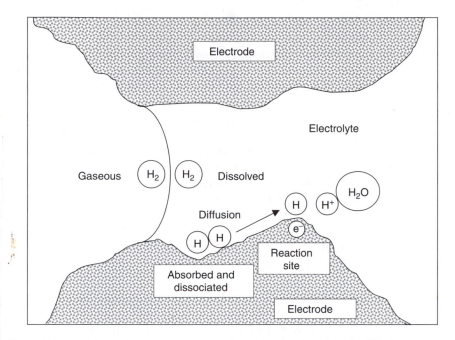

Figure 1.7 Physical and chemical processes around the three phase boundary of an H_2 electrode.

As an example, Figure 1.7 illustrates the various transport and chemical processes occurring around the three-phase boundary in a hydrogen electrode. Notice that the electrochemical reactions producing electric energy occur at this three-phase boundary (i.e., gas reactant, liquid electrolyte, and solid electrode surface). Normally any of these 13 processes can influence the performance of a fuel cell, exhibiting the complex nature of a fuel cell operation — thus one may appreciate the importance of understanding the electrochemical reactions and the fundamental transport processes in fuel cells. These aspects of fuel cells are described in detail in Chapters 3 and 4.

Function and Requirement of Electrodes   Well, let's not be overwhelmed and lose sight because the complex processes involved, the details of which depend on the components making up the cell. Simply put it, a fuel cell is made up of two electrodes and one electrolyte. The purpose of using electrodes is three-fold, that is to

- Provide a place for the electrochemical reaction to occur easily (typically referred to as reaction sites);
- Provide a flow path for reactant supply to, and product removal from, the reaction sites (reactant delivery and product removal or simply called mass transfer);
- Collect the electrons and provide a flow path for electron transfer (current collection, or electron conductor, and ion insulator).

As a result, it becomes necessary to make the electrodes in porous form so that the solid portion is used for reaction sites and electric current collection, while the void region is used for reactant flow and product removal. Since electrodes are operated in severe adverse conditions, they are required to be chemically stable in the reducing and oxidizing environment of the anode and cathode side, mechanically strong to resist deformation (sintering and creepage), among many other requirements. In the early development of fuel cell technology, precious metals such as platinum were routinely employed as electrode materials to provide reaction sites and which are too expensive for large-scale commercial applications. In order for performance enhancement and cost reduction, each porous electrode is typically subdivided into two layers: a catalyst layer adjacent to the electrolyte region for electrochemical reactions to occur and a backing layer for current collection and mass transport, which is also rightfully called a gas diffusion layer.

Function and Requirement of Electrolyte   Similarly, the function of electrolyte in a fuel cell is also three-fold, it is to act as:

- An ion conductor, so that ions can migrate from one electrode to another through the electrolyte, completing the cycle of mass transfer and electric circuit;
- An electron insulator, so that electrons are forced to migrate through the external circuit, thus providing electric power output. Otherwise, if electrons leak through the electrolyte, electric shorting occurs and no power output is attained;
- A barrier in order to separate the reactants (or prevent reactant from crossover from the anode to the cathode, or vice versa, and mixing together), so that controlled electrochemical reaction can proceed as expected. Otherwise, the crossover of the reactants from the anode or the cathode can form a combustible mixture, leading to fire hazards, and in severe cases, explosions may occur, not to even mention the loss of the electric power output!

A good effective electrolyte for fuel cells must possess a set of desirable characteristics, such as high dissolubility for reactants and high diffusion coefficients for reactant migration, in order to increase the concentrations of the reactant at the reaction sites. This is because the rate of electrochemical reaction for power generation is directly proportional to the reactant concentration available for the reaction.

Fuel Cell Performance   In fuel cell operation, the anode has a lower electric potential, the cathode has a higher electric potential, and the electric potential difference between the cathode and anode is often called **the actual cell potential**, E. The direction of electric current in the external circuit is from the cathode to the anode and it is from the anode to the cathode side in the electrolyte, as illustrated in Figure 1.6. One may recall that the positive direction of electric current is defined as the positive direction of motion for positively charged particles; hence it is just opposite to the direction of the motion by the electrons, which are negatively charged.

For a fuel cell with hydrogen oxidation reaction shown in Equation (1.2) at the anode, and oxygen reduction reaction shown in Equation (1.3) at the cathode, the best possible cell potential difference is 1.229 V when the fuel is pure hydrogen and the oxidant is pure oxygen, and both are at the standard state of 25 °C and 1 atm. Cell potential is achieved when the fuel cell is operated under thermodynamically reversible condition, hence it is often referred to as reversible cell potential, which is explained fully in next Chapter 2.

In practice, a fuel cell always has energy losses, exhibited in terms of electric potential losses. These losses are due to many irreversibilities which reduce the cell potential difference E to about 0.7~0.8 V at practical current densities for optimal fuel cell performance and realistic power output – a result of optimization (or compromise) between high efficiency and high power density requirement. This is because cell potential difference typically decreases as more current is drawn from the cell (i.e., more power is output from the cell). This conflicting variation of high-energy conversion efficiency and high power density constitutes a fundamental consideration in the fuel cell system design and optimization.

The voltage loss in a fuel cell is more often called overvoltage or overpotential, and the phenomena that give rise to voltage loss is referred to as polarization, hence a plot of cell voltage change as a function of cell current (or more often current density) is referred to as polarization curve. Figure 1.8 shows a typical polarization curve for an acid electrolyte fuel cell. As it is shown, the largest voltage loss is related to oxygen reduction reaction in the cathode due to the slow rate of oxygen reduction

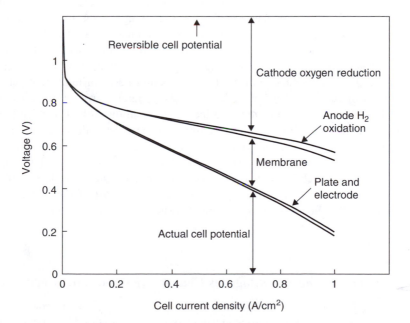

Figure 1.8 Typical cell polarization curve and breakdown of cell voltage losses for an acid electrolyte fuel cell.

reaction and mass transfer. As a result, the facilitation and acceleration of the slow cathode kinetics remains one of the key issues to be tackled. In comparison, hydrogen oxidation reaction in the anode proceeds very fast, at least orders of magnitude faster than the cathode counterpart, and thus the overpotential arising from it is very small and often negligible at the typical fuel cell operation condition. The second largest overpotential is due to resistance loss related to proton migration through the electrolyte; a search for an "ideal" electrolyte is continuing. Some other minor energy losses occur in the electrode and other cell components. As mentioned earlier, the actual cell potential decreases from the reversible cell potential when the current drawn from the cell is gradually increased to meet higher power output (i.e., higher load) demand.

Polarization has been classified as activation, concentration, and ohmic. They are due to resistance, respectively, to electrochemical reactions, to the transfer of ions through the electrolyte and electrons through the electrodes and other cell components, and to the mass transfer limitations in supplying reactants to the electrochemical reaction sites. A major effort in the development of fuel cell technology involves finding the optimal techniques to reduce these polarization effects. It has been identified[9] that effective approaches to achieve this include (i) selection of sufficiently active catalysts, often precious Noble metals, to promote electrochemical reactions, (ii) engineering of component and cell designs that reduce the active component thickness like electrolytes and electrodes, (iii) minimization of electrical resistance by suitable material selection and mechanical design, and (iv) enhancement of the rates of reactant and ionic diffusions.

On the other hand, the rather low electric potential of a single cell output (about 0.7~0.8 V) presents another major engineering challenge to the fuel cell development and practical applications. Because such low voltages are not feasible to power any realistic electric loading, it becomes necessary to combine many cells together in series to achieve required system voltages. This is accomplished by the so-called **cell stacking** — the technique of electric connection between individual cells. In earlier days of fuel cell development, electricity generated from each cell is harvested through the edge collection — a technique resulting in high voltage losses. Presently, bipolar arrangement is preferred by placing an electrically conducting bipolar plate separating each cell so that the cathode of one cell is in electrical contact with the anode of the next cell. The fuel and air supply to each cell is accomplished by the flow channels machined on the bipolar plate. The design of fuel cell stacks has a major impact on the performance and cost of the fuel cell systems; hence will influence to a large degree the success of fuel cells for commercial applications.

1.3.3   What Are the Advantageous Characteristics of Fuel Cells?

Due to its unique principle of operation with direct one-step energy conversion, a fuel cell power system has numerous advantageous characteristics. First and foremost, a fuel cell system has a high practical energy conversion efficiency at full and partial load. As pointed out earlier, fuel cell energy efficiency is higher than heat engine efficiency under practical operating conditions and at the design load which will be scrutinized carefully in the next chapter. For example, in space explorations, fuel cells operate on pure hydrogen and oxygen with over 70% chemical to electrical efficiency and the only byproduct water from the fuel cell constitutes the sole source of drinking water for the crew of the spacecraft.

One of the most important points concerning energy efficiency is that at partial load, the efficiency of a heat engine decreases due to increased mechanical and heat losses; whereas the efficiency of a fuel cell will normally increase from the full to partial loading because the energy conversion efficiency of a fuel cell is directly proportional to the cell voltage. At the reduced loading, power output, hence the cell current density, is reduced from the designed full load, and the cell potential increases. This property is especially useful when fuel cells are used as the power source for vehicular applications because a motor vehicle spends a considerable time, if not absolute majority of its lifetime, in partial load.

Another important advantage of a fuel cell system is its extremely low pollutant emissions. Since fuel cell operation does not involve high temperature environment as in combustion process, no harmful chemical pollutants are formed from the dissociative reactions at high temperatures such as nitride oxides (NO_x), sulphide oxides (SO_x), carbon monoxide, although a minute amount of them may be formed in the auxiliary subsystems.

Furthermore, because of its high energy efficiency, a fuel cell system produces, on a unit energy basis, less chemical, carbon dioxide and thermal emissions (or low waste heat disposal). For example, the commercially available fuel cell system, the International Fuel Cells' PC 25,[10] has a chemical to electrical energy efficiency of 40% and thermal energy efficiency of 50% for cogeneration with negligible chemical pollutant emissions and over 50% reduction in carbon dioxide emission when compared with the corresponding fossil fuelled power plant for the same amount of the electricity generated. Its waste heat rejection to the environment (only around 10%) is only a fraction of what is rejected from the conventional power plant. Excessive waste heat rejection into the rivers and lakes through cooling water may affect adversely the local aquatic life and ecosystem.

The advantage of a fuel cell system includes its excellent and fast transient load response. Since fuel cell reactions for power generation involve the transport of ions, electrons, and gas molecules and auxiliary subsystems operate at constant pressure and temperature, fuel cell response time to load changing is limited only by the response in the reactant flow, which is very fast for a properly designed fuel cell power plant. For example, low-temperature fuel cells can respond and adjust the power output requirement within milliseconds when operating with hydrogen and air, and MW-size fuel cell systems for utility application including a fuel processing subsystem have a response time of within a second ($0.3 \sim 1$ s reported for IFC's PC 23, the predecessor for PC 25). It has also been demonstrated[11] that fuel cell power plants meet the load-following requirement for stationary utility applications, the varying power delivery requirement as set in the U.S. Federal Urban Driving Schedule for passenger vehicles, and a load change rate of 20 kW/s for transit bus application.

Even though faster response is possible, passenger comfort puts limitations on transit bus acceleration rate. Further, the fast response is achieved whether fuel cell systems operate on pure hydrogen, or processed fuel from the reformation of methanol or natural gas.

Fuel cell systems operate quietly, with no vibration or noise because they are free of mechanical moving components (except perhaps in the reactant flow auxiliary equipment). Considering their low emissions and low waste heat disposal, stationary fuel cell systems can be installed at the locations where electricity is needed without intrusion. This makes onsite, distributed and dispersed power generation possible, without exclusive dependence on the high-voltage electric transmission networks (or grids). Quiet operation is also valuable for congested urban transport applications. The electrochemical reaction byproducts of heat and water are easily recoverable, making fuel cell operation without a need for external water supply, and co-generation is easily implemented. Combined heat and electricity generation further improves the overall energy efficiency of fuel cell systems. The high efficiency is achieved without dependence on the power plant size, from a few watts to MW size or beyond. Thus fuel cell systems of various sizes with the same high-energy conversion efficiency can be realized for various applications, ideally suited for onsite, distributed and dispersed stationary applications.

The electrochemical nature of fuel cell operation without moving parts also makes fuel cell systems highly reliable and suitable for unattended mode of operation. High reliability results in long scheduled maintenance, hence low operation and maintenance cost, making fuel cells well suited for economic applications to, for example, distributed power generation particularly in developing rural areas and for installations with low power demand. Experience indicates[11] that the first commercial fuel cell system has already exceeded the maintenance requirements for stationary power plants (continuous duty operation) of one-year scheduled maintenance interval and five-year overhaul period. In U.S. spacecraft application, only one failure has occurred over 80,000 hours of fuel cell operation in space. That failure occurred on the second space shuttle flight due to foreign particle contamination blocking an aspirator nozzle and thereby restricting the removal of fuel cell water. Since the corrective redesign, no failures have occurred in 70,000 hours of operation. Field trials and demonstrations show that the cell stack and fuel processing components are extremely reliable, electrical and mechanical ancillary components such as pumps, valves, and fans account for most of the failures with the rest due to miscellaneous reasons such as sensors and procedures.

Other advantages of fuel cell systems include the modular design and fabrication process, for easy scale-up of the power plant size. This is because many cells connected together form a stack, and one or a number of stacks connected together form a fuel cell system. Fuel cells mainly operate with hydrogen gas as fuel, which can be manufactured from renewable sources or conventional fossil fuels, and they can also operate on methanol or natural gas directly or indirectly, which may also be derived from renewable sources. Therefore, the multifuel capability of fuel cell systems is well suited for the diversification and development of a sustainable future energy system.

1.3.4   Why Do We Need Fuel Cells?

Fuel cells are needed because fuel cells possess the advantageous characteristics of high-energy conversion efficiency and environmental benign operation and help alleviate and may eventually eliminate dependence on the limited and fast depleting fossil-fuel reserves, health concerns and environmental degradation since fuel cell is virtually pollution free. Furthermore, fuel cell systems are flexible, fitting into a variety of different applications such as onsite, distributed and dispersed generation as well as cogeneration, without the need for a long-distance electric power transmission network. But above all, they are compatible with renewable energy sources and carriers for future energy security, economic growth, and sustainable development.

It must be pointed out that fuel cells are an energy conversion technology, not an energy resource, and thus its contribution to sustainable development depends on many other factors — in large part the process by which its fuel is procured because hydrogen is not abundantly available on Earth. It is the choice and procurement of fuel for fuel cell systems that has more or less plagued the development and deployment of the fuel cells and it seems that this plague is not going to be cured any time soon.

1.3.5   What Are the Applications of Fuel Cells?

Fuel cell systems produce electric power on demand; therefore, they can be used in a diverse situation whenever power is needed and where other methods of power generation are unable to meet the specific requirements of the particular applications. The largest potential commercial market for fuel cells is the stationary power generation for utility application, including

- Residential applications in the size range of around 1 to 5 kW, operating on natural gas which is available for many residential houses in cities today. Besides electricity generation, waste heat can be used for hot water or space heating. Imagine that every household is powered with such a fuel cell unit located inside the house (say, in the basement), which is connected to the electric grid. When the household needs more power for a short while, it can draw the electricity from the grid. On the other hand, it can sell the surplus power to the grid, so that the fuel cell system will be always operating at the designed full load. Such a concept is almost an exact analogy to many personal computers connected to the Internet.
- Onsite cogeneration power plants in the size range of about 200-kW to 1-MW capacity. This is perhaps the first- and near-term commercial application for fuel cell systems.
- Dispersed (or distributed) electric power generation in the size range of around 2-MW to 20-MW capacity ranges. This is expected to be the next step in the commercial application of fuel cells for electric utility industry. At present, electricity is typically generated in a central thermal power plant and then distributed through the grid. In the future, it is possible to have a fuel cell system generating electricity

and heat for a neighbourhood of a few blocks of streets with natural gas as the primary fuel for fuel cell operation, where the electric connection for each house could be easily buried underground to avoid the severe impact of natural disasters like the ice storm in Quebec in January 1998.

- Baseload electric power plants in the size range of about 100-MW to 300-MW capacity ranges while operating on coal or natural gas. This is potentially one of the most lucrative markets for fuel cells. Currently, 750-GW installed capacity can be found in the United States alone.

Fuel cell commercialization in transportation application is dawning! Urban transit buses and passenger vehicles have been successfully demonstrated with fuel cell power units. The most well known fuel cell powered buses are from Ballard Power Systems, running on the streets of Chicago and Vancouver daily for three years, and that of the passenger vehicles is the Necar series of fuel cell powered motor vehicles by Daimler/Chrysler. Transportation is a huge world market, considering 600 million vehicles are found on the road world wide and this number is increasing rapidly. The internal-combustion engine powered vehicles of today are very inefficient with typical energy efficiency of 30% or so and pollutant emissions. They consume about 6000-million liters of fuel daily. Now visualize that a fuel cell vehicle has an energy efficiency of 40% or more with zero emissions. A difference of 10% in the fuel efficiency means 600-million liters of fuel savings daily for the same output power or distance driven by these vehicles.

On the other hand, fuel cells have a long history of successful application in space explorations. The first operation of a fuel cell in space was in 1964. From then on to the late 1960s, 1-kW fuel cell power unit was used in Gemini program, representing the first practical use of fuel cells. The Apollo space program used 1.4-kW fuel cell power units from 1966 to 1978 for 18 flights. Since 1981, U.S. space shuttle flights have been employing three 12-kW fuel cell generators, totaling 70,500 hours of operation for 95 shuttle flights. Space applications constitute the first routine practical use of fuel cells.

The U.S. military has showed a huge and persistent interest in fuel cell power systems because of their unique operation and characteristics. For example, submarines with fuel cell propulsion systems will be very quiet with low noise and thermal signatures, easily hiding from enemy detection during combat. Large fuel cell power generators had also been envisioned as a power source for the Strategic Defense Initiative (the so-called "Star Wars" program) in early 1980s. Under consideration by U.S. Army is the small backpack fuel cell unit for solders in field combat. Other possible military applications abound.

Although fuel cell application considered in the past had been mostly for large power plants, fuel cells are being considered for applications with small power ratings, such as cellular phones, notebook computers, and so on. The DC power generated from fuel cells can be used directly for these applications, and no DC-to-AC or AC-to-DC conversion is needed. Because of their high efficiency, no pollutant emissions and quiet operation, fuel cells will find more applications where a power supply is required.

1.3.6   What Are the Objectives for...?

The ultimate objective for fuel cell analysis, research and development (R&D), as well as demonstration is to develop a mature technology (including manufacturing) for cost-effective reliable fuel cell systems with the aforementioned characteristics and sufficiently long lifetime for a given application. Notice here the cell systems must be cost effective (i.e., low capital cost) in order to be economically competitive in the market place, reliable for low maintenance and operation cost with a stable and high-quality power source, and long lifetime comparable with the competing alternative technology including the conventional thermal power plants. Above all, the fuel cell systems developed must possess the desirable characteristics exhibiting their advantages such as high energy efficiency and environmental benign operation.

However, this book covers the fundamental aspects of fuel cells to the practical various types of fuel cell technologies. It is hoped that the readers will be able to do the following two tasks after reading this book:

- A quick and rough estimate of the performance of existing fuel cell systems (performance evaluation) and
- A quick and rough design of a new fuel cell system (i.e., performance prediction).

The fundamental theories covered in the book allow the readers to do some simplified analysis, modelling, and simulation of phenomena occurring in fuel cells, so that this information may be used to substitute or complement difficult, expensive, and time-consuming experimental work for cost-effective research and development activities.

1.3.7   What Are the Backgrounds Needed for Fuel Cell Analysis and R&D?

The science and technology of fuel cells involve many branches and disciplines of science and engineering, including but not limited to the following:

- Thermodynamics,
- Electrochemistry,
- Chemistry and chemical engineering,
- Fluid mechanics,
- Heat and mass transfer,
- Material science (metallurgy) and materials engineering,
- Polymer science and specifically ionomer chemistry,
- Design, manufacturing and engineering optimization,
- Solid mechanics and mechanical engineering, and
- Electromagnetism and electrical engineering.

Therefore fuel cells require a multidisciplinary knowledge, interaction, and integration. Very often, one individual may not possess the necessary background and teamwork and collaboration is a must for a successful fuel cell R&D program. It may be said that fuel cells are truly multidisciplinary subject.

1.3.8   What Do Typical Fuel Cell Analysis, Modelling, and Simulation Look Like?

Theoretical analysis, modelling, and simulation may be performed for a particular phenomenon or a collection of phenomena that occur in a specific fuel cell system. The objective may be to understand and elucidate the phenomena and their impact on fuel cell performance, and how to design the various components of a fuel cell system to harvest the desirable effects or to avoid a detrimental influence. Hence, the ultimate objective is to carry out either performance evaluation of an existing fuel cell or attempt performance prediction for a new design. Because the relationship between cell voltage and current density drawn from the cell will constitute the primary measure of a fuel cell performance, and other performance measures such as cell energy conversion efficiency and power density can be easily derived from this information, the ultimate goal of any fuel cell analysis and modelling is to calculate the cell polarization curve under various operating and design conditions.

The calculation of cell polarization curve begins typically with the Nernst equation, which gives the reversible (or the best possible) cell potential, Er. Then all three modes of voltage losses, namely activation, ohmic, and concentration overpotential are calculated from the conservation equations for total mass (continuity equation), species (including both neutral and charged species), momentum, and energy along with the transport equations for the transport phenomena of mass (including neutral and charged species), momentum, and energy. The mechanism and rate of electrochemical reactions are employed for the rate of mass and electric current production.

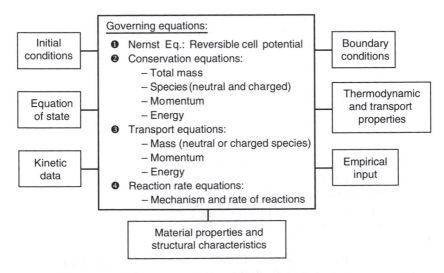

Figure 1.9 Elements of typical fuel cell analysis, modelling, and simulation.

Finally, the actual cell potential is determined by subtracting the overpotentials from the reversible cell potential. From cell voltage and current density, the energy efficiency, power density, and rate of waste heat generation (which dictates the cooling requirement) can be easily established. Figure 1.9 illustrates the elements of a typical fuel cell analysis, modelling, and simulation. Treatment of the conservation and transport equations gives rise to theoretical models of various degrees of complexity. With the advancement of computer science and numerical techniques, it may become possible to solve these equations for sophisticated geometrical designs.

1.4   CLASSIFICATION OF FUEL CELLS

Fuel cells may be classified in a variety of different ways, depending on the criteria used, which are typically the parameters related to fuel cell operation or construction. Fuel cell systems involve a vast number of variables, such as the type of electrolyte used, the type of ion transferred through the electrolyte, the type of reactants (e.g., primary fuels and oxidants), operating temperature and pressure, direct or indirect use of the primary fuels, and primary or regenerative systems. Because the choice of the electrolyte defines the properties of a fuel cell, including the operation principle, design and construction, as well as the materials that can be used for the cell and stack components, it is customary now that fuel cells are primarily named by the nature of their electrolyte used.

1.4.1   Classification by Electrolyte Used

By this naming convention, if a fuel cell contains as the electrolyte:

- An alkaline solution (normally potassium hydroxide, or KOH) in water, then it is referred to as an alkaline fuel cell (AFC);
- Phosphoric acid, then it is referred to as a phosphoric acid fuel cell (PAFC). (Other type of acids, such as sulfuric acid, have been attempted and the search for "super acids" with all the desirable attributes continues);
- A (solid) proton exchange (or proton-conducting) membrane, then it is called the proton exchange membrane fuel cell (PEMFC). (Often also referred to as solid polymer fuel cell (SPFC), solid polymer electrolyte fuel cell (SPEFC), polymer electrolyte fuel cell (PEFC), polymer electrolyte membrane fuel cell (PEMFC), etc.);
- A molten carbonate salt, then it is called a molten carbonate fuel cell (MCFC);
- A solid oxide ion-conducting ceramic, then it is called solid oxide fuel cell (SOFC).

1.4.2   Classification by Type of Ion Transferred Through the Electrolyte

As discussed earlier, fuel cells can also be classified by other criteria, such as the type of ion transferred through the electrolyte. This is because half-cell reactions in a fuel cell are directly related to the type of ions carrying electric current in the electrolyte, thus exerting significant influence on cell design, operation, and performance. For example, if the ions are positively charged, it may be referred to as cation transfer fuel cells, such as PAFC and PEMFC where H^+ is the ion transferred through the electrolyte. These cells may also be called **acid electrolyte fuel cells**, and they have following important characteristics:

- The oxygen reduction kinetics, or cathode reaction is slow, leading to high-activation polarization, thus necessitating the need of Noble metals as catalysts in order to provide reasonable electrocatalytic activity for fuel cell operation. The use of Noble metals increases the capital cost of the fuel cell.
- Electrochemical reaction byproduct water is formed at the cathode (i.e., on the side of oxidant). Since oxygen gas has much smaller diffusion coefficient compared to hydrogen, its transfer rate is limited, yielding a low concentration at the reaction sites. Water presence at the cathode requires a careful removal method to balance the water content there, which may be achieved at the designed operation condition but not quite easy to achieve over the wide range of conditions a fuel cell may encounter during the load varying situation. Therefore, the cathode is susceptible to water flooding, which often occurs in real fuel cells. Water flooding significantly reduces the rate of oxygen mass transfer, and lowers the oxygen concentration available for the reduction reaction, causing severe concentration polarization.

Clearly these two problems have been plaguing acid electrolyte fuel cells and even though methods have been developed to deal with them, an ideal solution is still waiting to be discovered.

On the other hand, anion transfer fuel cells refer to those cells with negatively charged ions transporting through the electrolyte, such as hydroxyl ion OH^- in alkaline fuel cells, carbonate ion CO_3^{2-} in molten carbonate fuel cells and oxide ion O^{2-} in solid oxide fuel cells described earlier. These cells may also be called "alkaline" type of fuel cells. They possess the following important characteristics:

- Fast oxygen reduction kinetics, or cathode reaction is fast, leading to low activation polarization. Hence, the Noble metals may be not needed necessarily as the catalyst.
- Product H_2O is formed at the anode (i.e., on the side of H_2 supply). Since hydrogen molecules are much smaller with a large diffusion coefficient, the presence of water will not affect the hydrogen mass transfer dramatically. Therefore, concentration polarization as a result of water flooding remains small due to high mobility of H_2 molecules.

As a result, alkaline fuel cells have better performance than acid electrolyte fuel cell under the same operating condition if pure hydrogen and oxygen are used as the reactants.

1.4.3   Classification by Type of Reactants Used

Fuel cells are often referred to by the primary fuel and oxidant used for their operation. The H_2–O_2 fuel cells imply that pure hydrogen and oxygen are used as the reactants, and H_2–air fuel cells denote that air instead of oxygen is used as the oxidant. Some other examples are ammonia–air fuel cells, hydrocarbon–air fuel cells, hydrazine–air fuel cells, hydrogen–chlorine fuel cells, hydrogen–bromine fuel cells, etc. The most important of this type of fuel cell is methanol fuel cell, in which methanol can be used indirectly through the steam reforming to produce the secondary fuel of hydrogen-rich gas mixture first — the so-called indirect methanol fuel cell. On the other hand, methanol can also be used directly in fuel cell electrochemical reactions, leading to the well-known direct methanol fuel cell (DMFC).

1.4.4   Classification by Operating Temperature

Fuel cells operating at high temperatures are being referred to as high-temperature fuel cells, both molten-carbonate ($\sim$650 °C) and solid-oxide ($\sim$1000 °C) fuel cells belong to this category. Whereas the low-temperature fuel cells include alkaline (60 °C$\sim$80 °C) and proton-exchange membrane (80 °C) fuel cells. The phosphoric-acid fuel cell ($\sim$200 °C) is sometimes also regarded as low-temperature fuel cell, however, it might be more appropriately called intermediate-temperature fuel cell, as its operating temperature is substantially higher than the low-temperature fuel cells. Chapter 6 explains that this operating temperature is not too high for bottoming cycles and not too low for easy cell material selection and long-operation lifetime, resulting in undesirable effects detrimental to the performance and cost reduction of the phosphoric acid fuel cell.

1.4.5   Other Methods of Fuel Cell Classification

Fuel cells may also be referred to by as many other methods as the number of variables related to the fuel cell system. For example, a pressurized fuel cell operates at pressures above the atmospheric pressure. A direct methanol or gasoline fuel cell indicates that the methanol or gasoline is fed directly into the fuel cell anode for electro-oxidation and electricity generation, as discussed in the previous section. For an indirect methanol or gasoline fuel cell, the primary fuel (methanol or gasoline) is converted into hydrogen-rich gas first through a reforming process, and the secondary fuel (hydrogen) is then supplied to the anode for hydrogen electrooxidation to produce electricity. If the same cell is used in the forward reaction to produce electricity (fuel cell reaction), and in the reverse reaction to produce the same reactants from the forward reaction products by electric energy input (electrolytic reaction), such a cell may be called a **regenerative fuel cell**, since it can regenerate the reactants itself. In that sense, a cell that can only be used for forward reaction may be called primary fuel cell. A hydrogen–chlorine (or hydrogen–bromine) fuel cell can be easily used as a regenerative fuel cell as well by using hydrochloric (or hydrobromic) acid as electrolyte; while a hydrogen–oxygen fuel cell forming water is not easy to implement as a regenerative system because of the inefficient reverse reaction and unstable electrode structure.

In summary, six major types of fuel cells have been developed or are being developed for practical applications. Five of them are classified based on their electrolytes used, including the alkaline fuel cell (AFC), phosphoric acid fuel cell (PAFC), proton exchange membrane fuel cell (PEMFC), molten carbonate fuel cell (MCFC) and solid oxide fuel cell (SOFC). Direct methanol fuel cells (DMFC) are classified based on the fuel used for the electricity generation. Table 1.1 provides a summary of the operational characteristics and application of the six major types of fuel cell.

1.5   COMPOSITION OF AIR

For terrestrial and commercial applications of fuel cells, air is almost invariably used as the oxidant. Atmospheric air contains some moisture (water vapor), reducing the amount of oxygen in humid air. On the other hand, air is often humidified on purpose before being fed into fuel cells, either for the humidification of the electrolyte as in PEM fuel cells or for the prevention of solid carbon formation in the reformed fuel stream for the high-temperature fuel cells. Therefore, both dry and humid air will be discussed in this section.

1.5.1   Dry Air

Dry air is a mixture of gases that has a representative composition by volume of 20.95% oxygen, 78.09% nitrogen, 0.93% argon, and trace amounts of other gases like carbon dioxide, neon, helium, methane, and so on. However, only oxygen in the air is electrochemically active and nitrogen is an inert species that only dilutes the mixture. The exact composition of air may vary slightly depending on the geographic location and weather conditions. Table 1.2 shows the standardized principle constituents of dry air by the U.S. Bureau of Standards.

Although many gas species are present in the air, their concentrations are too low to have any noticeable impact on fuel cell operation and performance. The only exception may be carbon dioxide — approximately 300-ppm concentration in the air is still too much for alkaline fuel cells and can severely degrade their performance and lifetime. With this exception in mind, it is sufficiently accurate to regard air as consisting of 21% O_2 (20.95% to be exact) and 79% N_2 (or 79.05%), the latter is also termed atmospheric or apparent nitrogen, which contains other trace gases.[12] Consequently, for every mole of oxygen in the air, there are $(1 - 0.2095)/0.2095 = 3.773$ moles of apparent nitrogen, thus 4.773 moles of air.

Therefore, in the following chapters, air will be assumed to have 21% oxygen and 79% nitrogen with the molecular weight of 28.964 and nitrogen will be implied as atmospheric nitrogen with a molecular weight of 28.160, which is slightly different from the molecular weight of the pure molecular nitrogen of 28.013, as shown in Table 1.2. As this difference is very small, well within the typical engineering accuracy requirement, it is often neglected without further specification.

Table 1.1 Operational Characteristics and Technological Status of Various Fuel Cells

Type of Fuel Cells	Operating Temperature (°C)	Power Density (mW/cm²) (Present) Projected	Projected Rated Power Level (kW)	Fuel Efficiency (Chemical to Electrical)	Lifetime Projected (hours)	Capital Cost Projected (U.S.$/kW)	Areas of Application
AFC	60–90	(100–200) >300	10–100	40–60	>10,000	>200	Space, Mobile
PAFC	160–220	(200) 250	100–5000	55	>40,000	3000	Dispersed and distributed power
PEMFC	50–80	(350) >600	0.01–1000	45–60	>40,000	>200	Portable, Mobile, Space, Stationary
MCFC	600–700	(100) >200	1000–100,000	60–65	>40,000	1000	Distributed power generation
SOFC	800–1000	(240) 300	100–100,000	55–65	>40,000	1500	Baseload power generation
DMFC	90	(230) ?	0.001–100	34	>10,000	>200	Portable, Mobile

Table 1.2   Principle Constituents of Dry Air

Gas	PPM by Volume	Molecular Weight	Mole Fraction	Mole Ratio
O_2	209,500	31.999	0.2095	1
N_2	780,900	28.013	0.7809	3.7274
Ar	9,300	39.948	0.0093	0.0444
CO_2	300	44.010	0.0003	0.0014
Air	1,000,000	28.964	1.0000	4.773 (rounded)

U.S. Bureau of Standards on air composition.

1.5.2   Humid Air

Atmospheric (or room) air usually contains a small amount of water vapor, which can be easily determined if the humidity ratio is known for the water vapor–air mixture. The humidity ratio, also called **absolute humidity** or **specific humidity**, is defined as the ratio of the mass of water vapor, m_w, to the mass of air, m_a, in the mixture:

$$\gamma = \frac{m_w}{m_a} \tag{1.5}$$

Therefore the humidity ratio is very useful in determining quickly the amount of water present in, or needed to humidify, a given amount of air. For example, $\gamma = 0.006$ represents that for every kg of dry air, 0.006 kg of water can be found in the mixture.

However, for a given condition a maximum amount of water can exist in the mixture in the vapor form, corresponding to the existence of a maximum value of γ. When this maximum value, γ_{max}, is exceeded, the excess water will exist in the liquid form or condense to the liquid state if the water is initially in the vapor form. Whereas liquid water in the mixture will vaporize when $\gamma < \gamma_{max}$. The value of γ_{max} depends primarily on the temperature of the mixture, and the tendency of water vaporization or condensation cannot be easily judged by a mere value of given γ. Normally another related parameter, namely relative humidity, is often used to represent conveniently the degree of water saturation in a particular mixture.

The relative humidity RH of a water vapor containing mixture is defined as the ratio of the partial pressure of the water vapor in the mixture, P_w, to the saturation pressure of water, $P_{sat}(T)$, corresponding to the mixture temperature, T, or

$$RH = \frac{P_w}{P_{sat}(T)} \tag{1.6}$$

Therefore, liquid water evaporates if RH < 1 (i.e., $P_w < P_{sat}(T)$), and condenses if RH > 1 (i.e., $P_w > P_{sat}(T)$). When $P_w = P_{sat}(T)$, the partial pressure of the water vapor is equal to the saturation pressure of water corresponding to the mixture temperature and the corresponding mixture is often referred to as water vapor saturated, or simply saturated. Hence, water saturation occurs at RH = 1, equivalent to the equal rate of two opposite processes occurring at the same time: water vaporization and

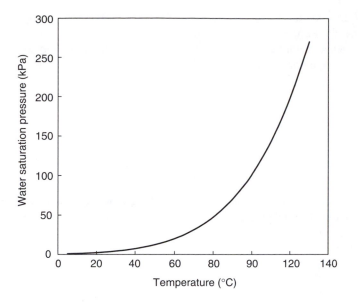

Figure 1.10 The dependence of water saturation pressure on temperature.

condensation. The relative humidity of 100% corresponds to the occurrence of the maximum value of humidity ratio γ_{max} for the given mixture. Consequently, the relative humidity has a numerical range of 0 to 1 for mixtures in equilibrium, corresponding to dry gas and fully humidified mixture, respectively, as two extreme cases.

It might be pointed out that the presence of inert gases in the mixture, in a strict sense, changes the saturation pressure of water by a very small amount, which is much less than 1% for conditions typically encountered in fuel cell operation, and thus negligible in engineering calculations including fuel cells. Figure 1.10 shows the variation of the water saturation pressure, P_{sat}, as a function of temperature. Clearly, P_{sat} is a highly nonlinear function of temperature T, and increases with T greatly. This increase becomes more and more rapid at higher temperatures. This accelerated increase of P_{sat} at higher temperatures has serious implications; in fact, it causes severe complications in the design and operations of PEM fuel cells because of the need to fully humidify the reactant streams around 80 °C in order to maintain complete hydration of the membrane electrolyte. The saturation pressure of water is typically tabulated as thermodynamic properties, and can be found in any thermodynamics textbook, such as Reference [13]. A closed-form equation is also available based on the latest information.[14] Appendix A shows the saturated vapor pressure of water at selected temperatures in tabular form as well as correlation equations with various accuracy. Knowledge of water saturation pressure is needed for the illustration of the significance of reactant gas humidification and the associated problems. In fact, humidification has the dominant impact on the selection of operating pressure for large PEM fuel cell stacks, as explained in Chapter 7.

Typically the amount of water vapor in the air is small, and water vapor at partial pressures below 1 atm may be approximated well as an ideal gas, of course the accuracy depends on the temperature as well. This is the case for typical fuel cell operations like PEM fuel cells. Therefore, the resulting water-vapor-containing-gas mixture, such as humid air, may still be considered as a mixture of ideal gases. Using the ideal gas equation of state, we obtain

$$\gamma = \frac{W_w P_w}{W_{air} P_{air}} = 0.662 \frac{P_w}{P_{air}} \tag{1.7}$$

or

$$RH = \frac{\gamma P_{air}}{0.622 P_{sat}(T)} \tag{1.8}$$

where W_w and W_{air} are the molecular weight of the water vapor and air, respectively. Also notice that the total pressure of the mixture, P, is the sum of the partial pressures of the water vapor and the air,

$$P = P_{air} + P_w \tag{1.9}$$

From Equations (1.7)–(1.9), it is clear that for a given mixture of air and water vapor (i.e., known P_w and P_{air}), the humidity ratio γ is fixed as long as $\gamma < \gamma_{max}$, and no matter how the mixture temperature T is being changed. But this is not the case for the relative humidity. Since water saturation depends sensitively on the mixture temperature, relative humidity will change appreciably with the mixture temperature as shown in Equation (1.6) or Equation (1.8). Especially, when the temperature is reduced while the mixture total pressure is maintained constant, the saturation pressure of water is decreased as shown in Figure 1.10 (or Appendix A), whereas the partial pressure of water vapor remains constant in the process. Therefore, relative humidity increases quickly. At RH = 100%, that is, when the mixture temperature T becomes equal to the saturation temperature, $T_{sat}(P_w)$, corresponding to the partial pressure of the water vapor P_w, the mixture becomes saturated by water vapor. Any attempt to reduce the mixture temperature further (i.e., $T < T_{sat}(P_w)$) results in the condensation of the excess water into the liquid state so that the remaining water vapor maintains RH = 100%. Therefore, the water saturation temperature corresponding to the partial pressure of the water vapor, $T_{sat}(P_w)$, is often called dew-point temperature, since at this temperature liquid droplets (dew) starts to form in the humid gas mixture. On the other hand, RH decreases quickly for a given mixture if the mixture temperature T is increased, because water saturation temperature increases rapidly as shown in Figure 1.10.

This description sheds light on the two critical issues in PEM fuel cells: water and thermal management which are coupled. Since heat is generated in fuel cells due to reversible and irreversible mechanisms, details of which are described in the next chapter, some cooling techniques are implemented for thermal management. However, if nonuniform temperature occurs in the fuel cell structure due to either nonuniform cooling or nonuniform heat generation, the local gas mixture may become unsaturated, and water evaporation occurs, leading to the dehydration of the membrane — the so-called formation of **local hot spots** and then **dry spots** in the membrane. If the local gas mixture is over-saturated, then the excess water will condense leading to the possibility of liquid water flooding the electrode pore region. Since water evaporation or condensation involves a large amount of heat absorbed or released as a result of the change in the thermal energy content of water, and this amount of heat absorption or release can affect the cell temperature distribution, hence the need to adjust cooling arrangement to improve the uniformity of the temperature distribution. To summarize, the nonlinear dependence of the water saturation pressure on the temperature imposes a significant technical challenge to the optimal design and operation of PEM fuel cells.

1.6   TYPICAL HYDROCARBON FUELS

Although almost all modern fuel cells under development consume hydrogen in the cell for power generation except the direct methanol fuel cell, the primary fuel to be used may have to be hydrocarbon fuels, depending on the specific application (i.e., electric utilities), due to a lack of access to cheap hydrogen at the present time.

Various families of hydrocarbon exist as shown in Table 1.3, and they differ according to their molecular structures. The cyclanes and aromatics have a closed ring-like structure for their carbon chain. The cyclanes, also called cycloparaffins, cycloalkanes, or naphthenes, have single carbon–carbon bond (C–C). While the aromatic or benzene family has a building block of six carbon atoms forming a closed ring, each carbon atom has a single and a double bond with its two neighboring carbon atoms ($-C=$) and the one additional bond for each carbon atom can be attached various side chains, as simple as one hydrogen atom, or as complex as any hydrocarbon group in various structural arrangements. The alkane, alkene, and alkyne families have an open carbon chain structure with the two ends of the carbon chain remaining unconnected. The alkane family has its molecules entirely made up of single carbon–carbon bonds (C–C), whereas alkene and alkyne families consist of one double or one triple carbon–carbon bonds (i.e., $C=C$ or $C\equiv C$), respectively, with the rest of carbon atoms connected in a single bond. The simplest species in the alkane family is methane (CH_4) and ethane (C_2H_6). If one hydrogen atom in the alkane family is

Table 1.3   Basic Hydrocarbon Families[15]

Family Name	Other Designations	Molecular Formula
Alkanes	Paraffins	C_nH_{2n+2}
Alkenes	Olefins	C_nH_{2n}
Alkynes	Acetylenes	C_nH_{2n-2}
Cyclanes	Cycloalkanes, cycloparaffins, Naphthenes	C_2H_{2n} or $(CH_2)_n$
Aromatics	Benzene family	C_nH_{2n-6}
Alcohols		$C_nH_{2n+1}OH$

replaced by one hydroxyl (—OH) group, it becomes the common alcohol family, for example, methyl alcohol (CH_3OH), or methanol, replaces methane (CH_4); and ethyl alcohol (C_2H_5OH), or ethanol, corresponds to ethane (C_2H_6), and so on. Therefore, alcohols can be generically designated as ROH, where R is the parent hydrocarbon radical in the alkane family.

The hydrocarbon fuels commonly encountered in practice are:

- Natural gas: Mostly methane (CH_4) with some ethane (C_2H_6), propane (C_3H_8), CO_2, and N_2. Typical heating value is about 54 MJ/kg.
- Gasoline: A mixture of several hundred hydrocarbons between hexane (C_6H_{14}) and decane ($C_{10}H_{22}$). Typical heating value is about 48 MJ/kg. Octane (C_8H_{18}) can be taken as representative.
- Diesel: A mixture of several hundred hydrocarbons, N–cetane (N–hexadecane, $C_{16}H_{34}$) is the representative; it can contain up to about 1% sulphur. Typical heating value is about 44 MJ/kg.
- Coal: A complex mixture of C, H, O, N, S and various organic compounds and trace metals. The exact composition changes significantly and the corresponding heating value ranges from about 15-to-35 MJ/kg, depending on the type of the coal.
- Wood: A complex cellulose with a typical formula of $C_6H_{10}O_5$. The heating value depends on the type of the wood, its moisture content and its resin content. Typical heating value of moisture and resin free wood is about 21 MJ/kg.

Hydrogen as a fuel occupies a uniquely important position in the fuel cells development, and five out of the six major types of fuel cells shown in Table 1.1 operate directly on hydrogen as fuel for electricity production. Methanol is perhaps the choice of primary fuels for transportation applications in the future due to its compact form and ability to be easily reformed into hydrogen rich gas. In contrast, the present choice of primary fuels for vehicles is gasoline. A summary of their physical and chemical properties is presented in Table 1.4 for comparison purpose.

1.7   A GLIMPSE OF FUEL CELL HISTORY

Fuel cells, the electric energy production through electrochemical reactions of chemically reacting substances, are an ancient phenomenon, existing in nature much longer than human history, let alone human civilization. They have existed in nature for millions of years in the form of the organs and muscles that supply the shock of the electric eel and other electric fish, such as, a Mediterranean variety, the torpedo ray, is capable of producing 300 V[16]! Acupuncture in Chinese medicine utilizes the strategic locations on human body, called *jingluo**, that have electric potential difference along them as well. These systems may be referred to as the biological fuel cells, whose

---

*They are, in Chinese medicine, main and collateral channels on human body regarded as a network of passages, through which vital energy circulates and along which the acupuncture points are distributed. A weak electric-current flow through these points during acupuncture can enhance the healing effect of acupuncture.

Table 1.4 Physical and Chemical Properties of Hydrogen, Methanol, and Gasoline

Property	Hydrogen	Methanol	Gasoline
Molecular Weight	2.02	32.04	~107.0
Density of Gas[a] (g/m^3)	84	790	~4400
Limits of Flammability in Air (Volume %)	4.0–75.0	6.7–36.0	1.0–7.6
Limits of Flammability in Oxygen (Volume %)	4.7–93.9		
Limits of Detonability in Air (Volume %)	18.3–59.0		1.1–13.3
Minimum Energy for Ignition in Air (mJ)	0.02	~0.3	0.24
Auto Ignition Temperature (°C)	585	385	228–501
Flash Point (°C)		11	−40
Flame Temperature in Air (°C)	2045		2197
Burning Velocity in Air (m/s)	2.7–3.2		0.37–0.43
Detonation Velocity in Air (m/s)	1480–2150		1400–1700
Heat of Combustion (low) (MJ/kg)	120		44.5
Heat of Combustion (high) (MJ/kg)	142		48

Note: All values are given at 1 atm pressure, 25 °C, and for dry gases.

[a]For comparison purpose, the density of dry air is $1.2 \, kg/m^3$.

power density is so low due to the slow biological reactions, that they do not really bear much potential for practical applications. Modern fuel cell technology, however, did not evolve from a fish, but was a result of human invention.

Sir William Robert Grove (1811–1896) is usually credited as the inventor of fuel cells when he convincingly demonstrated the principle of fuel cells in 1839.[17] However, before Grove, Sir Humphrey Davy (1778–1829), the founding father of electrochemistry, following the work of Luigi Galvani (1737–1798) on electric currents generated by contact of unlike metals (the Galvanic cell) and Alessandro Volta (1745–1827) on what are now known as Voltaic cells, had reported the construction of simple galvanic combinations on January 9, 1802.[18] Davy's cell seemed to have derived its power from the reaction of carbon at one electrode with oxygen taken from the nitric acid at the other. Although Davy did not succeed in constructing such a carbon (coal) — consuming cell for electricity generation in practical sense, he was able to give himself a feeble electric shock from a battery of such devices*. Despite this "failure" to invent a fuel cell consuming an economic fuel such as coal, he moved on to make great achievements in and contributions to electrochemistry.

---

*Or in Davy's original writing: "the powers of it are demonstrated by its agency upon the limbs of frogs, and by its effects upon the organs of sense."[18]

However, this started the aim and dream of the inventors of the mid and late years of the nineteenth century who sought to achieve, in one step, an electric cell which would act continuously and at the same time convert the energy of the reaction with air of a useful or economic fuel such as coal, directly to electric power — the so-called **direct coal fuel cell**. This is because coal was the only primary fuel in large scale use during the nineteenth century. The first use of the name "fuel cell" seems to have been in 1889 by Mond and Langer[19] who brought first marked improvement in Grove's cell and built the first technical fuel cell, in modern sense.

Significant effort has been spent and significant progress has been made since Grove's time, and one may venture to divide roughly the historical development of fuel cells into the following three waves (or eras) of fuel cell research and development, based on its popularity, the effort and amount of money spent on it[20]:

1.7.1   The First Wave (1839–1890s)

From the invention of fuel cells by Grove in 1839, the birth of industrial electricity and scientists' interest kept fuel cell development going, albeit the work on fuel cells appeared sporadic and less coordinated. Chemical to electrical energy efficiency was very low, on the order of only a few percentages, for the steam engine-driven dynamo power plants in those days. Almost all the developmental work was focused on developing a cell system that could consume the economic and practical fuel of the time — coal directly — and the goal was to achieve the potential high efficiency of fuel cell power plants, which showed the promise of well over 50% efficiency, at least theoretically. The direct coal fuel cell was eventually given up as an unachievable task.

1.7.2   The Second Wave (1950s–1960s)

Intensive fuel cell research and development activities started again from Bacon's pioneering engineering and success of his alkaline fuel cell systems. Highly publicized and successful fuel cell applications in space explorations, notably U.S. Gemini and Apollo programs, led to high hope and over-optimism in terrestrial and commercial fuel cell power systems. Many development programs were initiated with public and private funding, and almost all modern fuel cells and beyond were under investigation and development, including the alkaline fuel cell, the phosphoric fuel cell, the molten carbonate fuel cell, the solid oxide fuel cell, the proton exchange membrane fuel cell, direct hydrocarbon/alcohol fuel cell, as well as nitrogen compound-based fuel cells (e.g., hydrazine–ammonia fuel cell). During this period, fuel cell systems may be characterized by mobile and immobile electrolytes, depending on the specific application. For example, alkaline fuel cells had mobile electrolytes for terrestrial and military (submarine) applications, while immobile electrolytes were adopted for space programs for the alkaline fuel cell as well as PAFCs, MCFCs, and SOFCs for commercial applications. However, almost none of these development programs met their initially set targets, industrialists' hopes died, and expectations dimmed at the end of the 1960s for fuel cells as a potential terrestrial power source, although fuel cell faithfuls persevered and persisted, and fuel cells survived!

1.7.3 The Third Wave (1980s–Present)

Interest in fuel cells was renewed with high hopes once again due to many reasons, primarily due to the environmental and health concerns regarding pollution emissions at the local and global levels caused by the combustion of fossil fuels, diminishing natural resources, public concern on nuclear power safety, and the need and desire of sustainable development with renewable resources. All types of fuel cells are being intensively developed around the world, including, AFCs, PAFCs, MCFCs, SOFCs, and PEMFCs. Only direct methanol fuel cells survive in the family of direct hydrocarbon/alcohol fuel cells, due to slow rate of electrochemical reactions leading to poor performance. Except for DMFCs, all other fuel cells are being designed to operate on hydrogen as the fuel, and carbonaceous fuels are being reformed to hydrogen-rich gas mixtures first before being delivered to fuel cells. The goal is to improve the performance (i.e., energy efficiency, power density, reliability, lifetime, etc.), and reduce the capital cost for the second-wave fuel cells. Notable characteristics of the third-wave fuel cells are that the electrolytes are immobilized (or at least semi-immobile) for all applications, with increasing interest and focus on PEMFCs, MCFCs, SOFCs, and DMFCs.

Beyond the old saying of "Nothing goes beyond three times," there is good reason for considerable optimism at this time. With the lessons learned and experience amassed from the past, the advance and maturing of many related disciplines and technologies have been helping the understanding of almost every aspect of fuel cells, and solution to resolving technical barriers is fast approaching with worldwide effort. In fact, the routine practical application of fuel cells have lasted more than 20 years in space programs and commercial applications are dawning. The light at the end of tunnel is shining and the future of the fuel cells appears bright. A detailed account of historical development of fuel cells, especially during the last several decades, can be found in Reference [21].

1.8 TYPICAL FUEL CELL SYSTEMS

In general, a fuel cell power system involves more than just a fuel cell itself because fuel cells need a steady supply of qualifying fuel and oxidant as reactants for continuous generation of electric power. The oxidant is usually pure oxygen for specialized applications like in space and some military applications, and is almost invariably air for terrestrial and commercial applications. Depending on the specific types of fuel cells, both fuel and oxidant streams need to meet certain impurity requirements before being qualified as adequate for fuel cell operations. Therefore, a fuel cell power system is usually composed of a number of sub-systems for fuel processing, oxidant conditioning, electrolyte management, cooling or thermal management, and reaction product removal, etc. A schematic of a typical rudimentary fuel cell system is illustrated in Figure 1.11. Normally a power-conditioning unit is required to convert the DC electric power into AC electric power because fuel cells generate DC electric power while most of electric equipment operates on AC electricity. The waste heat produced in the fuel cell power section is often integrated through a series of heat

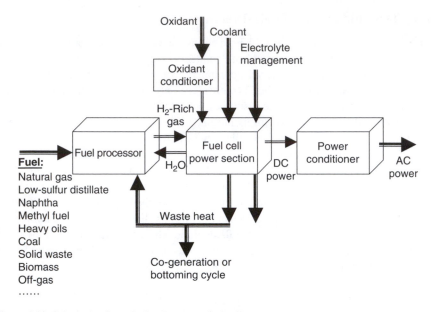

Figure 1.11 Schematic of a typical rudimentary, fuel cell system.

exchangers into the fuel cell system for better energy efficiency and it is also possible for some types of fuel cells to use the waste heat as the heat source for either cogeneration or bottoming cycles for additional electric energy generation. The cogeneration of heat and hot water (and sometimes steam) along with electricity generation increases the overall energy efficiency of the fuel cell system to as much as 85% or more. Heat is critical to human survival, for example, for space heating and household use. Both heat and steam are significantly important commodity in industrial processes, in addition to many other practical applications.

The DC-to-AC converter is a fairly mature technology due to the incorporation of semiconductor and integrated circuit technology, and its conversion energy efficiency is very high, as much as over 96% for megawatt-size power plants. The fuel processor converts the primary and/or portable fuel (e.g., such as natural gas, low-sulfur distillate, naphtha, methyl fuel — mostly methanol — heavy oils, coal, solid waste, biomass, etc.) into H_2 and CO. These secondary fuels (H_2 and CO) are considerably more electrochemically active in the electrochemical cell stack than the primary fuels. Even though fuel processing technology is highly advanced and efficient, it typically accounts for a third of power plant size, weight and cost for the hydrocarbon-fueled fuel cell power plants; roughly the electrochemical fuel cell stack accounts another third of the size, weight, and cost; while the ancillary components and subsystems associated with air supply, thermal management, water recovery and treatment, cabinet ventilation and system control and diagnostics (often referred to as the balance of the plant) accounts the remaining third. In fuel cell systems, the most important subsystem is the electrochemical fuel cell stack, and fuel processor is the second major sub-system if hydrocarbon fuels are used as the primary fuel. Therefore, we will focus on the fuel cell power section for this book, with some coverage on fuel processing and oxidant conditioning.

1.9 OBJECTIVES AND SCOPE OF THE BOOK

A major objective of this book is to provide a coherent and comprehensive treatment of fuel cells from basic fundamentals to state-of-the-art technology and knowledge so that both beginners and veterans in the field will appreciate. The subject matter is described in a sufficiently simple and gradual manner so that those with no prior experience and exposure to fuel cells, but with a basic engineering background, can read and understand, while those experienced in the field will also find this book a useful and valuable reference. It is hoped, therefore, that some readers may be motivated to join in and work on the exciting development of fuel cell technology.

The book is divided into 11 chapters, with the first chapter serving as the introduction and the next three chapters devoted to the key physical and electrochemical sciences that form the fundamental framework of fuel cells: chemical thermodynamic analysis for the best possible fuel cell performance in Chapter 2, electrochemical kinetics in Chapter 3, and transport phenomena and conservation principles for mass (including charged species), momentum and energy in Chapter 4. The latter two chapters focus on irreversible loss mechanisms in fuel cells, including activation, ohmic, and concentration polarization. In Chapters 5–10, each major type of fuel cells under research, development, and demonstration will be described, including alkaline fuel cells, phosphoric acid fuel cells, proton exchange membrane fuel cells, fuel processing and direct methanol fuel cells, molten carbonate fuel cells, and solid oxide fuel cells. The fundamentals are applied to each type of fuel cells in order to develop an understanding of the working principles and performances of these fuel cells. The final chapter (Chapter 11) is devoted to fuel cell system design, integration, and optimization.

Traditionally, in the past several decades, any book on fuel cells does not provide a complete coverage without some sort of discussion on the capital cost of fuel cell power plants. In fact, high capital costs have been plaguing the fuel cell industry and community for decades, if not centuries. Even today, the estimated capital costs of several thousand dollars U.S./kW or higher must be reduced substantially in order to be competitive in the market place. For example, a cost reduction to $30 \sim 50$ U.S./kW might be necessary for fuel cells to compete with the internal combustion engines in the transportation arena. In stationary power generation, the advanced gas turbine combined cycle power systems have fuel to electricity conversion efficiency approaching 60% with modest emissions and installed capital costs under $600 U.S./kW, and they are expected to be installed in the United States as central station bulk-power generators to meet the majority of new capacity for electricity over the next decade or so.[22] Transportation and stationary electricity generation are perhaps the two largest sectors of fuel cell applications. Therefore, it seems that fuel cell power systems have formidable, if surmountable, barriers to overcome before widespread commercial applications can be realized. Fuel cell development is currently technology driven, rather than market driven. However, it is the objective of this book to stay focused on the fundamental and technical aspects of fuel cells, rather than their economics, although the cost issue will be pointed out whenever the need arises.

1.10 SUMMARY

This chapter provides the introductory coverage on fuel cells. We begin with the motivation for fuel cells, followed by the fundamental concepts, ideas, and operational principles of fuel cells. A comparison of fuel cells is made with their main competitors: battery and heat engines. Important physical and chemical phenomena occurring in an electrode, especially around the three-phase zone, are briefly discussed that directly lead to the main functions and requirements for the electrode and electrolyte in fuel cells. We then provided a typical performance curve for fuel cells and outlined the advantages and practical applications of fuel cells as well. We point out the objective for and background needed for fuel cell RD&D. You should be able to describe the various ways fuel cells are classified and the reason for the names of the six major types of fuel cells as well as their typical operating characteristics. Commerical fuel cell applications invariably use the ambient air as the oxidant for fuel cell reaction, you should be able to describe the dry and humid air composition, the various representations of the moisture in the air and their relations. Although hydrogen is the direct fuel for almost all practically important fuel cells (except the direct methanol fuel cell), you should also know in general hydrocarbon fuels and their general characteristics. Towards the end of this chapter we provide a brief account of historical development of fuel cells and a typical fuel cell system. You should develop appreciation for the various system components when hydrocarbon fuels are used as the primary fuel.

BIBLIOGRAPHY

1.   Chalk, S. G., J. F. Miller, and F. W. Wagner, 2000. Challenges for fuel cells in transport applications. *J. Pow. Sour.*, 86: 40–51.
2.   Smith, W. 2000. The role of fuel cells in energy storage. *J. Pow. Sour.*, 86: 74–83.
3.   Maggetto, G. and J. Van Mierlo, 2000. Electric and electric hybrid vehicle technology: A survey. *IEE Seminar 2000 on Electric. Hybrid & Fuel Cell Vehicles*. London: IEE.
4.   Vincent, C. A. 2000. Battery systems for electric vehicles. *IEE Seminar 2000 on Electric. Hybrid & Fuel Cell Vehicles*. London: IEE.
5.   Mellor, P. H., N. Schofield, and D. Howe, 2000. Flywheel and supercapacitor peak power buffer technologies. *IEE Seminar 2000 on Electric. Hybrid & Fuel Cell Vehicles*. London: IEE.
6.   Blomen, L. J. M. J. and M. N. Mugerwa 1993. *Fuel Cell Systems*. New York: Plenum Press.
7.   Campbell, C. J. and J. H. Laherrere, 1998. The end of cheap oil. *Sci. Am.*, March 1998, pp. 78–83.
8.   Hirschenhofer, J. H., D. B. Stauffer, and R. R. Engleman, 1994. *Fuel Cells: A Handbook* (Revision 3), U.S. Dep. En.
9.   Schora, F. C. and E. H. Camara, 1991. *Fuel Cells: Power for the Future*. In Energy and the Environment in the 21st Century, eds. J. W. Tester, D. O. Wood and N. A. Ferrari. Cambridge, MA: MIT Press, pp. 959–971.
10.   International Fuel Cells' website: http://www.ifc.com. Accessed in 2002.

11. King, J. M. and M. J. O'Day, 2000. Applying fuel cell experience to sustainable power products. *J. Pow. Sour.*, 86: 16–22.
12. Heywood, J. B. 1988. *Internal Combustion Engine Fundamentals*. New York: McGraw Hill.
13. Cengel, Y. A. and M. A. Boles 2002. *Thermodynamics, An Engineering Approach*, 4th ed. New York: McGraw Hill.
14. Wagner, W., J. R. Cooper, A. Dittmann, J. Kijima, H.-J. Kretzschmar, A. Kruse, R. Mares, K. Oguchi, H. Sato, I. Stocker, O. Sifner, Y. Takaishi, I. Tanishita, J. Trubenbach. and Th. Willkommen, 2000. The IAPWS industrial formulation 1997 for the thermodynamic properties of water and steam. *J. Eng. Gas. Turb. Pow*. 122: 1–184.
15. Turns, S. R. 2000. *An Introduction to Combustion*, 2nd ed. New York: McGraw Hill.
16. Appleby, A. J. and F. R. Foulkes, 2000. *Fuel Cell Handbook*. New York: Van Nostrand Reinhold.
17. Grove, W. R. 1839. On Voltaic Series and the Combination of Gases by Platinum. *Phil. Mag.*, 14: 127–130.
18. Davy, H. 1802. An account of a method of constructing simple and compound Galvanic combinations, without the use of metallic substances, by means of charcoal and different fluids. *J. Nat. Phil. Chem. Arts*, 1: 144–145.
19. Mond, L. and C. Langer, 1889. A new form of gas battery. *Proc. R. Soc. London*, 46: 296–304.
20. Kivisaari, J. 1995. *Fuel Cell Lecture Notes*.
21. Stone, C. and A. E. Morrison, 2002. From Curiosity to Power to Change the World. *Sol. State Ionics*, 152–153: 1–13.
22. Rastler, D. 2000. Challenges for fuel cells as stationary power resource in the evolving energy enterprise. *J. Pow. Sour.*, 86: 34–39.

PROBLEMS

1.1. List the major similarities and differences between a fuel cell and a battery.
1.2. List the major similarities and differences between a fuel cell and a heat engine-driven generator.
1.3. Describe the working principle of an acid-electrolyte fuel cell and the major components of a fuel cell.
1.4. Discuss the function of and requirement for electrodes in a fuel cell.
1.5. Discuss the function of and requirement for electrolyte in a fuel cell.
1.6. Discuss the function of and requirement for catalyst in a fuel cell.
1.7. Discuss the function of and requirement for electrode backing layer in a fuel cell.
1.8. Describe the general characteristics of fuel cell systems.
1.9. Discuss the major advantages of fuel cell systems that make them popular among today's advanced energy conversion and power generation technologies.
1.10. Describe various classifications and types of fuel cells.
1.11. Describe the typical operating temperature of various major types of fuel cells that are under intense development today.
1.12. What parameters are used to represent the amount of water and the degree of humidification for water vapor containing gas mixtures?
1.13. Describe briefly the history of fuel cell development.

THERMODYNAMICS OF FUEL CELLS

2.1 INTRODUCTION

This chapter focuses on several thermodynamic concepts that are important in the study of fuel cells. Since in fuel cells reactants (fuel and oxidant) react to produce the useful electric energy output, we first review several thermodynamic concepts related to reacting systems that are specifically useful for fuel cell analysis: absolute enthalpy, enthalpy of reaction, heating values, Gibbs function, Gibbs function of formation, and Gibbs function of reaction. Next we examine the maximum possible performance for fuel cells, namely, the reversible cell potential by using the first and second laws of thermodynamics, and its variations with the operating conditions such as temperature, pressure, and reactant concentrations in the reactant streams. Finally, the issue of energy conversion efficiency is presented with the help of the first and second laws of thermodynamics. The maximum possible efficiency for fuel cells is investigated; a comparison is made with Carnot efficiency which is the maximum possible efficiency for heat engines against which fuel cells are competing for commercial success. Then the possibility of more than 100% efficiency for fuel cells is examined. The energy conversion efficiency for a fuel cell system comprised of fuel cells and auxiliary equipment is considered, and efficiency loss mechanism for operating fuel cells is also discussed. Examples are presented, where necessary, to illustrate the concepts and principles involved.

2.2 REVIEW OF THERMODYNAMIC CONCEPTS

This section focuses as on a brief review of a number of important concepts and properties in thermodynamics involving chemical reaction that will be needed in the analysis of fuel cell reactions. We begin with the concepts of absolute enthalpy and enthalpy of formation, including the definition of the standard temperature and pressure. The calculation of average specific heat and the sensible enthalpy change is illustrated. We then consider the enthalpy of reaction, enthalpy of combustion, as well as higher and lower heating values. The section ends with a brief discussion on the Gibbs function of reaction and Gibbs function of formation, and we emphasize the standard Gibbs function of formation and the Gibbs function of formation at other temperatures — the latter is convenient for analysis of chemical reactions occurring at elevated temperatures (including those in fuel cells) but it is not commonly used or available in many thermodynamics texts.

2.2.1   Absolute Enthalpy and Enthalpy of Formation

It is very useful to apply the concept of absolute enthalpy in the thermodynamic analysis involving chemically reacting systems. Absolute enthalpy at any state, which can be determined by system temperature T and pressure P, is often defined as the enthalpy that includes both chemical energy and sensible thermal energy. Chemical energy is associated with chemical bonds and often represented by the enthalpy of formation, h_f, at a given reference state defined by a temperature T_{ref} and a pressure P_{ref}, and the sensible thermal energy, Δh_s, represents the enthalpy difference between the values at the given state and the reference state defined. For any species i, its absolute enthalpy can therefore be expressed as

$$h_I(T, P) = h_{f,i}(T_{ref}, P_{ref}) + \Delta h_{s,i}(T, P) \tag{2.1}$$

where $h_{f,i}(T_{ref}, P_{ref})$ is usually measured by the amount of enthalpy changes in forming the species i, hence the name of enthalpy of formation, from the elements in their naturally occurring state at the reference state defined. Any state can be defined as the reference state; however, it is more convenient and practical to use 25 °C and 1 atm pressure for chemical thermodynamic tabulations. Therefore, the reference state defined by temperature, $T_{ref} = 25\,°C$ and pressure $P_{ref} = P^o = 1$ atm (or 101,325 Pa) is often called the standard reference state. Further, it is the convention in chemical thermodynamics that enthalpy of elements in their naturally occurring state is defined as zero at the standard reference state; consequently, enthalpy of formation for these elemental species is zero at the standard reference state. For example, hydrogen exists as diatomic molecules (i.e., hydrogen gas, H_2, the naturally occurring state) at 25 °C and 1 atm; hence,

$$h_{f,H_2}(T_{ref}, P_{ref}) = h_{f,H_2}(T_{ref}, P^o) = h^o_{f,H_2}(25\,°C) = 0$$

where the superscript "^{o}" is often used to denote that the value is evaluated at the standard-state pressure. Similarly, for oxygen and nitrogen

$$h^o_{f,O_2}(25\,°C) = 0 \quad \text{and} \quad h^o_{f,N_2}(25\,°C) = 0$$

The enthalpy of formation can be determined by laboratory measurements or by advanced methods of statistical thermochemistry. In laboratory measurements, the standard enthalpy of formation, $h^o_{f,i}(25\,°C)$, for a given substance is equal to the amount of heat absorbed or released when one mole of the substance is formed from its elemental substances in their standard reference state, for example,

$$C(s) + O_2(g) \longrightarrow CO_2 - 393,522\,J/mol$$

where the negative sign indicates that heat is released when CO_2 is formed from its elemental substances: solid carbon, $C(s)$, and oxygen gas, $O_2(g)$. The standard enthalpy of formation for many substances is available in many thermodynamic property tabulations and it is given in Appendix B for selected substances at the standard reference state. Note that the standard enthalpy of formation often has negative values, a result of the choice of the reference state.

The sensible enthalpy change, $\Delta h_{s,i}(T, P)$, can be determined by the thermodynamic property relations or tabulations. However, for solids and liquids (they are often approximated as incompressible substances) as well as ideal gases it is a function of temperature only, and it is convenient to calculate by using the specific heat at constant pressure, $c_p(T)$, as follows:

$$\Delta h_{s,i}(T, P) = h_i(T, P) - h_{f,i}(T_{ref}, P_{ref})$$
$$= \Delta h_{s,i}(T)$$
$$= \int_{T_{ref}}^{T} c_p(T)dT = \bar{c}_p(T - T_{ref}) \qquad (2.2)$$

where

$$\bar{c}_p = \frac{1}{T - T_{ref}} \int_{T_{ref}}^{T} c_p(T)dT \qquad (2.3)$$

is the average specific heat over the temperature range involved. The specific heat at constant pressure is also a thermodynamic property, available in many property tables, and it is given in Appendix A.3 for selected species. Based on laboratory measurements or advanced statistical thermodynamics, accurate analytical expressions are available for the specific heat. Appendix A.3 gives the specific heat at constant pressure for a number of gases (A) at 300 K, (B) as a function of temperature, and (C) as a third-degree polynomial fit to the data. In general, specific heat c_p does not vary with temperature appreciably; and it can be approximated as a linear function of temperature within a reasonable temperature range, say

$$c_p = a + bT$$

Then, we have

$$\bar{c}_p = \frac{1}{T - T_{ref}} \int_{T_{ref}}^{T} c_p(T)dT = \frac{1}{T - T_{ref}} \int_{T_{ref}}^{T} (a + bT)dT$$

$$= \frac{1}{T - T_{\text{ref}}} \left[a(T - T_{\text{ref}}) + \frac{b}{2} \left(T^2 - T_{\text{ref}}^2 \right) \right]$$

$$= a + b \left(\frac{T + T_{\text{ref}}}{2} \right) = c_p \left(\frac{T + T_{\text{ref}}}{2} \right)$$

Clearly, in this case the average specific heat is exactly the same as the specific heat evaluated at the average temperature $(T + T_{\text{ref}})/2$. This suggests that since c_p is a weak function of temperature over the typical temperature range encountered in most of engineering applications, the average specific heat $\bar{c}_p$ can be determined with any of the following three methods:

(a) Evaluate $\bar{c}_p$ at the average temperature $(T + T_{\text{ref}})/2$ from data given in Appendix A.3B, or $\bar{c}_p \approx c_p[(T + T_{\text{ref}})/2]$.
(b) Evaluate $\bar{c}_p$ as the average of specific heats $c_p(T)$ and $c_p(T_{\text{ref}})$ from data given in Appendix A.3B, or $\bar{c}_p \approx [c_p(T) + c_p(T_{\text{ref}})]/2$.
(c) Evaluate by integrating the third-order polynomial expression given in Appendix A.3C. Methods (a) and (b) require the system temperature T be known. If it is not known, the specific heats may be evaluated at T_{ref} or with an anticipated T. Once T is known, the calculation may be repeated until convergence (iteration). However, such iteration process is normally not required because of the weak dependence of specific heat on temperature. These three methods can yield results typically within a few percent range, acceptable for most engineering applications.

Note that enthalpy is an extensive thermodynamic property and the lower case symbol h represents the specific enthalpy either on a per-mole or per-unit mass basis. For the present notation, we will use the lower case symbol to denote the specific property on a per mole basis, unless otherwise stated. Specific properties on a mole basis are most convenient for thermodynamic analysis involving reacting systems.

Example 2.1 Determine the average specific heat at constant pressure for H_2 at temperatures between $T = 25\,°C$ and $T = 1000\,K$ and at the pressure of 1 atm by using (a) the average temperature; (b) the average of the specific heats; (c) the third-order polynomial expression; (d) then compare the results obtained from these three methods.

SOLUTION
H_2 can be considered as an idea gas for the specified conditions ranging from $T_{\text{ref}} = 25\,°C$ (298 K) to $T = 1000\,K$ at 1 atm. Hence, specific heat at constant pressure is a function of temperature only and the calculation methods discussed earlier can be used without modification.

(a) The specific heat at constant pressure, $c_p(T)$, for H_2 is given in Appendix A.3B as a function of temperature. At the average temperature

$$\bar{T} = \frac{T + T_{\text{ref}}}{2} = \frac{298 + 1000}{2} = 649\,K$$

Then Table A.3B gives (at 650 K),

$$\bar{c}_p = 14.571 \, \text{kJ}/(\text{kg} \cdot \text{K})$$

Since the molecular weight for H_2 is

$$W_{H_2} = 2.016 \, \text{kg/kmol}$$

We have the average specific heat at constant pressure on a per-mole basis

$$\bar{c}_p = 14.571 \, \text{kJ}/(\text{kg} \cdot \text{K}) \times 2.016 \, \text{kg/kmol} = 29.375 \, \text{kJ}/(\text{kmol} \cdot \text{K})$$

(b) From Appendix A.3b, we have

$$c_p(300 \, \text{K}) = 14.307 \, \text{kJ}/(\text{kg} \cdot \text{K}), \quad \text{and} \quad c_p(1000 \, \text{K}) = 14.983 \text{kJ}/(\text{kg} \cdot \text{K})$$

Therefore, the average specific heat, as the average of the specific heats at 298 K (approximated as 300 K for convenience) and 1000 K becomes

$$\bar{c}_p = \frac{1}{2}[c_p(300 \, \text{K}) + c_p(1000 \, \text{K})] = \frac{1}{2}(14.307 + 14.983) \, \text{kJ}/(\text{kg} \cdot \text{K})$$

$$= 14.645 \, \text{kJ}/(\text{kg} \cdot \text{K}) \times 2.016 \, \text{kg/kmol}$$

$$= 29.524 \, \text{kJ}/(\text{kmol} \cdot \text{K})$$

(c) The specific heat at constant pressure, $c_P(T)$, for H_2 is given in Appendix A.3C as a function of temperature in the form of a third-order polynomial expressed as

$$c_p(T) = a + bT + cT^2 + dT^3$$

where $a = 29.11$, $b = -0.1916 \times 10^{-2}$, $c = 0.4003 \times 10^{-5}$, and $d = -0.8704 \times 10^{-9}$. Then the average specific heat, by definition, is determined as

$$\bar{c}_p = \frac{1}{T - T_{\text{ref}}} \int_{T_{\text{ref}}}^{T} c_p(T) dT = \frac{1}{T - T_{\text{ref}}} \int_{T_{\text{ref}}}^{T} (a + bT + cT^2 + dT^3) \, dT$$

$$= \frac{1}{T - T_{\text{ref}}} \left[a(T - T_{\text{ref}}) + \frac{b}{2}(T^2 - T_{\text{ref}}^2) + \frac{c}{3}(T^3 - T_{\text{ref}}^3) \right.$$

$$\left. + \frac{d}{4}(T^4 - T_{\text{ref}}^4) \right]$$

Substituting all the numerical values into the above expression, we obtain

$$\bar{c}_p = 29.409 \, \text{kJ}/(\text{kmol} \cdot \text{K})$$

(d)   The result from method (c) is sufficiently accurate and can be viewed as the exact result for comparison purpose. Then the error for the method (a) is

$$\left| \frac{29.375 - 29.409}{29.409} \right| = 0.1\%$$

and the error for method (b) is

$$\left| \frac{29.524 - 29.409}{29.409} \right| = 0.4\%$$

Clearly, the approximate but simple methods (a) and (b) yield results that are in very close agreement with the result from much more elaborate calculation of method (c). This is not surprising since specific heat is a very weak function of temperature over the temperature ranges on the order of hundred degrees. Therefore, simple methods (a) and (b) can be used safely in all fuel cell calculations with errors well within the engineering tolerances.

COMMENT

1.   In thermodynamic calculations, temperature must be in Kelvin scale (absolute temperature) in order to obtain correct results.
2.   When using the tabulated data for calculations, care should be taken for units associated with the numerical values tabulated.
3.   If we use the specific heat at 300 K, instead of the preceding three averaging methods, the result would be 28.843 kJ/(kmol · K), which is in error by about 2%. Error of this magnitude would be acceptable for most engineering applications.

With this presentation on the enthalpy of formation and the calculation of sensible enthalpy through the average specific heat, we are ready to determine the absolute enthalpy for substances that may be encountered in fuel cell analysis.

Example 2.2   Determine the absolute enthalpy of H_2, O_2, and water (H_2O) at the pressure of $P = 1$ atm and temperature of (a) 25 °C; (b) 80 °C; (c) 200 °C; (d) 650 °C; and (e) 1000 °C. Notice that for conditions (a) and (b), water can be in the liquid form, $H_2O(\ell)$, or in the vapor form, $H_2O(g)$; calculate the absolute enthalpy for both cases.

SOLUTION

From Appendix A.2, the enthalpy of formation at the standard reference state (25 °C and 1 atm) is, for H_2, O_2, liquid water $H_2O(\ell)$, and water vapor $H_2O(g)$, respectively

$$h_{f,H_2} = 0; \quad h_{f,O_2} = 0; \quad h_{f,H_2O(\ell)} = -285{,}826 (\text{J/mol});$$

$$h_{f,H_2O(g)} = -241{,}826 (\text{J/mol})$$

Therefore, the emphasis now is to determine the sensible enthalpy change from the reference temperature of $T_{ref} = 25\,°C = 298$ K to the given temperature T.

(a) For $T = 25\,°C = 298$ K: Since the temperature is the same as the reference temperature, the sensible enthalpy is zero for all the species, and the absolute enthalpy is therefore equal to the enthalpy of formation in this case. That is,

$$h_{f,H_2} = 0; \quad h_{f,O_2} = 0; \quad h_{f,H_2O(\ell)} = -285{,}826(J/mol);$$

$$h_{f,H_2O(g)} = -241{,}826(J/mol)$$

(b) For $T = 80\,°C = 353$ K: The average temperature is $(298 + 353)/2 = 325.5$ K. From Appendix B, the average specific heat for H_2, O_2, and water vapor $H_2O(g)$ are, after interpolating the tabulated data and converting to per-mole basis through the use of molecular weight

$$\bar{c}_{p,H_2} = 14.367\,kJ/(kg \cdot K) \times 2.016\,kg/kmol = 28.964\,kJ/(kmol \cdot K)$$
$$= 28.964\,J/(mol \cdot K)$$
$$\bar{c}_{p,O_2} = 0.923\,kJ/(kg \cdot K) \times 31.999\,kg/kmol = 29.535\,kJ/(kmol \cdot K)$$
$$= 29.535\,J/(mol \cdot K)$$
$$\bar{c}_{p,H_2O(g)} = 33.860\,J/(mol \cdot K)$$

For liquid water $H_2O(\ell)$, Appendix B yields

$$\bar{c}_{p,H_2O(\ell)} = 4.182\,kJ/(kg \cdot K) \times 18.015\,kg/kmol$$
$$= 75.339\,kJ/(kmol \cdot K) = 75.339\,J/(mol \cdot K)$$

Therefore, the absolute enthalpy for H_2, O_2, $H_2O(\ell)$, and $H_2O(g)$ is determined as follows

$$h_{H_2} = h_{f,H_2} + \bar{c}_{p,H_2}(T - T_{ref}) = 0 + 28.964 \times (353 - 298) = 1{,}593.0\,J/mol$$
$$h_{O_2} = h_{f,O_2} + \bar{c}_{p,O_2}(T - T_{ref}) = 0 + 29.535 \times (353 - 298) = 1{,}624.4\,J/mol$$
$$h_{H_2O(\ell)} = h_{f,H_2O(\ell)} + \bar{c}_{p,H_2O(\ell)}(T - T_{ref}) = -285{,}826 + 75.339 \times (353 - 298)$$
$$= -281{,}682\,J/mol$$
$$h_{H_2O(g)} = h_{f,H_2O(g)} + \bar{c}_{p,H_2O(g)}(T - T_{ref}) = -241{,}826 + 33.860 \times (353 - 298)$$
$$= -239{,}964\,J/mol$$

(c) For $T = 200°C = 473$ K: At this temperature water will be in vapor form for the pressure of 1 atm. The average temperature is $(298 + 473)/2 = 385.5$ K. From Appendix A.3B, the average specific heat for H_2, O_2 and water vapor $H_2O(g)$ are, similar to the process shown in (b) above

$$\bar{c}_{p,H_2} = 14.462\,\text{kJ}/(\text{kg} \cdot \text{K}) \times 2.016\,\text{kg/kmol} = 29.155\,\text{kJ}/(\text{kmol} \cdot \text{K})$$
$$= 29.155\,\text{J}/(\text{mol} \cdot \text{K})$$
$$\bar{c}_{p,O_2} = 0.937\,\text{kJ}/(\text{kg} \cdot \text{K}) \times 31.999\,\text{kg/kmol} = 29.983\,\text{kJ}/(\text{kmol} \cdot \text{K})$$
$$= 29.983\,\text{J}/(\text{mol} \cdot \text{K})$$
$$\bar{c}_{p,H_2O(g)} = 34.343\,\text{J}/(\text{mol} \cdot \text{K})$$

Therefore, the absolute enthalpy for H_2, O_2, and $H_2O(g)$ becomes

$$h_{H_2} = h_{f,H_2} + \bar{c}_{p,H_2}(T - T_{\text{ref}}) = 0 + 29.155 \times (473 - 298)$$
$$= 5,102.1\,\text{J/mol}$$
$$h_{O_2} = h_{f,O_2} + \bar{c}_{p,O_2}(T - T_{\text{ref}}) = 0 + 29.983 \times (473 - 298)$$
$$= 5,247.0\,\text{J/mol}$$
$$h_{H_2O(g)} = h_{f,H_2O(g)} + \bar{c}_{p,H_2O(g)}(T - T_{\text{ref}}) = -241,826 + 34.343$$
$$\times (473 - 298) = -235,816\,\text{J/mol}$$

(d)   For $T = 650\,°\text{C} = 923\,\text{K}$: At this temperature water exists in vapor form for the pressure of 1 atm. The average temperature is $(298 + 923)/2 = 610.5\,\text{K}$. From Appendix A.3B, the average specific heat for H_2, O_2, and water vapor $H_2O(g)$ are, similar to the process shown in (b)

$$\bar{c}_{p,H_2} = 14.551\,\text{kJ}/(\text{kg} \cdot \text{K}) \times 2.016\,\text{kg/kmol} = 29.335\,\text{kJ}/(\text{kmol} \cdot \text{K})$$
$$= 29.335\,\text{J}/(\text{mol} \cdot \text{K})$$
$$\bar{c}_{p,O_2} = 1.006\,\text{kJ}/(\text{kg} \cdot \text{K}) \times 31.999\,\text{kg/kmol} = 32.191\,\text{kJ}/(\text{kmol} \cdot \text{K})$$
$$= 32.191\,\text{J}/(\text{mol} \cdot \text{K})$$
$$\bar{c}_{p,H_2O(g)} = 36.528\,\text{J}/(\text{mol} \cdot \text{K})$$

Therefore, absolute enthalpy for H_2, O_2, and $H_2O(g)$ becomes

$$h_{H_2} = h_{f,H_2} + \bar{c}_{p,H_2}(T - T_{\text{ref}}) = 0 + 29.335 \times (923 - 298)$$
$$= 18,334\,\text{J/mol}$$
$$h_{O_2} = h_{f,O_2} + \bar{c}_{p,O_2}(T - T_{\text{ref}}) = 0 + 32.191 \times (923 - 298)$$
$$= 20,119\,\text{J/mol}$$
$$h_{H_2O(g)} = h_{f,H_2O(g)} + \bar{c}_{p,H_2O(g)}(T - T_{\text{ref}}) = -241,826 + 36.528$$
$$\times (923 - 298) = -218,996\,\text{J/mol}$$

(e) For $T = 1000\,°C = 1273$ K: At this temperature and 1 atm water exists in vapor form. The average temperature is $(298 + 1273)/2 = 785.5$ K. From Appendix B, the average specific heat for H_2, O_2, and water vapor $H_2O(g)$ are,

$$\bar{c}_{p,H_2} = 14.681\,\text{kJ}/(\text{kg}\cdot\text{K}) \times 2.016\,\text{kg}/\text{kmol} = 29.597\,\text{kJ}/(\text{kmol}\cdot\text{K})$$

$$= 29.597\,\text{J}/(\text{mol}\cdot\text{K})$$

$$\bar{c}_{p,O_2} = 1.051\,\text{kJ}/(\text{kg}\cdot\text{K}) \times 31.999\,\text{kg}/\text{kmol} = 33.631\,\text{kJ}/(\text{kmol}\cdot\text{K})$$

$$= 33.631\,\text{J}/(\text{mol}\cdot\text{K})$$

$$\bar{c}_{p,H_2O(g)} = 38.518\,\text{J}/(\text{mol}\cdot\text{K})$$

Therefore, absolute enthalpy for H_2, O_2, and $H_2O(g)$ becomes

$$h_{H_2} = h_{f,H_2} + \bar{c}_{p,H_2}(T - T_{\text{ref}}) = 0 + 29.597 \times (1273 - 298)$$

$$= 28{,}857\,\text{J}/\text{mol}$$

$$h_{O_2} = h_{f,O_2} + \bar{c}_{p,O_2}(T - T_{\text{ref}}) = 0 + 33.631 \times (1273 - 298)$$

$$= 32{,}790\,\text{J}/\text{mol}$$

$$h_{H_2O(g)} = h_{f,H_2O(g)} + \bar{c}_{p,H_2O(g)}(T - T_{\text{ref}}) = -241{,}826 + 38.518$$

$$\times (1273 - 298) = -204{,}271\,\text{J}/\text{mol}$$

COMMENT

1. These calculations reveal that as the temperature increases, the specific heat at constant pressure also increases but very slowly, confirming our early statement that specific heat is a very weak function of temperature.
2. The increase in the specific heat is the smallest for H_2, which is the smallest molecule among the three species considered; is the largest for H_2O, the largest in molecular size.
3. For a compound substance such as H_2O, enthalpy of formation dominates in the numerical value of absolute enthalpy, whereas sensible enthalpy is small in comparison. For $H_2O(g)$ at 1273 K, the sensible enthalpy contributes only about 18% to absolute enthalpy.
4. In the determination of absolute enthalpy, almost all the effort is spent on the determination of sensible enthalpy. Therefore, it might be wise to devise a way to minimize, or even eliminate, the sensible enthalpy calculation (a method is given later in this chapter).

2.2.2 Enthalpy of Reaction, Enthalpy of Combustion, and Heating Values

For a chemically reacting system, enthalpy of reaction is defined as the difference between the enthalpy for the reaction products and that for the reactants. Hence, the enthalpy of reaction can easily be calculated once the absolute enthalpy for the mixture of products and mixture of reactants is known. In laboratory measurements, enthalpy of reaction can be easily determined as the amount of heat released or absorbed during an **isobaric** chemical reaction. The chemical reaction can occur either in a closed system for which the final system temperature must be the same as the initial system temperature, as shown in Figure 2.1, or in a steady flow reactor where the reactants entering and products exiting the reactor have the same temperature and pressure, as shown in Figure 2.2. For either system, the product mixture and reactant mixture need to have the same temperature and pressure, although the mixture composition, as specified by the number of moles n_i for each species i, is certainly different for the reactant and product because of the occurrence of the chemical reaction inside the system. Note that we have adopted the conventional notation that heat absorbed by the system is defined as positive.

Using energy balance (or the first law of thermodynamics), one can show that for both of the systems shown in Figures 2.1 and 2.2, the amount of heat absorbed by the system is related to the absolute enthalpies for the product and reactant mixtures as follows:

$$q = h_P - h_R = \Delta h_{\text{reaction}} \tag{2.4}$$

where the subscript "P" stands for the product and "R" for the reactant. The lower case q for the heat transfer and h for the absolute enthalpy represent the respective values

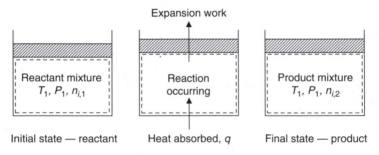

Figure 2.1 Closed-system reactor used for the determination of enthalpy of reaction. The dashed curve represents the system boundary.

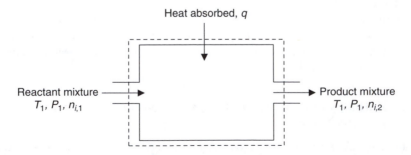

Figure 2.2 Steady-flow reactor used for the determination of enthalpy of reaction. The dashed curve represents the system boundary.

on a per unit mole (or mass) of one of the reactants, typically fuel for fuel cell reaction or combustion analysis (e.g., see Example 2.3). If enthalpy of reaction, $\Delta h_{reaction}$, is negative, heat is released during the reaction, consequently such a reaction is called **exothermic**; similarly, if the enthalpy of reaction, $\Delta h_{reaction}$, is positive, indicating that heat is absorbed by the system in order for the reaction to proceed, such a reaction is often termed **endothermic**. Therefore, endothermic reaction requires external means to provide the heat needed for the reaction to occur; whereas exothermic reaction can usually proceed by itself, once the reaction is initiated because the heat created during the reaction can usually be transferred to the surrounding medium. If heat generated is not transferred out of the system, then the system temperature will be increased; higher temperature will enhance the rate of reaction, hence even more heat generation, leading to a self-accelerated reaction process.

If the temperature and pressure for the reactant and product mixtures are 25 °C and 1 atm (the standard reference state), the resulting enthalpy change between the product and reactant is designated as standard enthalpy of reaction. If one of the reactants involved in the reaction is a fuel and the reaction is exothermic, such a reaction is often called combustion reaction. When a combustion process is **complete**, that is, when a fuel (usually hydrocarbon substances) reacts **completely** with an oxidant (usually oxygen) to form stable final products (usually CO_2 and H_2O) — meaning that all of the carbon in the fuel is converted to the stable final product CO_2 and all of the hydrogen in the fuel is converted to the stable final product H_2O, the enthalpy change for such a combustion reaction is called enthalpy of combustion. Obviously, enthalpy of combustion depends on the initial state of the fuel and oxidant and the final state of the products (because H_2O can be in liquid or vapor form with different energy content, or enthalpy values).

The enthalpy of combustion is always negative since the reaction is exothermic. Normally a positive value is desired for engineering application; hence heating value is defined as the absolute value of the standard enthalpy of combustion, that is, for reaction occurring at the standard reference temperature and pressure. When the H_2O in the product condenses to liquid, more heat is released during the reaction and the heating value is higher, hence it is called the higher heating value (HHV). Conversely, a lower heating value (LHV) results if the H_2O in the product remains in the vapor form. The difference between the HHV and LHV is equal to the enthalpy of condensation (or vaporization) for the amount of water in the product. The heating value can be expressed in terms of a per-mole fuel or per-unit mass fuel basis, and the latter is more convenient for practical use. For example, for methane (CH_4) at 25 °C and 1 atm,

$$HHV = 55,528 \, kJ/kg \, fuel$$

$$LHV = 50,016 \, kJ/kg \, fuel$$

Therefore, the higher heating value is approximately 11% larger than the lower heating value for methane.

Example 2.3 Determine the higher and lower heating values for gaseous (a) hydrogen, H_2; and (b) *n*-octane, $C_8H_{18}(g)$, reacting with O_2 to form product water. The reaction occurs at the standard reference state (25 °C and 1 atm).

SOLUTION

Heating value is equal to the magnitude of the enthalpy of reaction. We will use Equation (2.4) in conjunction with Equations (2.1)–(2.3) to determine the enthalpy of reaction first.

(a)   For 1 mol of H_2, the reaction equation can be written as

$$H_2(g) + \frac{1}{2}O_2 \longrightarrow H_2O$$

The enthalpy of reaction per mole of fuel (i.e., H_2 in this case) can be expressed, whether the product water is liquid or vapor, as

$$\Delta h_{reaction} = h_{H_2O} - \left(h_{H_2} - \frac{1}{2}h_{O_2}\right)$$

The sensible enthalpies are zero for all the species involved since the reaction occurs at the standard reference state. Further, the enthalpies of formation for H_2 and O_2 are also zero. Therefore, we obtain

$$\Delta h_{reaction} = h_{f,H_2O}$$

From Appendix A, we finally have

$$HHV = -\Delta h_{f,H_2O(\ell)} = 285{,}826\,J/mol\ H_2$$
$$LHV = -\Delta h_{f,H_2O(g)} = 241{,}826\,J/mol\ H_2$$

Therefore, for each mol H_2 burned completely with oxygen, there is 285,826 J of energy released as heat if the product water is liquid; otherwise the heat released is 241,826 J when the product water is in vapor state.

(b)   For 1 mole of n-octane, $C_8H_{18}(g)$, the reaction equation is

$$C_8H_{18}(g) + 12.5O_2 \longrightarrow 9H_2O + 8CO_2$$

The enthalpy of reaction can be written as,

$$\Delta h_{reaction} = 9h_{H_2O} + 8h_{CO_2} - \left(h_{C_8H_{28}(g)} + 12.5h_{O_2}\right)$$

Again, the sensible enthalpies are zero for all the species involved for the reaction taking place at the standard reference state. Hence, absolute enthalpy of each species is equal to their respective enthalpy of formation. From Appendix A.2, we have

$$\Delta h_{reaction} = 9 \times (-285{,}826) + 8 \times (-393{,}522) - [(-208{,}450) - 12.5 \times 0]$$
$$= -5{,}512{,}160\,J/mol\ C_8H_{18}(g)$$

when the product water is liquid, and

$$\Delta h_{reaction} = 9 \times (-241,826) + 8 \times (-393,522) - [(-208,450) - 12.5 \times 0]$$
$$= -5,116,160 \, J/mol \, C_8H_{18}(g)$$

when the product water is in vapor state. Therefore, the heating values are

$$HHV = -\Delta h_{reaction} = 5,512,160 \, J/mol \, C_8H_{18}(g)$$
$$LHV = -\Delta h_{reaction} = 5,116,160 \, J/mol \, C_8H_{18}(g)$$

COMMENT

1. In the analysis of chemically reacting systems, it is conventional to express the chemical reaction equation on a per-mole fuel basis.
2. As carbon content in the fuel molecule increases, the difference between the high and low heating values diminishes. For example, HHV is about 18% higher than LHV for H_2 and 11% higher for CH_4, but it is even less than 8% for n-octane, $C_8H_8(g)$.
3. This example indicates that heating value on a per-mole basis increases with the carbon content in the molecule. This is understandable because the molecular size increases with the carbon content, and a larger molecule is expected to have more energy contained within it. However, on a per-unit mass basis, the result will be different. For this example, the higher heating value for H_2

$$HHV = 141,779 \, kJ/kg \, H_2$$

Whereas for n-octane

$$HHV = 48,255 \, kJ/kg \, C_8H_{18}(g).$$

In fact on a per-unit mass basis, hydrogen has the highest energy content among all chemical fuels. It is interesting to point out that on a per-unit volume basis, hydrogen has the lowest energy content, or $1720 \, MJ/m^3$ at $25 \,°C$ and 1 atm for the higher heating value, which is equivalent to $0.4777 \, kW \cdot hr/L$. For a typical passenger vehicle, the gas tank is about 50 L, hence it can contain slightly less than $24 \, kW \cdot h$ energy of hydrogen, which lasts not even 12 min for a compact vehicle driven by 50 kW engine with 40% overall efficiency. This simple estimate allows us to appreciate the significance of the technical challenges related to hydrogen storage onboard vehicles.

It is useful to emphasize that enthalpy of formation for a given compound species is really enthalpy of reaction when that species is formed from elemental substances that form the species and the standard enthalpy of formation is the enthalpy of reaction when the species formation reaction occurs at the standard reference state ($25 \,°C$ and 1 atm). Then enthalpy of formation for a compound species i at any temperature and pressure, $h_{f,i}(T, P)$, can be determined by the above procedure for the determination of the enthalpy of reaction for the formation reaction of the compound species i from its constituent elemental substances.

Example 2.4   Determine the enthalpy of formation for H_2O at the pressure of $P = 1$ atm and temperature of (a) $80\,°C$; (b) $200\,°C$; (c) $650\,°C$; and (d) $1000\,°C$. Notice that for the condition (a), water can be liquid, $H_2O(\ell)$ or vapor, $H_2O(g)$; calculate for both cases.

SOLUTION

The formation reaction for H_2O can be written as

$$H_2 + \frac{1}{2}O_2 \longrightarrow H_2O$$

The enthalpy of formation for H_2O at any temperature and pressure is equal to the enthalpy of reaction for the preceding reaction and it can be written as

$$h_{f,H_2O}(T, P) = h_{H_2O}(T, P) - \left(h_{H_2}(T, P) + \frac{1}{2}h_{O_2}(T, P) \right)$$

$$= h_{f,H_2O}(T_{ref}, P_{ref}) + \Delta h_{s,H_2O}(T, P)$$

$$- \left(\Delta h_{s,H_2}(T, P) + \frac{1}{2}\Delta h_{s,O_2}(T, P) \right)$$

where Equation (2.1) has been used in the second equation and the standard enthalpy of formation is zero for H_2 and O_2 (at $T_{ref} = 25\,°C$ and $P_{ref} = 1$ atm). The standard enthalpy of formation for liquid and gaseous water is available in Appendix A.2, both absolute and sensible enthalpy for H_2, O_2, and H_2O have already been determined in Example 2.2. Utilizing those results for absolute enthalpy values, we obtain the enthalpy of formation for water at $P = 1$ atm and the temperature given as:

(a) For $T = 80\,°C = 353$ K:

$$h_{f,H_2O(\ell)}(T, P) = -281,682 - (1,593.0 + 0.5 \times 1,624.4) = -284,087\,J/mol$$

$$h_{f,H_2O(g)}(T, P) = -239,964 - (1,593.0 + 0.5 \times 1,624.4) = -242,369\,J/mol$$

(b) For $T = 200\,°C = 473$ K:

$$h_{f,H_2O(g)}(T, P) = -235,816 - (5,102.1 + 0.5 \times 5,247.0) = -243,542\,J/mol$$

(c) For $T = 650\,°C = 923$ K:

$$h_{f,H_2O(g)}(T, P) = -218,996 - (18,334 + 0.5 \times 20,119) = -247,390\,J/mol$$

(d) For $T = 1000\,°C = 1273$ K:

$$h_{f,H_2O(g)}(T, P) = -204,271 - (28,857 + 0.5 \times 32,790) = -249,523\,J/mol$$

COMMENT
1. Clearly the magnitude of the enthalpy of formation increases with temperature (or more heat is released when one mole of water is formed), because the reactants H_2 and O_2 need to be heated to the higher temperature first before the reaction, or they contain more energy in the first place.
2. Thermodynamic property evaluation is encountered all the time in the analysis of chemically reacting systems such as fuel cells. Skill is needed for property calculation.

Similar to the above example, enthalpy of formation at any temperature and pressure can be determined and is available in Appendix A.5 for CO, CO_2, H_2, H_2O, N_2 and O_2. Also tabulated are the specific heat at constant pressure, sensible enthalpy, absolute entropy and Gibbs function of formation as a function of temperature. These tabulations will help simplify greatly the determination of the property values needed for thermodynamic analysis. However, care must be taken in using the numerical values tabulated — as one may have already noticed in Appendix A.5, the enthalpy of formation has been defined as zero for all elemental substances such as H_2 and O_2 at 1 atm and any temperature, instead of at 25 °C as defined earlier in association with Equation (2.1). This is equivalent to choosing a new reference datum for property evaluation purpose at each temperature. Therefore, it is important to use consistent data in analysis and care must be taken when utilizing the tabulated data for property calculations.

Example 2.5   Determine the enthalpy of formation for $H_2O(g)$ at the pressure of $P = 1$ atm and temperature of (a) 80 °C; (b) 200 °C; (c) 650 °C; and (d) 1000 °C by using Appendix A.5.

SOLUTION
Since the desired property value is not tabulated directly in Appendix A.5, data interpolation will be used.

(a)   For $T = 80\,°C = 353$ K:

$$h_{f,H_2O(g)}(T, P) = -242{,}391 \text{ J/mol}$$

(b)   For $T = 200\,°C = 473$ K:

$$h_{f,H_2O(g)}(T, P) = -243{,}562 \text{ J/mol}$$

(c)   For $T = 650\,°C = 923$ K:

$$h_{f,H_2O(g)}(T, P) = -247{,}363 \text{ J/mol}$$

(d)   For $T = 1000\,°C = 1273$ K:

$$h_{f,H_2O(g)}(T, P) = -249{,}350 \text{ J/mol}$$

COMMENT

1. Data interpolation is almost always used when determining property values from a tabulated list of data.
2. The property for water listed in Appendix A.5 is for gaseous water only.
3. Even though the results obtained in this example are slightly different when compared to those obtained in the previous example, the difference is extremely small, less than a fraction of 1%. This small difference arises due to the approximate nature of calculation presented in both examples.
4. Clearly, it is significantly easier to determine the relevant properties from Appendix A.5.

2.2.3   Gibbs Functions of Reaction Formation

Gibbs function is also a thermodynamic property which is useful for analysis involving chemically reacting systems. It is defined as

$$g = h - Ts \tag{2.5}$$

where s is (absolute) entropy, another thermodynamic property. Both g and s can be expressed in a per-mole or per-unit mass basis. As pointed out earlier, the former is used throughout, unless stated otherwise.

Similar to enthalpy of reaction, Gibbs function of reaction is defined as the change in Gibbs function between the reaction products and reactants when the product and reactant have the same temperature and pressure. That is,

$$\Delta g_{\text{reaction}} = g_P - g_R \tag{2.6}$$

When 1 mol of a compound substance is formed from its elemental substances, the resulting Gibbs function of reaction is called Gibbs function of formation. If the formation reaction occurs at the standard reference state ($25\,°C$ and 1 atm), then it is called standard Gibbs function of formation, which is available from Appendix A.2. Appendix A.5 lists the Gibbs function of formation at other temperatures and it will become useful in the thermodynamic analysis of fuel cells presented in the rest of this chapter.

2.3   REVERSIBLE CELL POTENTIAL

In a fuel cell, the chemical energy of a fuel and an oxidant is converted directly into electrical energy, which is exhibited in terms of cell potential and electrical current output. The maximum possible electrical energy output and the corresponding electrical potential difference between the cathode and anode are achieved when the fuel cell is operated under the thermodynamically reversible condition. This maximum possible cell potential is called **reversible cell potential**, one of the significantly important parameters for fuel cells. We apply fundamental thermodynamic principle to derive the reversible cell potential in this section.

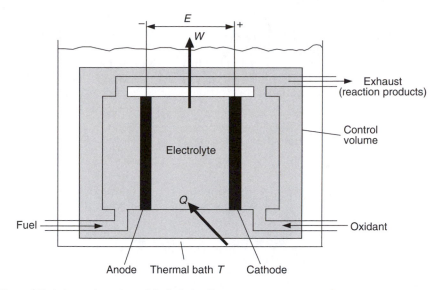

Figure 2.3 A thermodynamic model of a fuel cell system.

A thermodynamic system* model is shown in Figure 2.3 for the analysis of fuel cell performance. It is a control-volume system for the fuel cell to which fuel and oxidant streams enter and product or exhaust stream exits. The fuel cell is located inside a thermal bath in order to maintain the desired system temperature, T. The reactant streams (fuel and oxidant) and the exhaust stream are considered to have the same temperature, T, and pressure, P. It is assumed that the fuel and oxidant inflow and the exhaust outflow are steady; the kinetic and gravitational potential energy changes are negligible. Further, the overall electrochemical reactions occurring inside the fuel cell system boundary is described as follows:

$$\text{Fuel (e.g., } H_2) + \text{Oxidant (e.g., } O_2) \Rightarrow \dot{W} + \dot{Q} + \text{Product} \tag{2.7}$$

where $\dot{W}$ is the rate of work done by the system, and $\dot{Q}$ is the rate of heat transferred into the system from the surrounding constant temperature thermal bath, which may, or may not, be in thermal equilibrium with the fuel cell system at the temperature, T, and pressure, P. For hydrogen/oxygen fuel cells, the reaction product is usually water. Then, the first and second laws of thermodynamics can be written, respectively, for the present fuel cell system, as

$$\frac{dE_{C.V.}}{dt} = \left[\left(\dot{N}h + KE + PE\right)_F + \left(\dot{N}h + KE + PE\right)_{Ox}\right]_{in}$$
$$- \left[\left(\dot{N}h + KE + PE\right)_{Ex}\right]_{out} + \dot{Q} - \dot{W} \tag{2.8}$$

---

*A thermodynamic system, or simply system, is in thermodynamics a collection of matter under study (or analysis); whereas the jargon "fuel cell system" in fuel cell literature usually denotes the fuel cell power plant that consists of fuel cell stack(s) and auxiliary equipment (see Section 1.8 for more details). In this chapter, a fuel cell system may imply both meanings. However, the context will tell which it is meant to be.

$$\begin{pmatrix} \text{Increase In} \\ \text{System Energy} \end{pmatrix} = \begin{pmatrix} \text{Energy In} \\ \text{By Mass Flow} \end{pmatrix} - \begin{pmatrix} \text{Energy Out} \\ \text{By Mass Flow} \end{pmatrix}$$

$$+ \begin{pmatrix} \text{Energy In} \\ \text{As Heat} \end{pmatrix} - \begin{pmatrix} \text{Energy Out} \\ \text{As Work} \end{pmatrix}$$

$$\frac{dS_{C.V.}}{dt} = [(\dot{N}s)_F + (\dot{N}s)_{Ox}]_{in} - [(\dot{N}s)_{Ex}]_{out} + \frac{\dot{Q}}{T} + \dot{\wp}_s \tag{2.9}$$

$$\begin{pmatrix} \text{Increase In} \\ \text{System Entropy} \end{pmatrix} = \begin{pmatrix} \text{Entropy In} \\ \text{By Mass Flow} \end{pmatrix} - \begin{pmatrix} \text{Entropy Out} \\ \text{By Mass Flow} \end{pmatrix}$$

$$+ \begin{pmatrix} \text{Entropy In By} \\ \text{Heat Transfer} \end{pmatrix} + \begin{pmatrix} \text{Entropy} \\ \text{Generated} \end{pmatrix}$$

where $\dot{N}$ is the molar flow rate, h is the (absolute) enthalpy per unit mole, s is the specific entropy on a mole basis, and $\dot{\wp}_s$ is the rate of entropy generation due to irreversibilities. The subscript "F", "Ox" and "Ex" stand for fuel, oxidant and exhaust stream, respectively. "KE" and "PE" denote kinetic and gravitational potential energy that are being carried in and out of the system by the mass flow.

For a steady process, there are no temporal changes in the amount of energy $E_{C.V.}$ and entropy $S_{C.V.}$ within the control volume system, hence, $dE_{C.V.}/dt = 0$ and $dS_{C.V.}/dt = 0$. Further, the changes in the kinetic and gravitational potential energy are negligible for the present process, as assumed earlier. Therefore, Equations (2.8) and (2.9) can be simplified as follows:

$$\dot{N}_F(h_{in} - h_{out}) + \dot{Q} - \dot{W} = 0 \tag{2.10}$$

$$\dot{Q} = -T\dot{\wp}_s - \dot{N}_F T \, (s_{in} - s_{out}) \tag{2.11}$$

where

$$h_{in} = \left(h_F + \frac{\dot{N}_{Ox}}{\dot{N}_F} h_{Ox} \right)_{in} ; \quad \text{and} \quad h_{out} = \frac{\dot{N}_{Ex}}{\dot{N}_F} h_{Ex} \tag{2.12}$$

h_{in} is the amount of enthalpy-per-mole of fuel carried into the system by the reactant inflow, and h_{out} is the amount of enthalpy-per-mole of fuel taken out of the system by the exhaust stream. Similarly,

$$s_{in} = \left(s_F + \frac{\dot{N}_{Ox}}{\dot{N}_F} s_{Ox} \right)_{in} ; \quad \text{and} \quad s_{out} = \frac{\dot{N}_{Ex}}{\dot{N}_F} s_{Ex} \tag{2.13}$$

are the amount of entropy-per-mole of fuel brought into the system by the reactant inflow, and the amount of entropy-per-mole of fuel carried out of the system by the outgoing exhaust stream containing the reaction products, respectively.

Substitution of Equation (2.11) into Equation (2.10) yields

$$\dot{W} = \dot{N}_F \left(h_{in} - h_{out} \right) - \dot{N}_F T \left(s_{in} - s_{out} \right) - T \dot{\wp}_s \qquad (2.14)$$

Let

$$w = \frac{\dot{W}}{\dot{N}_F}; \quad q = \frac{\dot{Q}}{\dot{N}_F}; \quad \text{and} \quad \wp_s = \frac{\dot{\wp}_s}{\dot{N}_F} \qquad (2.15)$$

represent, respectively, the amount of work done, heat transferred, and entropy generated per-unit mole of fuel, Equations (2.11) and (2.14) then become

$$q = -T\wp_s - T \left(s_{in} - s_{out} \right) = T\Delta s - T\wp_s \qquad (2.16)$$

$$w = \left(h_{in} - h_{out} \right) - T \left(s_{in} - s_{out} \right) - T\wp_s \qquad (2.17)$$

Because the enthalpy and entropy change for the fuel cell reaction is defined as

$$\Delta h = h_{out} - h_{in}; \quad \text{and} \quad \Delta s = s_{out} - s_{in} \qquad (2.18)$$

Equation (2.17) can also be expressed as

$$w = -\Delta h + T\Delta s - T\wp_s = -\left[(h - Ts)_{out} - (h - Ts)_{in} \right] - T\wp_s \qquad (2.19)$$

From the definition of the Gibbs function (per-mole of fuel) $g = h - Ts$, Equation (2.17) or Equation (2.19) can also be written as

$$w = -(g_{out} - g_{in}) - T\wp_s = -\Delta g - T\wp_s \qquad (2.20)$$

Because by the second law of thermodynamics, entropy can be generated but can never be destroyed, we know $\wp_s \geq 0$, and also the absolute temperature (in Kelvin scale) $T > 0$ by the third law of thermodynamics, the maximum possible work (i.e., useful energy) output from the present system occurs when $\wp_s = 0$, or under the thermodynamically reversible condition, since the change in the Gibbs function is usually negative for useful fuel cell reaction. Therefore, from Equation (2.20) it is clear that the maximum possible work output from the present fuel cell system is equal to the decrease in Gibbs function, or

$$w_{max} = -\Delta g \qquad (2.21)$$

for all reversible processes, regardless of the specific type of fuel cells involved. In fact, it might be pointed out that in the derivation of Equations (2.20) and (2.21), no specifics about the control volume system have been stipulated, hence they are valid for any energy conversion systems.

For a fuel cell system, the electrical energy output is conventionally expressed in terms of the cell potential difference between the cathode and the anode. Since (electrical) potential is the (electrical) potential energy per unit (electrical) charge, its SI unit is J/C, which is more often called volt or simply V. Potential energy is defined as the work done when charge is moved from one location to another in the electrical field, normally refers to external circuits. For the internal circuit of fuel cells, such as the one shown in Figure 2.3, electromotive force is the terminology often used, which is also defined as the work done by transferring one Coulomb positive charge from a low to a high potential. Hence, electromotive force also has the SI unit of J/C, or V. We adopt the terminology of cell potential, instead of electromotive force, from now on; and we use the notation E to represent cell potential. Because normally electrons are the particles transferred that carry electrical charge, we express the work done by a fuel cell as follows

$$w(\text{J/mol fuel}) = E \times (\text{C/mol fuel})$$

or

$$w = E \times (nN_0e) = E \times (nF) \tag{2.22}$$

where n is the number of moles of electrons transferred per mole of fuel consumed, N_0 the Avogadro's number ($= 6.023 \times 10^{23}$ number of electrons/mole electron), and e the electric charge per electron ($= 1.6021 \times 10^{-19}$ Coulomb/electron). Since $N_0e = 96,487$ Coulomb/mole electron $= F$ is often known as the Faraday constant, the cell potential becomes, from Equation (2.20)

$$E = \frac{w}{nF} = \frac{-\Delta g - T\wp_s}{nF} \tag{2.23}$$

Hence, the maximum possible cell potential, or the reversible cell potential E_r, becomes

$$E_r = -\frac{\Delta g}{nF} \tag{2.24}$$

From the reversible cell potential given Equation (2.23) can also be rewritten as

$$E = E_r - \frac{T\wp_s}{nF} = E_r - \eta \tag{2.25}$$

where

$$\eta = \frac{T\wp_s}{nF} \tag{2.26}$$

is the cell voltage loss due to irreversibilities (or entropy generation). Clearly, the actual cell potential can be calculated by subtracting the cell voltage loss from the reversible cell potential. Alternatively, the amount of entropy generation per-mole fuel consumed can be determined as

$$\wp_s = \frac{nF\eta}{T} = \frac{nF(E_r - E)}{T} \tag{2.27}$$

Thus, the amount of entropy generation, representing the degree of irreversibilities (the degree of deviation from the idealized reversible condition), for the fuel cell reaction process can be measured once the cell potential E and the cell operating temperature T are known.

Note that the Gibbs function is a thermodynamic property, determined by state variables such as temperature and pressure. Hence, the change in the Gibbs function for the fuel cell reaction discussed here,

$$\Delta g = \Delta h - T\Delta s \tag{2.28}$$

is also a function of the system temperature T and pressure P, so is the reversible cell potential. The specific effect of the operating conditions, such as temperature, pressure, and reactant concentrations, on the reversible cell potential are presented in the next section. If the reaction occurs at the standard reference temperature and pressure (25 °C and 1 atm), the resulting cell potential is usually called the standard reversible cell potential $E_r^{o}*$, or

$$E_r^{o}(T_{\text{ref}}) = -\frac{\Delta g(T_{\text{ref}}, P_{\text{ref}})}{nF} \tag{2.29}$$

If pure hydrogen and oxygen are used as reactants to form product water, then $E_r^{o}(25\,°C) = 1.229\,V$ for the product water in liquid form, and $E_r^{o}(25\,°C) = 1.185\,V$ if the product water is in vapor form. The difference in E_r^{o} corresponds to the energy required for the vaporization of water. It might be pointed out that any fuel containing hydrogen (including hydrogen itself, hydrocarbons, alcohols, and to a lesser extent, coal) has two values for Δg and Δh, one higher and one lower, depending on whether the product water is in the form of liquid or vapor. Hence, care should be taken when referring to reversible cell potential and energy efficiency, which is discussed later in Section 2.5.

Example 2.6   Determine the standard reversible cell potential when H_2 is used as fuel and O_2 is used as oxidant, with the following cell reaction

$$H_2 + \frac{1}{2}O_2 \longrightarrow H_2O$$

Obtain the result by considering the product H_2O is (a) in the liquid state, $H_2O(\ell)$ and (b) in the vapor state, $H_2O(g)$, respectively.

SOLUTION
For the standard reversible cell potential, E_r^{o}, we know that the fuel cell reaction occurs at the standard reference temperature and pressure (i.e., $T = 25\,°C$ and $P = 1$ atm).

---

*In literature, the superscript "o" sometimes denotes the value at the standard reference condition of 25 °C and 1 atm; and sometimes it also refers to parameters evaluated at 1 atm. To avoid confusion, the latter meaning has been adopted here.

Whether the product is in the liquid or vapor state, the change in the Gibbs function for the fuel cell reaction can be written as

$$\Delta g = g_{out} - g_{in} = (h_{out} - h_{in}) - T(s_{out} - s_{in})$$

where the specific numerical values for h_{out} and s_{out} depend on whether the product H_2O is liquid or vapor. From the given reaction equation, we know H_2 is the fuel, O_2 is the oxidant, and H_2O is the exhaust, and hence

$$\frac{\dot{N}_{Ox}}{\dot{N}_F} = \frac{1}{2} \text{ mol } O_2/\text{mol fuel}; \quad \frac{\dot{N}_{Ex}}{\dot{N}_F} = 1 \text{ mol } H_2O/\text{mol fuel}$$

Then the absolute enthalpy and entropy at the inlet and outlet can be written, according to Equations (2.12) and (2.13)

$$h_{in} = h_{H_2} + \frac{1 \text{ mol } O_2}{2 \text{ mol fuel}} h_{O_2}; \quad h_{out} = \frac{1 \text{ mol } H_2O}{1 \text{ mol fuel}} h_{H_2O}$$

$$s_{in} = s_{H_2} + \frac{1 \text{ mol } O_2}{2 \text{ mol fuel}} s_{O_2}; \quad s_{out} = \frac{1 \text{ mol } H_2O}{1 \text{ mol fuel}} s_{H_2O}$$

We know that the sensible enthalpies for all species involved are 0 since the reaction occurs at the standard reference state; and the enthalpies of formation for H_2 and O_2 are also 0 at 25 °C and 1 atm by definition. Therefore, the numerical values for the enthalpy of all the species involved are equal to their respective values for the standard enthalpy of formation for these species. From Appendix A.2, we obtain the values for the enthalpy of formation and entropy for each species at the standard reference state as follows

$$h_{f,H_2} = 0 \text{ J/mol}; \quad s_{H_2} = 130.68 \text{ J/(mol} \cdot \text{K)}$$

$$h_{f,O_2} = 0 \text{ J/mol}; \quad s_{O_2} = 205.14 \text{ J/(mol} \cdot \text{K)}$$

$$h_{f,H_2O(\ell)} = -285,826 \text{ J/mol}; \quad s_{H_2O(\ell)} = 69.92 \text{ J/(mol} \cdot \text{K)}$$

$$h_{f,H_2O(g)} = -241,826 \text{ J/mol}; \quad s_{H_2O(g)} = 188.83 \text{ J/(mol} \cdot \text{K)}$$

(a)   When the product H_2O is in the liquid state, we have

$$\Delta h = h_{out} - h_{in} = \frac{1 \text{ mol } H_2O}{1 \text{ mol fuel}} h_{H_2O(\ell)} - \left(h_{H_2} + \frac{1 \text{ mol } O_2}{2 \text{ mol fuel}} h_{O_2} \right)$$

$$= \frac{1 \text{ mol } H_2O}{1 \text{ mol fuel}} h_{f,H_2O(\ell)} - \left(h_{f,H_2} + \frac{1 \text{ mol } O_2}{2 \text{ mol fuel}} h_{f,O_2} \right)$$

$$= 1 \text{ mol } H_2O/\text{mol fuel} \times (-285,826 \text{ J/mol } H_2O) - (0 + \frac{1}{2} \times 0) \text{ J/mol fuel}$$

$$= -285,826 \text{ J/mol fuel}$$

$$\Delta s = s_{\text{out}} - s_{\text{in}} = \frac{1 \text{ mol } H_2O}{1 \text{ mol fuel}} s_{H_2O(\ell)} - \left(s_{H_2} + \frac{1 \text{ mol } O_2}{2 \text{ mol fuel}} s_{O_2} \right)$$

$$= 1 \text{ mol } H_2O/\text{mol fuel} \times 69.92 \text{ J}/(\text{mol } H_2O \cdot K)$$

$$- \left[130.68 \text{ J}/(\text{mol fuel} \cdot K) + \frac{1}{2} \text{ mol } O_2/\text{mol fuel} \right.$$

$$\left. \times 205.14 \text{ J}/(\text{mol } O_2 \cdot K) \right]$$

$$= -163.25 \text{ J}/(\text{mol fuel} \cdot K)$$

Since the reaction temperature $T = 25\,°C = 298$ K, we have

$$\Delta g = \Delta h - T\Delta s = -285,826 \text{ J}/\text{mol fuel} - 298 \text{ K} \times (-163.25) \text{ J}/(\text{mol fuel} \cdot K)$$

$$= -237,177.50 \text{ J}/\text{mol fuel}$$

For the present reaction, 2 moles of electrons are transferred for each mole of fuel, H_2, consumed, or

$$n = 2 \text{ mol } e^-/\text{mol fuel}$$

and recall that Faraday constant is

$$F = 96,487 \text{ C}/\text{mol electron}$$

Now substituting all the known parameter values into Equation (2.24) or Equation (2.29), we obtain

$$E_r^o = -\frac{\Delta g(T, P)}{nF} = -\frac{-237,177.50 \text{ J}/\text{mol fuel}}{2 \text{ mol } e^-/\text{mol fuel} \times 96,487 \text{ C}/\text{mol } e^-}$$

$$= 1.229 \text{ J/C}$$

or

$$E_r^o = 1.229 \text{ V}$$

(b)   When the product H_2O is in the vapor state, we have

$$\Delta h = h_{\text{out}} - h_{\text{in}} = \frac{1 \text{ mol } H_2O}{1 \text{ mol fuel}} h_{H_2O(g)} - \left(h_{H_2} + \frac{1 \text{ mol } O_2}{2 \text{ mol fuel}} h_{O_2} \right)$$

$$= \frac{1 \text{ mol } H_2O}{1 \text{ mol fuel}} h_{f,H_2O(g)} - \left(h_{f,H_2} + \frac{1 \text{ mol } O_2}{2 \text{ mol fuel}} h_{f,O_2} \right)$$

$$= 1 \text{ mol } H_2O/\text{mol fuel} \times (-241{,}826 \text{ J/mol } H_2O) - (0 + \frac{1}{2} \times 0) \text{ J/mol fuel}$$

$$= -241{,}826 \text{ J/mol fuel}$$

$$\Delta s = s_{\text{out}} - s_{\text{in}} = \frac{1 \text{ mol } H_2O}{1 \text{ mol fuel}} s_{H_2O(g)} - \left(s_{H_2} + \frac{1 \text{ mol } O_2}{2 \text{ mol fuel}} s_{O_2} \right)$$

$$= 1 \text{ mol } H_2O/\text{mol fuel} \times 188.83 \text{ J/(mol } H_2O \cdot K)$$

$$- \left[130.68 \text{ J/(mol fuel} \cdot K) + \frac{1}{2} \text{ mol } O_2/\text{mol fuel} \right.$$

$$\left. \times 205.14 \text{ J/(mol } O_2 \cdot K) \right]$$

$$= -44.42 \text{ J/(mol fuel} \cdot K)$$

Since the reaction temperature $T = 25\,°C = 298$ K, we have

$$\Delta g = \Delta h - T\Delta s = -241{,}826 \text{ J/mol fuel} - 298 \text{ K}$$

$$\times (-44.42) \text{ J/(mol fuel} \cdot K)$$

$$= -228{,}588.84 \text{ J/mol fuel}$$

Now substituting all the known parameter values into Equation (2.24) or Equation (2.29), we obtain

$$E_r^o = -\frac{\Delta g(T, P)}{nF} = -\frac{-228{,}588.84 \text{ J/mol fuel}}{2 \text{ mol } e^-/\text{mol fuel} \times 96{,}487 \text{ C/mol } e^-}$$

$$= 1.185 \text{ J/C}$$

or

$$E_r^o = 1.185 \text{ V}$$

COMMENT

This example reveals a number of important issues worth mentioning.

1. First, for most of the useful fuel cell reactions (i.e., reactions potential for application in fuel cells), the changes in enthalpy, entropy, and Gibbs function are negative, that is,

$$\Delta h < 0; \quad \Delta s < 0; \quad \text{and} \quad \Delta g < 0$$

implying that reaction products have less energy content and less microscopic disorder (less entropy content) than reactants going into the reaction. Then, Equation (2.16) indicates that

$$q = -T\wp_s - T(s_{\text{in}} - s_{\text{out}}) = -T\wp_s + T\Delta s < 0$$

or heat is actually transferred from the fuel cell system to the surrounding, not the other way around, as Figure 2.3 might suggest. The amount of heat transfer from the fuel cell, which is often called heat generated in the fuel cell or waste heat, is attributable to irreversibilities and the reduction in entropy. Even at the idealized thermodynamically reversible condition, waste heat is generated because of the entropy change from reactants to products. This is an important point to remember, and we shall come back to this issue later when we discuss the energy conversion efficiency for fuel cells (Section 2.5).

2. Secondly, since gas (vapor) state contains more energy and entropy than liquid state does, reaction product in gas state results in less changes in enthalpy, entropy and Gibbs free energy, the corresponding cell potential is then smaller. This is clear when comparing $E_r^o = 1.229\,\text{V}$ for liquid water with $E_r^o = 1.185\,\text{V}$ for gaseous water as the reaction product. Similarly, one can infer that liquid fuel yields smaller cell potential than gaseous fuel.

3. Thirdly, similar to enthalpy shown in Equation (2.1), the Gibbs function can also be decomposed into two parts: the Gibbs function of formation representing chemical energy and the sensible Gibbs function representing thermal energy, that is

$$g_i(T, P) = g_{f,i}(T_{\text{ref}}, P_{\text{ref}}) + \Delta g_{s,i}(T, P)$$

The Gibbs function of formation is also tabulated and it is given in Appendix A.2 for selected substances at the standard reference state and Appendix A.5 at any temperature, but 1 atm. Gibbs function can be utilized to calculate the reversible cell potential, E_r, easily. Take for example the present example problem, we can easily have

$$\Delta g = g_{\text{out}} - g_{\text{in}} = \frac{1\ \text{mol}\ H_2O}{1\ \text{mol fuel}} g_{H_2O} - \left(g_{H_2} + \frac{1\ \text{mol}\ O_2}{2\ \text{mol fuel}} g_{O_2} \right)$$

$$= \frac{1\ \text{mol}\ H_2O}{1\ \text{mol fuel}} g_{f,H_2O} - \left(g_{f,H_2} + \frac{1\ \text{mol}\ O_2}{2\ \text{mol fuel}} g_{f,O_2} \right)$$

For liquid H_2O as product, from Appendix A.2 we have

$$\Delta g = 1\ \text{mol}\ H_2O/\text{mol fuel} \times (-237,180\ \text{J/mol}\ H_2O)$$

$$- (0 + \frac{1}{2}\ \text{mol}\ O_2/\text{mol fuel} \times 0)$$

$$= -237,180\ \text{J/mol fuel}$$

and

$$E_r^o = 1.229\ \text{V}$$

Similarly, for vapor H_2O as product,

$$\Delta g = 1 \text{ mol } H_2O/\text{mol fuel} \times (-228{,}590 \text{ J/mol } H_2O)$$

$$-(0 + \frac{1}{2} \text{ mol } O_2/\text{mol fuel} \times 0)$$

$$= -228{,}590 \text{ J/mol fuel}$$

and

$$E_r^o = 1.185 \text{ V}$$

Clearly, it is much easier to use the tabulated Gibbs function of formation directly for the determination of the reversible cell potential.

4.   Finally, we may mention that when H_2 and O_2 are used as the reactant, the fuel cell is often called H_2–O_2 fuel cell.

Table 2.1   Standard Enthalpy and Gibbs Function of Reaction for Candidate Fuels and Oxidants, and Corresponding Standard Reversible Cell Potential as well as Other Relevant Parameters (at 25 °C and 1 atm)[a]

Fuel	Reaction	n n	$-\Delta h$ (J/mol)	$-\Delta g$ (J/mol)	E_r^o (V)	η^b (%)
Hydrogen	$H_2 + 0.5O_2 \rightarrow H_2O(\ell)$	2	286.0	237.3	1.229	82.97
	$H_2 + Cl_2 \rightarrow 2HCl(aq)$	2	335.5	262.5	1.359	78.33
	$H_2 + Br_2 \rightarrow 2HBr(aq)$	2	242.0	205.7	1.066	85.01
Methane	$CH_4 + 2O_2 \rightarrow CO_2 + 2H_2O(\ell)$	8	890.8	818.4	1.060	91.87
Propane	$C_3H_8 + 5O_2 \rightarrow 3CO_2 + 4H_2O(\ell)$	20	2221.1	2109.3	1.093	94.96
Decane	$C_{10}H_{22} + 15.5O_2 \rightarrow 10CO_2 + 11H_2O(\ell)$	66	6832.9	6590.5	1.102	96.45
Carbon Monoxide	$CO + 0.5O_2 \rightarrow CO_2$	2	283.1	257.2	1.333	90.86
Carbon	$C(s) + 0.5O_2 \rightarrow CO$	2	110.6	137.3	0.712	124.18[c]
	$C(s) + O_2 \rightarrow CO_2$	4	393.7	394.6	1.020	100.22[c]
Methanol	$CH_3OH(\ell) + 1.5O_2 \rightarrow CO_2 + 2H_2O(\ell)$	6	726.6	702.5	1.214	96.68
Formaldehyde	$CH_2O(g) + O_2 \rightarrow CO_2 + H_2O(\ell)$	4	561.3	522.0	1.350	93.00
Formic Acid	$HCOOH + 0.5O_2 \rightarrow CO_2 + H_2O(\ell)$	2	270.3	285.5	1.480	105.62[c]
Ammonia	$NH_3 + 0.75O_2 \rightarrow 1.5H_2O(\ell) + 0.5N_2$	3	382.8	338.2	1.170	88.36
Hydrazine	$N_2H_4 + O_2 \rightarrow 2H_2O(\ell) + N_2$	4	622.4	602.4	1.560	96.77
Zinc	$Zn + 0.5O_2 \rightarrow ZnO$	2	348.1	318.3	1.650	91.43
Sodium	$Na + 0.25H_2O + 0.25O_2 \rightarrow NaOH(aq)$	1	326.8	300.7	3.120	92.00

[a] Adapted from A.J. Appleby, 1993. *Fuel Cell Systems*, Chapter 5, ed., L.J.M.J. Blomen and M.N. Mugerwa, New York: Plenum Press. With kind permission of Springer Science and Business Media.

[b] Energy conversion efficiency.

[c] There is a conceptual problem with these efficiency data; see Section 2.5 for explanation.

Similar to the preceding example, the standard reversible cell potential, E_r^o, can be determined for any other electrochemical reaction. Some of the potential fuel cell reactions and the resulting E_r^o are shown in Table 2.1 along with other relevant parameters. From this table, it might be noted that E_r^o should be approximately above 1 V in order for the reaction to be realistic for fuel cell application. This is because if E_r^o is much less than 1 V and considering the cell voltage loss that inevitably occur in practical fuel cells due to irreversibilities, the actual cell potential might become too small to be useful for practical applications. Therefore, the rule of thumb is, for any proposed fuel and oxidant, to calculate E_r^o and to see if E_r^o is on the order of 1 V or larger before proceeding to any further work on it.

The calculation of reversible cell potential may be expressed in a more general form that would be convenient for computer programming. For a generalized electrochemical reaction:

$$\underbrace{\sum_{i=1}^{N} v_i' M_i}_{\text{Reactant}} \longrightarrow \underbrace{\sum_{i=1}^{N} v_i'' M_i}_{\text{Product}} \tag{2.30}$$

where M_i is the chemical formula for species i; v_i' and v_i'' are the number of moles for species i in the reactant and product mixture, respectively; and N is the total number of species in the chemically reacting system. Take H_2 and O_2 reaction to form water for example,

$$H_2 + \frac{1}{2}O_2 \longrightarrow H_2O$$

Then, we have

$$M_1 = H_2; \quad M_2 = O_2; \quad M_3 = H_2O;$$
$$v_{H_2}' = v_F' = 1; \quad v_{O_2}' = 0.5; \quad v_{H_2O}' = 0$$
$$v_{H_2}'' = 0; \quad v_{O_2}'' = 0; \quad v_{H_2O}'' = 1$$

The enthalpy and entropy at the inlet and outlet are equal to those of reactant and product, respectively

$$h_{in} = h_R = \frac{1}{v_F'} \sum_{i=1}^{N} v_i' h_{M_i}; \quad h_{out} = h_P = \frac{1}{v_F'} \sum_{i=1}^{N} v_i'' h_{M_i} \tag{2.31}$$

$$s_{in} = s_R = \frac{1}{v_F'} \sum_{i=1}^{N} v_i' s_{M_i}; \quad s_{out} = s_P = \frac{1}{v_F'} \sum_{i=1}^{N} v_i'' s_{M_i} \tag{2.32}$$

The enthalpy and entropy changes for the generalized reaction, Equation (2.30), are

$$\Delta h = h_P - h_R = \frac{1}{\nu_{F'}} \sum_{i=1}^{N} \left(\nu_i'' - \nu_i' \right) h_{M_i};$$

$$\Delta s = s_P - s_R = \frac{1}{\nu_{F'}} \sum_{i=1}^{N} \left(\nu_i'' - \nu_i' \right) s_{M_i} \tag{2.33}$$

Then, the change in the Gibbs function can be determined as

$$\Delta g = \Delta h - T\Delta s \tag{2.34}$$

or alternatively

$$\Delta g = g_P - g_R = \frac{1}{\nu_{F'}} \sum_{i=1}^{N} \left(\nu_i'' - \nu_i' \right) g_{M_i} \tag{2.35}$$

Finally, the reversible cell potential can be calculated according to Equation (2.24).

2.4 EFFECT OF OPERATING CONDITIONS ON REVERSIBLE CELL POTENTIAL

The most important operating conditions that influence fuel cell performance are operating temperature, pressure, and reactant concentrations. Before analyzing these influences, we need some thermodynamic relations that will be useful for analysis. Recall that the Gibbs function and enthalpy are defined as

$$g = h - Ts \tag{2.36}$$

$$h = u + Pv \tag{2.37}$$

where u is the internal energy and v is the specific volume. Combining these equations together and taking the differential, we obtain

$$dg = du + Pdv + vdP - Tds - sdT \tag{2.38}$$

Another fundamental relation in thermodynamics is the Gibbs equation for a simple compressible substance, which is

$$Tds = du + Pdv = dh - vdP \tag{2.39}$$

Substituting Equation (2.39) into Equation (2.38) results in

$$dg = vdP - sdT \tag{2.40}$$

Therefore, we arrive at two important relations for fuel cell analysis

$$\left(\frac{\partial g}{\partial T}\right)_P = -s; \quad \left(\frac{\partial g}{\partial P}\right)_T = v \tag{2.41}$$

Applying the Equations (2.40) and (2.41) to Gibbs function change for a particular fuel cell reaction, such as the generalized one given in the Equation (2.30), we finally obtain

$$\left(\frac{\partial \Delta g}{\partial T}\right)_P = -\Delta s \tag{2.42}$$

$$\left(\frac{\partial \Delta g}{\partial P}\right)_T = \Delta v \tag{2.43}$$

where Δs and Δv are the respective changes in the entropy and specific volume between the products and reactants. It might be emphasized that these two relations are obtained without making any specific assumptions such as ideal gas approximation, hence they are valid for any substances undergoing chemical reactions. They are utilized for the following analysis of temperature and pressure effect on the reversible cell potential.

2.4.1   Effect of Temperature on Reversible Cell Potential E_r

The reversible cell potential, E_r, given in Equation (2.24), is a function of temperature because the change in the Gibbs function depends on the fuel cell operating temperature and pressure. Hence,

$$E_r(T, P) = -\frac{\Delta g(T, P)}{nF} \tag{2.24}$$

Then incorporating Equation (2.42), the change of the reversible cell potential with temperature can be expressed as

$$\left(\frac{\partial E_r(T, P)}{\partial T}\right)_P = -\frac{1}{nF}\left(\frac{\partial \Delta g(T, P)}{\partial T}\right)_P = \frac{\Delta s(T, P)}{nF} \tag{2.44}$$

Clearly, the variation of E_r with temperature depends on the change in entropy for the particular fuel cell reaction and three possible situations may arise:

1. If $\Delta s < 0$, like $H_2 + \frac{1}{2}O_2 \longrightarrow H_2O(g)$ shown in the previous example, the reversible cell potential decreases with cell operation temperature.
2. If $\Delta s > 0$, then the reversible cell potential increases with temperature, that is, for the reaction $C(s) + \frac{1}{2}O_2 \longrightarrow CO$, the entropy change is about $+89$ J/K.
3. If $\Delta s = 0$, then the reversible cell potential is independent of temperature, like the reaction $CH_4 + 2O_2 \longrightarrow CO_2 + 2H_2O(g)$.

For many useful electrochemical reactions, the entropy change is negative and is almost constant with the change of temperature to a good approximation, provided the temperature change $T - T_{ref}$f is not too large. Then Equation (2.44) may be integrated from the standard reference temperature, $T_{ref} = 25\,°C$, to the arbitrary fuel cell operating temperature T, while keeping pressure P constant,

$$E_r(T, P) = E_r(T_{ref}, P) + \left(\frac{\Delta s\,(T_{ref}, P)}{nF}\right)(T - T_{ref}) \qquad (2.45)$$

Alternatively, we can expand the reversible cell potential expression, Equation (2.24), in the Taylor series in terms of temperature, T, around the reference temperature, T_{ref}, keeping P constant

$$E_r(T, P) = E_r(T_{ref}, P) + \left(\frac{\partial E_r(T_{ref}, P)}{\partial T}\right)_P (T - T_{ref})$$

Now considering Equation (2.44), the preceding equation again reduces to Equation (2.45).

Example 2.7   For H_2 and O_2 reaction:

$$H_2 + \frac{1}{2}O_2 \longrightarrow H_2O$$

at $P = 1$ atm. Determine the reversible cell potential expression as a function of temperature based on Equation (2.45). Then, determine the reversible cell potential at $80\,°C$ and 1 atm. Take $T_{ref} = 25\,°C$ and water can be either (a) liquid or (b) vapor.

SOLUTION
The entropy change for the present fuel cell reaction at the standard reference temperature and pressure and the reversible cell potential have already been determined in the previous example, they are

$$\Delta s(25\,°C, 1\,atm) = -163.25\ \text{J/(mol fuel} \cdot \text{K)} \text{ for } H_2O(\ell); \quad E_r = 1.229\ \text{V}$$

$$\Delta s(25\,°C, 1\,atm) = -44.42\ \text{J/(mol fuel} \cdot \text{K)} \text{ for } H_2O(g); \quad E_r = 1.185\ \text{V}$$

and the number of electron transferred is $n = 2$ mol e^-/mol fuel H_2.

(a)   For liquid water as reaction product and $T_{ref} = 25\,°C$, $P = 1$ atm:

$$\left(\frac{\Delta s(T_{ref}, P)}{nF}\right) = \frac{-165,25\ \text{J/(mol fuel} \cdot \text{K)}}{2\ \text{mol } e^-/\text{mol fuel} \times 96,487\ \text{C/mol } e^-}$$

$$= -0.8460 \times 10^{-3}\ \text{V/K}$$

Therefore, we have the desired expression

$$E_r(T, 1\,\text{atm}) = E_r(25\,°\text{C}, 1\,\text{atm}) + \left(\frac{\Delta s(25\,°\text{C}, 1\,\text{atm})}{nF}\right)(T - T_{\text{ref}})$$

or

$$E_r(T, 1\,\text{atm}) = 1.229\,\text{V} - 0.8460 \times 10^{-3}\,\text{V/K}(T - T_{\text{ref}})$$

Hence, for every degree of temperature increase, reversible cell potential is reduced by 0.8460 mV. At temperature of 80 °C, we obtain

$$E_r(80\,°\text{C}, 1\,\text{atm}) = 1.229\,\text{V} - 0.8460 \times 10^{-3}\,\text{V/K} \times (80 - 25)\text{K}$$

$$= 1.229\,\text{V} - 0.4653\,\text{V}$$

$$= 1.182\,\text{V}$$

That is, the reversible cell potential is reduced by about 3.8% from an increase in temperature from 25 °C to 80 °C.

(b)   For gaseous water (vapor) as reaction product and $T_{\text{ref}} = 25°\text{C}$, $P = 1$ atm:

$$\left(\frac{\Delta s(T_{\text{ref}}, P)}{nF}\right) = \frac{-44.42\,\text{J/(mol fuel}\cdot\text{K)}}{2\,\text{mol e}^-/\text{mol fuel} \times 96,487\,\text{C/mol e}^-}$$

$$= -0.2302 \times 10^{-3}\,\text{V/K}$$

Therefore, the desired expression can be written as

$$E_r(T, 1\,\text{atm}) = E_r(25\,°\text{C}, 1\,\text{atm}) + \left(\frac{\Delta s(25\,°\text{C}, 1\,\text{atm})}{nF}\right)(T - T_{\text{ref}})$$

or

$$E_r(T, 1\,\text{atm}) = 1.185\,\text{V} - 0.2302 \times 10^{-3}\,\text{V/K} \times (T - T_{\text{ref}})$$

That is, for every degree of temperature increase, the reversible cell potential is reduced by 0.2302 mV. At temperature of 80 °C, we obtain

$$E_r(80\,°\text{C}, 1\,\text{atm}) = 1.185\,\text{V} - 0.2302 \times 10^{-3}\,\text{V/K} \times (80 - 25)\text{K}$$

$$= 1.185\,\text{V} - 0.01266\,\text{V}$$

$$= 1.172\,\text{V}$$

That is, the reversible cell potential is reduced by about 1.1% from an increase in temperature from 25 °C to 80 °C if the product water is vapor instead of in the liquid form.

COMMENT

1. The temperature difference in °C and in K is identical.
2. The reduction in the reversible cell potential is very small with temperature. This is because the sensible Gibbs function is very small compared to the Gibbs function of formation for reasonable temperature changes (up to a few hundred degrees Kelvin).
3. The reduction in the reversible cell potential is smaller for gaseous water as reaction product than for liquid water product, due to the smaller magnitude of the entropy change for the reaction in the former case.

It must be emphasized that the expression given in Equation (2.45) is an approximation. Strictly speaking, the reversible cell potential at any temperature and pressure should be determined from Equation (2.24) by calculating first the property changes for the particular fuel cell reaction involved. Such a procedure has been followed for the hydrogen and oxygen reaction to form gaseous water, and the results are presented in Figure 2.4. Clearly, the reversible cell potential indeed decreases almost linearly as temperature is increased over a large temperature range. However, it is noticed that the reversible cell potential is larger for product water as liquid at low temperatures, but it decreases much faster than the gaseous water as product when temperature is increased. So that at temperatures slightly above about 373 K, the reversible cell potential for liquid water product actually becomes smaller. This may seem curious, but it is because at such high temperatures pressurization is necessary to keep the product water in liquid form as the reactants, hydrogen and oxygen, are fed at 1 atm. Also notice that the critical temperature for water is about 647 K, beyond which distinct liquid state does not exist for water, hence the shorter curves for the liquid water case shown in Figure 2.4.

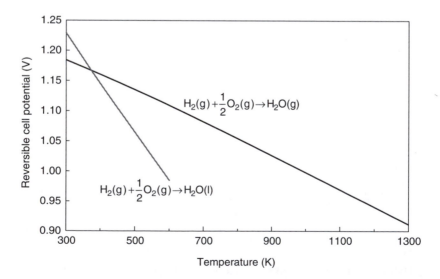

Figure 2.4 Effect of temperature on the reversible cell potential of a hydrogen–oxygen fuel cell for the reaction of $H_2 + \frac{1}{2}O_2 \longrightarrow H_2O$ at the pressure of 1 atm.

As pointed out earlier, the entropy change for most of fuel cell reactions is negative; consequently the reversible cell potential decreases as temperature is increased as shown in Figure 2.4. However, for some few reactions such as

$$C(s) + \frac{1}{2}O_2(g) \longrightarrow CO(g)$$

the entropy change is positive, for example, $\Delta s = 89 \, J/(mol \, fuel \cdot K)$ at the standard reference temperature and pressure. As a result, the reversible cell potential for this type of reactions will increase with temperature.

Suppose N_P and N_R represent the number of moles of products and reactants, respectively, which are in gaseous state and on a per mole fuel basis, and $\Delta N = N_P - N_R$ represents the change, per mole fuel, in the number of moles of gas species during the reaction, then as a rough rule of thumb, it might be stated that

1. $\Delta s > 0$ for $\Delta N > 0$ (due to increasing disorder because of more molecules in the product) and reversible cell potential increases with temperature;
2. $\Delta s < 0$ for $\Delta N < 0$ (due to decreasing disorder because of less molecules in the product) and reversible cell potential decreases with temperature;
3. $\Delta s \approx 0$ for $\Delta N = 0$ and reversible cell potential is almost independent of temperature.

Figure 2.5 shows the reversible cell potential as a function of temperature for a number of important fuel cell reactions. The aforementioned three trends of variation

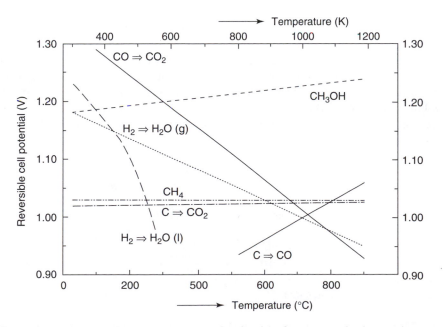

Figure 2.5 Standard reversible cell potential, E_r, as a function of temperature for the most important fuel cell reactions at the pressure of 1 atm. (From E. Barendrecht, 1993. *Fuel Cell Systems*, eds. L.J.M.J. Blomen and M.N. Mugerwa. New York: Plenum Press. With permission.)

for the reversible cell potential versus temperature can be clearly seen in Figure 2.5. For example, for methane reaction with oxygen,

$$CH_4(g) + 2O_2(g) \longrightarrow CO_2(g) + 2H_2O(g)$$

and for solid carbon, C(s), reaction with oxygen,

$$C(s) + O_2(g) \longrightarrow CO_2(g)$$

the change in the number of moles for the gaseous species is zero and the reversible cell potential for these two reactions is almost a horizontal line, independent of temperature.

2.4.2   Effect of Pressure on Reversible Cell Potential E_r

Taking partial derivative of Equation (2.24) with respect to pressure while keeping temperature constant, we obtain

$$\left(\frac{\partial E_r}{\partial P}\right)_T = -\frac{1}{nF}\left(\frac{\partial \Delta g}{\partial P}\right)_T = -\frac{\Delta v}{nF} \tag{2.46}$$

where

$$\Delta v = v_P - v_R \tag{2.47}$$

represents the volume change of all the gaseous species in the reaction on a per-mole fuel basis; the volume of the solids and liquids is much smaller than the volume of gas species, and can be neglected for the present purpose; v_P and v_R are the specific volume (per-mole fuel) of gas products and reactants, respectively. Consider all the reactant and product gases can be treated as ideal gases, Equation (2.47) can be expressed as

$$\Delta v = v_P - v_R = \frac{N_P \Re T}{P} - \frac{N_R \Re T}{P} = \frac{\Delta N \Re T}{P} \tag{2.48}$$

where $\Re$ is the universal constant. Combining Equations (2.46) and (2.48) yields

$$\left(\frac{\partial E_r}{\partial P}\right)_T = -\frac{\Delta N \Re T}{nF}\frac{1}{P} \tag{2.49}$$

Equation (2.49) indicates that

1. If $\Delta N > 0$, meaning product contains more moles of gas species than reactant, the reversible cell potential will decrease with pressure, such as for the reaction $C(s) + \frac{1}{2}O_2(g) \longrightarrow CO(g)$;

2. If $\Delta N < 0$, which is the case for most of useful fuel cell reactions, the reversible cell potential will increase with pressure;
3. If $\Delta N = 0$, the reversible cell potential will not change with pressure.

Note for (1) and (2), the amount of change for reversible cell potential decreases gradually when pressure is increased. Higher-pressure operation results in mechanical problems, such as mechanical strength of the cell components, cell sealing problem, corrosion, and so on. This implies that the performance gain in E_r at high-pressure operation diminishes, and may become undesirable from a system design point of view.

Now integrating Equation (2.49) from the standard reference pressure, $P_{ref} = 1$ atm, to an arbitrary pressure, P, while keeping temperature fixed, results in

$$E_r(T, P) = E_r(T, P_{ref}) - \frac{\Delta N \Re T}{n F} \ln\left(\frac{P}{P_{ref}}\right) \tag{2.50}$$

This equation indicates that the pressure dependence of the reversible cell potential is a logarithmic function, and hence the dependence becomes weaker as pressure P is increased.

Equation (2.50) can be easily derived based on Equation (2.24) and thermodynamic property relations without resorting to the above differentiation and integration process. The alternate derivation is given below. First we can write

$$\Delta g(T, P) = \Delta g(T, P_{ref}) + \left[\Delta g(T, P) - \Delta g(T, P_{ref})\right] \tag{2.51}$$

Since $\Delta g = g_P - g_R = \Delta h - T \Delta s$, and for liquids, solids and ideal gases, enthalpy is a function of temperature only (i.e., independent of pressure), the square bracket term in Equation (2.51) becomes

$$\left[\Delta g(T, P) - \Delta g(T, P_{ref})\right] = [-T \Delta s(T, P)] - [-T \Delta s(T, P_{ref})]$$
$$= \{-T [s_P(T, P) - s_R(T, P)]\} - \{-T [s_P(T, P_{ref}) - s_R(T, P_{ref})]\}$$
$$= \{-T [s_P(T, P) - s_P(T, P_{ref})]\} - \{-T [s_R(T, P) - s_R(T, P_{ref})]\} \tag{2.52}$$

Notice that entropy is independent of pressure for solids and liquids which can be approximated as incompressible, therefore, the entropy change for products and reactants over the pressure range from the reference pressure, P_{ref}, to the arbitrary pressure, P, can be attributed to the gas species in the reactant and product, and the gas species can be approximated as ideal gas. Recall the entropy change per mole of mixture for ideal gases can be written as

$$s(T_2, P_2) - s(T_2, P_2) = c_P \ln\left(\frac{T_2}{T_1}\right) - \Re \ln\left(\frac{P_2}{P_1}\right) \tag{2.53}$$

Since in Equation (2.52), the entropy is expressed on a per-mole fuel basis, instead of per-mole mixture, referring to the generalized electrochemical reaction shown in

Equation (2.30), Equation (2.52) can be rewritten as

$$\left[\Delta g(T, P) - \Delta g(T, P_{ref})\right] = N_P \left\{-T\left[-\Re \ln\left(\frac{P}{P_{ref}}\right)\right]\right\}$$

$$-N_R \left\{-T\left[-\Re \ln\left(\frac{P}{P_{ref}}\right)\right]\right\}$$

$$= (N_P - N_R)\, \Re T \ln\left(\frac{P}{P_{ref}}\right)$$

$$= \Delta N \Re T \ln\left(\frac{P}{P_{ref}}\right) \qquad (2.54)$$

Now combining Equations (2.51), (2.52), and (2.54) with Equation (2.24), we obtain once again Equation (2.50).

Example 2.8   For H_2 and O_2 reaction:

$$H_2(g) + \frac{1}{2}O_2(g) \longrightarrow H_2O$$

at $T = 25\,°C$. Determine the reversible cell potential expression as a function of pressure based on Equation (2.50). Then, determine the reversible cell potential at $25\,°C$ and 3, 5, and 10 atm, respectively. Take $P_{ref} = 1$ atm and water can be either (a) liquid or (b) vapor.

SOLUTION
Example 2.6 shows that

$$E_r(25\,°C, 1\text{ atm}) = \begin{cases} 1.229 \text{ V if water is liquid} \\ 1.185 \text{ V if water is vapor} \end{cases}$$

(a)   For liquid water as reaction product and at $T = 25\,°C = 298$ K, $P = 1$ atm:
The number of mole change between the gaseous product and gaseous reactant for the given reaction is

$$\Delta N = 0 - (1 + 0.5) = -1.5 \text{ mol/mol fuel}$$

and the number of electron transferred is $n = 2$ mol e^-/mol fuel H_2. Hence,

$$-\left(\frac{\Delta N \Re T}{nF}\right) = -\frac{-1.5 \text{ mol/mol fuel} \times 8.3143 \text{ J/(mol} \cdot \text{K)} \times 298 \text{ K}}{2 \text{ mol } e^-/\text{mol fuel} \times 96{,}487 \text{ C/mol } e^-}$$

$$= 19.26 \times 10^{-3} \text{ V}$$

Therefore, we have the desired expression

$$E_r(25\,°C, P) = E_r(25\,°C, 1\text{ atm}) + 19.26 \times 10^{-3}\ln\left(\frac{P}{P_{ref}}\right)$$

or

$$E_r(25\,°C, P) = 1.229\text{ V} + 19.26 \times 10^{-3}\text{ V} \times \ln\left(\frac{P}{P_{\text{ref}}}\right)$$

Hence, the reversible cell potential at the pressure of 3, 5, and 10 atm is

$$E_r(25\,°C, 3\text{ atm}) = 1.229\text{ V} + 19.26 \times 10^{-3}\text{ V} \times \ln(3) = 1.250\text{ V}$$

$$E_r(25\,°C, 5\text{ atm}) = 1.229\text{ V} + 19.26 \times 10^{-3}\text{ V} \times \ln(5) = 1.260\text{ V}$$

$$E_r(25\,°C, 10\text{ atm}) = 1.229\text{ V} + 19.26 \times 10^{-3}\text{ V} \times \ln(10) = 1.273\text{ V}$$

It is seen that the reversible cell potential at 10 atm is only about 3.6% higher than the corresponding value at 1 atm. Clearly the pressure effect on the reversible cell potential is minimal. But note that the pressure effect on the actual cell potential may be more significant due to enhanced electrochemical kinetics and mass transfer processes.

(b) For water vapor as reaction product and at $T = 25\,°C = 298$ K, $P = 1$ atm:
The number of mole changes for the given reaction now becomes

$$\Delta N = 1 - (1 + 0.5) = -0.5 \text{ mol/mol fuel}$$

because the water vapor should now be counted in the mole number change. Then

$$-\left(\frac{\Delta N \Re T}{nF}\right) = -\frac{-0.5\text{ mol/mol fuel} \times 8.3143\text{ J/(mol} \cdot \text{K)} \times 298\text{ K}}{2\text{ mol e}^-/\text{mol fuel} \times 96,487\text{ C/mol e}^-}$$

$$= 6.420 \times 10^{-3}\text{ V}$$

Therefore, we have the desired expression

$$E_r(25\,°C, P) = E_r(25\,°C, 1\text{ atm}) + 6.420 \times 10^{-3} \ln\left(\frac{P}{P_{\text{ref}}}\right)$$

or

$$E_r(25\,°C, P) = 1.185\text{ V} + 6.420 \times 10^{-3}\text{ V} \times \ln\left(\frac{P}{P_{\text{ref}}}\right)$$

Hence, the reversible cell potential at the pressure of 3, 5, and 10 atm is

$$E_r(25\,°C, 3\text{ atm}) = 1.185\text{ V} + 6.420 \times 10^{-3}\text{ V} \times \ln(3) = 1.192\text{ V}$$

$$E_r(25\,°C, 5\text{ atm}) = 1.185\text{ V} + 6.420 \times 10^{-3}\text{ V} \times \ln(5) = 1.195\text{ V}$$

$$E_r(25\,°C, 10\text{ atm}) = 1.185\text{ V} + 6.420 \times 10^{-3}\text{ V} \times \ln(10) = 1.200\text{ V}$$

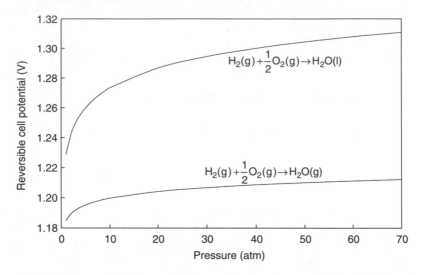

Figure 2.6 Standard reversible cell potential, E_r, as a function of pressure for the fuel cell reaction of $H_2\ (g) + \frac{1}{2}O_2 \longrightarrow H_2O$ at the temperature of 25 °C.

It is evident that the change in the reversible cell potential is even smaller compared to liquid water product at the low temperature of 25 °C.

COMMENT

1. It is important to emphasize that in calculating the number of mole change ΔN, only gaseous species are counted, and both liquid and solid species are not included in the calculation. The reason should be clear by now [refer to the alternate derivation for Equation (2.50).]

2. Operation at pressure as high as 10 atm only increases the reversible cell potential by less than 15 mV for the gaseous water as product and about 44 mV for the liquid water. In practice, it may not be wise to increase the cell pressure for the small gain of cell potential, especially considering the energy needed for the compression of the reactants to the high pressure. High pressure operation may promote corrosion of cell components by gases and electrolytes. The cell sealing difficulty and higher mechanical strength requirements for the cell and system components, although higher pressure may enhance reaction kinetics reduce other capital costs, such as smaller gas pipes and tanks due to the pressurization. It is clear that an optimal operating pressure exists and needs to be determined for specific fuel cell design.

Using the results in the preceding example, we can easily calculate the pressure effect for various pressure values, and the results are shown in Figure 2.6 for both liquid and vapor water as the reaction product. It is seen that the reversible cell potential increases with pressure, very fast for low-pressure values and gradually slow down for

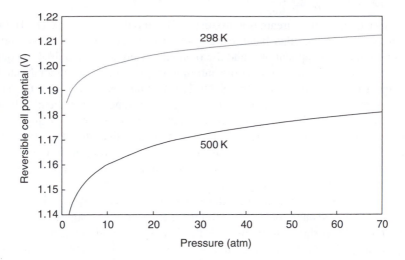

Figure 2.7 Standard reversible cell potential, E_r, as a function of pressure for the fuel cell reaction of H_2 (g) $+ \frac{1}{2}O_2 \longrightarrow H_2O$ at the temperature of 298 K and 500 K, respectively.

higher-pressure values. Also, the pressure effect is larger for liquid water as product because of the larger coefficient arising from the larger change in the number of moles between the product and reactant, ΔN.

It might be also emphasized that the pressure effect is small on the reversible cell potential at low temperatures, as shown in the preceding example. However, this effect increases significantly at high temperatures because the pressure effect coefficient, $-\Delta N \Re T/(nF)$, is directly proportional to temperature. Figure 2.7 shows the pressure dependence of the reversible cell potential at two different temperatures for the reaction $H_2(g) + \frac{1}{2}O_2(g) \longrightarrow H_2O(g)$.

It might be pointed out that for high temperature fuel cells the dependence of the actual cell potential E on the pressure follows closely the results given in Equation (2.50), whereas a significant deviation occurs for the low temperature fuel cells. The difference arises from the fact that at high temperatures, reaction kinetics are very fast and pressurization primarily increases the reactant concentration, hence better performance directly. At low temperatures, the reaction kinetics are slow and a higher reactant concentration does not yield a proportional increase in the cell potential due to the significant cell potential loss associated with the slow kinetics.

2.4.3   Effect of Concentration on Reversible Cell Potential

All the results obtained so far for the reversible cell potential are, strictly speaking, valid for pure fuel and pure oxidant in the reactant streams for the fuel cell reaction and pure reaction product in the exhaust stream. In reality, both fuel and oxidant streams are diluted by many other species for a variety of reasons. For example, H_2 as fuel is often humidified with water vapor before entering proton exchange membrane fuel cells for the hydration of the proton-conducting membrane; if H_2 is obtained from hydrocarbon fuel by reforming process (either steam reforming or partial oxidation reforming), then the fuel stream is a mixture of H_2 (about 50%–70%), CO_2, $H_2O(g)$, CO, and so on. On the other hand, oxygen in air is usually used as an oxidant for most of commercial applications, and the oxidant stream is also required to contain sufficient amount of water vapor before entering the proton exchange membrane fuel cells, again for the hydration of the electrolyte membrane. Fuel cell performance, including reversible cell potential, is affected by the presence of the chemically inert diluents which lowers the concentration of fuel and oxidant in the anode and cathode stream and increases resistance to mass transport to the active sites. The effect of the chemically active species on the cell performance (i.e., CO poisoning of the PEM fuel cells) is not considered for the analysis presented in this subsection and is described in later chapters for the specific type of fuel cells.

Consider that fuel, oxidant and exhaust streams, respectively, consist of a mixture of solids, liquids and ideal gases. Refer back to the schematic thermodynamic system for fuel cell shown in Figure 2.3, the first and second law analyses leading to Equations (2.20)–(2.29) are still valid because all inert diluents do not participate in the fuel cell reaction for energy conversion. However, the change for the Gibbs function, Δg, should be evaluated at the mixture temperature and the partial pressure of the fuel, oxidant and reaction product for the fuel, oxidant and exhaust stream, respectively.

Let P_i be the partial pressure of a component i in an ideal gas mixture of temperature T and total pressure P, then the Gibbs function of the component is $g_i(T, P_i)$ by definition, and also

$$g_i(T, P_i) = h_i(T) - Ts_i(T, P_i) \tag{2.55}$$

From Gibbs equation, Equation (2.39), we have for component i

$$ds_i = \frac{dh_i}{T} - \frac{v}{T}dP = \frac{dh_i}{T} - \frac{\Re}{P}dP \tag{2.56}$$

where the equation of state for ideal gases, $pv = \Re T$, has been used to obtain the second equation in Equation (2.56). Integration of Equation (2.56) from the partial pressure, P_i, to the mixture total pressure, P, while keeping the temperature fixed, yields

$$s_i(T, P) - s_i(T, P_i) = -\Re(\ln P - \ln P_i)$$

or

$$s_i(T, P_i) = s_i(T, P) - \Re \ln\left(\frac{P_i}{P}\right) \tag{2.57}$$

Substitution of Equation (2.57) into Equation (2.55) results in

$$g_i(T, P_i) = g_i(T, P) + \Re T \ln\left(\frac{P_i}{P}\right) \tag{2.58}$$

where

$$g_i(T, P) = h_i(T) - T s_i(T, P) \tag{2.59}$$

is the Gibbs function for component i at the mixture temperature T and total pressure P.

Now consider the generalized electrochemical reaction given in Equation (2.30)

$$\sum_{i=1}^{N} v_i' M_i \longrightarrow \sum_{i=1}^{N} v_i'' M_i \tag{2.30}$$

$$\underbrace{\qquad\qquad}_{\text{Reactant}} \qquad \underbrace{\qquad\qquad}_{\text{Product}}$$

where N is the total number of species in the reacting system, including solids, liquids, and gas species and both electrochemically reacting and inert species as well. Since solids and liquids are treated as incompressible, the pressure has no effect on the value of the Gibbs function, or

$$g_i(T, P_i) = g_i(T, P) \text{ for incompressible substances}$$

Then the Gibbs function change for the preceding generalized reaction can be written, according to Equations (2.35) and (2.58), as

$$\Delta g(T, P_i) = \frac{1}{v_{F'}} \sum_{i=1}^{N} \left(v_i'' - v_i' \right) g_i(T, P_i)$$

$$= \frac{1}{v_{F'}} \sum_{i=1}^{N} \left(v_i'' - v_i' \right) g_i(T, P) + \frac{1}{v_{F'}} \sum_{i=1}^{N_g} \left(v_i'' - v_i' \right) \Re T \ln\left(\frac{P_i}{P} \right)$$

or

$$\Delta g(T, P_i) = \Delta g(T, P) + \Re T \ln K \tag{2.60}$$

where $\Delta g(T, P)$ is the Gibbs function change for fuel, oxidant, and exhaust streams at the system temperature T and pressure P; similarly, $\Delta g(T, P_i)$ is the Gibbs function change for fuel, oxidant, and exhaust streams at the temperature T and respective partial pressure P_i; and

$$K = \prod_{i=1}^{N_g} \left(\frac{P_i}{P} \right)^{(v_i'' - v_i')/v_{F'}} \tag{2.61}$$

is defined similarly to the equilibrium constant for partial pressure (although they are not the same); N_g is the total number of gas species in the reacting system, excluding the solid and liquid species; and for ideal gases the pressure ratio can be expressed in terms of the molar fraction X_i as follows:

$$X_i = \frac{P_i}{P} \tag{2.62}$$

Now substituting Equation (2.60) into Equation (2.24), we obtain the expression for reversible cell potential when inert diluents exist in the fuel, oxidant, and exhaust streams

$$E_r(T, P_i) = E_r(T, P) - \frac{\Re T}{nF} \ln K \tag{2.63}$$

This is the general form of the Nernst equation, representing the effect of the reactant and product concentrations on the reversible cell potential. When the reactant streams contain inert diluents for a given operating temperature and pressure, the diluents cause a voltage loss for the reversible cell potential with the amount of which is generally called the Nernst loss, and its magnitude is equal to the second term on the right-hand side of Equation (2.63).

Example 2.9   For a given H_2–O_2 fuel cell, the following electrochemical reaction occurs

$$H_2(g) + \frac{1}{2}O_2(g) \longrightarrow H_2O(\ell)$$

at the standard reference state. It is known that the molar fraction of H_2 in the fuel stream is 0.5 and the molar fraction of O_2 in the oxidant stream is 0.21 (i.e., air). The remaining species are chemically inert. Determine the reversible cell potential for the preceding reaction.

SOLUTION
Since the fuel cell operates at the standard temperature and pressure, $T = 25\,°C$ and $P = 1$ atm, because the reactant streams are not pure, the reversible cell potential can be determined from the Nernst equation. From Equation (2.63),

$$E_r(T, P_i) = E_r(T, P) - \frac{\Re T}{nF} \ln K$$

The reactant molar fraction is given in the problem as:

$$X_{H_2} = 0.5; \quad X_{O_2} = 0.21$$

Hence, from Equations (2.61) and (2.62) we have

$$K = X_{H_2}^{(0-1)/1} X_{O_2}^{(0-1/2)/1} = X_{H_2}^{-1} X_{O_2}^{-1/2}$$

Note the product $H_2O(\ell)$ is in the liquid state, hence is not included in the calculation of K. Example 2.6 shows that the reversible cell potential at the standard temperature and pressure is, when pure H_2 and pure O_2 are used as reactants

$$E_r(T, P) = 1.229 \text{ V}$$

Substituting the numerical values into the symbolic expression leads to

$$E_r(T, P_i) = 1.229 \text{ V} - \frac{8.3143 \text{ J}/(\text{mol} \cdot \text{K}) \times 298 \text{ K}}{2 \text{ mol e}^-/\text{mol fuel} \times 96{,}487 \text{ C}/\text{mol e}^-}$$

$$\times \ln\left(0.5^{-1} \times 0.21^{-1/2}\right)$$

$$= 1.229 \text{ V} - 0.01892 \text{ V}$$

$$= 1.210 \text{ V}$$

COMMENT

Although the reversible cell potential is decreased due to the presence of inert diluents, the example shows the Nernst loss amounts to less than 20 mV, a fairly small value when considering the fact that only 21% of the oxidant stream is composed of O_2 and only 50% of the fuel stream is H_2. The next chapter shows that the voltage loss due to irreversibilities (e.g., slow oxygen electrochemical reduction reaction, the slow rate of mass transfer, and the reduction in the rate of mass transfer due to the presence of these inert diluents) is much larger. Also, if the reaction product is in the gaseous state, its effect will be negative as well when inert diluent is found in the exhaust stream.

2.5   ENERGY CONVERSION EFFICIENCY

2.5.1   Definition of Energy Conversion Efficiency

The efficiency for any energy conversion process or system is often defined as*

$$\eta = \frac{\text{Useful energy obtained}}{\text{Energy available for conversion that's an expense}} \tag{2.64}$$

Based on this definition, it is well known that 100% energy conversion efficiency is possible by the first law of thermodynamics. But is not possible by the second law of thermodynamics because many energy conversion systems produce power output by means of thermal energy, such as steam and gas turbines, internal combustion engines, which involve irreversible losses of energy. These thermal energy conversion systems are often referred to as heat engines.

---

*Note that in literature η is commonly used as efficiency in thermodynamics; whereas it is also conventionally used as overpotential, or voltage loss, for fuel cell analysis, as in electrochemistry. In this book, η is used for both in order to be consistent with the convention, and its meaning would become clear from the context.

On the other hand, there exist thermal energy conversion systems that have 100% or even higher efficiency. One of the examples is the refrigeration system such as heat pumps, air conditioners, etc. For these systems, often other performance measures are used, such as coefficient of performance (COP), instead of efficiency. Further, these systems are typically power consuming, instead of power generation.

Therefore, the rest of the book focuses on the energy conversion efficiency of power generation systems, especially fuel cells, and how it compares with the efficiency of heat engines, a widely used existing and mature technology against which fuel cells are being compared and competing for commercial application. This section derives the reversible energy conversion efficiency, the best possible efficiency for fuel cells. This efficiency is compared with the Carnot efficiency, which is the best possible efficiency for heat engines. Then we demonstrate that these forms of efficiency are really identical and they are the different forms of the best possible efficiency under the same conditions that are dictated by the second law of thermodynamics. The misconception that over 100% fuel cell efficiency is possible is clarified and other efficiencies associated with a fuel cell power plant will be introduced. Finally, additional energy loss mechanism in operating fuel cells is described briefly. During this process, the amount of waste heat generation in the fuel cells is determined, which is equal to the amount of cooling required for operating fuel cells.

2.5.2   Reversible Energy Conversion Efficiency for Fuel Cells

For the present fuel cell system described in Figure 2.3, the energy balance equation, Equation (2.8), can be written as, on a per-unit mole of fuel basis

$$h_{in} - h_{out} + q - w = 0 \tag{2.65}$$

which indicates that the enthalpy change, $-\Delta h = h_{in} - h_{out}$, provides the energy available for conversion into the useful energy exhibited as work here and it is the expense to be paid for the useful work output. At the same time, waste heat, q, is also generated which would represent a degradation of energy. The amount of waste heat generated can be determined from the second law expression, Equation (2.9) or Equation (2.16), as

$$q = T\Delta s - T\wp_s \tag{2.66}$$

and the useful energy output as work is, from Equation (2.20) or combining Equations (2.65) and (2.66)

$$w = -\Delta g - T\wp_s \tag{2.67}$$

Therefore, the energy conversion efficiency for the fuel cell system described in Figure 2.3 becomes, according to Equation (2.64)

$$\eta = \frac{w}{-\Delta h} = \frac{\Delta g + T\wp_s}{\Delta h} \tag{2.68}$$

Note that both Δh and Δg are negative for power generation systems, including fuel cells as it is clearly shown in Table 2.1. By the second law, the entropy generation per-unit mole of fuel is

$$\wp_s \geq 0 \tag{2.69}$$

and the equality holds for all reversible processes whereas entropy is always generated for irreversible processes. Therefore, the maximum possible efficiency allowed by the second law is, when the process is reversible (i.e., $\wp_s = 0$)

$$\eta_r = \frac{w_{max}}{-\Delta h} = \frac{\Delta g(T, P)}{\Delta h(T, P)} \tag{2.70}$$

Since both the enthalpy and Gibbs function change depend on the system temperature and pressure, so does the energy conversion efficiency. It should be pointed out that in the above derivation, no assumption specifically related to fuel cell has been made, the only assumption made is that the energy conversion system for power production is reversible for all processes involved. Thus, Equation (2.70) is valid for any power production system, be it electrochemical converter like fuel cells or conventional thermal energy converter like heat engines, as long as the process is reversible. Hence, it may be called the **second law efficiency**, since it is the maximum possible efficiency that is allowed by the second law of thermodynamics. In what follows we will demonstrate that the maximum possible efficiency for conventional heat engines, the well-known Carnot efficiency, is really the second law efficiency applied specifically to the conventional thermal power cycles, thus is equivalent to Equation (2.70).

2.5.3   Carnot Efficiency: The Reversible Energy Conversion Efficiency for Heat Engines

Consider a heat engine operating between two temperature thermal energy reservoirs (TERs), one at a high temperature T_H and the other at a low temperature T_L, as shown in Figure 2.8. The heat engine obtains energy from the high-temperature TER in the form of heat with the quantity q_H, a portion of this heat is converted to work output w and the remainder is rejected to the low-temperature TER in the amount of q_L as waste heat. Applying the first and second laws to the 2T heat engine, we have

$$1^{st} \text{ Law: } w = q_H - q_L \tag{2.71}$$

$$2^{nd} \text{ Law: } \wp_{s,HE} = \frac{q_L}{T_L} - \frac{q_H}{T_H} \tag{2.72}$$

where $\wp_{s,HE}$ represents the amount of entropy production during the energy conversion process by means of the heat engine. From Equation (2.72), the amount of heat rejection can be determined as

$$q_L = \frac{T_L}{T_H} q_H + T_L \wp_{s,HE} \tag{2.73}$$

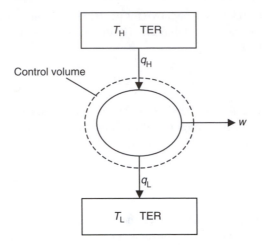

Figure 2.8 Thermodynamic system model of heat engines operating between two temperature thermal energy reservoirs (TERs).

The efficiency for the heat engine is, by the definition of Equation (2.64)

$$\eta = \frac{w}{q_H} \qquad (2.74)$$

Substituting Equations (2.71) and (2.73) into Equation (2.74) yields

$$\eta = 1 - \frac{T_L}{T_H} - \frac{T_L}{q_H} \wp_{s,HE} \qquad (2.75)$$

As pointed out earlier, the second law of thermodynamics dictates that entropy generation within the heat engine can never be negative; at most, it can vanish under the thermodynamically reversible condition. Therefore, the maximum possible efficiency for the heat engine is achieved if the process is reversible ($\wp_{s,HE} = 0$),

$$\eta_{r,HE} = 1 - \frac{T_L}{T_H} \qquad (2.76)$$

This is the familiar Carnot efficiency, giving the upper bound for efficiency of all 2T heat engines. Because $T_L < T_H$, low temperature $T_L \neq 0$ by the third law of thermodynamics, and the high temperature T_H is finite, 100% efficiency is not possible by the second law for any energy conversion system that produces power output using heat engines, such as steam and gas turbines, internal combustion engines, etc. The second law requires that the entropy generation term must never be negative. In contrast, 100% efficiency is always possible by the first law, which merely states the principle of energy conservation.

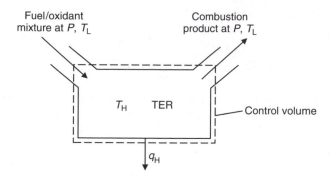

Figure 2.9 Thermodynamic system model of high-temperature thermal energy reservoir (TER) maintained by combustion process of a fuel/oxidant mixture.

2.5.4   Equivalency of Carnot and Fuel Cell Efficiency

As shown previously, both the Carnot efficiency and the reversible energy conversion efficiency for fuel cells Equation (2.70), are the maximum possible efficiency allowed by the second law, hence, they may be called the second law efficiency; the former is applied specifically to heat engines, while the latter is derived for fuel cells. Therefore, they must be related somehow as they both are the maximum possible efficiency dictated by the second law. In this subsection we demonstrate that they are actually equivalent, just expressed in a different form, under a suitable condition for the comparison.

Suppose for the heat engine the high-temperature TER is maintained at T_H by the combustion of a fuel with an oxidant, both reactants are originally at the temperature of T_L, as shown schematically in Figure 2.9. It is assumed that both the fuel and oxidant are the same as used in Figure 2.3 for the derivation of fuel cell performance, the combustion process is carried out at the same system pressure P in a controlled manner such that the combustion products leave the TER at the pressure P and temperature T_L. Neglecting the changes in the kinetic and gravitational potential energy, the first and second laws become for the high-temperature TER

$$\text{First law:} \quad q_H = h_R - h_P = -\Delta h(T_L, P) \tag{2.77}$$

$$\text{Second law:} \quad \wp_{s,\text{TER}} = (s_P - s_R) + \frac{q_H}{T_H} = \Delta s(T_L, P) + \frac{q_H}{T_H} \tag{2.78}$$

After rearranging, Equation (2.78) gives the temperature T_H resulting from the combustion process

$$T_H = \frac{q_H}{\wp_{s,\text{TER}} - \Delta s(T_L, P)} \tag{2.79}$$

Substitution of Equations (2.77) and (2.79) into Equation (2.75) leads to

$$\eta = \frac{\Delta g(T_L, P)}{\Delta h(T_L, P)} + \frac{T_L}{\Delta h(T_L, P)} \left(\wp_{s,\text{HE}} + \wp_{s,\text{TER}} \right) \tag{2.80}$$

where $\Delta g(T_L, P) = \Delta h(T_L, P) - T_L \Delta s(T_L, P)$ is the change in the Gibbs function between the reaction product and reactant. If all the processes within the heat engine and high-temperature TER are reversible ($\wp_{s,\text{HE}} = 0$ and $\wp_{s,\text{TER}} = 0$), then Equation (2.80) reduces to

$$\eta_r = \frac{\Delta g(T_L, P)}{\Delta h(T_L, P)} \tag{2.81}$$

which is exactly the same as Equation (2.70) — the efficiency expression derived for fuel cells. Note that in order for the combustion process to be reversible (i.e., $\wp_{s,\text{HE}} = 0$) theoretically, there should be no product dissociations and no incomplete combustion products or byproducts (e.g., pollutants) formed and perfect combustion products should consist of stable chemical species only, as would be obtained from an ideal and complete stoichiometric reactions. Therefore, it may be stated that any reversible heat engine operating under the maximum temperature limit allowed by a *perfect* combustion of a fuel/oxidant mixture has the same efficiency as that of a reversible isothermal fuel cell using the same fuel and oxidant and operating at the same temperature as that of the low-temperature TER. Or simply stated, the maximum possible efficiency is the same for both fuel cells and heat engines.

As a result, the often-heard statement that fuel cells are not subject to the Carnot efficiency and therefore they are more efficient is partly *correct* and partly *incorrect*, depending on the angle of viewpoint. It is "correct" because fuel cells do not require two temperature TERs to operate, and no T_H and T_L are involved. Clearly the Carnot efficiency does not apply to fuel cell operations — this statement is preferable rather than the statement that a fuel cell is not subjected to the Carnot efficiency limitation. On the other hand, it is "incorrect" because the Carnot efficiency is in essence a specific form of the second law efficiency which imposes upper limits on the performance of any energy conversion systems including both fuel cells and heat engines.

From the preceding analysis, it is clearly shown that both heat engines and fuel cells are subject to the same second law limitation. Then why do we often hear the statement that fuel cells have higher energy conversion efficiency than the corresponding heat engines? The reason may be several folds.

- The theoretical combustion temperature T_H is not achieved in practice: Since flame temperature is usually above 3000 K for a perfect combustion, product dissociation occurs and incomplete combustion products are formed, such as CO, NO, C (particulates), OH, O, N, etc., leading to a lower-temperature T'_H (usually about 2200 K for hydrocarbon fuels). In other words, the combustion process is invariably irreversible, resulting in the degradation of useful energy. That is, a perfect combustion is unachievable in practice.

- Due to material (metallurgical) consideration, even T'_H has to be lowered in heat engines so that metal components can have sufficient mechanical strength. For example, the maximum allowable temperature for gas turbines is about 1000 K* (with good cooling of turbine blades).
- The heat rejected to the lower temperature TER is not achieved in practice at the temperature T_L, rather at a temperature T'_L which is higher than the ambient atmospheric temperature T_L, e.g., $T'_L \cong 550$ K** for gas turbines instead of about 300 K. Referring to Figure 2.9, it is not possible in reality for the hot combustion product to leave the high-temperature TER at the low-temperature of T_L, instead it is more likely at the high-temperature T_H.

Therefore, the actual maximum efficiency

$$\eta'_r = 1 - \frac{T'_L}{T'_H} \cong 1 - \frac{550}{1000} = 45\%$$

which is only about half of the theoretical value $\eta_r = 1 - T_L/T_H \cong 1 - 300/3000 = 90\%$. Considering other related losses such as frictions, heat engines usually have lower than about 40% energy efficiency. For automobile engines, a rough estimate is that of chemical energy of fuels released as heat

- One-third is lost to the cooling water (metallurgical requirement),
- One-third is lost to the environment by exhaust stream,
- One-third is converted to useful work output.

that is, the actual efficiency is only about 33%, much less than the Carnot efficiency — the maximum possible efficiency allowed by the second law.

On the other hand, fuel cells can operate isothermally at a temperature sufficiently low enough so that no limitation is imposed by materials and sufficiently close to the atmospheric temperature so that the degree of irreversibility arising from cooling requirement is much less than the corresponding heat engines. For example, proton exchange membrane fuel cells can be operated at $T \cong 80\,°C$, at which q_L is rejected to the ambient atmosphere at $T_L \cong 298$ K (or 25 °C), compared to gas turbines of $T'_L \cong 550$ K! Even though irreversibilities do occur for fuel cells, but at a lesser degree, so that fuel cells have a higher practical efficiency than the conventional heat engines. Figure 2.10 shows a comparison of the practical energy conversion efficiency for different power generation technologies as a function of scale [2], which clearly illustrates the superior performance of fuel cells over the conventional heat engines. Therefore, the important point about fuel cells is not that they are not Carnot efficiency limited, rather they are (i) free from incomplete reaction or product dissociation

---

* The maximum temperature for gas turbines has increased significantly in the last decade to about 1300 °C–1400 °C.

** The exhaust temperature for gas turbines is typically in the range of 300 °C–600 °C.

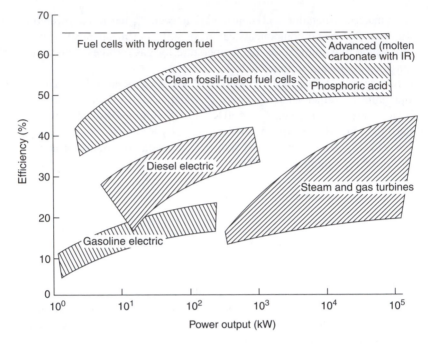

Figure 2.10 Comparison of practical energy conversion efficiency (based on lower heating value) of different technologies as a function of scale.[2] Note the commercially available combined cycle (based on natural gas) has the chemical-to-electrical energy efficiency of around 58% or higher and the total energy efficiency is much higher if co-generation is considered.

because of the much lower operating temperatures, (ii) free from the high temperature limit imposed by materials on any heat engines, and (iii) less irreversibilities associated with heat rejection process.

It should be emphasized that just like any other energy conversion system, fuel cells can never achieve, although it is quite possible to achieve very closely, the maximum possible efficiency allowed by the second law. The various mechanisms of irreversible losses in fuel cells is described later in this section.

2.5.5 Possibility of Over 100% Fuel Cell Efficiency: Is It Real or Hype?

It is well known that no heat engine could have efficiency of 100% or more, including the ideal Carnot efficiency, as discussed earlier. However, it has been reported that the ideal fuel cell efficiency, η_r, according to Equation (2.70), could be even over 100% in principle for some special fuel cell reactions (e.g. see References [3], and [4]), even though it is unachievable in practice. This has also sometimes been used as evidence that fuel cells could have higher energy efficiency than the competing heat engines. Is this realistic even under the thermodynamically reversible condition? The answer is negative! With the following analysis, we show that this is really due to a conceptual error in stretching the application of Equation (2.70) beyond its validity range.

Consider the thermodynamic model system used for fuel cell analysis, as shown in Figure 2.3. The amount of heat transfer from the surrounding thermal bath to the fuel cell system is given in Equation (2.66) for practical fuel cells. Under the thermodynamically reversible condition, the amount of heat transfer becomes

$$q = T\Delta s = \Delta h - \Delta g \tag{2.82}$$

For most of fuel cell systems, Δs is negative (i.e., $\Delta s < 0$ just like Δh and Δg), indicating that heat is actually transferred from the fuel cell to the ambient environment, or heat is lost from the fuel cell system, rather than the other way around. Hence, the second law efficiency, according to Equation (2.70)

$$\eta_r = \frac{\Delta g}{\Delta h} = \frac{\Delta h - T\Delta s}{\Delta h} = 1 - \frac{T\Delta s}{\Delta h} < 1 \tag{2.83}$$

is less than 100%, as it should be by the common perception of the parameter called efficiency.

However, for some special reactions, such as

$$C(s) + \frac{1}{2}O_2(g) \longrightarrow CO(g) \tag{2.84}$$

the entropy change Δs is positive. Physically it indicates that the fuel cell absorbs heat from the ambiance and converts it completely into electrical energy along with the chemical energy of the reactants. This is equivalent to that the less useful form of energy — heat is converted completely into the more useful form of energy — electric energy without the generation of entropy (i.e., reversible condition) during the conversion process when Equation (2.70) is used for the efficiency calculation — such a process is clearly a violation of the second law. Therefore, the second law efficiency for this particular fuel cell reaction becomes larger than 100%, that is, an unphysical result, when Equation (2.70) is utilized for the efficiency calculation for this type of fuel cell reactions. In fact, for the reaction shown in Equation (2.84) and according to Equation (2.70), the reversible fuel cell efficiency would be equal to $\eta_r = 124\%$ at the standard temperature and pressure, 163% at 500 °C and 1 atm, and 197% at 1000 °C and 1 atm!

The root of the problem from the straight forward application of Equation (2.70) leading to the physically impossible result of over 100% energy efficiency is as follows. At atmospheric temperature for fuel cell operations, the energy from the thermal bath (or the atmosphere) as heat may be free. But at elevated temperatures, external means must be employed to keep the thermal bath at temperatures above the ambient atmospheric temperature, which constitutes an expense. Therefore, the heat from the thermal bath to the fuel cell system is no longer a free energy input, rather it is part of the energy input which has to be paid. By the definition, Equation (2.64), the efficiency definition for fuel cells has to be modified accordingly, such that the ideal second law efficiency is no longer over 100% for fuel cells. Thus, we conclude that the reversible fuel cell efficiency shown in Equation (2.70) is only valid for fuel cell reactions where the entropy change between the product and reactant is negative (hence, heat is lost from the fuel cell) and it cannot be applied for reactions with positive entropy change, such as the one given in Equation (2.84).

2.5.6   Practical Fuel Cell Efficiency and Energy Loss Mechanisms

From the preceding analysis, it is clear that energy loss in fuel cells occurs under both reversible and irreversible conditions. We discuss each type of energy loss mechanisms and associated expression for energy conversion efficiency in fuel cells. Since this chapter is devoted to thermodynamic analysis, the reversible loss mechanism is presented in detail, whereas the irreversible loss mechanisms are only briefly described since the next two chapters are devoted to a detailed description of the various processes that occur under the thermodynamically irreversible condition.

Reversible Energy Loss and Reversible Energy Efficiency   The energy loss in fuel cells under reversible condition is equal to the heat transferred (or lost) to the environment, as given in Equation (2.82),

$$q = T\Delta s = \Delta h - \Delta g \tag{2.82}$$

because of the negative entropy change for the fuel cell reaction. The associated energy conversion efficiency, which has been called the second law efficiency or the reversible energy conversion efficiency, has been derived and given in Equation (2.70) or (2.83). Combining Equation (2.82) with Equation (2.83) yields

$$\eta_r = \frac{\Delta g}{\Delta h} = \frac{\Delta g}{\Delta g + T\Delta s} \tag{2.85}$$

Dividing the numerator and the denominator by the factor (nF), and utilizing Equations (2.24) and (2.44), Equation (2.85) becomes

$$\eta_r = \frac{E_r}{E_r - T\left(\frac{\partial E_r}{\partial T}\right)_P} \tag{2.86}$$

Therefore, when the entropy change is negative, as described earlier, the reversible efficiency, η_r, is less than 100% and the reversible cell potential decreases with temperature; and according to Equation (2.86), the reversible efficiency, η_r, also decreases with temperature. For example, for H_2 and O_2 reaction forming gaseous water at 1 atm pressure, Example 2.7 shows that

$$\left(\frac{\partial E_r}{\partial T}\right)_P = -0.2302 \times 10^{-3} \text{ V/K}$$

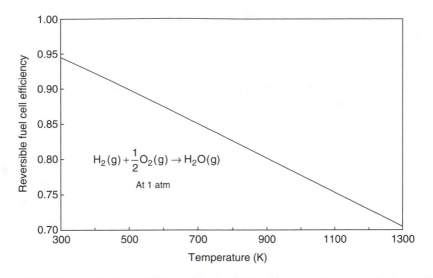

Figure 2.11 The reversible fuel cell efficiency (based on LHV) as a function of temperature for the reaction of $H_2 + \frac{1}{2}O_2 \longrightarrow H_2O(g)$ occurring at 1 atm pressure.

at 25 °C, and the reversible efficiency is about 95% at 25 °C, and it becomes 88% at 600 K and 78% at 1000 K. Figure 2.11 illustrates the reversible efficiency as a function of temperature for the hydrogen and oxygen reaction with gaseous water as the reaction product. It is seen that the reversible efficiency decreases almost linearly. For most fuel cell reactions,

$$\left(\frac{\partial E_r}{\partial T}\right)_P = -(0.1 \sim 1.0) \times 10^{-3}\ \text{V/K}$$

at 25 °C and 1 atm, hence the reversible efficiency is typically around 90%.

However, for the reaction of carbon and oxygen to form carbon monoxide, as shown in Equation (2.84), the entropy change is positive and the reversible cell potential increases with temperature, as presented previously; hence, the reversible efficiency also increases with temperature, according to Equation (2.86). But as discussed previously, the efficiency expression, Equation (2.85) or (2.86), is not really valid for such reactions.

From the reversible energy efficiency, Equation (2.70), and dividing the numerator and denominator by the factor (nF), we have, after utilizing the reversible cell potential, Equation (2.24)

$$\eta_r = \frac{E_r}{(-\Delta h/nF)} = \frac{E_r}{E_{tn}} \tag{2.87}$$

where

$$E_{tn} = -\frac{\Delta h}{nF} \tag{2.88}$$

is called **thermoneutral voltage** (or potential voltage), a voltage a fuel cell would have if all the chemical energy of the fuel and oxidant is converted to electric energy (i.e., 100% energy conversion into electricity). For example, for the reaction

$$H_2(g) + \frac{1}{2}O_2(g) \longrightarrow H_2O(\ell)$$

$E_{tn} = 1.48\,V$ and the corresponding reversible efficiency is $\eta_r = 83\%$ at 25 °C and 1 atm. Whereas at the same temperature and pressure, for the reaction

$$H_2(g) + \frac{1}{2}O_2(g) \longrightarrow H_2O(g)$$

$E_{tn} = 1.25\,V$ and the corresponding reversible efficiency is $\eta_r = 95\%$.

From the preceding discussion, it is noted that for hydrogen and oxygen reaction, the reversible cell efficiency can differ by as much as 14%, depending on whether the product water is liquid or vapor or whether the higher or lower heating value is used for the efficiency calculation under identical operating condition. Since for most of hydrocarbon fuels that contain hydrogen (including hydrogen itself, hydrocarbons, alcohols, and to a lesser extent coal), two values exist for the change in the enthalpy and Gibbs function, that is,

For natural gas (methane, CH_4): $\dfrac{LHV}{HHV} = 0.90;$

For coals of typical hydrogen and water content: $\dfrac{LHV}{HHV} = 0.95 \sim 0.98$

Therefore, different efficiency values result, depending on which heating value $(-\Delta h)$ is used for the efficiency calculation. Typically in fuel cell analysis, the HHV is used unless stated otherwise; this convention is used throughout this book unless explicitly stated otherwise.

It should be emphasized that from the preceding analysis it is known that for most fuel cell reactions, the reversible efficiency, η_r, decreases as the fuel cell operating temperature is increased. This effect is important in considering high-temperature fuel cells, namely molten carbonate fuel cells and solid oxide fuel cells. For example, Figure 2.11 indicates that the reversible cell efficiency is reduced to lower 70% (based on LHV) for hydrogen and oxygen reaction at the typical operating temperature of 1000 °C for solid oxide fuel cells, as opposed to around 95% at 25 °C as discussed here. This significant reduction in the reversible cell efficiency seems against high temperature fuel cells. However, the irreversible losses, to be described in the following text, decrease drastically as temperature is increased, so that the practical fuel cell performance (such as, efficiency and power output under practical operating condition) increases. Therefore, further analysis should be needed for efficiency under practical operating condition rather than the idealized reversible condition, which is the focus of the following discussion.

Irreversible Energy Losses   For fuel cells, the reversible cell potential and the corresponding reversible efficiency are obtained under the thermodynamically reversible condition, implying that there is no rigorous occurrence of continuous reaction or electrical current output. For practical applications, a useful amount of work (electrical energy) is obtained only when a reasonably large current I is drawn from the cells because the electrical energy output is through the electrical power output, which is defined as

$$\text{Power} = EI \quad \text{or} \quad \text{Power Density} = EJ \qquad (2.89)$$

However, both the cell potential and efficiency decrease from its corresponding (equilibrium) reversible values because of irreversible losses when current is increased. These irreversible losses, the subject of the next two chapters, are often called **polarization**, **overpotential** or **overvoltage**[*]in literature and they originate primarily from three sources: activation polarization, ohmic polarization, and concentration polarization. The actual cell potential as a function of current is the result of these polarizations; therefore, a plot of the cell potential versus current output is conventionally called a **polarization curve**. It should be noticed that the magnitude of electrical current output depends largely on the active cell area, therefore, a better measure is the current density J (A/cm^2) instead of current I, itself and the unit A/cm^2 is often used rather than A/m^2 as the unit for the current density because square meter is too large to be used for fuel cell analysis.

A typical polarization curve is illustrated in Figure 2.12 for the cell potential as a function of current density. The ideal cell potential–current relation is independent of the current drawn from the cell, and the cell potential remains equal to the reversible cell potential. The difference between the thermoneutral voltage and the reversible cell potential represents the energy loss under the reversible condition (the reversible loss). However, actual cell potential is smaller than reversible cell potential and decreases as the current drawn is increased due to the three mechanisms of irreversible losses: activation, ohmic, and concentration polarization. The activation polarization, η_{act}, arises from the slow rate of electrochemical reactions, and a portion of the energy is lost (or spent) on driving up the rate of electrochemical reactions in order to meet the rate required by the current demand. The ohmic polarization, η_{ohm}, arises due to electrical resistance in the cell, including ionic resistance to the flow of ions in the electrolyte and electronic resistance to the flow of electrons in the rest of the cell components. Normally, ohmic polarization is linearly dependent on the cell current. Concentration polarization, η_{conc}, is caused by the slow rate of mass transfer resulting in the depletion of reactants in the vicinity of active reaction sites and the overaccumulation of reaction products which block the reactants from reaching the reaction sites. It usually becomes significant, or even prohibitive, at high current density when

---

[*]The terms "polarization," "overpotential" (or "overvoltage") have been loosely used in literature to denote cell potential (or voltage) loss. Their subtle differences are pointed out in the next chapter.

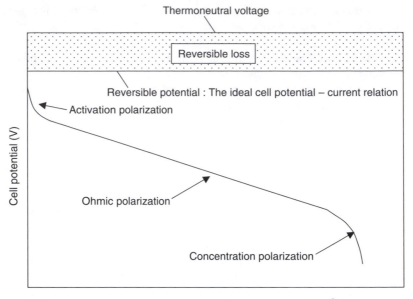

Figure 2.12 Schematic of a typical polarization curve. The cell potential for a fuel cell decreases as the current drawn from the cell is increased due to activation, ohmic, and concentration polarizations.

the slow rate of mass transfer is unable to meet the high demand required by the high current output. As shown in Figure 2.12, concentration polarization is often the cause of cell potential decrease rapidly to zero. The current (density) corresponding to the zero cell potential is often called the limiting current (density), and evidently it is controlled by the concentration activation. From Figure 2.12, it is also clear that activation polarization occurs at small current density, while concentration polarization occurs at high current density. The linear drop in the cell potential due to resistance loss occurs at immediate current density and practical fuel cell operation is almost always located within the ohmic polarization region.

Figure 2.12 also indicates that even at zero current output from the fuel cell, the actual cell potential is smaller than the idealized reversible cell potential. This small difference in cell potential is directly related to the chemical potential difference between the cathode and anode. So that even at zero external load current, electrons are delivered to the cathode, where oxygen ions are formed, and migrate through the electrolyte to the anode where they deionize to release electrons. The electron released migrates back to the cathode to continue the process or "exchange." The ionization/deionization reactions proceeding at a slow rate yield an extremely small current, often called exchange current I_0 or exchange current density J_0, and the cell potential is reduced below the reversible cell potential. Therefore, exchange current arises from the fact that electrons migrating through the electrolyte rather than through the external load, and about $0.1 \sim 0.2$ V of cell potential loss results from the exchange process. Consequently, the efficiency of a real fuel cell is about $8\% \sim 16\%$ lower than the reversible cell efficiency, η_r, even at close to zero current output.

The exchange current density J_0 is very small, it is at least about 10^{-2} A/cm^2 for H_2 oxidation at the anode, and about 10^{-5} times slower for O_2 reduction at the cathode. In comparison, the O_2 reduction process at the cathode is so slow that competing anodic reactions play a significant role, such as oxidation of electrocatalyst, corrosion of electrode materials, oxidation of organic impurities in the anode structure. All these anodic reactions result in the corrosion of electrodes, thereby limiting the cell life unless appropriate counter measures are taken.

It should be pointed out that the cell potential loss resulting from the exchange current diminishes when the current drawn through the external load is increased beyond a certain critical value. As the external current is increased, the cell potential decreases as shown in Figure 2.12, thus the driving force for the exchange current is reduced, leading to a smaller exchange current — this is the only form of energy losses that decreases when external current is increased.

From this discussion, it becomes clear that the actual cell potential E is lower than the reversible cell potential E_r and the difference is due to the potential losses due to the above irreversible loss mechanisms. Therefore,

$$E = E_r - (\eta_{act} + \eta_{ohm} + \eta_{conc}) \tag{2.90}$$

The irreversible energy loss as heat (or waste heat generation) per-mole fuel consumed can be easily obtained because the entropy generation is, according to Equations (2.27) and (2.90)

$$\wp_s = \frac{nF(E_r - E)}{T} = \frac{nF(\eta_{act} + \eta_{ohm} + \eta_{conc})}{T} \tag{2.91}$$

Then, Equation (2.16) becomes for the total heat loss from the fuel cell

$$q = T\Delta s - T\wp_s = \underbrace{T\Delta s}_{\text{Reversible loss}} \underbrace{- nF(\eta_{act} + \eta_{ohm} + \eta_{conc})}_{\text{Irreversible losses}} \tag{2.92}$$

Since the entropy change is negative ($\Delta s < 0$) for most fuel cell reactions, the heat generated is negative as well, implying that energy as heat is lost from the fuel cell shown in Figure 2.3 for both reversible and irreversible losses.

Because $T\Delta s = \Delta h - \Delta g$ by the definition of the Gibbs function change for fuel cell reactions, Equation (2.92) can be written as

$$\frac{q}{nF} = \frac{T\Delta s}{nF} - (\eta_{act} + \eta_{ohm} + \eta_{conc}) = \frac{\Delta h - \Delta g}{nF} - (\eta_{act} + \eta_{ohm} + \eta_{conc}) \tag{2.93}$$

Considering the definition for the thermoneutral voltage and the reversible cell potential, the above expression becomes

$$\frac{q}{nF} = -E_{tn} + E_r - (\eta_{act} + \eta_{ohm} + \eta_{conc}) \tag{2.94}$$

Combining with Equation (2.90), Equation (2.94) reduces to

$$\frac{q}{nF} = -(E_{tn} - E) \tag{2.95}$$

Hence, the equivalent cell potential loss due to the energy loss from the fuel cell as heat is equal to the difference between the thermoneutral voltage and the actual cell potential.

The rate of heat loss per mole fuel consumed in the fuel cell, Equation (2.92), can be expressed as an equivalent power loss:

$$P_{\text{HeatLoss}} = I\left(\frac{q}{nF}\right) = -I(E_{tn} - E) = I\left[\frac{T\Delta s}{nF} - (\eta_{\text{act}} + \eta_{\text{ohm}} + \eta_{\text{conc}})\right] \quad (2.96)$$

This expression is important in determining the cooling requirements of fuel cell stacks.

Various Forms of Irreversible Energy Efficiency   After the description of the irreversible energy losses, we can now introduce several forms of energy efficiency that would be useful in the analysis of fuel cell performance.

(1) Voltage Efficiency η_E: The voltage efficiency is defined as:

$$\eta_E = \frac{E}{E_r} \quad (2.97)$$

Because the actual cell potential E is compared with the maximum possible cell potential E_r allowed by the second law, the voltage efficiency is really a specific form of the energy efficiency, representing the degree of departure of the cell operation from the idealized thermodynamically reversible condition. As shown in Equation (2.90), $E < E_r$, hence $\eta_E < 1$. For example, for H_2 and O_2 reaction at the standard reference state forming liquid water, we know the reversible cell potential is equal to 1.229 V. If such a cell is operating at a cell potential of 0.7 V, the corresponding voltage efficiency can be determined as

$$\eta_E = \frac{E}{E_r} = \frac{0.7}{1.229} = 0.57\%$$

For the same cell, but if the reaction product water is gaseous, then the corresponding voltage efficiency becomes

$$\eta_E = \frac{E}{E_r} = \frac{0.7}{1.185} = 0.59\%$$

(2) Current Efficiency η_I: The current efficiency is a measure of how much current is produced from a give amount of fuel consumed in fuel cell reaction; and it is defined as

$$\eta_I = \frac{I}{nF\left(\frac{dN_F}{dt}\right)} \quad (2.98)$$

where dN_F/dt represents the rate of fuel consumption in the fuel cell (mol/s). The current efficiency would be less than 100% if part of the reactants participates in non current-productive side reactions, called parasitic reactions, such as reactants cross over the electrolyte region, incomplete conversion of reactants to desired products, reaction with the cell components, or even reactant leakage from the cell compartment due to sealing problem, etc. For most practical fuel cells, especially at operating conditions where the current output is sufficiently larger than zero (without the effect of exchange current discussed previously), the current efficiency is about 100%. This is because for practical fuel cells, all the parasitic reactions are undesirable and would have been removed by appropriate design.

However, for direct methanol fuel cells about 20% of the liquid methanol can cross over to the cathode side through the proton-conducting polymer membrane, implying the current efficiency is only about 80% for such cells. Similarly, if the polymer membrane is too thin, say 50 μm or thinner, hydrogen crossover in a proton exchange membrane fuel cell may not be negligible, depending on the operating current density drawn from the cell. On the other hand, hydrogen crossover is minimal and can be neglected if the membrane used is sufficiently thick, such as Nafion 117 — a commonly used polymer membrane electrolyte for proton exchange membrane fuel cells. Therefore, specific care should be taken for the current efficiency in practical fuel cells.

(3) Overall Free Energy Conversion Efficiency, η_{FC}: The overall free energy conversion efficiency is defined as the product of the reversible efficiency, voltage and current efficiency:

$$\eta_{FC} = \eta_r \times \eta_E \times \eta_I \tag{2.99}$$

If the current efficiency is 100% as is often the case for well-designed practical fuel cells, substituting the definitions for the various efficiencies into Equation (2.99) lead to

$$\eta_{FC} = \frac{\Delta g}{\Delta h} \times \frac{E}{E_r} \times 1 = \frac{E_r}{E_{tn}} \times \frac{E}{E_r} \times 1 = \frac{E}{E_{tn}}$$

or

$$\eta_{FC} = \frac{E}{E_{tn}} \tag{2.100}$$

Therefore, the overall free energy conversion efficiency is really the overall efficiency for energy conversion process occurring within the fuel cell. Because the thermoneutral voltage is a fixed value for a given fuel and oxidant under a given operating condition of temperature and pressure, the overall energy conversion efficiency for fuel cells is proportional to the actual cell potential

$$\eta_{FC} \sim E \tag{2.101}$$

This is a significantly important result. Once the actual cell potential is determined, the energy conversion efficiency of the fuel cell is known as well. This is the primary reason that in fuel cell literature, it is almost always that the cell polarization curve is given without specifically showing the cell energy efficiency as a function of the current. Further, Equation (2.101) implies that the fuel cell efficiency will depend on the current output in the same way as the cell potential, that is, decrease as the current output is increased.

(4) Fuel Cell System Efficiency η_S: As discussed in Section 1.2, a fuel cell system is composed of one or multiple fuel cell stacks and a number of auxiliary equipment, which would also have its own energy efficiency of η_{aux}. Therefore, the total fuel cell system efficiency is equal to the product of the efficiency values for the fuel cell and all the auxiliary equipment:

$$\eta_s = \eta_{FC} \times \eta_{aux} \tag{2.102}$$

Figure 2.10 shows a comparison of fuel cell system efficiency with the efficiency of a number of competing technologies as a function of scale. The efficiency is determined based on the lower heating value.

Example 2.10   For proton exchange membrane (PEM) fuel cells, pristine H_2 and O_2 are normally used as the reactants. At the operating temperature of 80 °C and pressure of 1 atm, the following electrochemical reaction occurs

$$H_2(g) + \frac{1}{2}O_2(g) \longrightarrow H_2O(\ell)$$

One of the design objectives is often to operate the PEM fuel cell with a cell voltage between 0.7~0.8 V. Determine the overall fuel cell efficiency, η_{FC}, (a) for the above reaction; (b) for the above reaction occurring at 80 °C and 3 atm.

SOLUTION
From Equation (2.100), we know that the overall fuel cell efficiency can be determined from

$$\eta_{FC} = \frac{E}{E_{tn}}$$

assuming 100% current efficiency, where the thermoneutral voltage can be calculated from

$$E_{tn} = -\frac{\Delta h}{nF}$$

provided that the enthalpy change for the given reaction under the given condition is known. Therefore, the focus here is to acquire the enthalpy change.

(a) The enthalpy change for the given reaction is equal to the enthalpy of formation for liquid water at 80 °C and 1 atm; the latter can be determined from the results obtained in Example 2.4 for the given condition of 80 °C and 1 atm. In Example 2.4 the enthalpy of formation for liquid water is shown as, at 80 °C and 1 atm

$$h_{f,H_2O(\ell)} = -284,087 \, J/(mol \cdot K)$$

Therefore, the enthalpy change for the present reaction is

$$\Delta h = -284,087 \, J/(mol \, fuel \cdot K)$$

Hence, the thermoneutral voltage is obtained as

$$E_{tn} = -\frac{-284,087}{2 \times 96,487} = 1.472 \, V$$

For the given cell potential of $E = 0.7 \sim 0.8 \, V$, the overall fuel cell efficiency is

$$\eta_{FC} = \frac{0.7 \sim 0.8}{1.472} = 0.476 \sim 0.543$$

(b) At the pressure of 3 atm, because enthalpy is independent of pressure for liquids (as incompressible substance) and ideal gases (a good approximation for H_2 and O_2), the enthalpy of reaction for the given reaction is also independent of pressure. This leads to the same enthalpy of reaction as at 1 atm. Since the actual cell potential of $0.7 \sim 0.8 \, V$ is also given in the problem, it turns out that the overall fuel cell efficiency in this example remains independent of pressure. Hence,

$$\eta_{FC} = \frac{0.7 \sim 0.8}{1.472} = 0.476 \sim 0.543$$

or the same as in (a).

COMMENT

Note that under the same operating temperature, the actual cell potential E increases when the operating pressure is increased because of the increased reversible cell potential and reduced cell voltage losses. In this case, the pressure affects the overall fuel cell efficiency. The pressure independency in this example is due to the fact that the cell potential E is given and fixed when the operating pressure is changed (or the operating cell potential E did not change with pressure). In practice, at the same cell voltage, the current drawn from the cell is increased as the operating pressure is raised, thus the output power is increased as well, even though the cell voltage and the efficiency remain the same.

2.5.7   Efficiency Loss in Operating Fuel Cells: Stoichiometry, Utilization, and Nernst Loss

In an operating fuel cell, reactant composition changes between the inlet and outlet of the fuel cell along the flow path over the electrode surface because reactants are consumed to yield current output and reaction products are formed along the way as well. The change in reactant composition results in additional loss of cell potential beyond those losses described in the preceding section. This potential loss arises from the fact that the cell potential E adjusts to the lowest electrode potential given by the Nernst equation, Equation (2.63), for the various reactant compositions at the exit of the anode and cathode chambers. This is because electrodes are usually made of good electronic conductors and consequently they are iso-potential surfaces. The cell potential E may not exceed the minimum local value set by the Nernst equation. This additional cell potential loss is often also called the Nernst loss, which is equal to the difference between the inlet and exit Nernst potentials determined based on the inlet and exit reactant compositions. According to Equation (2.63), this additional cell potential loss due to the consumption of reactants in the cell is, when the reactant streams are arranged in a concurrent flow

$$\eta_N = \frac{\Re T}{nF} \ln K_{\text{out}} - \frac{\Re T}{nF} \ln K_{\text{in}} = \frac{\Re T}{nF} \ln \frac{K_{\text{out}}}{K_{\text{in}}} \qquad (2.103)$$

where K_{in} and K_{out} are the equilibrium constant for partial pressure evaluated at the cell inlet and outlet gas compositions.

In the case of a fuel cell where both fuel and oxidant flow is in the same direction (concurrent), the minimum Nernst potential occurs at the flow outlet. When the reactant flows are in counterflow, crossflow, or more complex arrangements, it becomes difficult to determine the location of the minimum potential due to the reactant consumption. Appropriate flow channel design for the anode and cathode side can minimize the Nernst loss.

Equation (2.103) also implies that the Nernst loss will be extremely large, approach infinity if all the reactants are consumed in the in-cell electrochemical reaction leading to zero reactant concentration at the cell outlet. To reduce the Nernst loss to an acceptable level for practical fuel cell operations, reactants are almost always supplied more than the stoichiometric amount required for the desired current production. The actual amount of reactants supplied to a fuel cell is often expressed in terms of a parameter called stoichiometry, S_t:

$$S_t = \frac{\text{Molar flow rate of reactants supplied to a fuel cell}}{\text{Molar flow rate of reactants consumed in the fuel cell}}$$

$$= \frac{\dot{N}_{\text{in}}}{\dot{N}_{\text{consumed}}} \qquad (2.104)$$

For example, for proton exchange membrane fuel cells, typical operation uses

$$S_t \approx (1.1 \sim 1.2) \text{ for } H_2 \quad \text{and}$$

$$S_t \approx 2 \text{ for } O_2 \text{ (pure or in air)}$$

Therefore, stoichiometry really represents the actual flow rate for the reactant delivered to the fuel cell or how much reactant is consumed in the fuel cell for current production for a given amount of reactant supply. Because normally at least two types of reactant for fuel cell exist, one as fuel and another as oxidant, stoichiometry can be defined for either reactant. For molten carbonate fuel cell, there are two reactants in the oxidant stream, namely, CO_2 and O_2, then the stoichiometry for the oxidant is commonly defined for the species in deficiency.

Alternatively, reactant flow rate can be expressed in terms of a parameter called utilization, U_t

$$U_t = \frac{\text{Molar flow rate of reactants consumed in a fuel cell}}{\text{Molar flow rate of reactants supplied into the fuel cell}}$$

$$= \frac{\dot{N}_{\text{consumed}}}{\dot{N}_{\text{in}}} = \frac{1}{S_t} \qquad (2.105)$$

Clearly, stoichiometry and utilization are inversely proportional to each other. Although both parameters are used in practice, stoichiometry is more frequently used in literature for proton exchange membrane fuel cells, and utilization is often used for intermediate- and high-temperature fuel cells (PAFCs, MCFCs and SOFCs).

For properly designed practical fuel cells, no reactant crossover or leakage out of the cell may occur in general, therefore, the rate of reactant consumed within the cell is equal to the difference between the molar flow rate into and exiting the cell. Because there are two reactants for fuel cell reactions: a fuel and an oxidant, the parameter stoichiometry (or utilization) can be defined, respectively, for the fuel and oxidant. For example, the stoichiometry for the fuel may be expressed as

$$S_{t,F} = \frac{\dot{N}_{F,\text{in}}}{\dot{N}_{F,\text{consumed}}} = \frac{\dot{N}_{F,\text{in}}}{\dot{N}_{F,\text{rmin}} - \dot{N}_{F,\text{out}}} = \frac{1}{U_{t,F}} \qquad (2.106)$$

The stoichiometry or utilization for the oxidant can be written similarly.

Effect of reactant utilization on the reversible cell potential is illustrated in Figure 2.13 with the corresponding reactant composition at the cell outlet, in terms of mole fractions, as a function of utilization given in Table 2.2. It is seen that the reactant composition at the cell outlet decreases, hence the reversible cell potential decreases as well when the utilization factor is increased. The decrease is rapid when utilization goes beyond about 90%. In practical fuel cell operation, 100% utilization (or unity stoichiometry) will result in reactant concentrations vanishing at the cell exit, then the Nernst loss becomes dominant and the cell potential is reduced to zero — this is certainly undesirable situation, that needs to be avoided. Therefore, typical operation requires that the utilization be about 80%~90% for fuel, and 50% for oxidant in order to balance the Nernst loss with the parasitic losses associated with the reactant supply.

As shown in Equation (2.103), the additional Nernst loss due to the reactant depletion in the cell is directly proportional to the cell operating temperature. For

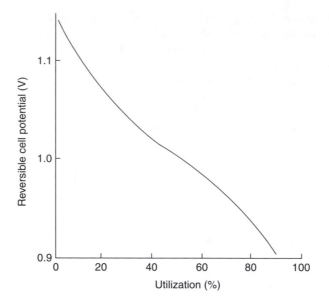

Figure 2.13 Reversible cell potential as a function of reactant utilization (both fuel and oxidant utilizations are set equal) for a molten carbonate fuel cell operating at 650 °C and 1 atm. Reactant compositions at the cell inlet: 80% H_2/20% CO_2 mixture saturated with $H_2O(g)$ at 25 °C for the fuel gas; and 60% CO_2/30% O_2/10% inert gas mixture for the oxidant gas.[5]

Table 2.2   Reactant Composition[a] at Cell Outlet as a Function of Utilization for a MCFC Operating at 650 °C and 1 atm [5]

| | Utilization[b] (%) | | | | |
Reactant	0	25	50	75	90
Anode[c]					
X_{H_2}	0.645	0.410	0.216	0.089	0.033
X_{CO_2}	0.064	0.139	0.262	0.375	0.436
X_{CO}	0.130	0.078	0.063	0.033	0.013
X_{H_2O}	0.161	0.378	0.458	0.502	0.519
Cathode[d]					
X_{CO_2}	0.600	0.581	0.545	0.461	0.316
X_{O_2}	0.300	0.290	0.273	0.231	0.158

[a] Reactant compositions (H_2 and CO for anode and O_2 and CO_2 for cathode) are given in mole fractions in the Table.

[b] Same utilization for fuel and oxidant is used for the calculation.

[c] Fuel gas composition at the anode inlet: 80% H_2/20% CO_2 saturated with H_2O at 25 °C; Fuel gas composition at the anode outlet: Calculated based on compositions for water–gas shift reaction in equilibrium (The details of calculations are deferred until Chapter 9).

[d] Oxidant gas composition at the cathode inlet: 30% O_2/60% CO_2/10% inert gas; Oxidant gas composition at the cathode outlet: Calculated based on the utilization or amount of consumption in the cell.

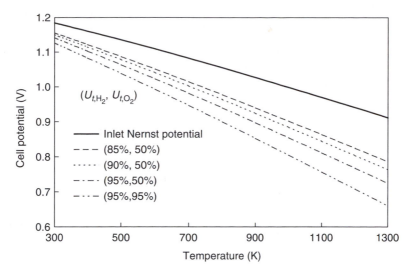

Figure 2.14 Inlet and outlet Nernst potential as a function of temperature and utilization for the reaction of $H_2(g) + O_2(g) \longrightarrow H_2O(g)$ at 1 atm.

example, for a H_2–air fuel cell, a change in the reactant gas composition from the cell inlet to the outlet that leads to a 60 mV cell potential loss at near room temperature (25 °C) would result in a loss of 300 mV at 1200 °C. Therefore, the additional Nernst loss arising from the reactant gas composition changes within the cell becomes more significant when the cell operating temperature is increased, and may become a serious loss mechanism for high temperature fuel cells. Figure 2.14 shows the reversible cell potential at the cell inlet and outlet for hydrogen and oxygen reaction forming gaseous water product at 1 atm as a function of temperature. The outlet Nernst potentials (i.e., the reversible cell potential at the cell outlet) are determined for oxygen utilization of 50%, and hydrogen utilizations of 85%, 90%, and 95%, respectively; as well as for the utilization of 95% for both hydrogen and oxygen. It is clearly seen that the outlet Nernst potential decreases when either utilization or temperature is increased.

If pure hydrogen is used as fuel, the anode compartment can be designed as a deadend chamber for hydrogen supply, as shown in Figure 2.15. Similarly, if pure oxygen is used as oxidant, a deadend cathode compartment can be employed. However, inert impurities in the reactant gas will accumulate at the anode and cathode compartment, and they must be removed either periodically or continuously in order to maintain a good fuel cell performance. Periodic purging or continuous bleeding can be implemented for this purpose, but this results in a small loss of fuel, and hence less than 100% utilization.

From the previous discussion, it is evident that 100% utilization for reactants is practically an unwise design. Since in-cell fuel utilization will never be 100% in practice, the determination of in-cell energy conversion efficiency and the cell potential must take utilization factor into consideration. If the fuel exiting the fuel cell is discarded (not recirculated back to the cell or not utilized for other useful purpose

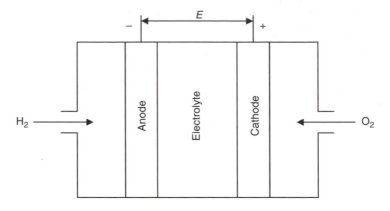

Figure 2.15 Schematic of deadend design for reactant supply to a fuel cell when pure hydrogen and pure oxygen are used as reactants.

such as providing heat for fuel preprocessing), then the overall energy conversion efficiency must be equal to the overall fuel cell efficiency given in Equation (2.99) multiplied by the utilization to take into account the fact that not all the fuel is being used for electric energy production.

Example 2.11   Determine the inlet and outlet Nernst potential for the following reaction

$$H_2 + \frac{1}{2}O_2 \longrightarrow H_2O(g)$$

at the temperature of 25 °C and 1000 K and pressure of 1 atm as a function of O_2 utilization. Assume the fuel is pure H_2, and O_2 is from the air supplied to the cell at the given temperature and pressure, H_2 utilization is zero, and reaction product water is formed on the oxidant side.

SOLUTION
The inlet and outlet Nernst potential are the reversible cell potential calculated based on the reactant composition at the cell inlet and outlet, respectively. Therefore, they can be calculated from the Nernst equation, Equation (2.63). Specifically, the reversible cell potential due to the change in O_2 concentration is, at the cell inlet

$$E_{r,\text{in}} = E_r(T, P_{O_2,\text{in}}) = E_r(T, P) - \frac{\Re T}{nF} \ln X_{O_2,\text{in}}^{-1/2}$$

At the cell inlet, the fuel is pure H_2, and the oxidant concentration is equal to O_2 in air, therefore we have the molar fractions for the reactants
   At the cell inlet:

$$X_{H_2,\text{in}} = 1; \quad X_{O_2,\text{in}} = 0.21; \quad X_{N_2,\text{in}} = 0.79; \quad P_{O_2,\text{in}} + P_{N_2,\text{in}} = P_{\text{in}} = 1 \text{ atm}$$

Since for the hydrogen and oxygen reaction at 25 °C and 1 atm, Example 2.6 shows that

$$E_r(25\,°C, 1\,atm) = 1.185\,V$$

At 1000 K and 1 atm, we have from Appendix A.5

$$E_r(1000\,K, 1\,atm) = -\frac{-192,652}{2 \times 96,487} = 0.9983\,V$$

The inlet Nernst potential at 25 °C and 1 atm is, respectively

$$E_r(25\,°C, 1\,atm) = 1.185 - \frac{8.3143 \times 298}{2 \times 96,487} \ln 0.21^{-1/2} = 1.175\,V$$

$$E_r(1000\,K, 1\,atm) = 0.9983 - \frac{8.3143 \times 1000}{2 \times 96,487} \ln 0.21^{-1/2} = 0.9647\,V$$

At the cell outlet, the fuel remains pure H_2 because of the zero utilization, whereas for the oxidant stream, the partial pressure for O_2 becomes $(1 - U_{t,O_2})P_{O_2,in}$ due to the in-cell O_2 consumption, the partial pressure for $H_2O(g)$ is $2U_{t,O_2}P_{O_2,in}$ because 2 mole $H_2O(g)$ is formed from every 1 mole O_2 consumed in the cell. The partial pressure for N_2 remains the same as the value at the inlet, $0.79P_{in}$, since it is inert species and does not participate in the cell reaction. Therefore, we have the molar fractions for the reactants at the cell outlet.

At the cell outlet:

$$X_{H_2,out} = 1;$$

$$X_{O_2,out} = \frac{(1 - U_{t,O_2})\,P_{O_2,in}}{(1 - U_{t,O_2})\,P_{O_2,in} + 2U_{t,O_2}\,P_{O_2,in} + 0.79P_{in}};$$

$$P_{O_2,out} + P_{H_2O,out} + P_{N_2,out} = P_{out} = P_{in} = 1\,atm$$

Since $P_{O_2,in} = X_{O_2,in}P_{in} = 0.21P_{in}$, and $P_{N_2,in} = X_{N_2,in}P_{in} = 0.79P_{in}$, simplifying the above expressions results in

$$X_{O_2,out} = \frac{0.21\,(1 - U_{t,O_2})}{1 + 0.21U_{t,O_2}}$$

Then the outlet Nernst potential becomes

$$E_{r,out} = E_r(T, P_{O_2,out}) = E_r(T, P) - \frac{\mathfrak{R}T}{nF} \ln X_{O_2,out}^{-1/2} = E_r(T, P)$$

$$-\frac{\mathfrak{R}T}{nF} \ln \left[\frac{1 + 0.21U_{t,O_2}}{0.21(1 - U_{t,O_2})} \right]^{1/2}$$

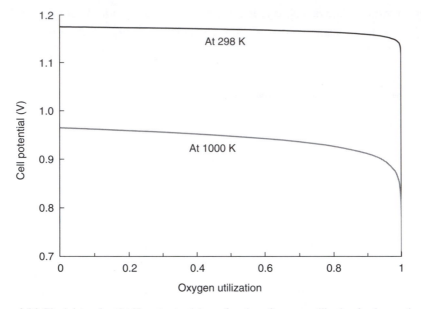

Figure 2.16 The inlet and outlet Nernst potential as a function of oxygen utilization for the reaction of $H_2(g) + \frac{1}{2}O_2(g) \longrightarrow H_2O(g)$ at the pressure of 1 atm and the temperature of 298 K and 1000 K, respectively.

or

$$E_{r,out} = 1.185 - 0.01284 \ln\left[\frac{1 + 0.21U_{t,O_2}}{0.21(1 - U_{t,O_2})}\right]^{1/2} \text{(V)}; \quad \text{at } T = 25\,^\circ\text{C} = 298\text{ K}$$

$$E_{r,out} = 0.9983 - 0.04309 \ln\left[\frac{1 + 0.21U_{t,O_2}}{0.21(1 - U_{t,O_2})}\right]^{1/2} \text{(V)}; \quad \text{at } T = 1000\text{ K}$$

Hence the Nernst potential at the outlet can be calculated for various oxygen utilizations at the two specified temperatures. Such results are shown in Figure 2.16. Clearly, the Nernst potential is hardly affected until the oxygen utilization is beyond 90%, at very close to 100% then the decrease is abrupt. Also it is evident that the Nernst loss is more significant at high temperature, as is suggested by the Nernst equation, Equation (2.63) itself.

COMMENT

1. Zero utilization can be approached in practice by providing excessively high reactant flow to the cell. According to Equation (2.105), if the reactant in-flow is far larger than what is needed for fuel cell reaction producing electric energy, i.e., $\dot{N}_{in} \gg \dot{N}_{consumed}$, the corresponding utilization becomes extremely small, so that the reactant concentration may be assumed constant from the cell inlet to the outlet because the amount of reactant in-cell consumption may be negligibly small in comparison.

2. Although the Nernst potential decreases with temperature, the actual cell potential E may increase with temperature because the irreversible cell potential loss may be reduced significantly when the temperature is increased. Details are found in the next chapter.

3. The actual cell potential will decrease much faster than the Nernst potential as the reactant utilization is increased because the cell potential loss will increase quickly due to an increase in the activation and concentration polarization arising from the higher resistance to the mass transfer and a low reactant concentration at the reaction sites.

4. In the present example, the total pressure loss from the cell inlet to the cell outlet has been neglected. In reality, pressure loss will occur due to the frictional and minor loss mechanisms associated with reactant flow and reactant in-cell consumption.

2.6   SUMMARY

This chapter is largely devoted to the analysis of fuel cell performance (cell potential and energy conversion efficiency) under the idealized reversible condition. The determination of the cell reversible potential and reversible cell efficiency depends strongly on the evaluation of thermodynamic properties, which are usually acquired from tabulated data or analytical property relations. The properties of importance include absolute enthalpy, enthalpy of reaction, heating values, Gibbs function, Gibbs function of formation, Gibbs function of reaction, entropy change for a reaction, and so on. One must become skilled at the property evaluation, as it is one of the basic skills needed for fuel cell analysis. The comparison of fuel cell efficiency with Carnot efficiency is then presented in length, and possibility of over 100% fuel cell efficiency is ruled out from the strict consideration of definition for efficiency and the second law of thermodynamics. Both reversible and irreversible energy loss mechanisms are discussed, waste heat generation in a fuel cell is provided, and various forms of efficiency are defined. Finally, the Nernst potential loss arising from the reactant consumption in practical cell is considered, and issues related to reactant utilization are outlined. From this chapter, you should be able to calculate the reversible cell potential and efficiency for a given fuel cell reaction, and be able to calculate the amount of heat generation which is equal to the amount of cooling required for practical fuel cells. You should also be familiar with the various energy loss mechanisms, including the Nernst loss. Finally, readers are encouraged to consult [6]–[9] for basic thermodynamic analysis presented in this chapter.

BIBLIOGRAPHY

1. Barendrecht, E. 1993. *Electrochemistry of Fuel Cells.* In Fuel Cell Systems, eds. L. J. M. J. Blomen and M. N. Mugerwa. New York: Plenum Press.
2. Kordesch, K. and G. Simader 1996. *Fuel Cells and Their Applications.* New York: VCH.
3. Appleby, A. J. and F. R. Foulkes 1989. *Fuel Cell Handbook.* New York: Van Nostrand Reinhold.

4. Appleby, A. J. 1993. *Characteristics of Fuel Cell Systems*. In Fuel Cell Systems, eds. L. J. M. J. Blomen, and M. N. Mugerwa. New York: Plenum Press.
5. Hirschenhofer, J. H., D. B. Stauffer and R. R. Engleman 1994. *Fuel Cells A Handbbok* (Rev. 3). U.S. Department of Energy.
6. Black, W. Z. and J. G. Hartley, 1991. *Thermodynamics*, 2nd ed. New York: HarperCollins Publishers.
7. Cengel, Y. A. 1997. *Introduction to Thermodynamics and Heat Transfer*. New York: McGraw-Hill.
8. Reynolds, W. C. and Perkins, H. C. 1977. *Engineering Thermodynamics*. New York: McGraw-Hill.
9. Turns, S. R. 2000. *An Introduction to Combustion*, 2nd Ed. New York: McGraw-Hill.

REVIEW QUESTIONS

2.1. Describe the concept of absolute enthalpy, enthalpy of formation, and sensible enthalpy. Describe their dependence on temperature and pressure for ideal gases.

2.2. What is the difference between absolute enthalpy and enthalpy of formation? Do you think there are conditions under which the two would become the same? What are these conditions? Why?

2.3. Describe the concept of Gibbs function, Gibbs function of formation, and Gibbs function of reaction. Describe their dependence on temperature and pressure for ideal gases.

2.4. What are the enthalpy of reaction and heating values? Why are there two heating values for common fuels? Is one value more appropriate than the other? Why?

2.5. In Equation (2.9), the entropy flow associated with energy transfer as heat is shown; however, the entropy flow associated with energy transfer as work is not included in the equation. Why?

2.6. Describe the relation between the entropy generation in a fuel cell with the cell potential loss. Explain whether cell potential loss would occur when no entropy generation exists.

2.7. Explain why entropy generation in a fuel cell would always lead to a loss in the cell potential? Does situation arise where cell potential loss does not lead to entropy generation?

2.8. Describe how waste heat generation in a fuel cell is related to the cell potential loss and the entropy generation in a fuel cell? Why are they related to each other?

2.9. Explain how reversible cell potential depends on temperature, pressure and reactant concentrations. Can the concentration of reaction products affect the reversible cell potential? Why?

2.10. Explain why the reversible cell efficiency is less than 100% for some fuel cell reactions, whereas it could be equal to or even more than 100% for some other reactions? Why could we build and operate fuel cells with over 100% efficiency?

2.11. Explain what is the reversible (second law) efficiency for fuel cell and for heat engines, and the conditions under which the two become equivalent. Is it possible to construct both fuel cell and heat engine to operate with the reversible efficiency? Why?

2.12. Why is fuel cell commonly regarded to have better performance than heat engine?

2.13. What is the significance of waste heat generation in a fuel cell? Will it cause operational or design problems for the fuel cell? Why?

2.14. What is the Nernst loss? What are the inlet and outlet Nernst potentials?

2.15. Explain why fuel utilization cannot be 100% for practical fuel cells? How can operation with zero utilization be achieved in practice?

2.16. Describe how Nernst loss could be minimized through design and operation strategies? How realistic are these strategies for practical fuel cells?

PROBLEMS

2.1   Calculate the reversible cell potential E_r and the second law efficiency η_r for the following reactions occurring at the temperature of 25 °C and pressure of 1 atm:

$$\text{(1)} \quad H_2 + \frac{1}{2}O_2 \longrightarrow H_2O(\ell)$$

$$\text{(2)} \quad H_2 + \frac{1}{2}O_2 \longrightarrow H_2O(g)$$

$$\text{(3)} \quad CH_4 + 2O_2 \longrightarrow CO_2 + 2H_2O(\ell)$$

$$\text{(4)} \quad C_3H_8 + 5O_2 \longrightarrow 3CO_2 + 4H_2O(\ell)$$

$$\text{(5)} \quad NH_3 + \frac{3}{4}O_2 \longrightarrow \frac{1}{2}N_2 + \frac{3}{2}H_2O(\ell)$$

$$\text{(6)} \quad C(s) + \frac{1}{2}O_2 \longrightarrow CO$$

$$\text{(7)} \quad C(s) + O_2 \longrightarrow CO_2$$

$$\text{(8)} \quad CH_3OH(g) + \frac{3}{2}O_2 \longrightarrow CO_2 + 2H_2O(\ell)$$

$$\text{(9)} \quad CH_3OH(\ell) + \frac{3}{2}O_2 \longrightarrow CO_2 + 2H_2O(\ell)$$

$$\text{(10)} \quad CO + \frac{1}{2}O_2 \longrightarrow CO_2$$

2.2   Calculate the reversible cell potential E_r and the second law efficiency η_r for the following reactions at the pressure of 1 atm and the temperature of 600 K and 1000 K, respectively

$$\text{(1)} \quad C(s) + O_2 \longrightarrow CO_2$$

$$\text{(2)} \quad C(s) + \frac{1}{2}O_2 \longrightarrow CO$$

$$(3) \quad CO + \frac{1}{2}O_2 \longrightarrow CO_2$$

$$(4) \quad H_2 + \frac{1}{2}O_2 \longrightarrow H_2O(g)$$

$$(5) \quad CH_4 + 2O_2 \longrightarrow CO_2 + 2H_2O(g)$$

2.3 Repeat the calculation for the reactions listed in the previous problem, but at the pressure of 5 atm and temperature of 600 K, and 10 atm and 1000 K, respectively.

2.4 Calculate the reversible cell potential E_r and the second law efficiency η_r for the reaction

$$N_2 + 3H_2 \longrightarrow 2NH_3(g)$$

at the temperature of 25 °C and pressure of 1 atm.

2.5 Since air contains 79% of N_2 (by volume) as compared to O_2 (which is only 21%), then if N_2 is used as oxidant, the N_2 concentration is almost four times of the oxygen concentration. Higher reactant concentration can reduce the Nernst loss and the concentration polarization. Further, the reaction product, NH_3 (ammonia), can be used as a fertilizer for farming and other applications. Calculate the reversible cell potential E_r and the second law efficiency η_r for the reaction

$$3H_2 + ?air(0.79N_2 + 0.21O_2) \longrightarrow 2NH_3(g) + 0.21?O_2$$

at the temperature of 25 °C and the total pressure of 1 atm. Then judge whether such a fuel cell reaction is practically useful. Explain the reasoning leading to your conclusion. The question mark (?) in the above reaction is the unknown amount of air needed to balance such reaction stoichiometrically.

2.6 Determine the inlet and outlet Nernst potential as well as the associated Nernst loss for the following reaction

$$H_2 + \frac{1}{2}O_2 \longrightarrow H_2O(\ell)$$

at the temperature of 80 °C and pressure of 1 and 3 atm, respectively, as a function of H_2 and O_2 utilization. Assume pure H_2 and O_2, and the reaction product water is formed at the oxidant side. Formulate the problem and present the results in 3D graphical form with H_2 and O_2 utilization as the two independent coordinates.

2.7 Recalculate the results in Problem 2.6 if the oxidant is O_2 in air instead of pure O_2.

2.8 Recalculate the results in Problem 2.6 if the oxidant is O_2 in air, and both the fuel and oxidant streams are fully saturated with water vapor at the cell inlet.

2.9 Consider the following hydrogen–oxygen fuel cell reaction

$$H_2 + \frac{1}{2}O_2 \longrightarrow H_2O(\ell)$$

at the standard temperature and pressure (1 atm and 25 °C). If the cell operates at the cell voltage of $E = 0.7$ V, then determine for each mole fuel consumed in the cell (a) the amount of electrical work done; (b) the amount of entropy generation; (c) the amount of cell potential loss; (d) the amount of waste heat generation.

2.10   Repeat the calculation from Problem 2.9 if the hydrogen-oxygen reaction occurs at 80 °C and 1 atm.

2.11   Repeat the calculation from Problem 2.9 if the reaction occurs reversibly, instead of at the cell potential of $E = 0.7$ V, for the condition of
(a) 25 °C and 1 atm,
(b) 80 °C and 1 atm,
(c) 80 °C and 3 atm.

2.12   Consider the following hydrogen–oxygen fuel cell reaction

$$H_2 + \frac{1}{2}O_2 \longrightarrow H_2O(\ell)$$

at the standard temperature and pressure (1 atm and 25 °C). If the cell operates at the cell voltage of $E = 0.7$ V and the current of $I = 1$ A with a fuel flow of 10 mL/min and measurements indicate that 2.3 mL/min of fuel leaks out of the cell due to a sealing problem, then determine the
(a) reversible cell efficiency,
(b) cell voltage efficiency,
(c) cell current efficiency, and
(d) overall free energy-conversion efficiency.

ELECTROCHEMISTRY OF FUEL CELLS

3.1 INTRODUCTION

In the preceding chapter, thermodynamic analysis tells us whether particular reactants will react to form products without the need of help from external means — that is, the feasibility of a given reaction for fuel cell application; and if feasible, it allows calculation of the cell potential difference E_r between the cathode and the anode, and the corresponding cell energy conversion efficiency η_r under the thermodynamically reversible condition — that is, the idealized best possible performance for a given fuel cell reaction. However, thermodynamic analysis of a given fuel cell cannot provide the important information regarding:

(i)   How fast reaction occurs, or the rate of electrochemical reaction that yields the electric current and power production. It is well known that fuel cell power density, either on a per-unit volume or mass basis, is an extremely important performance measure for some fuel cell applications, particularly for mobile (transportation) application.

(ii) How reactants react to form products, or the actual reaction pathway. The mechanism of electrode reaction is required for the prediction and analysis of reaction rate associated with the electric current/power production in a cell.

(iii) How much energy losses occur under the actual (irreversible) condition for the electrochemical reaction instead of the idealized reversible condition? How can energy losses be minimized under a given set of conditions?

These issues are dealt with within the specialized field called electrochemical kinetics, a study of the elementary electrochemical reactions and their rates. Clearly, it is essential to understand the underlying electrochemical reaction processes occurring at the anode and cathode for the study of fuel cells. In many types of fuel cells, especially the low temperature fuel cells, the rate of electrochemical reactions controls the rate of electrical power generation, and is the predominant factor for the cell voltage (energy) losses. Therefore, the cell performance is closely related to the electrochemical processes at each electrode. Although a significant progress has been made in the past few decades, our knowledge of detailed electrode processes is still quite limited. The prediction of fuel cell performance based on the detailed reaction kinetics for a practical fuel cell with a complex flow field is still unavailable, even though such effort has been made from first principles towards a complete solution combining transport phenomena and electrochemical processes. However, analyses of various complexities have been carried out based on existing knowledge, albeit limited, and useful results have been obtained for the fuel cell design and performance.

This chapter reviews some basic electrochemical kinetics concepts and discusses in more detail one of the three polarization types, namely, the activation polarization, while deferring the discussion of other two types of polarization (concentration and ohmic) to the next chapter because they are related to the transport of species, ions, and electrons. Next we describe the mechanisms of electrode reactions in fuel cells and electrocatalysis associated with the heterogeneous catalyzed reactions that occur invariably in fuel cells.

3.2   ELECTRODE POTENTIAL AND CELL POLARIZATION

A fuel cell typically consists of three major components, as shown schematically in Figure 3.1, they are

- An anode (or negative, or fuel) electrode;
- A cathode (or positive, or oxidant) electrode; and
- An electrolyte.

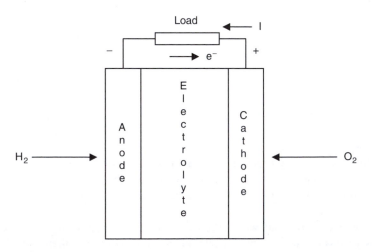

Figure 3.1 Schematic of a fuel cell.

Note that in this classification, the thin catalyst layer for each electrode has been regarded as a part of the electrode itself (otherwise, an electrode may be considered having a substrate, or more often called gas diffusion layer or electrode backing layer, and a catalyst layer where electrochemical reaction occurs). For hydrogen–oxygen fuel cells, the overall half-cell reaction for acid electrolyte fuel cell is

$$\text{At the anode: } H_2 \Longrightarrow 2H^+ + 2e^- \tag{3.1}$$

$$\text{At the cathode: } \frac{1}{2}O_2 + 2H^+ + 2e^- \Longrightarrow H_2O \tag{3.2}$$

In alkaline electrolyte fuel cell, the overall half-cell reaction becomes

$$\text{At the anode: } H_2 + 2OH^- \Longrightarrow 2H_2O + 2e^- \tag{3.3}$$

$$\text{At the cathode: } \frac{1}{2}O_2 + H_2O + 2e^- \Longrightarrow 2OH^- \tag{3.4}$$

Summarizing this anode and cathode half-cell reactions, the overall single-cell reaction results

$$H_2 + \frac{1}{2}O_2 \Longrightarrow H_2O \tag{3.5}$$

for both acid and alkaline electrolyte fuel cells.

For a fuel cell, the cell output voltage is equal to the electric potential difference between the cathode and anode. However, single-electrode potentials cannot be measured because the absolute zero of the electric potential is unknown. Since only the potential difference between the two electrodes matters, an arbitrary reference potential could be used for the determination of the electrode potential for the convenience of the study of electrochemical kinetics at each individual electrode. This is very similar to the determination of many thermodynamic properties such as energy, enthalpy, etc. where a reference datum is defined, such as the standard reference state defined in the previous chapter. In electrochemical study, it is conventional to define the reference potential as the potential of a hydrogen electrode undergoing the following reaction reversibly (so as to avoid the overpotential which would be unknown and variable, depending on the degree of irreversibility)

$$\frac{1}{2}H_2 \Longleftrightarrow H^+ + e^- \tag{3.6}$$

Such a reference electrode is often called the **reference hydrogen electrode** (RHE) and other single-electrode potentials are measured with respect to the potential of this RHE; the electric current is measured in another circuit, that is, between the experimental electrode of concern and a suitable counter electrode. However, one must realize that this reference potential (i.e., reference energy datum) associated with an RHE does not coincide with the energy datum defined at the standard reference state and care must be taken when energy balance is made for an individual electrode reaction.

Let $\phi_{a,r}$ and $\phi_{c,r}$ represent the anode and cathode potential, respectively, under the reversible condition, and ϕ_a and ϕ_c are the respective anode and cathode potential under the practical fuel cell working condition, then the reversible and the actual cell potential (difference between the cathode and the anode) can be written as

$$E_r = \phi_{c,r} - \phi_{a,r} \tag{3.7}$$

$$E = \phi_c - \phi_a \tag{3.8}$$

From Equation (2.25), the total cell overpotential* can be determined as

$$\eta = E_r - E = (\phi_{c,r} - \phi_{a,r}) - (\phi_c - \phi_a)$$

$$= (\phi_a - \phi_{a,r}) - (\phi_c - \phi_{c,r})$$

or

$$\eta = \eta_a - \eta_c \tag{3.9}$$

where

$$\eta_a = \phi_a - \phi_{a,r} > 0 \tag{3.10}$$

$$\eta_c = \phi_c - \phi_{c,r} < 0 \tag{3.11}$$

are the anode and cathode overpotential, respectively. It should be emphasized that for fuel cells, the anode overpotential is positive, while the cathode overpotential is negative. This arises from the convention that the electrode overpotential is defined conventionally in electrochemistry as the difference between the actual and the reversible electrode potential.

Therefore, the actual cell potential can be determined from the reversible cell potential and knowledge of the overpotentials as follows

$$E = E_r - \eta_a + \eta_c = E_r - \sum |\eta_i| \tag{3.12}$$

where $\sum |\eta_i|$ represents the total overpotentials for a working cell. Equation (3.12) denotes that the actual cell potential is the difference between the reversible cell potential and all the overpotentials occurring in the working cell. This is easily understood because the reversible cell potential represents the best idealized cell performance, and the overpotentials represent the energy losses due to irreversibilities. Figure 3.2 illustrates the various electrode potentials.

From this discussion, the following conditions apply for a H_2/O_2 fuel cell operating at 25 °C and 1 atm with the half-cell reactions shown in Equations (3.1)

---

*as noted in Chapter 2, η is commonly used as efficiency in thermodynamics, while it is also conventionally used as overpotential in electrochemistry and in fuel cell literature. Therefore, in this book, η is used for both efficiency as in Chapter 2 and overpotential as in this chapter in order to be consistent with the convention. The specific meaning would become clear from the context.

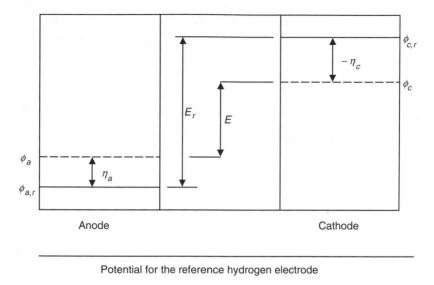

Anode   Cathode

Potential for the reference hydrogen electrode

Absolute zero of potential (unknown)

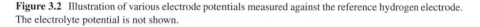

Figure 3.2 Illustration of various electrode potentials measured against the reference hydrogen electrode. The electrolyte potential is not shown.

and (3.2), that is, in acid electrolyte fuel cells,

Reversible Conditions	Irreversible Conditions
$E_r = 1.229\,V$	$E_r < 1.229\,V$
$\phi_{a,r} = 0.000\,V$	$\phi_a > 0.000\,V$
$\phi_{c,r} = 1.229\,V$	$\phi_c < 1.229\,V$

On the other hand, if the half-cell reactions are as shown in Equations (3.3) and (3.4), that is, for alkaline electrolyte fuel cells, the electrode potential becomes for the same H_2/O_2 cell

Reversible Conditions	Irreversible Conditions
$E_r = 1.229\,V$	$E_r < 1.229\,V$
$\phi_{a,r} = -0.828\,V$	$\phi_a > -0.828\,V$
$\phi_{c,r} = 0.401\,V$	$\phi_c < 0.401\,V$

As described in the previous chapter, the maximum cell potential (or electromotive force), E_{max}, is obtained when the cell is operated under the thermodynamically reversible condition, which, in practical term, is operated when there is no electric current is drawn from the cell (i.e., open circuit). Hence, ideally

$$E_{max} = E_r = E_{oc} \tag{3.13}$$

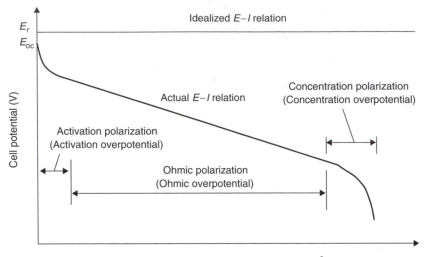

Figure 3.3 Cell polarization curve illustrating the idealized and actual cell potential–current relation with various forms of overpotential.

where the reversible cell potential, E_r, is determinable theoretically based on the analysis given in the previous chapter, and the open circuit cell potential, E_{oc}, is measurable experimentally. In reality, the open circuit cell potential is always slightly less than the reversible cell potential due to the practical irreversibilities as explained in the previous chapter (due to the exchange current to be explained later). The degree of deviation from E_r depends on the other electrode activities such as electrode corrosion reactions.

For the cell to provide useful electric energy output in the form of electric power, electric current I (or current density J) has to be drawn from the cell. The associated power output or power density is, as given in Equation (2.89)

$$\text{Power} = EI \quad \text{or} \quad \text{Power Density} = EJ \tag{3.14}$$

As the current drawn is increased, the degree of irreversibility increases as well, and the overpotentials are increased accordingly. Hence, the actual cell potential becomes lower and lower than the reversible cell potential due to irreversible processes that cause the loss in the cell potential. The difference between the reversible cell potential and the measured cell potential under working conditions is often called **overpotential** or **overvoltage**. The processes, which could be physical or chemical in nature, that give rise to overpotential, are often referred to as polarization. Hence, the actual cell potential and current relation curve is often called cell polarization curve, as shown schematically in Figure 3.3.

Overpotential may be classified according to the phenomenon believed to be responsible. For fuel cells, it is often divided into three types: activation, ohmic, and concentration overpotential (polarization), due to the resistance to electrochemical

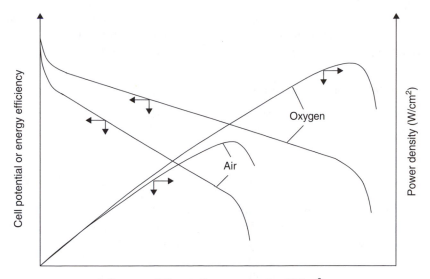

Figure 3.4 Schematic of cell potential (or cell energy-conversion efficiency) and power density as a function of current density for an H_2–O_2 and H_2–air fuel cell.

reactions, to the transport of electrons and ions in the cell components, and to the limitations in the mass transfer. This division may not always be realistic since these phenomena may be cooperative and division may not be justified. For example, in the analysis and simulation of fuel cell performance, as shown later, the activation and concentration overpotential may be so interrelated that they may be considered as a single, rather than separate, entity. However, overall it is still useful for the qualitative and quantitative examination of the various influences of physical and chemical processes occurring within the cell. Figure 3.4 shows the schematic of cell potential and power density when pure oxygen and air are used as the oxidant, respectively. As discussed in Chapter 2, the cell energy-conversion efficiency is proportional to the actual cell potential, thus cell potential also represents the cell energy efficiency. It is seen that cell power density increases with the current density almost linearly until reaching the regime of concentration polarization, where the cell potential is reduced so rapidly that the power density is reduced as well. Therefore, the power density often exhibits a peak located at the beginning of the concentration polarization regime. The activation polarization is described later in this chapter, while the other two forms of polarization is considered in Chapter 4.

3.3   SOME ELECTROCHEMICAL KINETICS CONCEPTS

Before describing the activation polarization in a fuel cell, a few useful kinetic concepts are reviewed here for easy presentation of materials that follow. These include the global and elementary reactions, the reaction rate expression for heterogeneous catalyzed reaction, the rate-limiting reaction step or rate-determining step within a series of reactions, and the steady-state approximation for reaction intermediaries. A brief description of each concept is given in the following text.

3.3.1   Global and Elementary Reactions

The reactions given in Equations (3.1)–(3.4) are often called the global or overall reaction mechanism occurring at the anode and cathode, respectively, which are also referred to as the overall half-cell reactions; and Equation (3.5) is the global or overall reaction mechanism for the entire cell reaction. Half-cell reactions are easily recognized in fuel cell reactions because they involve electrons and ions (charged species) as the reactants or products. However, this does not mean that the actual half-cell reaction follows the simplistic one-step process shown in Equations (3.1)–(3.4). The actual reaction may proceed through many sequential or simultaneous processes that are often called elementary reactions involving many intermediate species. For example, for the anodic global reaction shown in Equation (3.1), the following elementary reactions are known to be important

$$H_2 \rightleftharpoons H_{ad} + H_{ad} \quad \text{Tafel reaction} \tag{3.15}$$

$$H_{ad} \rightleftharpoons H^+ + e^- \quad \text{Volmer reaction} \tag{3.16}$$

Equation (3.15) is a dissociative chemisorption step, often referred to as Tafel reaction, and Equation (3.16) represents a charge transfer reaction, named after Volmer.

In this partial mechanism for molecular hydrogen oxidation reaction in the presence of acid electrolyte, reaction (3.15) suggests that H_2 does not yield to protons (H^+) and electrons (e^-) directly, as Equation (3.1) may have suggested. Instead it first adsorbs on the electrode surface, and then dissociates into two H atoms that remain adsorbed on the electrode, as indicated by the subscript "ad". The adsorbed H atom is the intermediate species during the oxidation process of the molecular hydrogen, H_2, and it is formed before the charge transfer reaction, Equation (3.16), that actually produces the proton and the electron — the final product for the anodic reaction. The global reaction, Equation (3.1), is obtained when reactions (3.15) and (3.16) are summarized and the reaction intermediate species, H_{ad}, is cancelled out from the equation. In general, a reaction mechanism denotes a series of elementary reactions that are necessary to describe a global reaction. It can therefore be said that the mechanism for anodic H_2 oxidation in the presence of acid electrolyte includes the dissociative chemisorption process followed by charge transfer reaction represented by reactions (3.15) and (3.16). In reality, the elementary reactions for the H_2 oxidation are more than just the Tafel–Volmer reactions, and they will be introduced later in this chapter. A reaction mechanism usually contains many elementary reactions, ranging from a few steps to several hundreds. However, the identification of a minimum number of elementary reactions needed to describe a given global reaction is of practical importance and remains under active investigation because the minimum number and the set of elementary reactions required may change with the specific operating condition, such as temperature, pressure, reactant composition and electrode potential, and so on.

In reality, many elementary reactions occurring in parallel or in sequence, involving many intermediate species and steps, may underlie a particular global reaction. As shown later in this chapter, the simple half-cell reaction for the oxidation of molecular hydrogen and the reduction of molecular oxygen in both acid and alkaline electrolyte, as shown in Equations (3.1)–(3.4), may proceed through a much more complex pathway, for example, involving either a two-electron or four-electron reaction steps for the reduction reaction of molecular oxygen. A reaction mechanism is often referred to a collection of elementary reactions necessary to describe an overall reaction. It is also quite possible that a different set of elementary reactions may be able to describe reasonably for a particular global reaction, adding to the complexities and uncertainties associated with this kind of study because it is difficult to measure the reaction intermediate species and figure out the exact reaction pathways. Therefore, determining reaction mechanisms, even for the simplest reaction, is not an easy task and it still remains a field of active research.

3.3.2   The Reaction Rate

In fuel cells, electrochemical reaction occurs at the electrode–electrolyte interface where the reactants are available for the reaction. For such a heterogeneous reaction, the rate of reaction, ω'', is often expressed in terms of the number of mole changes per-unit electrode surface area per-unit time, or mol/(m$^2 \cdot$ sec). Experimental measurements indicate that for an elementary reaction, the reaction rate is proportional to the product of the concentrations of the reactants present raised to a power, which is equal to the corresponding stoichiometric coefficient of the reaction. This statement is often referred to as the law of mass action.

Consider an elementary electrochemical reaction written generally as follows

$$\sum_{i=1}^{N} v_i' M_i \xrightarrow{k} \sum_{i=1}^{N} v_i'' M_i \tag{3.17}$$

Then the rate of reaction for species i (the rate of production) can be expressed as, according to the law of mass action

$$\omega_i'' = (v_i'' - v_i')k \prod_{i=1}^{N} [M_i]^{v_i'} \tag{3.18}$$

where the square bracket [ ] represents the molar concentration for the chemical species shown inside the bracket, mol/m^3; molar concentration for species i is also conventionally denoted by C_i, or

$$C_i = [M_i] = \frac{P_i}{RT} \tag{3.19}$$

where the last equality only holds for ideal gases. The exponent v_i' for $[M_i]$ is called the reaction order with respect to the species i, or reaction (3.17) is said to be v_i'-th order with respect to the species i, and the total reaction order for the given reaction is

$$\nu = \sum_{i=1}^{N} \nu_i' \tag{3.20}$$

The proportionality constant k is often referred to as the reaction rate constant, however, it only remains a constant under a given condition of reaction; and in reality, it is a strong function of temperature, the electrode overpotential, the kind of catalyzed surface used for the reaction facilitation, and the type of electrolyte used, e.g., whether the electrolyte is acid or alkaline, and the concentration of the acid or alkaline solution, etc.

The rate of heterogeneous reaction given in Equation (3.18) is more often expressed in terms of the current generated because electric current is much more convenient to measure experimentally. According to Faraday's law of electrochemical reaction, the current generated over a unit electrode surface area, or the current density as often used in fuel cell literature, can be written as follows

$$J = -nF\omega_F'' = -(\nu_F'' - \nu_F')k \prod_{i=1}^{N} [M_i]^{\nu_i'} \tag{3.21}$$

where k is the modified reaction rate constant, n is the number of the mole electrons transferred per mole of fuel consumed and F is the Faraday constant. By our definition earlier (in Chapter 2), $\nu_F' = 1$. The minus sign indicates that the rate of current generation is a direct result of the fuel consumption in the fuel cell, since ω_i'' represents the rate of production for the species i involved.

It might be pointed out that the reaction orders are always integers for elementary reactions; however, they may not be integers for global reactions, as a global reaction is the summarized representation of many elementary reactions occurring either simultaneously or in sequence. Therefore, for a global reaction the reaction rate constant and the reaction orders are determined by curve fitting the experimental data for the current density measurements. This global approach to the electrochemical reactions is useful in tackling practical problems where only the global feature of the reactions is needed. However, it does not represent what actually occurs in a fuel cell, and such a lack of understanding in the details of the reaction process may limit our ability to optimize the design and performance of a fuel cell. In reality, a reaction can proceed forward and backward simultaneously, as shown in the Tafel–Volmer reaction in Equations (3.15) and (3.16). For example, the Tafel reaction really includes the following two separate reactions occurring simultaneously

$$H_2 \longrightarrow H_{ad} + H_{ad} \tag{3.22}$$

$$H_{ad} + H_{ad} \longrightarrow H_2 \tag{3.23}$$

The first reaction indicates that molecular hydrogen is consumed, while in the second reaction molecular hydrogen is produced. These two reactions are conventionally and conveniently combined together into a single expression shown in Equation (3.15). The rate of reaction can be written separately for each reaction and the net rate of reaction (production) can be obtained by combining the two expressions together.

Similarly, for a general elementary reaction, such as the one shown in Equation (3.17), which is also undergoing reverse backward reaction, it is often written as

$$\sum_{i=1}^{N} v_i' M_i \underset{k_b}{\overset{k_f}{\rightleftharpoons}} \sum_{i=1}^{N} v_i'' M_i \tag{3.24}$$

where k_f and k_b are the reaction rate constant for the forward and backward reaction, respectively. Then the rate of reaction for the forward reaction, $\omega_{i,f}''$, is identical to the one given in Equation (3.18) with k being replaced by k_f. The rate of reaction for the backward reaction can be written similarly as

$$\omega_{i,b}'' = (v_i' - v_i'') k_b \prod_{i=1}^{N} [M_i]^{v_i''} \tag{3.25}$$

because for the backward reaction, the right-hand side of Equation (3.24) now represents the reactants and the left-hand side the products.

The net rate of the production for the species i becomes

$$\omega_i'' = \omega_{i,f}'' + \omega_{i,b}''$$

$$= (v_i'' - v_i') \left[k_f \prod_{i=1}^{N} [M_i]^{v_i'} - k_b \prod_{i=1}^{N} [M_i]^{v_i''} \right] \tag{3.26}$$

At equilibrium condition, the rate of reaction for the forward and backward reaction becomes identical, thus the net rate of reaction vanishes or $\omega_i'' = 0$. Then, Equation (3.26) yields

$$\frac{k_f}{k_b} = \prod_{i=1}^{N} [M_i]^{(v_i'' - v_i')} \equiv K_c(T, P) \tag{3.27}$$

where K_c is the equilibrium constant for concentration, which is in general a function of both temperature and pressure; and it is related to the equilibrium constant for partial pressure as follows, because $P_i = C_i RT$ for ideal gases

$$K_P = \prod_{i=1}^{N} \left(\frac{P_i}{P_0} \right)^{(v_i'' - v_i')} = \left(\frac{RT}{P_0} \right)^{\sum_{i=1}^{N} (v_i'' - v_i')} \prod_{i=1}^{N} C_i^{(v_i'' - v_i')} \tag{3.28}$$

The equilibrium constant for partial pressure is a function of temperature only, and can be determined from the Gibbs function of reaction for the reaction (3.24),

$\Delta g(T, P_0)$, as follows

$$K_P(T) = \exp\left(-\frac{\Delta g(T, P_0)}{RT}\right) \tag{3.29}$$

where $P_0 = 1$ atm is the reference pressure. When the number of moles for the gaseous reactants and the gaseous products are the same, Equation (3.28) indicates that

$$K_c = K_P \tag{3.30}$$

Therefore, K_c will be a function of temperature only. In general, the backward reaction rate constant can be determined from a knowledge of the forward reaction rate constant and the equilibrium constant according to Equations (3.27)–(3.29).

3.3.3   Surface Coverage

As already emphasized previously, the heterogeneous half-cell reaction occurs at the electrode surface, and hence the reacting species must be absorbed on the electrode surface before the electrochemical reaction can occur. In other words, an electrode surface is covered by adsorbed species, although the adsorption need not be complete (that is, the electrode surface need not be completely covered by adsorbed species), and a variety of species, other than the reactant, may also adsorb on the electrode surface. It is known that the rate of an electrode reaction may be substantially influenced by the diverse adsorbed species and the vacant sites (uncovered electrode surface). The importance of adsorption and desorption is represented by the degree of the electrode surface coverage by the adsorbed species — a variable may be simply called surface coverage θ_i, which is included as an additional parameter in the reaction rate equations.

The surface coverage for a species i may be defined as the fraction of an electrode surface covered by the adsorbed species i,

$$\theta_i = \frac{C_{i,\text{ad}}}{\sum_j (C_{j,\text{ad}})_s} \tag{3.31}$$

where the subscript "ad" represents the species i (or j) that is adsorbed on the electrode surface and "s" in the denominator denotes the concentration, $C_{j,\text{ad}}$, is at the saturation of the electrode surface, not necessarily at complete coverage of all the surface with the diverse adsorbed species. For example, for the half-cell reaction shown in Equation (3.1), the oxidation of the molecular hydrogen on a metal electrode surface (say, platinum) at the presence of acid electrolyte may occur actually through the Tafel–Volmer mechanism, as shown in Equations (3.15) and (3.16). However, in order to write the rate of reaction correctly, the Tafel–Volmer reaction may be more vividly expressed as

$$H_2 + 2M \overset{k_{1,f}}{\underset{k_{1,b}}{\rightleftharpoons}} 2(H - M) \quad \text{(Tafel reaction)} \tag{3.32}$$

$$\text{H} - \text{M} \underset{k_{2,b}}{\overset{k_{2,f}}{\rightleftharpoons}} \text{H}^+ + \text{e}^- + \text{M} \quad \text{(Volmer reaction)} \tag{3.33}$$

where M is a metal adsorption site. From Equation (3.31), θ_H measures $C_{\text{H}_{ad}}$. Hence, the expression for the rate of reaction, Equation (3.26), for the reaction (3.33) is more often written as follows

$$\omega''_{\text{H}_{ad}} = -k_{2,f}\theta_\text{H} + k_{2,b}C_{\text{H}^+}C_{\text{e}^-}(1 - \theta_\text{H}) \tag{3.34}$$

where $(1 - \theta_\text{H})$ represents the metal surface that is not covered by the adsorbed hydrogen, H_{ad}, for the Tafel–Volmer reaction under consideration. Further, $d\theta/dt \sim \omega''_{\text{H}_{ad}}$. Hence, Equation (3.34) may be simply written for $d\theta/dt$ with the proportionality constant being absorbed by the reaction rate constants.

Now we start to see the difference between the reaction rate expressions for the heterogeneous electrochemical reaction in fuel cells and the homogeneous single-phase chemical reaction. Note that for reactions occurring at the metal electrode–liquid electrolyte interface, the concentration of electrons may be taken as a constant, and then incorporated into the reaction rate constant, as is often the case in fuel cell literature.

3.3.4   Rate-Limiting Reaction Step or Rate-Determining Step

As discussed previously, there are many elementary reactions occurring either simultaneously or in sequence in an electrode reaction, or in a complex series-parallel sequence, some reactions proceed fast and some slow. For example, in addition to the Tafel–Volmer reaction for the anodic reaction of hydrogen shown earlier, the following reactions may occur with significance

$$\text{H}_{2,ad} \rightleftharpoons \text{H}_{ad} + \text{H}^+ + \text{e}^- \quad \text{(Heyrovsky reaction)} \tag{3.35}$$

$$\left.\begin{array}{l} \text{H}_{2,ad} \rightleftharpoons \text{H}^+_{2,ad} + \text{e}^- \\ \text{H}^+_{2,ad} \rightleftharpoons \text{H}_{ad} + \text{H}^+ \end{array}\right\} \quad \text{(Horiuti mechanism)} \tag{3.36}$$

Notice that Horiuti mechanism may be regarded as a special case of the Heyrovsky reaction with $\text{H}^+_{2,ad}$ as a postulated intermediate. Hence, Heyrovsky reaction and Horiuti mechanism represent the other electrochemical processes occurring at the hydrogen electrode, in addition to the Volmer reaction. The postulation of these additional mechanisms as alternatives to the Tafel reaction for the formation of H_{ad} from $\text{H}_{2,ad}$ thus provides additional degrees of freedom for the explanation of experimental observations.

Since it is generally a difficult task to determine a complex mechanism for a given electrode reaction under various conditions, the concept of **rate-determining step** of an overall reaction has been frequently used for the calculation of electrode reaction rate. The rate-determining step may be defined as the reaction step that determines the rate of the overall reaction. This concept holds both in the case of consecutive and of parallel reactions or a combination of the two types of the reactions. Further, it is known that many electrochemical reactions proceed by a consecutive mechanism, and few by a parallel-path mechanism. In other word, it may be regarded that the rate of an overall reaction is mainly influenced, or determined, by one step among many elementary reactions, called the rate-determining step. Such a concept considerably simplifies the analysis and calculation of the rate of an overall electrode reaction, and have been used extensively in fuel cell literature, although it is over-simplistic in the description of the complex electrode reactions. It is known that the most important factor in determining the power and efficiency of electrochemical energy conversion is the reaction rate of the rate-determining step, although other factors may be also important such as the adsorptive properties of reactants and intermolecular forces among the species adsorbed on the electrode.

For example, it is generally believed that the rate-determining step in the oxidation of molecular hydrogen at the anode of an acid electrolyte fuel cell is the charge transfer reaction, or Volmer reaction, given in Equation (3.33). However, for most of hydrocarbon fuels such as methanol, the electrode process is rate limited by the slow adsorption process of fuel molecules on the electrode surface. The slowness of the electrode process leads to high energy losses at the electrode and thus often determines the feasibility of the particular fuel for direct energy conversion in fuel cells.

3.3.5   Steady-State Approximation for Reaction Intermediaries

In many electrode reactions, the use of a single step — the rate-determining step for the determination of the rate of reaction is oversimplifying, and it may require the use of multistep reaction mechanism for a better representation of the reaction process. The anodic reaction of hydrogen, as shown in the previous section, is a multistep reaction, including Tafel–Volmer reaction, Heyrovski reaction, and Horiuti mechanism. The carbon monoxide poisoning of the anode catalyst, platinum, in the acid electrolyte, requires the simultaneous consideration of the reactions of the carbon monoxide and hydrogen at the anode electrode.

In multistep reaction mechanism, some highly reactive intermediate species may be formed that are short-lived with very low concentrations, while other species in the reaction system may have considerably longer lifetime with much higher concentrations. Although the analysis of such a reaction process can be dealt with by writing the reaction rate expression, as shown earlier, for each species involved, numerical solution for the species concentration and the rate of overall reaction is often difficult to obtain due to the small time step needed to resolve the shortest-lived intermediate species in the reaction process. Therefore, the analysis of such reaction systems can sometimes be greatly simplified by applying the steady state approximation (or assumption) for the highly reactive intermediate species. Physically, what happens in the reaction process is that the intermediate species is rapidly produced at the beginning of the reaction and its concentration rapidly approaches a steady-state value. This occurs because the intermediate species is rapidly consumed as soon as it is produced after the steady-state value is achieved, that is, its rate of formation becomes equal to the rate of destruction after the initial rapid buildup in its concentration. This situation arises when the reaction producing the intermediate is slow, and the reaction consuming the intermediate is very fast, leading to a very low level of the intermediate concentrations. A good example of electrode reaction would be the slow chemisorption of reactants (e.g., methanol or other hydrocarbon fuels) on anode electrode, followed by fast charge transfer reaction, to be described later in this chapter. The following example illustrates the principle and the application of the approximation described here as well as the great simplification brought by the approximation.

Example 3.1   Take the following reactions as the mechanism for the anodic reaction of hydrogen in an acid electrolyte fuel cell:

$$H_2 + 2M \underset{k_{1,b}}{\overset{k_{1,f}}{\rightleftharpoons}} H - M + H - M \qquad \text{(Tafel reaction)}$$

$$H_2 + M \underset{k_{2,b}}{\overset{k_{2,f}}{\rightleftharpoons}} H - M + H^+ + e^- \quad \text{(Heyrovsky reaction)}$$

$$H - M \underset{k_{3,b}}{\overset{k_{3,f}}{\rightleftharpoons}} M + H^+ + e^- \qquad \text{(Volmer reaction)}$$

Both forward and backward reactions are assumed possible with respective reaction rate constant, k_i's, are given for each reaction. The reaction is further assumed to occur at the metal electrode–liquid electrolyte interface. Use the steady-state approximation to derive an expression for the current density produced by the above reactions in terms of the partial pressure for H_2 and the concentration of the product, H^+.

SOLUTION

Since the surface coverage, θ_H, is a measure of the concentration of the adsorbed hydrogen atom, $H - M$, it will be used in the reaction rate expression instead of its concentration. The concentration of electrons at the reaction sites may be regarded as constant for the present metal electrode–liquid electrolyte system, hence it may be incorporated into the reaction rate constant term. Then the amount of current density production is related to the rate of electron (or proton) produced through the Faraday's law of electrochemistry, or

$$J = nF\omega''_{e^-}$$

where $n = 1$, and the rate of electron production due to the Heyrovsky and Volmer reactions are

$$\omega''_{e^-} = k_{2,f} P_{H_2}(1 - \theta_H) - k_{2,b}C_{H^+}\theta_H + k_{3,f}\theta_H - k_{3,b}C_{H^+}(1 - \theta_H)$$

Note the partial pressure of molecular hydrogen is used in the preceding rate equation, instead of the usual concentration, since molecular hydrogen can be taken as following ideal gas behavior, and the partial pressure is much easier to measure than the concentration itself. The difference between the concentration and the partial pressure is absorbed into the reaction rate constant term. In the above expression, the surface coverage, θ_H, is an unknown quantity, hence the current density produced cannot be determined without determining this parameter first. However, the adsorbed hydrogen atom, $H-M$, is a reaction intermediate species between the reactant, H_2, and the product, H^+ and e^-, and the net rate of $M-H$ production is, according to the Tafel and Heyrovsky reaction given

$$\frac{d\theta_H}{dt} = 2k_{1,f}P_{H_2}(1-\theta_H)^2 - 2k_{1,b}\theta_H^2 + k_{2,f}P_{H_2}(1-\theta_H) - k_{2,b}C_{H^+}\theta_H$$
$$- k_{3,f}\theta_H + k_{3,b}C_{H^+}(1-\theta_H)$$

After a short initial transient for the concentration of the adsorbed hydrogen atom, $H-M$, to increase quickly to a small but steady value, $d\theta_H/dt$ quickly reduces to zero, or the right-hand side of the preceding equation fast approaches zero. Thus with the steady-state approximation, we obtains

$$\frac{d\theta_H}{dt} = 0 = 2k_{1,f}P_{H_2}(1-\theta_H)^2 - 2k_{1,b}\theta_H^2 + k_{2,f}P_{H_2}(1-\theta_H) - k_{2,b}C_{H^+}\theta_H$$
$$- k_{3,f}\theta_H + k_{3,b}C_{H^+}(1-\theta_H)$$

The preceding equation is a quadratic algebraic equation and can be solved for θ_H. Then substituting the expression for θ_H into the equation for the rate of electron production, an expression for the current density produced by the previous reactions can be obtained, albeit it is lengthy and complicated.

COMMENT

1. Even for the simplest possible electrode reaction (i.e., for hydrogen anodic reaction), the reaction rate expression, after the simplification employing steady-state approximation, is still very complex so as to deter us from writing its exact expression in this example, implying the complexities associated with any analysis of fuel cell reaction process. However, the above formulation is sufficiently general to be able to explain almost any experimental observation related to hydrogen anodic reaction. Simpler results can be obtained for specific conditions. For example, if the reaction follows the Heyrovsky–Volmer mechanism, both the surface coverage and the current density produced can be shown to take the following form

$$\theta_H = \frac{k_{2,f}P_{H_2} + k_{3,b}C_{H^+}}{k_{2,f}P_{H_2} + k_{2,b}C_{H^+} + k_{3,b}C_{H^+} + k_{3,f}}$$

$$J = 2F\frac{k_{2,f}k_{3,f}P_{H_2} - k_{2,b}k_{3,b}C_{H^+}^2}{k_{2,f}P_{H_2} + k_{2,b}C_{H^+} + k_{3,b}C_{H^+} + k_{3,f}}$$

Further simplification is possible if the associated backward reactions are negligible under suitable condition (i.e., sufficiently large overpotential to drive the reaction to proceed in the forward direction), then we have

$$\theta_H = \frac{k_{2,f} P_{H_2}}{k_{2,f} P_{H_2} + k_{3,f}}$$

$$J = 2F\frac{k_{2,f} k_{3,f} P_{H_2}}{k_{2,f} P_{H_2} + k_{3,f}} = 2F k_{3,f} \theta_H$$

This result indicates that the Heyrovsky–Volmer mechanism together is equivalent to the Volmer reaction as the rate-determining step when the respective backward reactions are negligible — a condition true for most of practical operating fuel cells.

An exception is that when the Volmer reaction proceeds much faster than the Heyrovsky reaction because the partial pressure of H_2 is very small so that

$$k_{2,f} P_{H_2} \ll k_{3,f}$$

Then $\theta \ll 1$ and the current density production reduces to

$$J = 2F k_{2,f} P_{H_2}$$

This expression for the current density is now consistent with the overall reaction shown in Equation (3.1) if the overall reaction could be regarded as an elementary reaction with the reaction rate constant equal to $k_{2,f}$. In this case, the Heyrovsky reaction may be taken as the rate-determining step in the oxidation of hydrogen at the anode.

Finally, the stoichiometric number for the rate-determining step (e.g., the Volmer reaction here) can be determined, which is an important parameter in the electrochemistry and is defined as follows: if n electrons are transferred in a rate-determining step and N electrons in the overall electrochemical reaction, then

$$vn = N$$

where v is the stoichiometric number. Hence, $v = 2$ for the Volmer reaction. For the reduction of oxygen at the cathode, $v = 4$ usually, but $v = 2$ can occur when hydrogen peroxide forms as the reaction intermediate in the rate-determining step. The stoichiometric number can be used to help in the interpretation of the rate law.

2. Although the steady-state approximation for the reaction intermediate species may imply that the concentration of the intermediate is a constant, and may not change with time. In reality, the change may occur if the reaction condition is changed, however, the change will be very fast in order to reach a new steady state value consistent with the changed condition, and the readjustment process is very similar to the initial transient. Therefore, the new steady state value can still be determined based on the steady-state approximation.

3. One may have already realized that the effect of the steady-state approximation effectively reduces the number of the first-order differential equations, governing the concentration of the species present in the reaction system (including the reactants, products and all the intermediate species), that need to be solved for a particular reaction mechanism. For the present example, only one reaction intermediate is present in the reaction, hence only one equation is reduced from the differential equation to the algebraic equation, which is usually easy to solve analytically and this often results in the final solution to be expressed analytically as well.

3.4   ACTIVATION POLARIZATION FOR CHARGE TRANSFER REACTION

Activation overpotential for an electrode reaction arises from the resistance to the electrochemical reaction and the physicochemical processes associated with the adsorption of reactant molecules or atoms on the electrode surface and it is directly related to the activation energy of the rate-determining process. As pointed out earlier, the rate-determining step is the step with the slowest rate of reaction and the rates of all the other reactions occurring either simultaneously or in sequence at the same electrode will depend on the rate of the rate-determining step. It might be mentioned that it may be possible to have more than just one rate-determining step, which is excluded from discussion here for it is far less common. It is also possible that the rate-determining step may be changed from one reaction to another when the reaction condition such as temperature and electrode overpotential is changed, hence the rate-determining step is not an invariable for a given overall reaction.

Consider that the following simple electron transfer reaction is the rate-determining step, represented by

$$R \rightleftharpoons O + ne^- \tag{3.37}$$

It is assumed that both oxidation of R to O and reduction of O to R occur at the electrode surface simultaneously. The forward and backward reactions take place at the same time. The current (or current density) produced in association with the forward reaction is denoted as I_f (or J_f), and with the backward reaction is I_b (or J_b). Then,

$$J_f = k_f C_R \tag{3.38}$$

$$J_b = k_b C_O \tag{3.39}$$

It might be pointed out that J_f is associated with the oxidation of R to O, hence it is normally referred to as the anodic current density because the anode is defined as the seat of oxidation. Similarly, J_b is often called the cathodic current density. However, the term of forward-and-backward current density is used continuously here to avoid the confusion with the actual reactions occurring at the anode and cathode of a fuel cell. The net current density at the electrode is conventionally defined as the difference between the forward-and-backward current density, or

$$J = J_f - J_b = k_f C_R - k_b C_O \tag{3.40}$$

For the convenience of further analysis, we focus on the production of current density under the reversible and irreversible condition in the following sections.

3.4.1   Charge Transfer Reaction Under Reversible Condition

According to the transition state theory, for the reaction given in Equation (3.37), whether it is the forward or the backward reaction, there is an energy barrier to be overcome in order for the reaction to proceed successfully. A schematic of the variation of the Gibbs function (free energy) with the reaction coordinate is shown in Figure 3.5 for an elementary electrochemical reaction. The reaction coordinate represents the degree of the reaction completed, and the activated complex is the reaction intermediate species that sits on top of the energy mountain. Hence, the activated complex can move forward to form the reaction product or fall back to the reactant. For the reactant to become the product, or vice versa, the energy mountain must be climbed over or the energy barrier must be overcome. The magnitude of the energy barrier to be overcome is equal to the Gibbs function change between the activated complex and the reactant R or the product O, respectively, which is called

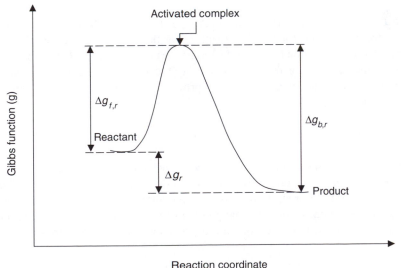

Figure 3.5 Schematic of the Gibbs function change with the reaction coordinate for an elementary electrochemical reaction under the reversible condition.

the Gibbs function of activation for the forward and backward reaction, respectively. They are*

$$\Delta g_{f,r} = g_{AC} - g_{R,r} \tag{3.41}$$

$$\Delta g_{b,r} = g_{AC} - g_{O,r} \tag{3.42}$$

The difference between the Gibbs function of activation for the forward and backward reaction is equal to the Gibbs function change for the overall reaction

$$\Delta g_r = \Delta g_{f,r} - \Delta g_{b,r} = g_{O,r} - g_{R,r} \tag{3.43}$$

where the subscript, R, O, and AC represent the reactant R, product O and the activated complex, AC, respectively; and the subscript r denotes the parameters under the reversible condition. For the reaction shown in Figure 3.5, the Gibbs function change for the forward reaction is negative, hence the forward reaction represents fuel cell (or battery) reaction and the backward reaction represents electrolytic reaction.

The current density (i.e., the rate of reaction) associated with the forward-and-backward reaction is given in Equations (3.38) and (3.39), respectively. For the Gibbs function change during the reaction shown in Figure 3.5 and under reversible (equilibrium) condition, the reaction rate constant may be expressed in the form, according to the transition state theory

$$k_f = B_f T \exp\left(-\frac{\Delta g_{f,r}}{RT}\right) \tag{3.44}$$

$$k_b = B_b T \exp\left(-\frac{\Delta g_{b,r}}{RT}\right) \tag{3.45}$$

where B's are the pre-exponential factor, $\Delta g_{f,r}$ is the molar Gibbs function of activation for the oxidation (or ionization) reaction for the present simple charge transfer reaction given in Equation (3.37), and similarly $\Delta g_{b,r}$ is the molar Gibbs function of activation for the reduction reaction (or discharge of ions). In general, the temperature dependence of the reaction rate constant is primarily due to the exponential term, and the linear dependence in the pre-exponential term is very weak in comparison and hence may be neglected, then the resulting expression becomes similar to the empirical Arrhenius equation for the reaction rate constant of a homogeneous chemical reaction.

When the reaction is reversible, in practice it means that the external circuit is open without net current being drawn from the electrode. In principle, an equal number of charged particles (ions or electrons) crosses the interface between the metal electrode and liquid electrolyte, so that an equal number of reactant R is oxidized and product O reduced. Under such a condition, the forward and backward reactions then occur

---

*In traditional electrochemical literature, the Gibbs function of the reactant, product and for the reaction, etc. is almost always written as ΔG_f, ΔG_b, and ΔG. However, the lower case symbol "g" is used here to emphasize that the Gibbs function change is on a per mole reactant (fuel) basis and to be consistent with the notation used in the previous chapter.

at the exactly same rate such that

$$J_f = J_b = J_0 \tag{3.46}$$

$$J_0 = B_f C_R T \exp\left(-\frac{\Delta g_{f,r}}{RT}\right) = B_b C_O T \exp\left(-\frac{\Delta g_{b,r}}{RT}\right) \tag{3.47}$$

where J_0 is the exchange current density (A/cm^2). It is a measure of the amount of electron transfer activity at the equilibrium electrode potential ϕ_r, and also represents how easy an electrochemical reaction can take place. Numerical values of the exchange current densities can change as much as by about 20 orders of magnitude, such as ranging from 10^{-18} A/cm^2 for the oxygen–evolution reaction on gold to 10^2 A/cm^2 for some metal–deposition reactions. For electrode reactions important for fuel cell applications, the exchange current density can vary from about 10^{-3} A/cm^2 at a platinum electrode to 10^{-12} A/cm^2 at a mercury electrode for H$_2$ electrode reaction; and from 10^{-10} A/cm^2 at a platinum electrode to 10^{-8} A/cm^2 at a copper electrode for O$_2$ electrode reaction. For acid electrolyte fuel cells, platinum is almost invariably used as catalyst, and the exchange current density is about seven orders of magnitude different between the H$_2$ oxidation and O$_2$ reduction reaction, therefore, O$_2$ reduction is a much slower process and incurs a much higher voltage loss than the H$_2$ oxidation reaction. In fact, for low to intermediate temperature fuel cells, voltage loss associated with O$_2$ reduction reaction in the cathode represents one single largest source of voltage losses. As a result, more effort has been spent on improving the cathodic process for the O$_2$ reduction reaction there.

Some typical values of exchange current density are given in Tables 3.1–3.3. Table 3.1 shows the exchange current density, J_0, for the hydrogen electrode reaction

Table 3.1   Exchange Current Density for Hydrogen–Electrode Reaction on Some Metals in H$_2$SO$_4$ at 25 °C[1]

Metal	Normality of H$_2$SO$_4$ Electrolyte	Exchange Current Density, J_0, A/cm^2
Pt	0.5	1×10^{-3}
Rh	0.5	6×10^{-4}
Ir	1.0	2×10^{-4}
Pd	1.0	1×10^{-3}
Au	2.0	4×10^{-6}
Ni	0.5	6×10^{-6}
Nb	1.0	4×10^{-7}
W	0.5	3×10^{-7}
Cd	0.5	2×10^{-11}
Mn	0.1	1×10^{-11}
Pb	0.5	5×10^{-12}
Hg	0.25	8×10^{-13}
Ti	2.0	6×10^{-9}

Table 3.2   Exchange Current Density, J_0, for Oxygen–Electrode Reaction on Some Metals at $25\,^\circ C^1$

Metal	J_0 in 0.1 N HClO$_4$ (pH $\sim$ 1), A/cm^2	J_0 in 0.1 N NaO (pH $\sim$ 12), A/cm^2
Pt	1×10^{-10}	1×10^{-10}
Pd	4×10^{-11}	1×10^{-11}
Rh	2×10^{-12}	3×10^{-13}
Ir	4×10^{-13}	3×10^{-14}
Au	2×10^{-12}	4×10^{-15}
Ag		4×10^{-10}
Ru		1×10^{-8}
Ni		5×10^{-10}
Fe		6×10^{-11}
Cu		1×10^{-8}
Re		4×10^{-10}

Table 3.3   Exchange Current Density, J_0, for Oxidation of Ethylene on Some Metals in 1N H$_2$SO$_4$ at $80\,^\circ C^1$

Metal	J_0, A/cm^2
Pt	1×10^{-10}
Pd	1×10^{-10}
Rh	5×10^{-11}
Ir	8×10^{-11}
Au	2×10^{-10}
Ru	5×10^{-11}

on a number of metals in acid solutions. It is seen that J_0 changes by as much as nine orders of magnitude, and its value is the largest for Noble metal electrodes, intermediate for the other transition metals and the lowest for mercury electrode. Table 3.2 provides the exchange current densities for the oxygen–dissolution reaction on some metal electrodes, and it shows that J_0 changes by about seven orders of magnitude from Ru to Au. Noble metals do not yield the highest exchange current density, in contrast to the hydrogen reaction shown in Table 3.1.

Table 3.3 presents the exchange current density for the oxidation of ethylene to carbon dioxide in acid electrolyte. Clearly, J_0 is persistently small even at the Noble metal surfaces and at the higher temperature of $80\,^\circ C$ when compared with the oxidation of hydrogen shown in Table 3.1. This rather low electrochemical reactivity of ethylene is representative of all electroorganic oxidation, such as methanol.

Consequently, this rather small rate of reaction limits the power density achievable at the acceptable energy conversion efficiency for practical applications.

From Equation (3.47) the following expression can be obtained, considering Equations (3.38)–(3.40), and (3.43)

$$\frac{C_O}{C_R} = \frac{k_f}{k_b} = \frac{B_f \exp\left(-\frac{\Delta g_{f,r}}{RT}\right)}{B_b \exp\left(-\frac{\Delta g_{b,r}}{RT}\right)} = \left(\frac{B_f}{B_b}\right) \exp\left(-\frac{\Delta g_r}{RT}\right) \tag{3.48}$$

Since the equilibrium constant for concentration is defined, for reaction (3.37), as

$$K_c = \frac{C_O}{C_R} \tag{3.49}$$

further $K_c = K_P$, the equilibrium constant for partial pressure, for Equation (3.37), and if the ratio of the pre-exponential factor is on the order of one, or

$$\frac{B_f}{B_b} \cong 1$$

then the familiar expression results for the equilibrium constant for concentration

$$K_c = \frac{k_f}{k_b} = \exp\left(-\frac{\Delta g_r}{RT}\right) \tag{3.50}$$

This indicates that the ratio of the forward-and-backward reaction rate constant is equal to the equilibrium constant for concentration, and it can be determined by the knowledge of the Gibbs function of the reaction and the temperature at which the reaction occurs.

On the other hand, it might be expected that the Gibbs function of activation for the activated complex, which represents the energy barrier to the given reaction, should be related to the Gibbs functions of the reactant and product. Therefore, the following relations may hold

$$\Delta g_{f,r} = c + a_f \Delta g_r \tag{3.51}$$

$$\Delta g_{b,r} = c - a_b \Delta g_r \tag{3.52}$$

where a_f and a_b are constant coefficient relating the Gibbs function changes and c is a constant. The minus sign represents the fact that the Gibbs function for the forward and backward reaction is positive and the Gibbs function of the reaction is negative for the case illustrated in Figure 3.5.

In electrochemistry a reversible (or equilibrium) potential of electrode for a given electrode reaction is defined as the electric potential difference between the metal electrode and the electrolyte solution when the net electric current is zero for the reaction. In analogy with the reversible cell potential presented in Equation (2.24), a reversible electrode potential, ϕ_r, which is the potential an electrode would have when the electrode reaction given in Equation (3.37) occurs reversibly, may be defined as follows

$$\phi_r = -\frac{\Delta g_r}{nF} \tag{3.53}$$

where the Gibbs function of reaction is given in Equation (3.43). An electrode potential, ϕ_r, so defined is consistent with the electrode potential measured with respect to a reference hydrogen electrode presented in Section 3.2. This is because the Gibbs function of species, whether charged or not, is measured in electrochemistry with respect to the reference hydrogen electrode. On the other hand, if the reduced species R in reaction (3.37) is regarded to be in the electrolyte bulk region, the reversible electrode potential, ϕ_r, defined in Equation (3.53) may also be interpreted as the potential difference between the metal electrode and the electrolyte at equilibrium.

Now combining Equation (3.53) with Equations (3.51) and (3.52) respectively results in

$$\Delta g_{f,r} = c - a_f nF\phi_r \tag{3.54}$$

$$\Delta g_{b,r} = c + a_b nF\phi_r \tag{3.55}$$

Substituting Equations (3.53)–(3.55) into Equation (3.43) yields the following relation for the two coefficients

$$-a_f - a_b = -1$$

or

$$a_b = 1 - a_f \tag{3.56}$$

Therefore the reaction rate constants presented in Equations (3.44) and (3.45) may be written as

$$k_f = B_f T \exp\left(-\frac{c}{RT}\right) \exp\left(\frac{a_f nF\phi_r}{RT}\right) \tag{3.57}$$

$$k_b = B_b T \exp\left(-\frac{c}{RT}\right) \exp\left(-\frac{(1-a_f)nF\phi_r}{RT}\right) \tag{3.58}$$

and the exchange current density given in Equation (3.47) may also be expressed as

$$J_0 = B_f C_R T \exp\left(-\frac{c}{RT}\right) \exp\left(\frac{a_f nF\phi_r}{RT}\right)$$

$$= B_b C_O T \exp\left(-\frac{c}{RT}\right) \exp\left(-\frac{(1-a_f)nF\phi_r}{RT}\right) \tag{3.59}$$

It is emphasized that the reversible electrode potential, ϕ_r, appeared in the Equation (3.59) may also be regarded as the metal electrode–electrolyte potential difference at equilibrium.

From Equation (3.59) the reversible electrode potential can be obtained as

$$\phi_r = \frac{RT}{nF} \ln\left(\frac{B_b}{B_f}\right) + \frac{RT}{nF} \ln\left(\frac{C_O}{C_R}\right) \tag{3.60}$$

Equation (3.60) may also be obtained thermodynamically by considering the reaction (3.37) under the reversible condition, or it is the specific form of the Nernst equation for the reaction (3.37). The first term on the right-hand side of the Equation (3.60) represents the reversible electrode potential when the concentration of the reactant and the product is not diluted, and the second term denotes the effect of concentration deviating from the undiluted values (which in practice may be taken as 1 mol/L for the purpose of experimental measurements). One may also verify that the reversible electrode potential determined from Equation (3.60) is identical to the definition given in Equation (3.53).

It may be pointed out that for any electrode reaction with exchange current density of $J_0 < 10^{-7}$ A/cm^2, it is unlikely that the equilibrium (or reversible) condition at that electrode (that is, reversible electrode potential, ϕ_r, as determined above) can be established for the reaction, because impurities likely to be present in the electrode material probably yields higher values of the exchange current density (i.e., quicker electrochemical reaction). As a result, the observed electrode potential is due to the impurities' reaction occurring at the electrode. Therefore, the experimentally measured open circuit voltage is in reality (about $0.1 \sim 0.2$ V) less than the reversible cell potential for H_2–O_2 fuel cell due to slow O_2 reduction reaction.

3.4.2   Charge Transfer Reaction Under Irreversible Condition

When a net current is produced from the electrode reaction, the electrode reaction shown in Equation (3.37) is no longer reversible (or becomes irreversible), and an imbalance of electron transfers exists (i.e., the forward-and-backward reactions take place at the different rates). The degree of the imbalance is represented by a net flow of electrons taking place either into the electrolyte (accompanying reduction at the cathode) or into the electrode (accompanying oxidation at the anode). The net amount of current flow to the electrode, as presented in Equation (3.40), depends on the extent to which the potential at the electrode differs from its equilibrium value, ϕ_r; this electrode potential difference has been defined as the overpotential in Equations (3.10) and (3.11), that is,

$$\eta = \phi - \phi_r \tag{3.61}$$

Therefore, the objective for this section is to establish the relation between the net current produced and the electrode overpotential incurred.

If an electric potential difference ϕ, other than the reversible value ϕ_r, exists between the reactant and the product, the forward and backward reaction are not

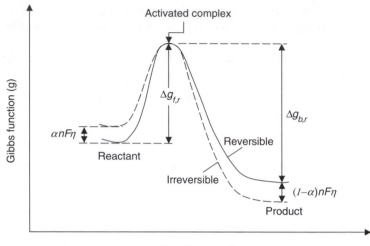

Figure 3.6 The effect of an electrode overpotential on the Gibbs function of the reactant and product as well as its variation with the reaction coordinate for an elementary electrochemical reaction under the irreversible condition with a net current flow.

balanced, resulting into a net current flow, the Gibbs function of the reactant R and the product O is then also different from their respective reversible values because of the influence by the overpotential between them, as shown in Figure 3.6. Physically, it is known that the overpotential developed will promote the forward reaction, presumably by raising the energy level of the reactant, and at the same time will hinder the backward reaction by lowering the energy level of the product. In general, the effect of the overpotential is not the same for the forward and backward reaction, that is, it is not symmetrically (or equally) distributed between the forward and backward reaction. Rather a proportion α is spent on raising the Gibbs function of the reactant, and the remaining proportion, $1 - \alpha$, is on reducing the Gibbs function of the product, as illustrated in Figure 3.6.

In the previous chapter it is shown that the Gibbs function change, Δg, corresponding to a potential difference, E, is related by

$$\Delta g = -nFE$$

such as the reversible cell potential difference given in Equation (2.24). Similarly, the Gibbs function change corresponding to an overpotential η can be expressed as equal to $-nF\eta$. Then the actual Gibbs function of activation for the forward-and-backward reactions become

$$\Delta g_f = \Delta g_{f,r} - \alpha nF\eta \tag{3.62}$$

$$\Delta g_b = \Delta g_{b,r} + (1 - \alpha)nF\eta \tag{3.63}$$

In analogy with the reaction rate constant given in Equations (3.44) and (3.45) under the reversible condition, the reaction rate constant under the irreversible condition may be expressed as

$$k_f = B_f T \exp\left(-\frac{\Delta g_f}{RT}\right) \tag{3.64}$$

$$k_b = B_b T \exp\left(-\frac{\Delta g_b}{RT}\right) \tag{3.65}$$

Taking into account of Equations (3.62)–(3.65), the current density associated with the forward-and-backward reaction becomes

$$J_f = k_f C_R = B_f C_R T \exp\left(-\frac{\Delta g_{f,r} - \alpha n F \eta}{RT}\right)$$

$$= B_f C_R T \exp\left(-\frac{\Delta g_{f,r}}{RT}\right) \exp\left(\frac{\alpha n F \eta}{RT}\right) \tag{3.66}$$

$$J_b = k_b C_O = B_b C_O T \exp\left(-\frac{\Delta g_{b,r} + (1-\alpha) n F \eta}{RT}\right)$$

$$= B_b C_O T \exp\left(-\frac{\Delta g_{b,r}}{RT}\right) \exp\left(-\frac{(1-\alpha) n F \eta}{RT}\right) \tag{3.67}$$

Considering the exchange current density given in Equation (3.47), the Equations (3.66) and (3.67) become

$$J_f = J_0 \exp\left(\frac{\alpha n F \eta}{RT}\right) \tag{3.68}$$

$$J_b = J_0 \exp\left(-\frac{(1-\alpha) n F \eta}{RT}\right) \tag{3.69}$$

Clearly the electrode overpotential enhances the forward reaction and hinders the backward reaction. The net current density arising from the electrode overpotential then becomes, following Equation (3.40)

$$J = J_f - J_b = J_0 \left\{ \exp\left(\frac{\alpha n F \eta}{RT}\right) - \exp\left(-\frac{(1-\alpha) n F \eta}{RT}\right) \right\} \tag{3.70}$$

which is known as the Butler–Volmer equation, representing the general relation between the net current density produced and the activation overpotential η. Where the parameter α is called the transfer coefficient (or symmetry factor), theoretically its numerical value lies between zero and one, or

$$0 < \alpha < 1 \tag{3.71}$$

Table 3.4 Experimentally Determined Values of Transfer Coefficient, α, for a Few Anode Reactions[2]

Reaction	Electrode	Transfer Coefficient α
$Fe^{2+} \longrightarrow Fe^{3+} + e^-$	Pt	0.58
$Ce^{3+} \longrightarrow Ce^{4+} + e^-$	Pt	0.75
$Ti^{3+} \longrightarrow Ti^{4+} + e^-$	Hg	0.42
$H_2 \longrightarrow 2H^+ + 2e^-$	Hg	0.50
$H_2 \longrightarrow 2H^+ + 2e^-$	Ni	0.58
$Ag \longrightarrow Ag^+ + e^-$	Ag	0.55

Experimentally it is often found to be in the vicinity of 0.5. Table 3.4 presents the measured value of the transfer coefficient for a few typical electrode reactions.

The Butler–Volmer equation shown in Equation (3.70) may be obtained in a different but easy way. In analogy with Equations (3.57) and (3.58), the reaction rate constant may be written as, when the electrode potential is ϕ and according to Equation (3.61), $\phi = \eta + \phi_r$

$$k_f = B_f T \exp\left(-\frac{c}{RT}\right) \exp\left(\frac{a_f n F\phi}{RT}\right)$$

$$= B_f T \exp\left(-\frac{c}{RT}\right) \exp\left(\frac{a_f n F(\eta + \phi_r)}{RT}\right)$$

$$= B_f T \exp\left(-\frac{c}{RT}\right) \exp\left(\frac{a_f n F\phi_r}{RT}\right) \exp\left(\frac{a_f n F\eta}{RT}\right) \quad (3.72)$$

and similarly

$$k_b = B_b T \exp\left(-\frac{c}{RT}\right) \exp\left(-\frac{(1 - a_f)n F\phi_r}{RT}\right) \exp\left(-\frac{(1 - a_f)n F\eta}{RT}\right) \quad (3.73)$$

Then the net current density becomes, taking into account Equation (3.59)

$$J = J_f - J_b = k_f C_R - k_b C_O$$

$$= J_0 \left\{ \exp\left(\frac{a_f n F\eta}{RT}\right) - \exp\left(-\frac{(1 - a_f)n F\eta}{RT}\right) \right\} \quad (3.74)$$

which is identical with Equation (3.70) if the coefficient a_f is set equal to the transfer coefficient α.

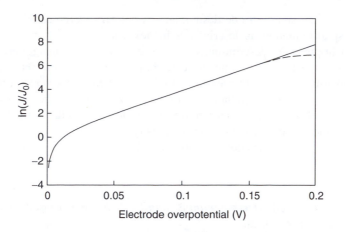

Figure 3.7 The dependence of current density as a function of electrode overpotential according to the Butler–Volmer equation for the condition of temperature $T = 25\,^\circ$C, transfer coefficient $\alpha = 0.5$, and the number of electron transferred $n = 2$.

Note that the Butler–Volmer equation may also be written as

$$J = J_0 \left\{ \exp\left(\frac{\alpha_a n F \eta}{RT} \right) - \exp\left(-\frac{\alpha_c n F \eta}{RT} \right) \right\} \tag{3.75}$$

where α_a is called the anodic transfer coefficient, α_c is the cathodic transfer coefficient, and

$$\alpha_a + \alpha_c = 1 \tag{3.76}$$

Recall that in fuel cell reactions,

$\eta > 0$, and hence $J > 0$ for the anodic oxidation reaction;
$\eta < 0$, and hence $J < 0$ for the cathodic reduction reaction.

Figure 3.7 illustrates the dependence of the current density on the electrode over-potential η at the temperature of 25 °C for the transfer coefficient of $\alpha = 0.5$ and the number of electron transfer $n = 2$. It is seen that almost a straight-line relation holds between the logarithm of the current density and the electrode overpotential for reasonably large values of η. The linear relation is deviated for small values of the overpotential, and at the very large overpotentials the current density is so large that the limited rate of mass transfer will be unable to supply the needed reactant species to the reaction sites for the current production. Therefore, the mass transport limitation will cause the current–overpotential relation to depart from the linear relation at the very large values of η, as shown by the dashed curve in Figure 3.7. However, the exact overpotential value at which deviation occurs and the degree of the deviation will depend on the mass transfer characteristics for a given cell design, which depend in turn on a number of factors including the reactant flow conditions, the flow channel, and electrode design, and as on. The overpotential related to the mass transport limitations is described in the next chapter.

From the preceding description, it becomes clear that the exchange current density can be measured for a given electrode structure, electrolyte and reactant used as follows: measure and plot the measured $\ln(J)$ vs η, then the intercept at the vertical coordinate for a linear extrapolation from the linear portion of the curve represents the logarithm of the exchange current density $\ln(J_0)$. This is the normal practice in the measurement of the exchange current density.

For the Butler–Volmer equation derived in Equation (3.70), two special cases can be considered:

(1) When the Electrode Overpotential is Small: When the electrode overpotential is small so that the exponent

$$\frac{\alpha n F \eta}{RT} < 0.1 \quad \text{and} \quad \frac{(1-\alpha)n F \eta}{RT} < 0.1 \tag{3.77}$$

then the exponential function in the Bulter–Volmer equation can be expanded in Taylor series according to the relation:

$$\exp(x) \approx 1 + x$$

and the Butler–Volmer equation can be simplified to

$$
\begin{aligned}
J &= J_0 \left\{ \left(1 + \frac{\alpha n F \eta}{RT}\right) - \left[1 - \frac{(1-\alpha)n F \eta}{RT}\right] \right\} \\
&= J_0 \left(\frac{n F \eta}{RT}\right)
\end{aligned} \tag{3.78}
$$

Therefore, a linear relation results between the current density and the electrode overpotential when the overpotential is sufficiently small. To satisfy the inequality (3.77), the electrode overpotential has to be smaller than 0.003 V for reactions occurring near the room temperature because the coefficient of η is approximately 39 V^{-1} and is strongly temperature dependent, if the transfer coefficient is taken as 0.5 and the number of electrons transferred as 2.

(2) When the Electrode Overpotential is Large: When the electrode overpotential is large so that the exponent

$$\frac{\alpha n F \eta}{RT} > 1.2 \quad \text{and} \quad \frac{(1-\alpha)n F \eta}{RT} > 1.2 \tag{3.79}$$

equivalent to $\eta > 0.05$ V for electrode reactions near the room temperature, the backward reaction is negligible compared with the forward reaction because the second term in the Butler–Volmer equation is at least an order of magnitude smaller than the first term. Therefore Equation (3.70) can be reduced to

$$J = J_0 \exp\left(\frac{\alpha n F \eta}{RT}\right) \tag{3.80}$$

or

$$\eta = \frac{RT}{\alpha n F} \ln\left(\frac{J}{J_0}\right) \tag{3.81}$$

which is the well-known Tafel equation, one of the fundamental relations in electrochemistry, first established in 1905 empirically.

The Tafel equation is conventionally written in the following form:

$$\eta = a \log(bJ) \tag{3.82}$$

where a and b are constants, and normally determined from a particular set of experimental data. Comparing with Equation (3.81), it is clear that

$$a = \frac{2.303 RT}{\alpha n F} \quad \text{and} \quad b = \frac{1}{J_0} \tag{3.83}$$

Therefore, both the transfer coefficient α and the exchange current density J_0 can be determined from the experimental measurements for a particular electrode reaction under a specified condition. Table 3.5 shows the Tafel Slope, a, and the exchange current density for a number of electrode reactions and Table 3.6 presents the Tafel slope, α, and the exchange current density, J_0, for hydrogen evolution reaction at various electrodes in different electrolyte solutions. Both tables are for the data measured at room temperature.

It should be emphasized that the Tafel slope, a, depends on the reaction mechanism involved, while the exchange current density, as shown in Equation (3.47) or

Table 3.5 Values of Tafel Slope, a, and Exchange Current Density, J_0, for Various Electrode Reactions at the Room Temperature[2]

Electrode	Electrolyte	Reaction	a (V)	J_0 (mA/cm^2)
Cu	1 mol/ℓ CuSO$_4$(aq)	Deposition of Cu	−0.051	2×10^{-2}
Ni	1 mol/ℓ NiSO$_4$(aq)	Deposition of Ni	−0.051	2×10^{-6}
Pt	0.05 mol/ℓ H$_2$SO$_4$(aq)	Evolution of O$_2$	−0.044	2×10^{-7}
Pt	1 mol/ℓ NaN$_3$(aq)	Evolution of N$_2$	−0.026	1×10^{-73}
Hg	0.1 mol/ℓ KOH(aq)	Evolution of H$_2$	−0.093	4×10^{-12}
Ag	7 mol/ℓ HCl(aq)	Evolution of H$_2$	−0.090	1.3×10^{-3}
Pd	0.5 mol/ℓ H$_2$SO$_4$(aq)	Evolution of H$_2$	−0.080	1

Table 3.6 Values of Tafel Slope, a, and Exchange Current Density, J_0, for Hydrogen Evolution Reaction at Various Electrodes in Different Electrolyte Solutions at the Room Temperature[2]

Electrode	Electrolyte	Concentration (mol/ℓ)	$-\log(J_0, \text{A/cm}^2)$	a (V)
Pt	HCl	0.5	2.6	0.028
Pd	H_2SO_4	0.5	3.0	0.080
Cu	HCl	0.1	6.0	0.117
Cu	NaOH	0.15	6.0	0.117
Ag	HCl	1.0	4.0	0.130^a
Ag	HCl	1.0	5.0	0.060^b
Au	HCl	0.1	5.0	0.097^a
Au	HCl	0.1	6.0	0.071^b
Cd	H_2SO_4	0.85	12.0	0.120
Hg	HCl	1.0	12.0	0.119
Hg	LiOH	0.1	12.0	0.102
Al	H_2SO_4	1.0	10.0	0.100
Sn	HCl	1.0	8.0	0.140
Pb	HCl	1.0	13.0	0.119
Pb	H_2SO_4	10.0	13.0	0.119
Mo	HCl	0.1	6.0	0.104^a
Mo	NaOH	0.1	7.0	0.116^a
Mo	HCl	0.1	7.0	0.080^a
Mo	NaOH	0.1	7.0	0.087^b
W	HCl	5.0	5.0	0.110
Fe	HCl	1.0	6.0	0.130
Fe	NaOH	0.1	6.0	0.120
Ni	HCl	1.0	5.0	0.109
Ni	NaOH	0.1	5.0	0.101

^{a} High current density, $J = 0.01 - 0.1 \text{A/cm}^2$.

^{b} Low current density, $J = 0.01 \text{A/cm}^2$.

Equation (3.59) earlier, is a function of the reactant concentration and temperature. In fact, it is a strong function of the temperature. Further, for exchange current density less than 10^{-7}A/cm^2, a mixed electrode potential may occur due to the electrode impurities' reactions. Therefore, care should be taken for measurements and data analysis under such conditions.

Example 3.2   For the hydrogen reaction at the platinum electrode surface immersed in a 0.5 N H_2SO_4 electrolyte solution, the following Volmer reaction is assumed to occur

$$H - M \rightleftharpoons M + H^+ + e^- \quad \text{(Volmer Reaction)} \tag{3.84}$$

Determine the current density associated with the forward-and-backward reaction as well as the net current density at the electrode overpotential of 0.01 V, 0.02 V, 0.05 V, 0.1 V, and 0.2 V, respectively, at the temperature of 25 °C by taking the transfer coefficient $\alpha = 0.5$.

SOLUTION

From Table 3.1 we find that the exchange current density for the given condition is

$$J_0 = 1 \times 10^{-3} \, \text{A/cm}^2 \tag{3.85}$$

For the Volmer reaction, the number of electrons transferred is $n = 1$. Therefore,

$$\frac{\alpha n F}{RT} = \frac{(1-\alpha)nF}{RT} = \frac{0.5 \times 1 \times 96,487}{8.3143 \times 298} = 19.471 \frac{1}{V} \tag{3.86}$$

since the transfer coefficient $\alpha = 0.5$ is given for the problem and the Butler–Volmer equation, Equation (3.70), becomes

$$J = J_0[\exp(19.471\eta) - \exp(-19.471\eta)]$$
$$= 1 \times 10^{-3}[\exp(19.471\eta - \exp(-19.471\eta)]; \quad \text{A/cm}^2$$

which provides the net current density. The current density associated with the forward-and-backward reaction is, respectively

$$J_f = J_0[\exp(19.471\eta)]$$
$$= 1 \times 10^{-3}[\exp(19.471\eta)]; \quad \text{A/cm}^2$$
$$J_b = J_0[\exp(-19.471\eta)]$$
$$= 1 \times 10^{-3}[\exp(-19.471\eta)]; \quad \text{A/cm}^2$$

Then the numerical values of the current density for the forward-and-backward reaction as well as the net current density can be calculated for the stated electrode overpotentials under the given condition. For brevity, the final results are presented in the following table:

	Electrode Overpotential η(V)				
	0.01	**0.02**	**0.05**	**0.1**	**0.2**
J_f(A/cm^2) ($\times 10^{-3}$)	1.2150	1.4761	2.6473	7.0083	49.117
J_b(A/cm^2) ($\times 10^{-3}$)	0.8231	0.6774	0.3777	0.1427	0.02036
J(A/cm^2) ($\times 10^{-3}$)	0.3919	0.7987	2.270	6.866	49.10

COMMENT

1. It is seen that the current density for the forward reaction increases quickly, while it decreases rapidly for the backward reaction, so that the net current density increases significantly as the electrode overpotential is increased.

2. It is known that among the acid electrolyte fuel cells, the rate of reaction is the fastest for hydrogen oxidation at the Pt electrode with the presence of H_2SO_4 electrolyte. Even so, the current density generated is still small, as shown in this example. At the electrode potential of 200 mV, the current density generated is only about 0.0491 A/cm². Therefore, for practical fuel cells measures need to be taken to increase the current density, hence the power output, at anode overpotentials as small as possible. Note that anode overpotential of 200 mV is a very large value in H_2–O_2 fuel cells.

3. The relation between the electrode overpotential and the net current density generation can be illustrated graphically. Because hydrogen oxidation usually occurs at the anode of a fuel cell, the corresponding electrode overpotential arising from the resistance to the electrochemical reaction in the anode is often referred to as the anode activation overpotential, as calculated in this example. It is conventional for fuel cell performance to express the overpotential as a function of the net current density generated. Therefore, the anode activation overpotential versus the net current density for this example can be presented in graphical form as shown in Figure 3.8. It is seen in the figure that the relation between the overpotential and the current density is almost linear at low current density values, and is exponential at high current densities, as the two extreme cases analyzed earlier. Whereas for the intermediate values of the current density, the relation is more complex, described by the hyperbolic sine function

$$J = 2 \times 10^{-3} \sinh(19.471\eta); \quad A/cm^2$$

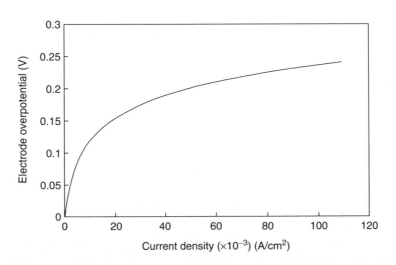

Figure 3.8 The anode activation overpotential as a function of the net current density generation for the conditions given in Example 3.2.

Example 3.3   For the oxygen reduction reaction at the platinum electrode surface immersed in a 0.05 mol/ℓ H_2SO_4 electrolyte solution, the following overall reaction occurs, in consistency with Equation (3.37)

$$H_2O \rightleftharpoons O_2 + 4H^+ + 4e^- \tag{3.87}$$

The actual reaction mechanism may involve a four-electron or a two-electron reaction pathway, which is described later in this chapter. Determine the current density associated with the forward-and-backward reaction as well as the net current density at the electrode overpotential of $-0.01\,V$, $-0.02\,V$, $-0.05\,V$, $-0.1\,V$, and $-0.2\,V$, respectively. Assume the reaction takes place at the temperature of $25\,°C$, and the transfer coefficient is $\alpha = 0.5$.

SOLUTION

Table 3.5 indicates that the exchange current density for the given condition is

$$J_0 = 2 \times 10^{-7}\,mA/cm^2$$

and the Tafel slope is

$$a = 0.044\,V$$

It should be noted the negative sign in Table 3.5 results from the convention of a positive value for the overpotential in the Tafel equation, or Equation (3.82); whereas in the use of the Butler–Volmer equation, the overpotential is considered as positive for the anode and negative for the cathode as discussed earlier.

From the given parameter values and according to Equation (3.83), we obtain from the Tafel slope

$$\frac{\alpha n F}{RT} = \frac{(1-\alpha)n F}{RT} = \frac{2.303}{0.044} = 52.34\frac{1}{V}$$

Assume that the reaction mechanism is the same at the high or low overpotentials, leading to the same exponent and the exchange current density for all overpotential values (notice that this assumption may not be true, in fact, usually not quite true for the oxygen reduction reaction). Then, the Butler–Volmer Equation (3.70) may be written as for the present problem

$$J = J_0 \left[\exp(52.34\eta) - \exp(-52.34\eta)\right]$$
$$= 2 \times 10^{-7} \left[\exp(52.34\eta) - \exp(-52.34\eta)\right]; \quad mA/cm^2$$

which represents the net current density. The current density associated with the forward-and-backward reaction may therefore be written as, respectively

$$J_f = J_0 \left[\exp(52.34\eta)\right]$$
$$= 2 \times 10^{-7} \left[\exp(52.34\eta)\right]; \quad mA/cm^2$$

$$J_b = J_0\left[\exp(-52.34\eta)\right]$$
$$= 2 \times 10^{-7}\left[\exp(-52.34\eta)\right]; \quad mA/cm^2$$

Then the current density for the forward-and-backward reaction as well as the net current density can be calculated for the stated electrode overpotential values under the given condition in this example. For brevity, the final results are calculated and presented in the following table

	Electrode Overpotential η(V)				
	−0.01	**−0.02**	**−0.05**	**−0.1**	**−0.2**
J_f (mA/cm^2) ($\times 10^{-7}$)	1.1850	0.7021	0.1460	0.01066	5.686×10^{-5}
J_b (mA/cm^2) ($\times 10^{-7}$)	3.3755	5.6970	27.389	375.08	7.034×10^4
J (mA/cm^2) ($\times 10^{-7}$)	−2.1905	−4.9949	−27.243	−375.07	-7.034×10^4

COMMENT

1. The results given in the table above indicate that the current density for the forward reaction decreases quickly, while it increases rapidly for the backward reaction; further more, the current density for the forward reaction is smaller than the backward reaction, so that the net current density is negative, indicating that the net reaction is in the backward direction. That is, H_2O is formed as the reaction product under the specified condition, as it should be for the cathode reaction in a fuel cell. It is also seen that the net current density decreases significantly as the magnitude of the electrode overpotential is increased.

2. Since the net current density is always negative for the present cathodic reaction, which would take place as, in a fuel cell

$$O_2 + 4H^+ + 4e^- \rightleftharpoons H_2O$$

and for the above reaction, the forward reaction is faster than the backward reaction so that the net current density associated with this reaction is positive, along with a positive value of the overpotential. It should be emphasized that the magnitude of the electrode overpotential, from its definition, represents the voltage (or energy) loss, irrespective of its sign as discussed here.

3. A comparison with the tabulated results shown in the previous example for the hydrogen oxidation at the anode reveals that the oxygen reduction reaction in the cathode is much slower; correspondingly the net current density generated

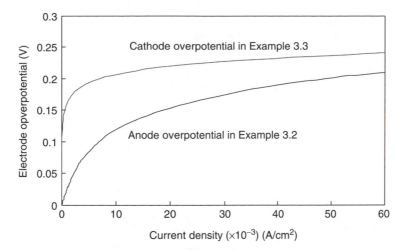

Figure 3.9 The cathode overpotential as a function of the net current density generated for the oxygen reduction reaction under the condition given in Example 3.3. The anode overpotential for the hydrogen oxidation in Example 3.2 is also shown for comparison purpose. Note that the current density corresponding to the cathode overpotential has been increased by a factor of 1000 for the clarity of presentation.

is much smaller for the cathodic reaction at the same electrode overpotential. Or alternatively, for the same net current density generated as in an operating fuel cell, the electrode overpotential is much smaller for the hydrogen oxidation reaction at the anode than the oxygen reduction reaction at the cathode. This much larger overpotential at the cathode suggests that the cathode process imposes severe performance limitation for a fuel cell; hence it is usually the focus of the attention for the performance enhancement.

4. Oxygen reduction reaction occurs at the cathode of a fuel cell and the corresponding electrode overpotential arising from the resistance to the electrochemical reaction in the cathode is conventionally called the cathode activation overpotential, as calculated in this example. The cathode activation overpotential can be regarded as a function of the net current density, and is shown in Figure 3.9. For comparison purpose, the anode activation overpotential determined from the previous example is also presented in the same figure. In order to show both results in the same plot, the current density corresponding to the cathode overpotential has been increased by a factor of 1000. It is evident that the cathode activation overpotential is significantly larger than the anode counterpart, as noticed earlier. This is representative of the relative magnitude of the activation overpotentials in an operating fuel cell.

5. In presenting the fuel cell performance data, it is sometimes the practice to plot the anode and cathode electrode potential, ϕ_a and ϕ_c according to its relation with the respective electrode overpotential as defined in Equations (3.10) and (3.11). That is, the potential of the anode and cathode electrode is, respectively

$$\phi_a = \phi_{a,r} + \eta_a = \eta_a$$
$$\phi_c = \phi_{c,r} + \eta_c = 1.229\text{V} - |\eta_c|$$

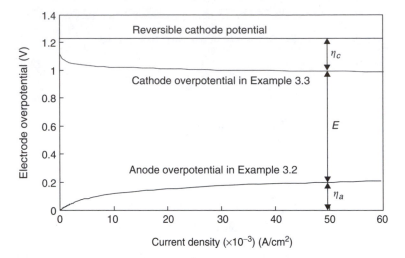

Figure 3.10 Electrode potential and the cell potential difference as a function of the net current density generated corresponding to the conditions stated in Examples 3.2 and 3.3. Note that the current density corresponding to the cathode overpotential has been increased by a factor of 1000 for the clarity of presentation.

This is because the reversible potential is 0 for the anode and 1.229 V for the cathode for an acid electrolyte H_2–O_2 fuel cell under the standard temperature and pressure as given in Section 3.2 earlier. If we use the anode overpotential as determined from Example 3.2 and the cathode overpotential as determined from Example 3.3, the anode and cathode overpotential can be presented graphically as shown in Figure 3.10 where the current density corresponding to the cathode overpotential has been increased by a factor of 1000. Clearly, the potential difference between the cathode and the anode represents the cell potential, as stated in Equation (3.8). Note that in an operating fuel cell other modes of overpotential occurs as well, which are not included in Figure 3.10 yet.

3.4.3 Generation of Net Current Density and Entropy

In Chapter 2, it is shown that the total overpotential for a cell, which arises from all the irreversibilities within the cell, is related to the total generation of entropy within the cell, as given in Equation (2.26). Similarly, an entropy generation can be defined corresponding to the anode and cathode activation overpotential, respectively. Specifically, the anode entropy generation corresponding to the anode activation overpotential is defined as

$$\wp_{s,a} = \frac{nF\eta_a}{T} \tag{3.88}$$

and the cathode entropy generation is

$$\wp_{s,c} = \frac{nF|\eta_c|}{T} = -\frac{nF\eta_c}{T} \tag{3.89}$$

The absolute value of the cathode overpotential is used in the Equation (3.89) because the entropy generation must be positive all the time required by the second law of thermodynamics, while the cathode overpotential is conventionally expressed as a negative value. Then the net current density generated in the anode and cathode can be written as, from the Butler–Volmer equation given in Equation (3.70)

$$J_a = J_{0,a} \left\{ \exp(\alpha \wp_{s,a}^*) - \exp\left[-(1-\alpha)\wp_{s,a}^*\right] \right\} \tag{3.90}$$

$$J_c = J_{0,c} \left\{ \exp(-\alpha \wp_{s,c}^*) - \exp\left[+(1-\alpha)\wp_{s,c}^*\right] \right\} \tag{3.91}$$

where the dimensionless entropy generation in the anode and cathode is normalized by the universal gas constant as follows

$$\wp_{s,a}^* = \frac{\wp_{s,a}}{R} \quad \text{and} \quad \wp_{s,c}^* = \frac{\wp_{s,c}}{R} \tag{3.92}$$

Since for most of fuel cell reactions the transfer coefficient α is about 0.5, Equations (3.90) and (3.91) can be written simply as

$$J_a = 2J_{0,a} \sinh(0.5\wp_{s,a}^*) \tag{3.93}$$

$$J_c = -2J_{0,c} \sinh(0.5\wp_{s,c}^*) \tag{3.94}$$

The preceding relation indicates that the generation of the net current density is related to the amount of the entropy generation. This is understandable because the entropy generation represents the degree of the irreversibility or departure from the thermodynamic equilibrium condition; and the net current density for a fuel cell reaction also represents the degree of the departure from the thermodynamic equilibrium at which the forward and backward reaction proceeds at the same rate with zero net current density generated. Therefore, the Butler–Volmer equation as written in Equations (3.90) and (3.91) or Equations (3.93) and (3.94) relates the electrochemical reaction to the thermodynamic analysis.

From this analysis and discussion, it becomes clear that the energy efficient approach to high current density (hence, high output power density) operation for practical fuel cells is to increase the exchange current density, because the generation of entropy represents the degradation of the useful energy. Increasing the cell operation temperature is commonly most effective to increase the exchange current density because of its strong temperature dependence. However, for each type of fuel cells under development limitations are found on the highest temperatures that a cell can be operated or tolerated. Therefore, other measures have been explored to increase the exchange current density. Some common techniques are to find the most effective catalyst or electrolyte or their combinations to facilitate the heterogeneous electrochemical reaction, and these techniques have been met with considerable difficulty for further significant progress. Since exchange current density is also a function of the reactant concentration, increasing the reactant concentration at the reaction sites is also effective to increase the current density without the increase in the generation of entropy; this can be achieved by increasing the operation pressure, by using the purified reactants (pristine hydrogen or oxygen), and by deploring electrolytes with high reactant solubilities, and so on. Due to the various practical limitations, the approach to minimize transport resistances to mass transfer is presently most feasible for each type of fuel cell, that includes the improved design of gas flow fields on the bipolar plates, electrode backing, and the catalyst layer.

3.4.4   Roughness Factor and Specific Reactive Surface Area of Electrode

As noticed earlier, especially in Examples 3.2 and 3.3, the current density generated is small even at significant electrode overpotentials. This is because of the extremely small values of the exchange current density for the electrode reactions involved. Therefore, the electrode is often designed and made in porous form. The porous (or void) region allows the reactant mass transfer to the reaction sites. But more important is the increase of the reactive surface areas for a same geometrical (or flat) electrode surface so that the current density measured in terms of the geometrical electrode surface is enhanced considerably without the increase in the generation of entropy. The amount of the reactive electrode surface area is often measured in terms of a roughness factor, defined as

$$\text{Roughness factor} = \frac{\text{Actual reactive surface area}}{\text{Geometrical (flat) electrode area}} = a \qquad (3.95)$$

The roughness factor of about 2000 can be achieved for platinized platinum. Typical value of about 600 or larger is often found in practical fuel cells.

The actual reactive surface area in a practical porous electrode is often difficult to determine. For this reason, the current density for a fuel cell is almost always defined in terms of geometrical area of the electrode, rather than the actual reactive surface area. That is, for a fuel cell

$$\text{Current Density } J_{FC} = \frac{\text{Current } I}{\text{Geometrical electrode area}} \qquad (3.96)$$

However, the exchange current density is conventionally measured experimentally in terms of the actual reactive surface area. Therefore, the current density commonly referred to for a fuel cell can be expressed as follows

$$\text{Current Density } J_{FC} = \frac{\text{Current } I}{\text{Actual reactive surface area}}$$
$$\times \frac{\text{Actual reactive surface area}}{\text{Geometrical electrode area}}$$

The second term on the right-hand side of the preceding equation is the roughness factor defined in Equation (3.95), and the first term is the current density presented in Equation (3.70) derived earlier in Section 3.4.2. Therefore, the current density expression useful for fuel cell analysis becomes

$$J_{FC} = a J_0 \left\{ \exp\left(\frac{\alpha n F \eta}{RT} \right) - \exp\left(-\frac{(1-\alpha) n F \eta}{RT} \right) \right\} ; \quad \text{A/cm}^2 \qquad (3.97)$$

It is clear from the expression (3.97) that the current density in a fuel cell can be increased by a factor equal to the value of the electrode roughness factor, if compared to an electrode with smooth flat surface, without degrading the cell potential or increasing the entropy generation. In practice, the roughness factor may be decreased in an operating fuel cell due to the crushing of the pore region under the strong cell compression force, resulting in some active surface area inaccessible by the reactant, or due to the catalyst particle agglomeration merging into larger particle sizes. The electrode sintering (plastic deformation) under compression may also restrict reactant mass transfer to the reactive sites. All these factors tend to degrade the long-term cell performance, limiting the cell lifetime.

For practical fuel cells, the roughness factor may not be a best measure for the effectiveness of the electrode structure design. This is because the mass transfer to the reactive sites is a very slow process. This fact, coupled with the fast electrochemical reaction within the porous volume of the electrode structure, often results in only a small fraction of the electrode structure is accessible by reactant species. Therefore, sometimes the reactive surface area per-unit volume of the electrode structure becomes important, which can be defined as

$$\text{Reactive surface area density} = \frac{\text{Actual reactive surface area}}{\text{Volume of electrode}} = A_v \qquad (3.98)$$

which is also referred to as the specific reactive surface area, m^2/m^3. Then a volumetric current density, J'_{FC}, may be defined as follows

$$\text{Volumetric current density } J'_{FC} = \frac{\text{Current } I}{\text{Volume of electrode}}$$

$$= \frac{\text{Current } I}{\text{Actual reactive surface area}}$$

$$\times \frac{\text{Actual reactive surface area}}{\text{Volume of electrode}}$$

$$= J \times A_v$$

Or it becomes after the substitution of Equation (3.70)

$$J'_{FC} = A_v J_0 \left\{ \exp\left(\frac{\alpha n F \eta}{RT} \right) - \exp\left(-\frac{(1-\alpha) n F \eta}{RT} \right) \right\} ; \quad \text{A/cm}^3 \qquad (3.99)$$

This volumetric current density is most useful in the analysis and simulation of the electrochemical reaction processes occurring in the electrode and the cell.

Based on their definitions, the electrode roughness factor and the specific reactive surface area are interrelated according to the following expression

$$A_v = \frac{a}{\delta}$$
(3.100)

where δ is the thickness of the active portion of the electrode, that is, it is the thickness of the catalyst layer.

3.4.5   Final Expression for the Butler–Volmer Equation

The exchange current density, as described in Section 3.4.1 and given in Equation (3.47), depends directly on the reactant concentration, C_R. It may be simply written as

$$J_0 = kC_R$$
(3.101)

For a practical fuel cell, the concentration of the reactant species at the reactive sites is influenced by many design and operation parameters and is most likely unknown. Whereas the exchange current density, as a fundamental parameter in electrochemistry, is usually measured for a known reactant concentration under a specific reference condition such as the operation temperature T. Therefore, we would have the exchange current density at the reference condition

$$J_{0,\text{ref}} = kC_{R,\text{ref}}$$
(3.102)

Since k is a proportionality constant, it is obtained as

$$k = \frac{J_{0,\text{ref}}}{C_{r,\text{ref}}}$$
(3.103)

Therefore, the exchange current density, Equation (3.101), becomes

$$J_0 = J_{0,\text{ref}} \frac{C_R}{C_{R,\text{ref}}}$$
(3.104)

In general, the reaction mechanism is difficult to determine and the reaction steps involved are often a summary of several or many elementary reactions. For such a globalized reaction, the exchange current density should be written as

$$J_0 = J_{0,\text{ref}} \left(\frac{C_R}{C_{R,\text{ref}}} \right)^{\gamma}$$
(3.105)

where γ is the reaction order with respect to the reactant R. If a number of species participates in the charge transfer reaction, then the concentration of each species should be included in this exchange-current density expression. Therefore, the generalized

Butler–Volmer equation typically used for fuel cell analysis and simulation becomes

$$J_{FC} = a J_{0,\text{ref}} \left(\frac{C_R}{C_{R,\text{ref}}} \right)^{\gamma} \left\{ \exp\left(\frac{\alpha n F \eta}{RT} \right) - \exp\left[-\frac{(1-\alpha)n F \eta}{RT} \right] \right\}; \quad \text{A/cm}^2$$

(3.106)

for the current density in terms of the geometrical electrode surface area and

$$J'_{FC} = A_v J_{0,\text{ref}} \left(\frac{C_R}{C_{R,\text{ref}}} \right)^{\gamma} \left\{ \exp\left(\frac{\alpha n F \eta}{RT} \right) - \exp\left[-\frac{(1-\alpha)n F \eta}{RT} \right] \right\}; \quad \text{A/cm}^3$$

(3.107)

for the current density expressed per-unit volume of the active electrode. Or in terms of the dimensionless entropy generation defined as

$$\wp_s^* = \frac{n F |\eta|}{RT} = \pm \frac{n F \eta}{RT}$$

(3.108)

where the plus sign for the anode reaction and the negative sign for the cathode reaction, the Butler–Volmer equation, Equations (3.106) and (3.107), becomes, respectively

$$J_{FC} = a J_{0,\text{ref}} \left(\frac{C_R}{C_{R,\text{ref}}} \right)^{\gamma} \left\{ \exp(\pm \alpha \wp_s^*) - \exp[\mp(1-\alpha)\wp_s^*] \right\}; \quad \text{A/cm}^2$$

(3.109)

$$J'_{FC} = A_v J_{0,\text{ref}} \left(\frac{C_R}{C_{R,\text{ref}}} \right)^{\gamma} \left\{ \exp(\pm \alpha \wp_s^*) - \exp[\mp(1-\alpha)\wp_s^*] \right\}; \quad \text{A/cm}^3$$

(3.110)

3.5   ACTIVATION POLARIZATION FOR SURFACE CHEMISORPTION

In the oxidation of hydrogen (H_2) and reduction of oxygen (O_2), the rate-determining step can be taken as the process of electron transfer, as described in the previous section. However, as mentioned at the beginning of the previous section, activation polarization at an electrode may also arise from the rate limiting step in the electrode reaction that is associated with the adsorption of reactant species on the electrocatalytic surface of the electrode (often called **reaction sites**). Surface adsorption reaction (or surface chemisorption) as the rate-determining step is more likely to be appropriate for the oxidation of hydrocarbon fuels than for hydrogen, that is, for the direct electrooxidation of methanol.

Because electron transfer reaction (ionization) is essential for the electric current production, although surface adsorption of reactant may be the rate limiting process, a two-step consecutive process is considered to occur at the electrode surface:

1. Adsorption of reactant on an electrode surface reaction site and its reverse reaction, desorption of reactant from the reaction site, occur simultaneously. This step is considered to occur very slowly, hence it is the rate-determining step.
2. Ionization of the adsorbed reactant at that reaction site (or electron transfer reaction). This step is considered to occur very fast, so that the reverse reaction, discharge of ion, may be neglected.

This two-step surface reaction may be represented as follows:

$$F(g) + M \overset{k_{1,f}}{\underset{k_{1,b}}{\rightleftharpoons}} F_{ad} \tag{3.111}$$

$$F_{ad} \overset{k_2}{\longrightarrow} F^{(+n)} + ne^- \tag{3.112}$$

if θ represents the fraction of electrode surface reaction sites that is covered by the adsorbed species, F_{ad}, at any particular time, then the rate of reaction for the formation of F_{ad} can be expressed as

$$\frac{d\theta}{dt} = k_{1,f} P_F (1 - \theta) - k_{1,b}\theta - k_2\theta \tag{3.113}$$

and the production of the current density, according to the charge transfer reaction of Equation (3.112) becomes

$$J = F\omega_{e^-}'' = nFk_2\theta \tag{3.114}$$

where P_F is the partial pressure of the gaseous species, F, representing the concentration of $F(g)$, k_2 is the reaction rate constant for the charge transfer reaction given in Equation (3.112), and is related to the change in the Gibbs function, as derived in Equation (3.64). On the other hand, the reaction rate constant for the adsorption and desorption reaction of Equation (3.111), $k_{1,f}$ and $k_{1,b}$, may be taken as a constant or to be dependent on the change in the Gibbs function, similar to the charge transfer reaction described in the previous section, thus leading to different result of varying complexities. We consider these two cases in the following two subsections.

3.5.1   Langmuir Model for Surface Chemisorption Reaction

The easiest and simplest assumption is that the reaction rate constants, $k_{1,f}$ and $k_{1,b}$, are constant, independent of surface coverage, θ. This is equivalent to assuming that during the adsorption and desorption process, all surface sites are equally reactive, irrespective of the surface coverage θ, and absorbed species can cover a surface until forming a complete unimolecular layer, or monolayer, after which further adsorption is no longer possible.

With $k_{1,f}$ and $k_{1,b}$ being taken as constant, the degree of surface coverage can be obtained by invoking the steady-state approximation for reaction intermediaries, or $d\theta/dt = 0$. Then, Equation (3.113) yields

$$\theta = \frac{k_{1,f} P_F}{k_{1,f} P_F + k_{1,b} + k_2} \tag{3.115}$$

Substituting θ into Equation (3.114), the current density becomes

$$J = \frac{n F k_2 k_{1,f} P_F}{k_{1,f} P_F + k_{1,b} + k_2} \tag{3.116}$$

In order to determine the electrode overpotential η that is needed to produce the current density shown above, the reaction rate constant, k_2, can be written, similar to Equation (3.64) along with Equation (3.62) for the charge transfer reaction

$$k_2 = BT \exp\left(\frac{\alpha n F \eta}{RT}\right) = \frac{J_0}{n F} \exp\left(\frac{\alpha n F \eta}{RT}\right) \tag{3.117}$$

where $BT = J_0/(nF)$ is the pre-exponential factor, and is a constant; and J_0 is the exchange current density when the surface coverage $\theta = 1$.

Substituting Equation (3.117) into Equation (3.116) and solving for η result in

$$\eta = \frac{RT}{\alpha n F} \ln\left[\frac{J(k_{1,f} P_F + k_{1,b})n F}{(n F k_{1,f} P_F - J) J_0}\right] \tag{3.118}$$

The Equation (3.118) clearly indicates that when a critical current density

$$J_L = n F k_{1,f} P_F \tag{3.119}$$

is reached, the electrode overpotential $\eta \rightarrow \infty$, which implies in practice that the cell potential is reduced to zero. Therefore, this critical current density represents the upper limiting value of the cell operation beyond which the cell is no longer operable and it is often called the limiting current density. Equation (3.119) indicates that the limiting current density J_L is reached when the surface coverage θ is reduced to zero due to the fast electron transfer reaction driven by a large electrode overpotential, so that as soon as the reactant F is adsorbed on the surface sites, it is consumed immediately by the electron transfer reaction of Equation (3.112) with negligible desorption reaction taking place.

Combining Equations (3.118) and (3.119) together yields

$$\eta = \frac{RT}{\alpha n F} \ln\left[\frac{J}{J_0} \frac{J_L + k_{1,b} n F}{(J_L - J)}\right] \tag{3.120}$$

since in general

$$\frac{J_L + k_{1,b} n F}{(J_L - J)} > 1$$

Equation (3.120) suggests that when comparing with the Tafel equation given by Equation (3.81), which was obtained for an electron transfer controlled process, the overpotential η for the adsorption process being the rate-determining step would increase faster, when the current density J is increased, than for the electron transfer being the rate-determining step.

It should be pointed out that the simplistic assumption of constancy for $k_{1,f}$ and $k_{1,b}$ in the Langmuir model is inadequate for many electrode reactions and a more complex model is required for proper description of the reactions in practice. One such model is described briefly in the next subsection.

3.5.2   Temkin Model for Surface Chemisorption Reaction

Similarly to the electron transfer reaction described in the precious subsection, the reaction rate constant for the adsorption and desorption process may be related to the Gibbs function change for the corresponding process, namely

$$k_{1,f} = B_f T \exp\left(-\frac{\Delta g_{ad}}{RT}\right) \tag{3.121}$$

$$k_{1,b} = B_b T \exp\left(-\frac{\Delta g_{de}}{RT}\right) \tag{3.122}$$

when Bs are the constant pre-exponential factor and Δg_{ad} and Δg_{de} are the Gibbs function change associated with the adsorption and desorption reaction, respectively. If Δg_{ad} and Δg_{de} are taken as constant, then the above equations essentially reduce to the Langmuir model.

However, as species become adsorbed, the fraction of surface covered by the adsorbed species is increased, and the associated Gibbs function change is increased as well, so that the continual process of the species adsorption becomes more difficult. For this case, it is the simplest to assume that the associated Gibbs function change depends on the surface coverage θ linearly, or

$$\Delta g_{ad} = a_{ad} + b_{ad}\theta \tag{3.123}$$

$$\Delta g_{de} = a_{de} + b_{de}(1 - \theta) \tag{3.124}$$

where a_{ad} and a_{de} are the Gibbs function change corresponding to $\theta = 0$ for adsorption process and $\theta = 1$ for the desorption process, respectively; and the a's and b's are constants, independent of the surface coverage θ.

Substitution of Equations (3.123) and (3.124) into Equations (3.121) and (3.122), respectively, yields

$$k_{1,f} = B_f T \exp\left(-\frac{a_{ad} + b_{bd}\theta}{RT}\right) \tag{3.125}$$

$$k_{1,b} = B_b T \exp\left(-\frac{a_{de} + b_{de}(1-\theta)}{RT}\right) \tag{3.126}$$

These two equations can be substituted into Equation (3.113) and the surface coverage θ can be determined by invoking the steady state approximation. Then combining with Equations (3.114) and (3.117), the electrode overpotential η can be obtained as a function of the current density generated. Due to the θ dependence of $k_{1,f}$ and $k_{1,b}$, η can no longer be expressed, analytically in a closed form, as a function of J.

However, when the surface coverage is very high ($\theta \to 1$) or very low ($\theta \to 0$), these results can be reduced to those of the Langmuir model described earlier. Temkin model has been found appropriate for the adsorption and desorption of carbon monoxide in polymer electrolyte membrane fuel cells.[5]

3.6   CELL POTENTIAL DISTRIBUTION AND ELECTRIC DOUBLE LAYER

Consider for the acid electrolyte fuel cell with the half-cell reaction given in Equations (3.1) and (3.2), proton (H^+) is produced at the anode and migrates through the electrolyte to the cathode to complete the electric circuit. Then a question arises: How could proton, which is positively charged, transfer from the anode at a lower electric potential to the cathode which has a higher potential, without external means?

Well, the answer may be very complicated. To put it simple, consider what happens in the electrolyte near the surface of an anode electrode as shown in Figure 3.11. Because an anode has a lower potential, positively charged ions tend to build up on the electrode surface. Thus between the electrode surface and the positive ions an equivalent capacitor is being set up, and the electrical field is very strong over the distance roughly on the order of the radius of the ions, which is very small, typically in the nm range. So that the electrical potential increases significantly over such a small distance. This small distance is often called **an electrical double layer** and the exact structure of the double layer is still under active research. Similarly near the cathode electrode, which has a higher potential, negatively charged ions tend to accumulate over its surface, with a correspondingly sharp drop in the potential over a small distance, again called the (electric) double layer. The end effect is that in the electrolyte the proton indeed is transported from a higher to a lower potential, as shown in Figure 3.12, when it migrates from the anode to the cathode. The corresponding potential drop across the electrolyte region follows the conventional Ohm's law, to be described in more detail in the next chapter.

Therefore, the actual potential distribution across the entire cell may be schematically depicted, as shown in Figure 3.12. The direction of the electric current I is

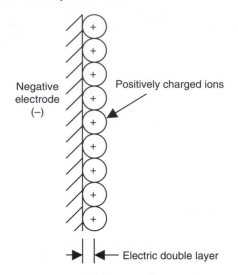

Figure 3.11 A simplified view of an electric double layer near the surface of a negative (anode) electrode.

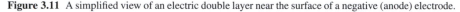
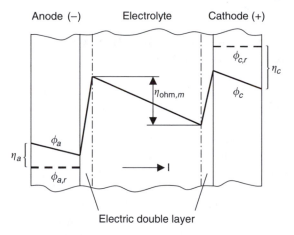

Figure 3.12 Schematic of potential distribution across the entire cell: solid line for the actual potential distribution and dashed lines for the reversible electrode potential. Also shown are the anode overpotential η_a, cathode overpotential η_c, and the ohmic overpotential in the electrolyte, $\eta_{ohm,m}$.

also indicated in the figure, along with the various overpotentials. Although the exact potential distribution in and around the double layers is still under active research, it may suffice to say that the function of the double layers is to enable the ion transfer through the electrolyte to complete the electric circuit. In this sense, it is very convenient to neglect the double layers, and to calculate the potential drop from the anode to the cathode. This potential drop across the entire cell represents the total voltage loss through the cell, including the anode and cathode activation overpotential as well as the ohmic overpotential through the electrode backing layers and the electrolyte region. Figure 3.13 illustrates the corresponding potential distribution when the double layers are neglected.[6]

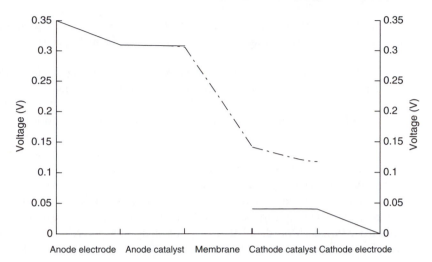

Figure 3.13 Potential drop across the entire cell when the electric double layers are not considered. The total cell potential drop is 0.35 V for the results shown here.[6]

3.7   ELECTRODE MECHANISMS IN FUEL CELLS

In this section, the major elementary reactions involved in the H_2–O_2 fuel cells are outlined because of their great practical significance and because almost all modern fuel cells are designed to employ H_2 and O_2 as the reactants, except direct methanol fuel cells. Both the anodic mechanism for hydrogen oxidation reaction and the cathodic mechanism for oxygen reduction reaction is described. It should be noted that the elementary steps involved in an electrode mechanism depends not only on the temperature, pressure and concentrations of the reactant, but maybe more importantly on the kind of electrode surface, the electrolyte (i.e., acid or alkaline), and the potential difference between the electrode surface and the bulk electrolyte. Due to the complexities involved, it is possible that a different set of elementary steps may be able to explain the same experimental observations and the complex electrode mechanisms believed to occur at the electrode surface may also change, or be improved over time as new observations become available and new insights are developed. For detailed discussions of electrode mechanisms, the reader should consult the original literature, reviews and more advanced texts with focus on the electrochemical kinetics such as.[1]

Furthermore, in this section, we only consider that H_2 and O_2 reactants are supplied from the gas phase to the electrode for reaction, excluding the situation of in-situ generation of H_2 and O_2 at the electrode as the reactant feeding mechanism, and limit our discussion to the presence of alkaline or acid electrolyte at the electrode.

3.7.1   Process at the Anode Electrode

The processes involved in the oxidation of molecular hydrogen to hydrogen ion are significantly easier to comprehend than the oxygen reduction reaction in the cathode electrode. This relative ease arises from the fact that the anode electrode potential is very low and remains almost constant (slow change) during the hydrogen oxidation process.

Based on the experimental observations, the processes occurring in the anode electrode are in general considered to involve mainly the following steps[3,7]:

1. Delivering of molecular hydrogen to the electrode surface (i.e., reaction sites): This step includes a number of physical and chemical processes preceding the electrochemical reaction and it generally includes the transport of molecular hydrogen from the gas phase supply outside the anode electrode structure to the liquid electrolyte surface, typically through the electrode structure. Then the dissolution of molecular hydrogen into the liquid electrolyte, followed by the diffusion through the electrolyte to the electrode surface and finally the adsorption on the electrode surface. More detailed description on the transport related phenomena is given in the next chapter. This step may be concisely expressed as follows:

$$H_2 \longrightarrow H_{2,dis} \longrightarrow H_{2,ad} \qquad (3.127)$$

where $H_{2,dis}$ represents the dissolved H_2 in the liquid electrolyte and $H_{2,ad}$ is the H_2 adsorbed on the electrode surface.

2. Hydration and ionization of the adsorbed molecular hydrogen at the electrode surface: This is the step for electrochemical reactions that produce electrons (i.e., electric current). As mentioned earlier, two reaction mechanisms are most likely to occur among the many possible elementary reactions:

    2.1. Dissociation of the adsorbed molecular hydrogen followed by the hydration and ionization of the adsorbed atomic hydrogen, or the so-called Tafel–Volmer reaction mechanism. This may be written as

$$H_{2,ad} \longrightarrow 2H_{ad} \quad \text{(Tafel reaction)} \qquad (3.128)$$

and

$$H_{ad} + OH^- \longrightarrow H_2O + e^- \quad \text{(Alkaline)} \qquad (3.129)$$

if the electrolyte is alkaline or

$$H_{ad} + H_2O \longrightarrow H_3O^+ + e^- \quad \text{(Acid)} \qquad (3.130)$$

if the electrolyte is acid.

    2.2. Hydration and ionization occur simultaneously, the so-called Heyrovsky–Volmer or Horiuti–Volmer mechanism. This may be written as

$$H_{2,ad} + OH^- \longrightarrow H_{ad} \cdot H_2O + e^- \longrightarrow H_{ad} + H_2O + e^- \quad \text{(Alkaline)} \qquad (3.131)$$

$$H_{ad} + OH^- \longrightarrow H_2O + e^- \quad \text{(Alkaline)} \qquad (3.132)$$

in the presence of alkaline electrolyte and

$$H_{2,ad} + H_2O \longrightarrow H_{ad} \cdot H_3O^+ + e^- \longrightarrow H_{ad} + H_3O^+ + e^- \quad (Acid)$$
$$(3.133)$$

$$H_{ad} + H_2O \longrightarrow H_3O^+ + e^- \quad (Acid) \qquad (3.134)$$

in the presence of acid electrolyte.

3. Removal of the reaction products (H_3O^+ or H_2O as well as e^-) from the electrode surface, thus the regeneration of the reaction sites. This includes the desorption of the products at the electrode surface and transfer of the products away from the surface into the electrolyte. The product H_3O^+ is transported through the acid electrolyte to the cathode to complete the electric circuit in acid electrolyte fuel cells, while H_2O is partly removed from the cell and partly transferred to the cathode for OH^- ion regeneration in alkaline electrolyte fuel cells. The electrons produced are transported through the anode electrode and the external circuit to the cathode, constituting the electric energy output from the cell reaction.

It might be pointed out again that the actual reaction pathways described in step (2) depends on the kind and surface structure of the electrode as well as the electrode operation conditions such as temperature, current drawn and electrode potential, and so on.

3.7.2   Process at the Cathode Electrode

The processes involved in the reduction reaction of molecular oxygen at the cathode electrode are much more complex than the hydrogen oxidation processes at the anode electrode described in Section 3.7.1. The associated complexity arises from the high cathode–electrode potential, the significant variations of the electrode potential during the oxygen reduction reaction, and the possibility of a number of simultaneous parallel and consecutive reaction mechanisms occurring at the cathode electrode.

From past studies and experimental observations, it is becoming clear[3,7] that hydrogen peroxide is the most important reaction intermediate in the oxygen reduction processes and it has also identified two overall reaction pathways in the oxygen reduction reaction at the presence of aqueous electrolytes. The identified overall pathways are the four-electron and peroxide pathways during the electrochemical reactions occurring at the cathode electrode, where the overall processes for the reduction of oxygen include the following major steps:

1. Delivery of molecular oxygen to the cathode electrode surface: This step involves a number of physical and chemical processes preceding the electrochemical reaction, and it generally includes the transport of molecular oxygen from the gas phase supply outside the cathode electrode structure to the liquid electrolyte surface, typically through the porous electrode structure, then the dissolution of molecular oxygen into the liquid electrolyte, followed by the diffusion of

the dissolved oxygen through the electrolyte–electrode surface, and finally the adsorption of the oxygen on the electrode surface. Or simply

$$O_2 \longrightarrow O_{2,\text{dis}} \longrightarrow O_{2,\text{ad}} \tag{3.135}$$

where $O_{2,\text{dis}}$ represents the dissolved O_2 in the liquid electrolyte and $O_{2,\text{ad}}$ the O_2 adsorbed on the electrode surface. As shown in the next chapter, the transport of molecular oxygen to the electrode surface is much more difficult, and hence much slower, than the corresponding process for the molecular hydrogen at the anode due to the much larger molecular size of the oxygen.

It might be pointed out that both H_2O (or H^+) and electrons are the reactants participating in the oxygen reduction reaction. H_2O (or H^+) is transported from the anode through the electrolyte and electrons are transferred from the anode through the external circuit. Therefore, the delivery of H_2O (or H^+) and electrons to the cathode reaction sites is as important as the delivery of oxygen itself.

2. Reduction of the adsorbed molecular oxygen at the electrode surface in the presence of aqueous electrolyte: This is the step for electrochemical reactions that combine with electrons produced at the anode and transported through the external circuit. The reactions only occur at the electrode surface that is covered by the electrolyte, hence, often referred to as the three phase boundary. Although the number of possible elementary reactions is larger than hydrogen oxidation process in the anode, the mechanism for the cathodic oxygen reduction reaction may involve the following two overall pathways:

   2.1. Four-Electron Pathway: The overall reaction for the reduction of oxygen in this mechanism is

$$O_2 + 2H_2O + 4e^- \longrightarrow 4OH^- \quad \phi_r = 0.401 \text{ V} \tag{3.136}$$

   in alkaline solutions, and

$$O_2 + 4H^+ + 4e^- \longrightarrow 2H_2O \quad \phi_r = 1.229 \text{ V} \tag{3.137}$$

   in acid solutions, where ϕ_r indicates the reversible electric potential, the reaction creates with respect to the reference hydrogen electrode.

   2.2. Peroxide Pathway: For this pathway, the following reactions occur in alkaline solutions

$$O_2 + H_2O + 2e^- \longrightarrow HO_2^- + OH^- \quad \phi_r = -0.065 \text{ V} \tag{3.138}$$

$$HO_2^- + H_2O + 2e^- \longrightarrow 3OH^- \quad \phi_r = 0.867 \text{ V} \tag{3.139}$$

$$HO_2^- \longrightarrow OH^- + \tfrac{1}{2}O_2 \quad \text{Decomposition reaction} \tag{3.140}$$

$$M - HO_2^- \longrightarrow OH^- + M - O \quad \text{Intermediate reaction} \tag{3.141}$$

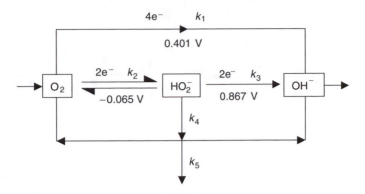

Figure 3.14 Global mechanism of the cathodic oxygen reduction in alkaline electrolyte. (From L.J.M.J. Blomen and M.N. Mugerwa, 1993. *Fuel Cell Systems*, New York: Plenum Press. With permission.)

whereas in acid solution the following reactions occur

$$O_2 + 2H^+ + 2e^- \longrightarrow H_2O_2 \qquad \phi_r = 0.67\,V \qquad\qquad (3.142)$$

$$H_2O_2 + 2H^+ + 2e^- \longrightarrow 2H_2O \qquad \phi_r = 1.77\,V \qquad\qquad (3.143)$$

$$H_2O_2 \longrightarrow H_2O + \tfrac{1}{2}O_2 \qquad \text{Decomposition reaction} \quad (3.144)$$

$$M - H_2O_2 \longrightarrow H_2O + M - O \qquad \text{Intermediate reaction} \qquad (3.145)$$

The reaction intermediate, the chemisorbed oxygen, M–O, can be desorbed via the reaction

$$2M - O \longrightarrow 2M + O_2 \qquad\qquad\qquad (3.146)$$

and the desorbed molecular oxygen, O_2, can be reduced again via these two overall reaction mechanisms.

3. Removal of the reaction products (OH^- or H_2O) from the electrode surface for the regeneration of the reaction sites and for the continuous reduction of molecular oxygen. This is especially important for low-temperature acid electrolyte fuel cells because product water (H_2O) is formed at the cathode and is in the liquid form. Liquid water accumulation in the porous electrode structure may block the transport of molecular oxygen to the reaction sites, severely hindering the process described in step 1, thereby degrading the cell performance due to oxygen starvation at the reaction sites. Such a phenomenon is often referred to as the water flooding of electrodes.

The global mechanism of the electroreduction of oxygen described in step 2 above may be represented concisely in Figures 3.14 and 3.15 for alkaline and acid electrolyte, respectively. The direct four-electron mechanism may involve many elementary reaction steps with an adsorbed peroxide as the reaction intermediate. However, peroxide does not desorb and show its presence in the electrolyte solution. On the other hand, the peroxide mechanism shown in step (2.2) involves peroxide species that are present in the electrolyte solution.

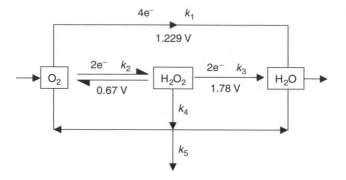

Figure 3.15 Global mechanism of the cathodic oxygen reduction in acid electrolyte. (From L.J.M.J. Blomen and M.N. Mugerwa, 1993. *Fuel Cell Systems*, New York: Plenum Press. With permission.)

It may be cautioned that the oxygen reduction mechanism is very complex, involving many parallel and consecutive elementary reactions and it also depends on the electrolyte and the electrode involved. Table 3.7 provides a summary of reaction pathways for oxygen reduction,[3,8] indicating the complexity and the possibility of various reduction reaction mechanisms. Thus, readers are referred to current research result for the latest development in this important area.

3.8   ELECTROCATALYSIS

Electrocatalysis means the facilitation or acceleration of an electrode reaction by a substance without any change in its physical and chemical properties in the overall reaction. Such a substance is often referred to as electrocatalyst or simply a catalyst. A catalyst is used to enhance the rate and selectivity of a particular reaction and it is regenerated cyclically during the reaction such that it can be used repeatedly and continuously for the acceleration of the reaction. In fuel cells, a catalyst is normally made part of the electrode surface or an electrode surface is usually made active catalytically.

It should be noted that the catalytic properties of a catalyst, that is, the reaction rate enhancement and the selectivity of (or preference for) a particular reaction, are also affected by the kind of electrolyte present, such as alkaline or acid solution described in the previous section. This is because the catalytical effect occurs at the catalyst surface wetted by electrolyte (or at the catalyst–electrolyte interface), where the reaction proceeds. Thus, electrocatalysis can be considered a specific type of heterogeneous catalysis whereby reactants and products adsorb onto the surface of the catalyst during the reaction process. Adsorbed reactants, activated by interaction with the catalyst surface, are rapidly and selectively transformed to adsorbed products. Finally, the adsorbed products leave the catalyst surface in a desorption step, cleaning the surface for reactant adsorption, and the next cycle of reaction thus continues. The reactant adsorption and product desorption process corresponds to step 1 and 3 for the processes occurring at the anode and cathode electrode, respectively, described in Section 3.7.2.

Table 3.7   A Summary of Reaction Pathways for Cathodic Oxygen-Reduction Reaction[3,8]

	Acid[a]	Base[b]
1. The "Oxide" Path		
1. $O_2 + 2M \rightarrow 2MO$	—	—
2. $MO + MH_2O \rightarrow MOH$	—	—
3. $MOH + H^+ + e \rightarrow M + H_2O$	Au	—
2. The "Electrochemical Oxide" Path		
1. $O_2 + 2M \rightarrow 2MO$	—	—
2. $MO + MH_2O + H^+ + e \rightarrow MOH + M + H_2O$	Au	—
3. $MOH + H^+ + e \rightarrow M + H_2O$	—	—
3. The "Hydrogen Peroxide" Path		
1. $O_2 + M + MH_2O \rightarrow MOH + MO_2H$	—	—
2. $MOH_2 + MO_2H \rightarrow MOH + MH_2O_2$	Pt	Pt
3. $M + MH_2O_2 \rightarrow 2MOH$	—	—
4. $MOH + H^+ + e \rightarrow M + H_2O$	Au	—
4. The "Metal Peroxide" Path		
1. $O_2 + M + MH_2O \rightarrow MHO_2 + MOH$	—	—
2. $M + MHO_2 \rightarrow MO + MOH$	Pt	Pt
3. $MO + MH_2O \rightarrow 2MOH$	—	—
4. $MOH + H^+ + e \rightarrow M + H_2O$	Au	—
5. The "Electrochemical Metal Peroxide" Path		
1. $O_2 + M + MH_2O \rightarrow MOH + MHO_2$	—	—
2. $MHO_2 + H^+ + e \rightarrow MO + H_2O$	—	—
3. $MO + MH_2O \rightarrow 2MOH$	—	—
4. $MOH + H^+ + e \rightarrow M + H_2O$	Pt	—
6. Hoar Alkaline Path		
1. $M + O_2 + 2e \rightarrow MO_2^{-2}$	Au	—
2. $M + MO_2^{-2} + 2H_2O \rightarrow 2MH_2O_2^-$	Pt	Pt
3. $MH_2O_2^- \rightarrow MOH + OH^-$	—	—
4. $MOH + e \rightarrow M + OH^-$	—	—
7. Conway and Bourgault Reaction Path		
1. $M + MH_2O + O_2 \rightarrow MHO_2 + MOH$	—	—
2. $MHO_2 \rightarrow MOH + MO$	Pt	Pt
3. $MO + H^+ + e \rightarrow MOH$	—	—
4. $MOH + H^+ + e \rightarrow M + H_2O$	Pt	—
8. Alternative Conway and Bourgault Reaction Path		
1. $M + MH_2O + O_2 \rightarrow MOH + MHO_2$	—	—
2. $MHO_2 + H^+ + e \rightarrow MO + H_2O$	—	—
3. $MO + H^+ + e \rightarrow MOH$	—	—
4. $MOH + H^+ + e \rightarrow MO + H_2O$	—	—
9. Riddiford Path		
1. $O_2 + MH_2O + H^+ + e \rightarrow MHO_2 + H_2O$	Au	—
2. $MHO_2 + H^+ + e \rightarrow MO + H_2O$	—	—

(Continued)

Table 3.7   (*Continued*)

	Acid[a]	Base[b]
3. $MO + MH_2O \rightarrow 2MOH$	—	—
4. $MOH + H^+ + e \rightarrow M + H_2O$	—	—
10. Krasilshchikov Path (Ni electrodes)		
1. $O_2 + 2M \rightarrow 2MO$	—	—
2. $MO + e \rightarrow MO^-$	—	—
3. $MO^- + H^+ \rightarrow MOH$	—	—
4. $MOH + H^+ + e \rightarrow M + H_2O$	—	—
11. Wade and Hackerman's Path		
1. $O_2 + 2e + 2M + MH_2O \rightarrow 2MOH^- + MO$	—	—
2. $MO + MH_2O + 2e \rightarrow 2MOH^-$	—	—
12. 1. $O_2 + H^+ + e + M \rightarrow MO_2H$	Au	—
2. $MO_2H + H^+ + e \rightarrow MO + H_2O$	—	—
3. $MO + H^+ + e \rightarrow MOH$	—	—
4. $MOH + H^+ + e \rightarrow M + H_2O$	—	—
13. 1. $M + O_2 \rightarrow 2MO$	—	—
2. $MO + H_2O \rightarrow MO - H - OH$	—	—
3. $MO - H - OH + e \rightarrow MO - H - OH^-$	—	—
4. $MO - H - OH^- + H^+ \rightarrow MOH + H_2O$	—	—
5. $MOH + H^+ + e \rightarrow M + H_2O$	—	—
14. 1. $O_2 + H^+ + e + M \rightarrow MHO_2$	Au	—
2. $MHO_2 + e \rightarrow MO + OH^-$	—	—
3. $MO + H_2O \rightarrow MO - H - OH$	—	—
4. $MO - H - OH + e \rightarrow MO - H - OH^-$	—	—
5. $MO - H - OH^- \rightarrow MOH + OH^-$	—	—
6. $MOH + H^+ + e \rightarrow M + H_2O$	—	—
15. Hoare Path		
1. $O_2 \text{ (aq)} \rightarrow MO_2$	—	—
2. $MO_2 + e \rightarrow MO_2^-$	—	—
3. $MO_2^- + H^+ \rightarrow MHO_2$	—	—
4. $MHO_2 + e \rightarrow MHO_2^-$	—	Pt, Au
5. $MHO_2^- + H^+ \rightarrow MH_2O_2$	—	—
6. $2H_2O_2 \xrightarrow{\text{(catalytic)}} 2H_2O + O_2 \text{ ads)}$	—	—
16. [Quoted by Ives]		
1. $M + O_2 + e \rightarrow MO_2^-$	—	—
2. $MO_2^- + H^+ \rightarrow MO_2H$	—	—
3. $MO_2H + e \rightarrow MO_2H^-$	—	Pt, Au
4. $MO_2H^- + H^+ \rightarrow MH_2O_2$	—	—
5. $MH_2O_2 + e \rightarrow MOH + OH^-$	—	—
6. $MOH + e \rightarrow M + OH^-$	—	—

[a] Rate-determining step for Pt and Au in H_2SO_2.
[b] Rate-determining step for Pt and Au in NaOH.

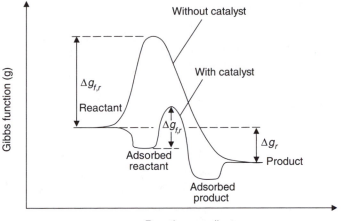

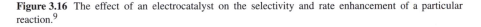

Reaction coordinate

Figure 3.16 The effect of an electrocatalyst on the selectivity and rate enhancement of a particular reaction.[9]

Figure 3.16 illustrates graphically the effect of an electrocatalyst on the reaction rate enhancement and the selectivity favoring a particular reaction, in analogy with Figure 3.6. The catalyst lowers the Gibbs function of activation by providing a surface on reaction site for adsorption and dissociation of the reactants, whereby the adsorbed reactants are much more easily reacted to form the products. The difference between the Gibbs function of the products and the reactants is the Gibbs function of reaction, Δg_r, and it is not changed by the presence of the catalyst. That is, the catalyst does not change the Gibbs function of the reactants and the products, rather it reduces the Gibbs function of activation for the particular reaction favored, thus enhancing the rate of and the preference for the reaction. Clearly, the catalyst only facilitates the thermodynamically favored reaction, but it cannot make a thermodynamically unfavorable reaction to occur.

As discussed earlier, the catalyzed electrochemical reaction occurs at the interface between the catalyst surface and the electrolyte, the rate of reaction depends on the catalyst surface-electrolyte potential difference, as well as the kind and surface morphology of the catalyst. The effect of the catalyst materials is clearly shown in Table 3.1, where the exchange current density, a measure of kinetic activity, for hydrogen–electrode reaction changes by as much as 10 orders of magnitude from the mercury-to-platinum electrode.

The catalytic effect of a catalyst can be correlated with the enthalpy of adsorption, Δh_{ad}, of a reactant on the catalyst surface, as shown in Figure 3.17 for H_2. Such a plot is often called a Volcano curve. A small value of Δh_{ad} indicates a very slow adsorption kinetics, and the adsorption process becomes the rate-determining step, limiting the rate of the overall reaction. A large value of Δh_{ad} will affect the desorption process, thereby limiting the overall reaction. Consequently, an intermediate value of Δh_{ad} would indicate a good catalytic effect, like platinum shown in Figure 3.17.

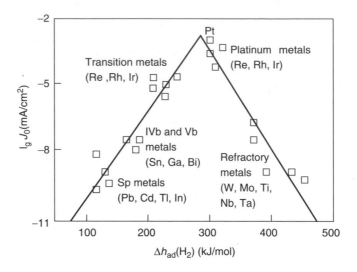

Figure 3.17 Volcano curve in the form of hydrogen (H_2) exchange current density J_0 as a function of the enthalpy of adsorption of hydrogen on various catalysts.[10]

The selectivity of a catalyst may be illustrated with the oxygen reduction-reaction mechanism described in the previous section. For example, the four-electron reaction pathway appears to be favored on Noble-metal electrocatalysts (i.e., Pt, Pd, Ag), metal oxides (such as perovskites, pyrochlores), and some transition-metal macrocycles. On the other hand, the peroxide pathway is predominant on graphite, carbon, gold, mercury, most oxide-covered metals such as Ni and Co, most transition-metal oxides (e.g., NiO), and some transition-metal macrocycles.

Regular transition elements are good candidate as electrocatalysts, especially, Fe, Co, Ni, and platinum metals which include Ru, Rh, Pd, Os, Ir, and Pt because these six elements resemble each other very closely. These elements are characterized by having an inner d level of electron energy levers not filled to capacity with electrons, and the unfilled inner d orbitals are readily available for bond formation with adsorbed species, the so-called d character. Thus, these elements are good substrates for adsorption. However, the best-known electrocatalysts may be the platinum metals, especially platinum, or the commonly called Noble metals.

Electrocatalysts are required for low-temperature fuel cells in order to achieve a reasonably fast reaction for both anode and cathode (i.e., a reasonable power output). For hydrogen–oxygen fuel cells, platinum or its alloys are typically used as catalysts for both anode and cathode reactions, whether the electrolyte is acid or alkaline. In addition, nickel and silver are sometimes used as effective catalyst for cathode oxygen reduction reaction in alkaline electrolytes. However, platinum can be easily poisoned by trace amounts of carbon monoxide and sulfur compounds, thus losing its effectiveness as a catalyst. On the other hand, non-Noble metal catalysts are more tolerant to these impurities. In general, a high temperature operation tends to alleviate the effect of poisoning, cell operation temperature may be limited due to other reasons for specific type of fuel cells.

High catalytic activity of a catalyst is usually achieved by a high catalyst surface area per-unit mass of the catalyst, which is in turn made possible by reducing the particle size of the powdered catalyst. The catalyst particle size commonly ranges from submicrons to nanometers. To maintain adequate catalyst layer porosity for the catalyst surface to remain wet by the electrolyte and accessible to the reactants, an electrocatalyst is usually made of at least two components: an active catalyst of fine size deposited on a high-surface-area carrier, often called support. Such a catalyst is often referred to as supported catalyst. Catalyst supports are porous, with significant port volume and capacity to maintain stable well-dispersed fine catalyst particles. Activated carbon is commonly used as support in low-temperature fuel cells. Often a third component, called a **promoter**, is used to increase the activity and/or stability of the catalyst, such as polytetrafluroethylene (PTFE) in low-temperature fuel cells for bondage and hydrophobicity, which is needed to prevent the liquid electrolyte from flooding the electrode structure or for water removal to prevent water flooding phenomena.

3.9   SUMMARY

This chapter is focused on the kinetics of electrochemical reactions occurring at the electrode surface, directly relevant to the electrode reactions in fuel cells. The potential of an electrode is defined first along with the various polarizations occurring in practical operating fuel cells, followed by an introduction of various fundamental concepts associated with the heterogeneous electrochemical reactions encountered in fuel cells, such as global and elementary reactions, the rate of reaction, the surface coverage, the rate-determining reaction step, steady-state approximation for reaction intermediaries, and so on. This paves the way for the description of activation polarization associated with the charge transfer reaction as the rate-determining step, the analysis under the reversible condition leads to the concept of exchange current density, and the irreversible reaction results in the Butler–Volmer equation for the rate of charge transfer reaction, an important relation to be remembered in the study of fuel cells. Tafel equation is shown as a special form of the Butler–Volmer equation at relatively large activation overpotentials. Then, the production of net current density from the charge transfer reaction is shown to be directly related to the production of entropy — a measure of thermodynamic irreversibility. The Butler–Volmer equation is derived for a flat electrode surface, and is converted to a form incorporating the porous nature of catalyst layer in practical fuel cells, then the rate of production for the volumetric current density in fuel cell catalyst layers is provided, and the concentration dependence and an overall reaction order are introduced for the Butler–Volmer equation. Activation polarization for surface chemisorption reaction as the rate-determining step is also described briefly with Langmuir and Temkin kinetics. The charge transfer reaction considered earlier may be viewed as the one-step mechanism for the generation of current as a result of the electrode reaction, while the surface chemisorption reaction is a three-step mechanism for the current production. Section 3.6 describes briefly the distribution of electric potential across an entire fuel cell, that brings about the concept of electric double layer, which is mainly introduced in the present context as the driving mechanism for the ion transport through the electrolyte. Section 3.7 provides an overview of the mechanism of anode and cathode electrode reaction processes in a fuel cell, taking examples of H_2 oxidation at the anode and O_2 reduction reaction at the cathode with the presence of either alkaline or acid electrolyte at the electrode surface. The complex reaction of cathodic reduction of oxygen is discussed mainly in terms of either four-electron or two-electron pathway as the primary mechanism. Finally, the concept of electrocatalysis for the acceleration of heterogeneous electrode reaction is touched upon in Section 3.8, a typical Volcano curve is introduced as the means of determining the effectiveness of a particular catalyst for fuel cell application, and the need for effective catalysts in fuel cells becomes clear due to the small values of the exchange current densities described in this chapter. It is hoped that the readers can now understand the basic concepts, ideas and terminologies of electrochemical kinetics that are often encountered in fuel cell literature.

BIBLIOGRAPHY

1. Bockris, J. O'M. and S. Srinivasan. 1969. *Fuel Cells: Their Electrochemistry*. New York: McGraw-Hill.
2. McDougall, A. 1976. *Fuel Cells*. London: MacMillan.
3. Kordesch, K. and G. Simader. 1996. *Fuel Cells and their Applications*. New York: VCH.
4. Liebhafsky, H. A. and E. J. Cairns. 1968. *Fuel Cells and Fuel Batteries*. New York: Wiley.
5. Baschuk, J. J. and X. Li 2003. *Int. J. En. Res.*, 27: 1095–1116.
6. Hum, B. and X. Li. 2004. *J. Appl. Electrochem.*, 34: 205–215.
7. Blomen, L. J. M. J and M. N. Mugerwa 1993. *Fuel Cell Systems*. New York: Plenum Press.
8. Gnanamuthu, D. S. and J. V. Petrocelli 1967. *J. Electrochem. Soc.*, 114: 1036.
9. Somorjai, G. A. 1994. *Introduction to Surface Chemistry and Catalysis*. New York: Wiley.
10. Kivisari, J. 1995. Fuel Cell lecture notes.

PROBLEMS

3.1   Describe how the electrode potential and overpotential are defined.

3.2   What are the various forms of irreversible losses for the cell voltage?

3.3   Discuss the concept of surface coverage and its usage.

3.4   What is activation polarization? Explain the cause of the activation polarization.

3.5   What is the rate-determining step in electrode reactions?

3.6   State the concept of steady-state approximation for reaction intermediaries, its validity, and limitation in practice.

3.7   Describe Faraday's law for electrochemical reactions.

3.8   Describe the various representations of an electrode reaction and point out which form is more suitable for writing an expression for the rate of reaction.

3.9   List all conditions under which the Bulter–Volmer equation is valid and derive the equation step by step by considering the reversible and irreversible charge transfer-reaction process. Describe the activation overpotential-current density relation for both anode and cathode.

3.10   What is Tafel equation? What is the Tafel slope? Describe the condition under which Tafel equation is valid.

3.11   List some important reasons for the use of porous electrodes in fuel cells.

3.12   Describe surface chemisorption reaction, the Langmuir and Temkin kinetics.

3.13   Following the derivation presented in this chapter, use the Temkin kinetics to develop equations that govern the activation overpotential as a function of the current density.

3.14   Describe the origin of concentration polarization.

3.15   Sketch the electrical potential distribution throughout a cell.

3.16   What is electric double layer in the electrolyte?

3.17   Describe the main reaction steps for hydrogen–oxidation reaction at the anode.

3.18   Describe the main reaction steps for oxygen reduction reaction at the cathode. What are the four-electron and two-electron mechanisms for the oxygen reduction reaction at the presence of acid and alkaline aqueous electrolyte?

3.19   What is the electrocatalysis and electrocatalyst? Describe the Volcano curve.

TRANSPORT PHENOMENA IN FUEL CELLS

4.1 INTRODUCTION

Transport of mass, heat, and momentum as well as electricity occurs simultaneously in fuel cells and the rates of these transport processes can have considerable, sometimes predominant, impact on the performance of the fuel cells. Although the generation of electric energy in fuel cells occurs through electrochemical reactions at electrodes, as described in Chapter 3, the reactants must be supplied to and the products must be removed from the sites of electrochemical reactions in order to maintain the continuous and steady generation of electricity. Similarly, the electricity generated and the waste heat accompanying the electricity generation and transport must be taken away properly as well. Therefore, the transport processes described here are equally important, as the electrochemical reactions, in determining the fuel cell performance. Increasing output power can be achieved by increasing the current density, which is in turn achieved by increasing the rates of the electrochemical reactions at the electrodes, as long as the rate of reactant supply and product removal can meet the increasing demand of the electrode reactions. Otherwise, a limiting (or maximum) current density will be approached and attempts to increase the output power further will fail, as in the region of concentration polarization in the cell polarization curve.

The phenomena associated with the transport of mass, heat, momentum, and electricity are quite complex and interlinked, involving multicomponent, multiphase and multi-dimensional transport processes. Thus, we are interested in presenting these phenomena in a manner as simple as possible, while retaining the essential physics involved. The objective of this chapter is to present the transport processes of mass, heat, momentum, and electricity, their fundamental rate laws as well as their impact on the cell performance; while the general equations governing their transports based on the conservation of mass, species, momentum and energy as well as the electrical charge are not described due to limitations and the complexities associated with the multicomponent and multiphase transport processes. Similarity and difference for these transport processes are also examined, and examples will be given to illustrate the effect on the fuel cell performance.

Through this chapter of studies on the transport processes involved in fuel cells, a better and more complete understanding of fuel cells can be developed and equations that are useful for calculating operation and design parameters are available.

4.2   SOME BASIC DEFINITIONS FOR MULTICOMPONENT MIXTURES

Consider a multicomponent mixture of volume $\mathcal{V}$, consisting of a total of N species. The concentration of a species i in such a mixture may be described conveniently by any one of the following methods:

- *Molar concentration C_i* defined as

$$C_i = \frac{n_i}{\mathcal{V}}; \quad \text{mole/m}^3 \tag{4.1}$$

or the number of moles for the species i per-unit volume.
- *Mass concentration ρ_i* defined as

$$\rho_i = \frac{m_i}{\mathcal{V}}; \quad \text{kg/m}^3 \tag{4.2}$$

or the mass for the species i per-unit volume, which is more commonly referred to as the partial density of the species i, and

$$\rho_i = C_i W_i \tag{4.3}$$

where W_i is the molecular weight of the species i.
- *Mole fraction X_i* defined as

$$X_i = \frac{n_i}{n} = \frac{C_i}{C} \tag{4.4}$$

which is the ratio of the number of the moles for the species i, n_i, to the total number of moles for all species in the mixture, $n = \sum_{i=1}^{N} n_i$. Similarly, $C = \sum_{i=1}^{N} C_i$ is the total molar concentration of all the species in the mixture.
- *Mass fraction Y_i* defined as

$$Y_i = \frac{m_i}{m} = \frac{\rho_i}{\rho} \tag{4.5}$$

which is the ratio of the mass for the species i, m_i, to the total mass in the mixture, $m = \sum_{i=1}^{N} m_i$. Similarly, $\rho = \sum_{i=1}^{N} \rho_i$ is the total mass concentration (or density) of all the species in the mixture.

From these definitions, the following identities are valid for the mixture:

$$\sum_{i=1}^{N} X_i = 1 \quad \text{and} \quad \sum_{i=1}^{N} Y_i = 1 \tag{4.6}$$

The molecular weight of the mixture may be expressed as

$$W_{\text{mix}} = \sum_{i=1}^{N} X_i W_i = \left[\sum_{i=1}^{N} \frac{Y_i}{W_i} \right]^{-1} \tag{4.7}$$

and the mass fraction is related to the mole fraction as follows

$$Y_i = X_i \left(\frac{W_i}{W_{\text{mix}}} \right) \quad \text{or} \quad X_i = Y_i \left(\frac{W_{\text{mix}}}{W_i} \right) \tag{4.8}$$

4.2.1   Average and Diffusion Velocity

Let $\mathbf{v}_i$ represents the absolute velocity of the species i with respect to a stationary coordinate and realize that each species in the mixture may have different absolute velocity. Then an average velocity for all the species in the mixture can be defined; however, depending on how the averaging is performed, we have

- Mass-average velocity $\mathbf{v}$ defined as

$$\mathbf{v} = \frac{\sum_{i=1}^{N} \rho_i \mathbf{v}_i}{\sum_{i=1}^{N} \rho_i} = \sum_{i=1}^{N} Y_i \mathbf{v}_i \tag{4.9}$$

or
- Molar-average velocity $\mathbf{v}^*$ defined as

$$\mathbf{v}^* = \frac{\sum_{i=1}^{N} C_i \mathbf{v}_i}{\sum_{i=1}^{N} C_i} = \sum_{i=1}^{N} X_i \mathbf{v}_i \tag{4.10}$$

where the bold symbols represent vector quantities.

Clearly the mass-average velocity $\mathbf{v}_i$ is averaged by the proportional amount of mass for each species in the mixture, while the molar-average velocity $\mathbf{v}^*$ is averaged by the proportional amount of the number of moles for each species in the mixture. In general, $\mathbf{v}$ is used in fluid mechanics by convention. It is important to note that the two average velocities, $\mathbf{v}$ and $\mathbf{v}^*$, are not equal, unless the molecular weight is identical for all the species in the mixture.

Species diffusion is the relative motion of a species with respect to the average (mean) motion of the mixture as a whole. Since the average motion of the mixture can be represented by either the mass-average velocity $\mathbf{v}$ or molar-average velocity $\mathbf{v}^*$, diffusion velocity of a species i in the mixture can be defined as follows

- mass-diffusion velocity:

$$\mathbf{V}_i = \mathbf{v}_i - \mathbf{v} \tag{4.11}$$

- molar-diffusion velocity:

$$\mathbf{V}_i^* = \mathbf{v}_i - \mathbf{v}^* \tag{4.12}$$

Therefore, the motion of the species i with respect to a stationary coordinate is contributed by the average (mean) flow, relative to the stationary coordinate, represented by either $\mathbf{v}$ or $\mathbf{v}^*$, and the relative motion of the species i with respect to the local mean motion of the fluid stream, represented by $\mathbf{V}_i$ or $\mathbf{V}_i^*$. The former (mean motion) is conventionally called **convection** or **bulk motion**, while the latter is called **diffusion**.

4.2.2   Rate of Mass Transfer

The rate of mass transfer, or the rate of the transport for species i in the mixture, is often expressed in terms of the amount of mass (or mole) transferred per-unit time and per-unit surface area that is normal to the direction of the mass transfer or the mass (molar) fluxes. Thus, the total mass and molar fluxes of the species i relative to the stationary coordinate become

$$\dot{\mathbf{m}}_i^{''} = \rho_i \mathbf{v}_i; \quad kg/(s \cdot m^2) \quad - \text{total mass flux} \tag{4.13}$$

$$\dot{\mathbf{N}}_i^{''} = C_i \mathbf{v}_i; \quad mol/(s \cdot m^2) \quad - \text{total molar flux} \tag{4.14}$$

Similarly, the mass and molar fluxes of the species i relative to the local mean fluid motion, called the **diffusional mass (or molar) fluxes**, are defined as

$$\dot{\mathbf{m}}_{i,d}^{''} = \rho_i \mathbf{V}_i; \quad kg/(s \cdot m^2) \quad - \text{diffusional mass flux} \tag{4.15}$$

$$\dot{\mathbf{N}}_{i,d}^{''} = C_i \mathbf{V}_i^*; \quad mol/(s \cdot m^2) \quad - \text{diffusional molar flux} \tag{4.16}$$

The mass and molar fluxes due to the bulk motion can be defined similarly as follows

$$\dot{\mathbf{m}}_{i,b}^{''} = \rho_i \mathbf{v}; \quad kg/(s \cdot m^2) \quad - \text{convectional mass flux} \tag{4.17}$$

$$\dot{\mathbf{N}}_{i,b}^{''} = C_i \mathbf{v}^*; \quad mol/(s \cdot m^2) \quad - \text{convectional molar flux} \tag{4.18}$$

From the diffusion velocity defined in Equations (4.11) and (4.12), the total mass and molar fluxes can be written as

$$\dot{\mathbf{m}}_i'' = \rho_i(\mathbf{v} + \mathbf{V}_i) = \rho_i\mathbf{v} + \rho_i\mathbf{V}_i = \dot{\mathbf{m}}_{i,b}'' + \dot{\mathbf{m}}_{i,d}'' \tag{4.19}$$

$$\dot{\mathbf{N}}_i'' = C_i(\mathbf{v}^* + \mathbf{V}_i^*) = C_i\mathbf{v}^* + C_i\mathbf{V}_i^* = \dot{\mathbf{N}}_{i,b}'' + \dot{\mathbf{N}}_{i,d}'' \tag{4.20}$$

Clearly, the total rate of mass transfer $\dot{\mathbf{m}}_i''$ or $\dot{\mathbf{N}}_i''$ is contributed by the amount of mass carried by the bulk motion of the fluid as well as the relative motion of the species with respect to the bulk motion, or simply put, by the convection and diffusion effect. The bulk fluid motion can be determined by the fluid continuity and the Newton's second law of motion, often called the conservation of mass and momentum equations, as described in fluid mechanics. Therefore, we describe the mass transfer by diffusion process in the next section.

The total rate of mass transfer for the mixture (i.e., for all the species in the mixture) is often defined as

$$\dot{\mathbf{m}}'' = \sum_{i=1}^{N} \dot{\mathbf{m}}_i'' \quad - \text{ total mass flux} \tag{4.21}$$

$$\dot{\mathbf{N}}'' = \sum_{i=1}^{N} \dot{\mathbf{N}}_i'' \quad - \text{ total molar flux} \tag{4.22}$$

Considering the definitions given in Equations (4.13) and (4.14) with the average velocities in Equations (4.9) and (4.10), the previous two equations become

$$\dot{\mathbf{m}}'' = \rho\mathbf{v} \quad - \text{ total mass flux} \tag{4.23}$$

$$\dot{\mathbf{N}}'' = C\mathbf{v}^* \quad - \text{ total molar flux} \tag{4.24}$$

which are more familiar to most of readers.

On the other hand, from the definitions of $\dot{\mathbf{m}}_{i,d}''$, $\mathbf{V}_i$, and $\mathbf{v}$, we have

$$\dot{\mathbf{m}}_{i,d}'' = \rho_i\mathbf{V}_i = \rho_i(\mathbf{v}_i - \mathbf{v}) = \rho_i\mathbf{v}_i - \frac{\rho_i}{\rho}\sum_{j=1}^{N}\rho_j\mathbf{v}_j$$

and from the definitions of $\dot{\mathbf{m}}_i''$ and Y_i, we can simplify the previous equation to

$$\dot{\mathbf{m}}_{i,d}'' = \dot{\mathbf{m}}_i'' - Y_i\sum_{j=1}^{N}\dot{\mathbf{m}}_j'' \tag{4.25}$$

where the first term on the right-hand side of the Equation (4.25) is the total mass flux for the species i and the second term on the right-hand side is the mass flux of the species i carried by the bulk fluid motion, $\dot{\mathbf{m}}_{i,b}''$. Similarly, we can derive for the molar flux

$$\dot{\mathbf{N}}_{i,d}'' = \dot{\mathbf{N}}_i'' - X_i\sum_{j=1}^{N}\dot{\mathbf{N}}_j'' \tag{4.26}$$

Example 4.1   A mixture consisting of hydrogen and water vapor has the composition in terms of the mole fraction of $X_{H_2} = 0.80$ and $X_{H_2O} = 0.20$. If the velocity of hydrogen and water vapor is in the same direction, with the magnitude of $v_{H_2} = 2$ m/s and $v_{H_2O} = 1$ m/s, determine the following:

(a) mass- and molar-average velocity;
(b) mass- and molar-diffusion velocity for both hydrogen and water vapor;
(c) total mass and molar flux for hydrogen and water vapor, respectively, if the mixture is at 1 atm and 80 °C;
(d) Diffusional mass and molar flux for both hydrogen and water vapor at 1 atm and 80 °C;
(e) The sum of the diffusional mass (and molar) flux as calculated in (d).

SOLUTION

(a) From the given conditions and Equation (4.10), the molar-average velocity is easily calculated as follows:

$$v^* = X_{H_2} v_{H_2} + X_{H_2O} v_{H_2O} = 0.80 \times 2 + 0.20 \times 1 = 1.8 \, \text{m/s}$$

To determine the mass-average velocity, mass fraction for hydrogen and water vapor need to be determined first. Since the molecular weight for hydrogen and water are known:

$$W_{H_2} \cong 2 \, \text{kg/kmol} \quad \text{and} \quad W_{H_2O} \cong 18 \, \text{kg/kmol}$$

The molecular weight for the hydrogen–water vapor mixture is, according to Equation (4.7)

$$W_{mix} = X_{H_2} W_{H_2} + X_{H_2O} W_{H_2O} = 0.80 \times 2 + 0.20 \times 18 = 5.2 \, \text{kg/kmol}$$

From Equation (4.8), the mass fraction is then obtained as

$$Y_{H_2} = X_{H_2} \left(\frac{W_{H_2}}{W_{mix}} \right) = 0.80 \times \left(\frac{2}{5.2} \right) = 0.3077$$

$$Y_{H_2O} = X_{H_2O} \left(\frac{W_{H_2O}}{W_{mix}} \right) = 0.20 \times \left(\frac{18}{5.2} \right) = 0.6923$$

According to Equation (4.9), the mass-average velocity becomes

$$V = Y_{H_2} V_{H_2} + Y_{H_2O} V_{H_2O}$$

$$= 0.3077 \times 2 + 0.6923 \times 1 = 1.3077 \, \text{m/s}$$

Clearly, the molar-average velocity is not equal to the mass-average velocity and their difference tends to decrease when the molecular weight for the species involved approaches each other, as this is evident from the previous calculations.

(b) From the definitions given in Equations (4.11) and (4.12), the diffusion velocity for hydrogen and water vapor, relative to the mass- and molar-averaged mean motion of the mixture, is

$$V_{H_2} = v_{H_2} - v = 2 - 1.3077 = 0.6923 \, \text{m/s}$$

$$V_{H_2}^* = v_{H_2} - v^* = 2 - 1.8 = 0.2 \, \text{m/s}$$

for the hydrogen, and

$$V_{H_2O} = v_{H_2O} - v = 1 - 1.3077 = -0.3077 \, \text{m/s}$$

$$V_{H_2O}^* = v_{H_2O} - v^* = 1 - 1.8 = -0.8 \, \text{m/s}$$

for the water vapor.

(c) The total density for the mass in the mixture may be determined from the equation of state by assuming that the mixture follows the ideal gas behavior, that is

$$\rho = \frac{P W_{mix}}{RT} = \frac{101,325 \, \text{Pa} \times 5.2 \, \text{kg/kmol}}{8314 \, \text{J/(kmol} \cdot \text{K)} \times (80 + 273) \, \text{K}} = 0.1795 \, \text{kg/m}^3$$

Then Equation (4.5) yields the partial density for hydrogen and water vapor

$$\rho_{H_2} = Y_{H_2}\rho = 0.3077 \times 0.1795 = 0.05523 \, \text{kg/m}^3$$

$$\rho_{H_2O} = Y_{H_2O}\rho = 0.6923 \times 0.1795 = 0.1243 \, \text{kg/m}^3$$

The corresponding total molar concentration for the mixture is obtained by Equation (4.3):

$$C = \frac{\rho}{W_{mix}} = \frac{0.1795 \, \text{kg/m}^3}{5.2 \, \text{kg/kmol}} = 34.52 \, \text{mol/m}^3$$

Then the molar concentration for hydrogen and water vapor is

$$C_{H_2} = X_{H_2}C = 0.80 \times 34.52 = 27.62 \, \text{mol/m}^3$$

$$C_{H_2O} = X_{H_2O}C = 0.20 \times 34.52 = 6.904 \, \text{mol/m}^3$$

Finally the total mass and molar fluxes become, from Equations (4.13) and (4.14)

$$\dot{m}_{H_2}'' = \rho_{H_2}v_{H_2} = 0.05523 \, \text{kg/m}^3 \times 2 \, \text{m/s} = 0.1105 \, \text{kg/(s} \cdot \text{m}^2)$$

$$\dot{N}_{H_2}'' = C_{H_2}v_{H_2} = 27.62 \, \text{mol/m}^3 \times 2 \, \text{m/s} = 55.24 \, \text{mol/(s} \cdot \text{m}^2)$$

for hydrogen, and

$$\dot{m}_{H_2O}'' = \rho_{H_2O}v_{H_2O} = 0.1243 \, \text{kg/m}^3 \times 1 \, \text{m/s} = 0.1243 \, \text{kg/(s} \cdot \text{m}^2)$$

$$\dot{N}_{H_2O}'' = C_{H_2O}v_{H_2O} = 6.904 \, \text{mol/m}^3 \times 1 \, \text{m/s} = 6.904 \, \text{mol/(s} \cdot \text{m}^2)$$

for water vapor.

(d)   From Equations (4.15) and (4.16), the diffusional fluxes are obtained as follows:

$$\dot{m}^{''}_{H_2,d} = \rho_{H_2} V_{H_2} = 0.05523 \, \text{kg/m}^3 \times 0.6923 \, \text{m/s} = 0.03824 \, \text{kg/(s} \cdot \text{m}^2)$$

$$\dot{N}^{''}_{H_2,d} = C_{H_2} V^*_{H_2} = 27.62 \, \text{mol/m}^3 \times 0.2 \, \text{m/s} = 5.524 \, \text{mol/(s} \cdot \text{m}^2)$$

$$\dot{m}^{''}_{H_2O,d} = \rho_{H_2O} V_{H_2O} = 0.1243 \, \text{kg/m}^3 \times (-0.3077) \, \text{m/s}$$

$$= -0.03825 \, \text{kg/(s} \cdot \text{m}^2)$$

$$\dot{N}^{''}_{H_2O,d} = C_{H_2O} V^*_{H_2O} = 6.904 \, \text{mol/m}^3 \times (-0.8) \, \text{m/s} = -5.523 \, \text{mol/(s} \cdot \text{m}^2)$$

(e)   The sum of the diffusional fluxes for both hydrogen and water vapor are

$$\dot{m}^{''}_{H_2,d} + \dot{m}^{''}_{H_2O,d} = -0.00001 \, \text{kg/(s} \cdot \text{m}^2) \cong 0 \quad \text{Within the Roundoff}$$
$$\text{Accuracies}$$

$$\dot{N}^{''}_{H_2,d} + \dot{N}^{''}_{H_2O,d} = 0.001 \, \text{mol/(s} \cdot \text{m}^2) \cong 0 \quad \text{Within the Roundoff}$$
$$\text{Accuracies}$$

COMMENT

1. From Equation (4.3), it is clear that the total mass flux differs from the total molar flux by a factor which is equal to the molecular weight of the species involved, that is

$$\dot{m}^{''}_i = \dot{N}^{''}_i W_i$$

However, this relation does not hold for the fluxes due to the bulk motion and due to the diffusion because the average and the diffusion velocity are not the same when based on the mass or the number of moles of the species involved. These observations can easily be verified by the results shown in this example.

2. The previous example suggests that all the diffusional fluxes for the species in the mixture, when summarized, are equal to zero. This is true in general. In fact, from Equations (4.11), (4.12), (4.15), and (4.16) we have

$$\sum_{i=1}^{N} \dot{m}^{''}_{i,d} = \sum_{i=1}^{N} \rho_i V_i = \sum_{i=1}^{N} \rho_i (v_i - v) = \sum_{i=1}^{N} \rho_i v_i - \sum_{i=1}^{N} \rho_i v = 0$$

$$\sum_{i=1}^{N} \dot{N}^{''}_{i,d} = \sum_{i=1}^{N} C_i V^*_i = \sum_{i=1}^{N} C_i (v_i - v^*) = \sum_{i=1}^{N} C_i v_i - \sum_{i=1}^{N} C_i v^* = 0$$

when Equations (4.9) and (4.10) are considered. This is understandable since species diffusion is the motion of species relative to the local mean motion of the fluid.

4.3   FICK'S LAW OF DIFFUSION

Consider a nonreacting mixture consisting of two molecular species: species i and j. The rate of mass transfer for species i diffusing through species j follows the following phenomenological equation

$$\dot{m}_{i,d}^{''} = -\rho \mathcal{D}_{ij} \nabla Y_i \qquad (4.27)$$

which is called Fick's law of diffusion, expressed in terms of the mass flux. This equation states that the diffusional mass flux arises from the mass concentration gradient and is in the direction of decreasing concentration as represented by the negative sign. Thus, species diffusion results in the preferential transfer of species i from a region of high concentration to a region of low concentration, similar to the transfer of heat by conduction in the direction of decreasing temperature. The proportionality constant $\mathcal{D}_{ij}$ is called **binary mass diffusivity**, or the diffusion coefficient of species i with respect to species j. It is a property of the binary mixture and has the unit of m^2/s. Values for some binary diffusion coefficients are given in Appendix B.

Combining Equations (4.15) and (4.27), we obtain

$$Y_i \mathbf{V}_i = -\mathcal{D}_{ij} \nabla Y_i \qquad (4.28)$$

or the diffusion velocity for species i, $\mathbf{V}_i$, can be determined if the mass fraction distribution is known.

Therefore, the total mass flux for species i relative to a stationary coordinate becomes, following Equation (4.19)

$$\dot{m}_i^{''} = \rho_i \mathbf{v} - \rho \mathcal{D}_{ij} \nabla Y_i \qquad (4.29)$$

This equation indicates that the total mass flux of species i relative to a stationary coordinate is contributed by two mechanisms: One is the species i carried by the bulk motion of the mixture relative to the stationary coordinate and the other is due to the species diffusion arising from the concentration gradient (relative to the bulk motion of the mixture). Considering Equations (4.5) and (4.23), Equation (4.29) can also be written as

$$\dot{m}_i^{''} = Y_i \dot{m}^{''} - \rho \mathcal{D}_{ij} \nabla Y_i \qquad (4.30)$$

where

$$\dot{m}^{''} = \dot{m}_i^{''} + \dot{m}_j^{''}$$

represents the total mass flux for the binary mixture.

In many situations, it is more convenient to express various aforementioned equations in terms of molar flux. As such, Fick's law of diffusion becomes

$$\dot{N}_{i,d}^{''} = -C \mathcal{D}_{ij} \nabla X_i \qquad (4.31)$$

or

$$X_i \mathbf{V}_i^* = -\mathcal{D}_{ij} \nabla X_i \tag{4.32}$$

The total molar flux of species i relative to a stationary coordinate is

$$\dot{\mathbf{N}}_i^{''} = C_i \mathbf{v}^* - C\mathcal{D}_{ij} \nabla X_i \tag{4.33}$$

or

$$\dot{\mathbf{N}}_i^{''} = X_i \dot{\mathbf{N}}^{''} - C\mathcal{D}_{ij} \nabla X_i \tag{4.34}$$

Example 4.2   For a binary mixture composed of species i and j, show that the binary mass diffusivity, $\mathcal{D}_{ij}$, is symmetrical to the subscript indices, that is

$$\mathcal{D}_{ij} = \mathcal{D}_{ji}$$

SOLUTION

From Equation (4.27), the diffusional mass flux for species i and j are, respectively

$$\dot{\mathbf{m}}_{i,d}^{''} = -\rho \mathcal{D}_{ij} \nabla Y_i$$

$$\dot{\mathbf{m}}_{j,d}^{''} = -\rho \mathcal{D}_{ji} \nabla Y_j$$

Because the total diffusional fluxes vanish, or

$$\dot{\mathbf{m}}_{i,d}^{''} + \dot{\mathbf{m}}_{j,d}^{''} = 0$$

we have

$$\mathcal{D}_{ij} \nabla Y_i + \mathcal{D}_{ji} \nabla Y_j = 0$$

For the present binary mixture composed of species i and j only, Equation (4.6) yields

$$Y_i + Y_j = 1 \quad \text{or} \quad Y_j = 1 - Y_i$$

Then we have

$$\mathcal{D}_{ij} \nabla Y_i + \mathcal{D}_{ji}(-\nabla Y_i) = 0$$

which results in

$$\mathcal{D}_{ij} = \mathcal{D}_{ji}$$

COMMENT

The symmetry of the binary mass diffusivity, $\mathcal{D}_{ij} = \mathcal{D}_{ji}$, is only valid for binary mixtures, but not valid for multicomponent mixtures of more than two species, even though similar diffusion coefficients can be defined for each pair of species. In fact, Fick's law of diffusion is only valid strictly for binary mixtures and more complex equation is described later in this chapter for diffusion in multicomponent mixtures.

4.4   SIMILARITY BETWEEN THE TRANSPORT OF MASS, ENERGY AND MOMENTUM

Analogy between the transport of mass, thermal energy (or commonly referred to as heat) and momentum can be drawn under certain conditions and is described in this section to show the similarity among these transport phenomena. The diffusion of thermal energy, or heat conduction as it is commonly known, follows the following phenomenological expression:

$$\dot{\mathbf{q}}_d'' = -k\nabla T \tag{4.35}$$

where $\dot{\mathbf{q}}_d''$ denotes heat flux, the rate of heat transfer by conduction per-unit time per-unit surface area which is normal to the direction of heat flux, with the unit of J/(s · m^2) or W/m^2; T is temperature with the unit of K; and the proportionality constant k is a property of the conducting medium, often called thermal conductivity, with the unit of W/(m · K). Equation (4.35) states that the rate of heat transfer by conduction is proportional to the gradient of temperature and the direction of heat transfer is in decreasing temperature, or from a high-temperature to low-temperature region. This expression is known as Fourier's law of conduction.

The diffusion of momentum is in general much more complex than the diffusion of species and thermal energy. It can occur in the direction of fluid flow as well as in the cross-stream direction, and it can depend on the velocity gradient both linearly for Newtonian fluids and nonlinearly for non-Newtonian fluids. Fluids commonly encountered, such as water, air, hydrogen, oxygen, nitrogen, carbon monoxide, and carbon dioxide, are Newtonian. For Newtonian fluids, the diffusion of momentum in terms of momentum flux $\dot{\mathbf{M}}_d''$ follows the following phenomenological expression:

$$\dot{\mathbf{M}}_d'' = -\mu[\nabla\mathbf{v} + (\nabla\mathbf{v})^T] \tag{4.36}$$

where the superscript "T" denotes the transpose of the velocity gradient $\nabla\mathbf{v}$, and it arises from the fact that the driving potential for momentum diffusion is the gradient of velocity, which is a vector quantity, whereas the driving potential for the diffusion of mass and heat is the gradient of concentration and temperature, respectively, which are scalar. The proportionality constant μ is a property of the fluid with the unit of kg/(m · s), and is called the **coefficient of shear viscosity**, or simply viscosity. Since momentum has the unit of kg · m/s, the unit for the momentum flux, or the rate of momentum transfer by diffusion per-unit time per-unit surface area, becomes

$$\frac{kg \cdot m}{s} / (s \cdot m^2) = \frac{N}{m^2}$$

That is, the momentum flux, $\dot{\mathbf{M}}_d''$, is equivalent to a force acting on a unit area, or stress. Thus, $\dot{\mathbf{M}}_d''$ is commonly replaced by τ representing shear and normal stress arising from the gradient of velocity or the transport of momentum. Equation (4.36) is known as Newton's law of viscosity.

Closer similarity between the transport of mass, momentum, and energy by diffusion can be obtained by considering the following situation: The flow of a binary mixture is smooth and orderly (laminar) in the x direction, the velocity, temperature, and concentration of species i change in the cross-stream y direction with negligible variation in the streamwise x direction, that is

$$Y_i = Y_i(y), \quad T = T(y), \quad v_x = v_x(y), \quad v_y \cong 0$$

Such a flow situation arises, for example, for a flow over a solid surface (wall). Then, the diffusional flux for the transport of mass, momentum and energy in the cross-stream y direction becomes

$$\dot{m}_{i,d,y}'' = -\mathcal{D}_{ij}\frac{\partial}{\partial y}(\rho Y_i) \quad \text{– Fick's Law for Constant } \rho \qquad (4.37)$$

$$(\text{or } \dot{N}_{i,d,y}'' = -\mathcal{D}_{ij}\frac{\partial}{\partial y}(CX_i) \quad \text{– Fick's Law for Constant } C) \qquad (4.38)$$

$$\dot{\mathbf{M}}_{d,y}'' (= \tau_{xy}) = -\nu\frac{\partial}{\partial y}(\rho v_x) \quad \text{– Newton's Law for Constant } \rho \qquad (4.39)$$

$$\dot{q}_y'' = -\alpha\frac{\partial}{\partial y}(\rho C_p T) \quad \text{– Fourier's Law for Constant } \rho C_p \qquad (4.40)$$

where $\nu = \mu/\rho$ is the kinematic viscosity with the unit of m²/s, and $\alpha = k/(\rho C_p)$ is the thermal diffusivity with the unit of m²/s. Both are the property of the mixture. The specific heat at constant pressure, C_p, has the unit of J/(kg · K), representing the property of the mixture as the thermal capacity of the mixture on the basis of unit-mass and unit-degree temperature change.

We shall see later the importance of the cross-stream transport phenomena due to diffusion in the design and operation of fuel cells, especially in connection with the nonuniform distribution of current density over the surface of the electrodes.

4.4.1   Transport Coefficients at Low Densities

The coefficient for the transport of mass, momentum and energy, $\mathcal{D}_{ij}$, ν, and α (or $\rho\mathcal{D}_{ij}$, μ and k), as described earlier, are properties of the transporting mixture, hence they are a function of the state of the mixture. For gas mixtures at low densities (ideal gases), kinetic theory shows that

$$\{\rho \mathcal{D}_{ij}, \mu, k\} \sim T^{1/2} \tag{4.41}$$

or

$$\{\mathcal{D}_{ij}, \nu, \alpha\} \sim \frac{T^{2/3}}{Pd^2} \tag{4.42}$$

where d is the molecular diameter.

Experimental results indicate stronger temperature dependence:

$$\{\rho \mathcal{D}_{ij}, \mu, k\} \sim T^n; \quad 0.5 \leq n \leq 1 \tag{4.43}$$

$$\{\mathcal{D}_{ij}, \nu, \alpha\} \sim \frac{T^m}{Pd^2}; \quad 1.5 \leq m \leq 2 \tag{4.44}$$

Clearly, these transport coefficient $\mathcal{D}_{ij}$, ν and α have the same dependence on temperature and pressure, and inversely proportional to the square of the molecular diameter. Therefore, hydrogen has the smallest molecular size, consequently, the largest diffusion coefficient, larger than oxygen diffusion coefficient under identical condition. This difference in the transport coefficients has significant impact on the performance of the hydrogen-fed anode and oxygen- (or air-) fed cathode, as we see soon in this chapter.

4.4.2   Ratios of Transport Coefficients

The similarity between the diffusional transport of mass, momentum and energy as well as among their corresponding transport coefficients has been described. In order to measure their relative importance and impact, it is conventional to define various dimensionless parameters in terms of their ratios. For example, the Prandtl number is defined as

$$\Pr \equiv \frac{\nu}{\alpha} \sim \frac{\text{Rate of momentum transport}}{\text{Rate of energy transport}} \tag{4.45}$$

The Lewis number is defined as

$$\text{Le}_{ij} \equiv \frac{\alpha}{\mathcal{D}_{ij}} \sim \frac{\text{Rate of energy transport}}{\text{Rate of mass transport}} \tag{4.46}$$

The Schmidt number is defined as

$$\text{Sc}_{ij} \equiv \frac{\nu}{\mathcal{D}_{ij}} \sim \frac{\text{Rate of momentum transport}}{\text{Rate of mass transport}} \tag{4.47}$$

Clearly, these three dimensionless numbers are related by

$$\text{Sc}_{ij} \equiv \Pr \times \text{Le}_{ij} \tag{4.48}$$

For many gases, these dimensionless numbers are very close to unity (as predicted by the kinetic theory); they are often slightly less than unity. This fact has not been explored in the analysis of fuel cells. For multicomponent mixtures, Le_{ij} and Sc_{ij} may be defined for each pair of species i and j in the mixture, based on the binary diffusion coefficient $\mathcal{D}_{ij}$. Therefore, the actual number of the dimensionless parameters appearing in a particular analysis depends on the number of species in the mixture.

4.5   TRANSPORT PHENOMENA FOR FLOW OVER A WALL

In fuel cells as well as many other practical applications, the transport phenomena occurring when fluid flows over a solid surface have particular importance because they can have dominant influence on, for example, the pumping power requirement for the fluid flow and the overall performance of heat (mass) transfer devices. A brief description is given below for the transport of momentum, energy and species between a fluid and the wall over which the fluid flows.

4.5.1   Velocity Boundary Layer: Momentum Transport

The description of transport phenomena over a wall is inevitably related to the concept of boundary layer. For illustration, consider a flow over a flat plate as shown in Figure 4.1. The fluid particles in contact with the wall take up the velocity of the wall, the so-called **no slip boundary condition** for viscous fluids ($\mu \neq 0$). Normally the wall is considered stationary ($v_s = 0$) with respect to the inertial frame of coordinate attached to the wall, such as x-y coordinate shown in Figure 4.1. The zero-velocity fluid particles at the wall slow down the motion of the fluid in the adjoining layer, which in turn slows down the motion of the fluid in the next layer, and so on. This retardation effect is weakened as the distance from the wall is increased such that at a sufficient distance away, say, $y = \delta$, the effect can be neglected, and the fluid velocity approaches the free stream velocity v_∞. The retardation of fluid motion near the wall results in the variation of fluid velocity, or momentum flux in the y direction, as indicated by Equation (4.36) or its simplified form, Equation (4.39). This momentum flux leads to a shear stress τ acting on the fluid in planes parallel to the fluid velocity (in the x direction), as shown in Figure 4.1, and the shear stress can be estimated by Equation (4.39).

It is evident that the fluid flow over the wall is characterized by two distinct regions: a thin region near the wall over which the velocity varies ($\tau \neq 0$) and a region away from the wall where the velocity variation is negligible ($\tau = 0$). The

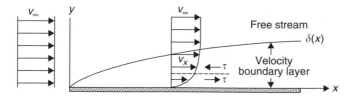

Figure 4.1 The development of velocity boundary layer over a flat plate.

thin region near the wall is often called the **velocity boundary layer**, whose thickness δ is commonly defined as the distance from the wall where $v_x = 0.99v_\infty$, or $v_x - v_s = 0.99(v_\infty - v_s)$ if the wall velocity v_s does not vanish. The effect of viscosity dominates inside the velocity boundary layer and becomes negligible outside the layer due to the increased distance away from the wall. As the fluid flows downstream (larger x), the effect of viscosity (or the wall) penetrates deeper into the free stream, the velocity boundary layer becomes thicker and δ increases with increasing x.

The significance of the momentum diffusion inside the velocity boundary layer gives rise to a shear stress acting on the wall by the fluid, τ_s. According to Newton's third law of mechanics and Equation (4.39), wall shear stress is determined by

$$\tau_s = \mu \frac{\partial u}{\partial y}\bigg|_{y=0} \tag{4.49}$$

or in dimensionless form for external flows

$$C_f \equiv \frac{\tau_s}{\frac{1}{2}\rho v_\infty^2} \tag{4.50}$$

where C_f is also called the local friction coefficient, and decreases with x due to the growth of the velocity boundary layer. It can be shown that for a given geometry, C_f depends on a dimensionless parameter, named Reynolds number:

$$\text{Re} \equiv \frac{\rho v_\infty x}{\mu} \tag{4.51}$$

which represents the relative importance of viscosity effect. Such that we have

$$C_f = f_1(\text{Re}) \tag{4.52}$$

Note that the above definition of the Reynolds number is typically for external flows such as the flow over a flat plate under discussion here, and the location in the flow direction, x, represents the length scale for the location under consideration. For flows inside ducts (internal flows), the length scale is normally chosen as the representative cross-sectional size of the flow passage, such as the diameter of a circular duct. For flows inside noncircular ducts, hydraulic diameter is the most often used length scale, defined as

$$D_h \equiv \frac{4 \times \text{Cross-Sectional Area}}{\text{Wetted Perimeter}} \tag{4.53}$$

where the wetted perimeter should include all solid surfaces acted upon by the shear stress τ_s for the given cross section. Sometimes, the effective laminar diameter may be used for flows through noncircular ducts, which makes the functional dependence for any internal flow, Equation (4.52), identical to that of the flow inside a circular duct.

4.5.2   Thermal Boundary Layer: Energy Transport

Similar to the development of a velocity boundary layer for a flow over a wall when the velocity of the free stream and the wall differs, a thermal boundary layer develops when the temperature of the free stream and the wall is not identical. Consider a fluid at temperature T_∞ flows over an isothermal flat plate at T_s, as shown in Figure 4.2. The fluid particles in contact with the wall quickly reaches local thermal equilibrium with the wall and assume the wall temperature (the continuity of temperature). These particles influence the temperature of the fluid in the adjoining layer through the transport of heat (thermal energy). Such process continues until at a sufficient distance away from the wall, say $y = \delta_t$, the wall effect becomes negligible and the fluid temperature no longer changes in the y direction, and approaches the free stream temperature. Again, the thermal field in the flow is characterized by two regions: a thin layer near the wall, called the thermal boundary layer, over which temperature varies from the wall temperature T_s to the free stream temperature T_∞ ($\dot{q}_y'' \neq 0$) and the region outside the thermal boundary layer where temperature variation is negligible ($\dot{q}_y'' \cong 0$). The heat flux in the y direction, $\dot{q}_y''$, can be determined from Equation (4.40). The thickness of the thermal boundary layer, δ_t, is often defined as the distance from the wall where $T - T_s = 0.99(T_\infty - T_s)$. As the fluid flows downstream, the effect of the heat transport (or the wall) is felt at a larger distance from the wall, thus δ_t increases with x.

The effect of the heat transfer in the y direction ($\dot{q}_y''$) over the thermal boundary layer is the heat transfer between the free stream and the wall, which is often determined by the Newton's law of cooling

$$\dot{q}_s'' = h_{\mathrm{H}}(T_s - T_\infty) \tag{4.54}$$

where h_{H} is termed the convection heat transfer coefficient with the unit of $\mathrm{W/(m^2 \cdot K)}$. It is important to emphasize that h_{H} is not a property of the fluid, rather it depends on the flow conditions in the thermal boundary layer, the wall geometry, and the type of fluid involved. Normally, h_{H} decreases in the flow direction due to the growth of the thermal boundary layer. Combining with Equation (4.40), we have

$$h_{\mathrm{H}} = \frac{-k\,\dfrac{\partial T}{\partial y}\Big|_{y=0}}{T_s - T_\infty} \tag{4.55}$$

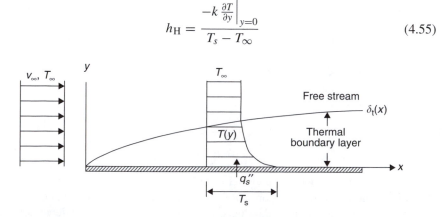

Figure 4.2   The development of thermal boundary layer over an isothermal flat plate.

A dimensionless convection heat transfer coefficient is commonly defined as

$$\text{Nu} \equiv \frac{h_H x}{k} \tag{4.56}$$

where Nu is commonly termed the Nusselt number and it may increase with increasing x due to the growth of the thermal boundary layer, although h_H will decrease with x. For a given geometry, it can be established that

$$\text{Nu} = f_2(\text{Re}, \text{Pr}) \tag{4.57}$$

that is, Nu is a function of the Reynolds and Prandtl numbers.

4.5.3   Concentration Boundary Layer: Species Transport

A concentration boundary layer develops near a wall when the concentration of a species in the fluid flowing over the wall differs from that at the wall surface, in almost exact analogy with the velocity and thermal boundary layers. Consider a binary mixture of chemical species i and j flows over a flat plate and the concentration of the species i is $C_{i,\infty}$ in the free stream and $C_{i,s}$ at the wall surface, as shown in Figure 4.3. Again the concentration distribution in the fluid flow over the wall is characterized by two regions: a thin layer near the wall, called the **concentration boundary layer**, over which the concentration of species i varies from $C_{i,s}$ to $C_{i,\infty}$ ($\dot{N}''_{i,d,y} \neq 0$) and the region outside the concentration boundary layer where the concentration variation is negligible ($\dot{N}''_{i,d,y} \cong 0$).

The mass flux in the y direction, $\dot{m}''_{i,d,y}$, can be determined by Equation (4.37). The thickness of the concentration boundary layer, δ_C, is similarly defined as the distance from the wall where $C_i - C_{i,s} = 0.99(C_{i,\infty} - C_{i,s})$. The effect of species transport penetrates further into the free stream as the fluid moves downstream, hence δ_C increases with x.

The rate of mass transfer for the species i between the free stream and the wall is often expressed similarly to Equation (4.54), as

$$\dot{N}''_{i,s} = h_m(C_{i,s} - C_{i,\infty}) \tag{4.58}$$

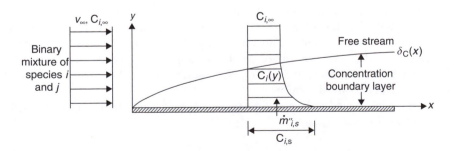

Figure 4.3 The development of concentration boundary layer over a flat plate.

in terms of the molar flux, or

$$\dot{m}_{i,s}'' = h_m(\rho_{i,s} - \rho_{i,\infty}) \tag{4.59}$$

in terms of the mass flux. Where h_m is called the convection mass transfer coefficient with the unit of m/s. Similar to h_H, h_m is not a property of the fluid mixture, rather a function of the flow conditions in the concentration boundary layer, the wall geometry and the fluid mixture. Considering Equation (4.37), we have, depending on whether molar or mass flux expression is utilized

$$h_m = -\frac{\mathcal{D}_{ij}\left.\frac{\partial C_i}{\partial y}\right|_{y=0}}{C_{i,s} - C_{i,\infty}} = -\frac{\mathcal{D}_{ij}\left.\frac{\partial \rho_i}{\partial y}\right|_{y=0}}{\rho_{i,s} - \rho_{i,\infty}} \tag{4.60}$$

A dimensionless convection mass-transfer coefficient is often defined as

$$\text{Sh} \equiv \frac{h_m x}{\mathcal{D}_{ij}} \tag{4.61}$$

which is also commonly known as Sherwood number. For a given geometry, Sh depends on the flow condition as represented by Re and the relative effectiveness of the momentum and mass transfer by diffusion in the velocity and concentration boundary layer, respectively, as represented by Sc, or

$$\text{Sh} \equiv f_3(\text{Re, Sc}) \tag{4.62}$$

Similarly, Sh may increase (or h_m may decrease) with increasing x arising from the concentration boundary layer development, and may be invariant for flow situations where δ_C remains constant.

4.5.4   Boundary Layer Analogies

The Boundary Layer Thickness   As described before, the Reynolds number represents the significance of viscosity effect. At small Re values, viscosity effect, which is like the frictional effect between the adjacent fluid layers, is dominant, and can diffuse away (or dissipate) sufficiently fast any disturbance developed in the flow field, thus preventing the disturbance being amplified that may alter the smooth, ordered and layered fluid motion (laminar) into fluctuating, random-like motion (turbulent). Therefore, a critical Reynolds number may exist, Re_c, such that laminar flow is maintained if $\text{Re} < \text{Re}_c$, and is destroyed due to the amplification of disturbance and the flow may become turbulent if $\text{Re} > \text{Re}_c$. For most of the fluid flows encountered in fuel cells, laminar flow dominates.

For laminar boundary layers, diffusional transport dominates in the cross-stream (y) direction, so that we may have

$$\frac{\delta}{\delta_t} \cong \text{Pr}^n \tag{4.63}$$

because the Prandtl number represents the relative effectiveness of momentum and heat transfer by diffusion in the respective velocity and thermal boundary layers. Similarly,

$$\frac{\delta}{\delta_C} \cong Sc^n \tag{4.64}$$

$$\frac{\delta_t}{\delta_C} \cong Le^n \tag{4.65}$$

where $n = 1/3$ for most flow situations of common fluids. For most gases, Pr, Sc, and Le are very close to unity, hence $\delta \cong \delta_t \cong \delta_C$. For turbulent flows, boundary layer development is predominantly influenced by the random fluctuations in the flow, rather than by the molecular diffusion. Therefore, the thickness for the velocity, thermal and concentration boundary layers is approximately the same for turbulent flows, or

$$\delta \cong \delta_t \cong \delta_C \tag{4.66}$$

It becomes clear that once the velocity boundary layer thickness, δ, is known, the thermal and concentration boundary layer thickness, δ_t and δ_C, can be determined easily. For example, for flow over a flat plate (parallel flow), Blasius obtained in 1908[2,3]

$$\frac{\delta}{x} \approx \frac{5.0}{Re^{1/2}} \tag{4.67}$$

For axisymmetric flow normal to a flat plate (stagnation flow),[4]

$$\delta \approx 2.0\sqrt{\frac{\nu}{B}} \tag{4.68}$$

where ν is the kinematic viscosity, and B is a positive constant, proportional to the stream velocity approaching the surface and inversely proportional to the characteristic length of the surface, with the unit of 1/s. Clearly for this case the boundary layer thickness is a constant, independent of the specific location along the surface. However, it is decreased by an increase in the stream velocity towards the surface. The boundary layer for stagnation flow is established as a balance between the outward viscous diffusion of the wall effect into the flow field, and the convection effect due to the flow towards the wall surface. Stagnation flow is important for the fundamental study of mass transfer effect on fuel cell performance, and such flow is often induced by the rotation of the electrode, or the so-called rotating disk electrode in electrochemistry. The significance of the stagnation flow is that its boundary layer thickness is identical every where, irrespective of the specific location on the surface, resulting in the uniform mass flux anywhere over the electrode surface since the transport of mass across the boundary layer is by diffusion. Furthermore, a comparison between Equations (4.67) and (4.68) indicates that the boundary layer thickness for the flow normal to a plate (stagnation flow) is about 2.5 times thinner than the flow parallel to the flat plate, thus the stagnation flow offers a higher rate of mass transfer, leading to a higher limiting current density for the electrode.

The Reynolds Analogy   The similarity among the development of the velocity, thermal, and concentration boundary layers suggests that the important engineering parameters of these boundary layers, the rate of transfer for momentum (shear stress), heat and mass (at the wall), must be related. Indeed, it can be shown that[5]

$$C_f \left(\frac{\text{Re}}{2} \right) = \text{NuPr}^{-1/3} = \text{ShSc}^{-1/3} \tag{4.69}$$

which is valid when $0.6 < \text{Pr} < 60$ for the analogy between the momentum and heat transfer (the first equality), and $0.6 < \text{Sc} < 3000$ for the second equality, or the analogy between the momentum and mass transfer. Equation (4.69) is known as the modified Reynolds analogy and it provides a method for the determination of the rate of transfer at the wall for the momentum, heat and species if any one of them is known.

However, Equation (4.69) is restricted to the following flow conditions: First, the boundary layer approximations must be valid; second, the pressure variation in the dominant flow direction (x) must be very small (negligible). This second condition is particularly important for laminar flows, while it can be relaxed considerably for turbulent flows. Especially when $\text{Pr} = \text{Sc} = 1$, Equation (4.69) reduces to

$$C_f \left(\frac{\text{Re}}{2} \right) = \text{Nu} = \text{Sh}$$

which is known as the Reynolds analogy.

For steady laminar flow over a flat plate, Blasius solution[2] yields the local friction coefficient as

$$C_f = 0.664 \text{Re}^{-1/2} \tag{4.70}$$

Then from the modified Reynolds analogy, Equation (4.69), the Nusselt and Sherwood number for the convective heat and mass transfer over a flat plate can be obtained immediately:

$$\text{Nu} \equiv \frac{h_H x}{k} = 0.332 \text{Re}^{1/2} \text{Pr}^{1/3}; \quad \text{Pr} \gtrsim 0.6 \tag{4.71}$$

$$\text{Sh} \equiv \frac{h_m x}{\mathcal{D}_{ij}} = 0.332 \text{Re}^{1/2} \text{Sc}^{1/3}; \quad \text{Sc} \gtrsim 0.6 \tag{4.72}$$

Note that Equation (4.70) was obtained for a constant wall velocity (usually taken as zero), hence Equations (4.71) and (4.72) are valid for constant wall temperature T_s and constant wall concentration $C_{i,s}$, respectively. If the condition at the plate surface is a uniform heat flux, instead of a uniform temperature, it has been shown that for steady laminar flows[6]:

$$\text{Nu} \equiv \frac{h_H x}{k} = 0.453 \text{Re}^{1/2} \text{Pr}^{1/3}; \quad \text{Pr} \gtrsim 0.6 \qquad (4.73)$$

Hence the Nusselt number is about 36% larger than the case for constant surface temperature shown in Equation (4.71). Using the modified Reynolds analogy, the Sherwood number for the mass transfer can be easily obtained as

$$\text{Sh} \equiv \frac{h_m x}{\mathcal{D}_{ij}} = 0.453 \text{Re}^{1/2} \text{Sc}^{1/3}; \quad \text{Sc} \gtrsim 0.6 \qquad (4.74)$$

Similarly, this expression for the Sherwood number for the constant mass flux at the surface is about 36% larger than its counterpart for constant surface concentration given in Equation (4.72). These differences for the different surface conditions can have significant implication on the fuel cell operation as we shall see in the next section. It might be pointed out that the thermophysical properties in Equations (4.70)–(4.74) should be evaluated at the so-called film condition, that is, the average of the value in the bulk gas phase and at the surface.

The analogy between the heat and mass transfer is also apparent when examining Equations (4.57) and (4.62). These equations imply that if a correlation for convection heat transfer coefficient Nu is available for a given flow, then a correlation for convection mass transfer for the same flow can be obtained simply by replacing Nu with Sh and Pr with Sc in the correlation for convection heat transfer, and vice versa. This heat and mass transfer analogy will be very useful since correlation for Nu are extensively available for a wide variety of flows, while correlations for Sh are relatively limited. We later use this analogy for the determination of mass transfer rates in a fuel cell environment.

It should be emphasized that the analogy between the cross-stream transport of momentum, heat, and species for boundary layer flows described previously is only valid when the mass transfer rate at the wall, $\dot{m}''_{i,s}$, is very small such that the wall mass transfer does not alter the boundary-layer flow field (momentum transfer) appreciably. If $\dot{m}''_{i,s} > 0$, the mass is transferred from the surface into the flow above the surface and such a flow is often called the **wall blowing**. On the other hand, the case of $\dot{m}''_{i,s} < 0$ implies that the mass is transferred from the flow into the wall, and is commonly referred to as the **wall suction**. Wall suction thins the boundary layer, thus enhancing the rate of transport; while the wall blowing thickens the boundary layer, reducing the rate of transport. Further, a strong blowing effect can blow the flow away from the surface, and the boundary layer flow concept breaks down. Our discussion in this chapter is limited to the situation of weak wall blowing/suction effect in order to simplify the analysis presented in the next two sections without a loss in the generality of the methodology presented. For strong effect of wall blowing/suction, the readers are referred to more advanced texts on the topics such as.[1,3,4]

4.6   CONCENTRATION POLARIZATION η_{conc}

With the preceding discussion on the mass transport, we now turn our attention to the concentration polarization in fuel cells, which is one of the three forms of polarization resulting in irreversible voltages (energy) losses. The activation polarization was described in the previous chapter, while the ohmic polarization is described in this chapter as well.

During the derivation of activation overpotential in the previous chapter, the concentration of the reactants at the reaction sites was implicitly taken as constant — independent of the cell current density (or the rate of electrochemical reaction at the electrodes). In reality, the finite rate of mass transfer sets up a concentration gradient along the direction of the mass transport and the reactant concentration at the reaction sites is invariably lower than what is available in the bulk reactant region, as electric current is drawn from the cell. The lower reactant concentration at the reaction sites due to the mass transfer limitations leads to a corresponding loss in the output cell voltage and such voltage loss is often referred to as the concentration overpotential.

As the current drawn from the cell is increased, the rate of electrochemical reactions at the electrodes is increased as well in order to meet the increased current output requirement. Consequently, the reactants are rapidly consumed at the electrodes by the reactions. The limited rate of mass transfer imposes an upper limit on the availability of the reactants in the vicinity of the reaction sites. When the reactant concentration is reduced to zero at the reaction sites, all the reactant transported in is completely consumed by the electrode reaction producing the current output, the maximum current output from the cell has thus been reached. Since further increase in the cell current output cannot be supported by the limited rate of mass transfer which provides the reactants for energy conversion in the cell (at the electrodes). This maximum current density from the cell is known as the limiting current density, at which the available (reversible) cell potential is completely offset by the overpotential, principally the concentration overpotential, so that the cell output potential reduces to zero. Clearly, the limiting current density J_L is predominantly influenced by the maximum rate of reactant transport, and a knowledge of J_L is of practical importance since it represents the allowable range for the cell operation.

Concentration overpotential is caused by a number of processes that hinder the transport of mass. For the anode- and cathode-electrode processes described in Section 3.7 especially for the processes described in association with Equations (3.127) and (3.135), the primary contributing processes are regarded as:

* The convective mass transfer from the gas phase in the flow channel to the porous electrode;
* The slow diffusion in the gas phase in the electrode pores;
* The solution/dissolution of reactants/products into/out of the electrolyte;
* The slow diffusion of reactants/products through the electrolyte to/from the electrochemical reaction sites.

In general, the low solubility of reactants in the electrolyte and the slow diffusion of reactants through the electrolyte constitute the major contribution to the concentration overpotential, η_{conc}. η_{conc} usually becomes significant (or even prohibitive) at high current density, as commonly familiar from the polarization curve for fuel cells because high rate of mass transfer is required to meet the demand of high current-density generation.

On the other hand, concentration overpotential can also be caused by the overaccumulation of reaction products in the electrode structure, which block the reactants from reaching the reaction sites. This phenomenon is especially serious for low temperature H_2/O_2 fuel cells because the reaction product water is in liquid state. This is often known as water flooding of electrodes, a severe performance-degrading mechanism for PEM fuel cells.

The supply of various species to the reaction sites for electrochemical reaction involves both reactants provided to the cell in the flow channel and the charged species (ions) from the bulk electrolyte. The latter is also related to the transport of electricity. In the following, we analyze the concentration overpotential by following the classic treatment first, then the engineering approach. These set the stage for modern comprehensive treatment of the subject.

4.6.1 Transport of Reactant Supply: Classic Approach

Consider an electrode as shown in Figure 4.4. The electrode-backing layer is much thicker (at least an order of magnitude) than the catalyst layer. Suppose the reactant

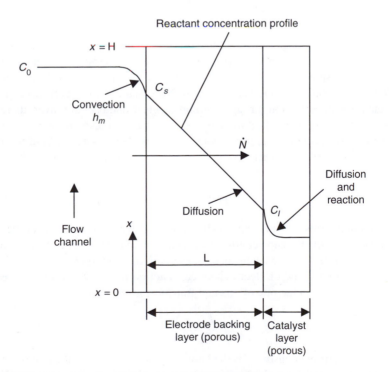

Figure 4.4 Typical process of reactant transport to the reactant sites.

is supplied to the flow channel at the concentration C_0. The reactant is transported from the flow channel to the electrode surface through convection with a known mass transfer coefficient h_m and through the backing layer predominantly by diffusion. If the concentration of the reactant at the interface between the backing layer and the catalyst layer is denoted as C_I, the rate of mass transfer under steady state becomes, according to Equation (4.58)

$$\dot{N} = \dot{N}'' A = A h_m (C_0 - C_s) \tag{4.75}$$

for the convection mass transfer at the electrode surface. Where A is the electrode surface area and C_s is the reactant concentration at the electrode surface. By Fick's law, the diffusional transport through the electrode-backing layer can be written as

$$\dot{N} = \dot{N}'' A = A \mathcal{D}^{\text{eff}} \frac{C_s - C_I}{L} \tag{4.76}$$

where L is the thickness of the electrode-backing layer. Combining Equations (4.75) and (4.76) together yields

$$\dot{N} = \frac{C_0 - C_I}{\sum R_m} \tag{4.77}$$

where

$$\sum R_m = \frac{1}{h_m A} + \frac{L}{\mathcal{D}^{\text{eff}} A} \tag{4.78}$$

represents the total resistance to the transport of reactant to the reaction sites since the term in the previous expression, $1/(h_m A)$, represents the resistance to the convective mass transfer; while the other term, $L/(\mathcal{D}^{\text{eff}} A)$, is the resistance to the diffusional mass transfer through the electrode-backing layer. Note that $\mathcal{D}^{\text{eff}}$ is the effective diffusion coefficient for the diffusion through the porous electrode backing layer, therefore it depends not only on the bulk diffusion coefficient $\mathcal{D}$, but also on the pore structure of the electrode. Assuming uniform pore size with the porosity (or void fraction) ϕ for the electrode, we may write, when the backing layer is free from the flooding of water and/or liquid electrolyte

$$\mathcal{D}^{\text{eff}} = \mathcal{D}\phi^{3/2} \tag{4.79}$$

The correction factor, $\phi^{3/2}$, is empirically based, known as Bruggenman type of correction and the bulk diffusion coefficient $\mathcal{D}$ is a function of temperature, pressure and the molecular size of the reactant species, as given in Equation (4.44).

Then Faraday's law of electrochemical reaction provides the link between the current generated and the rate of reactant transport as follows

$$\frac{J}{nF} = \frac{\dot{N}}{A} \tag{4.80}$$

where J is the current density, F is the Faraday constant, and n is the number of mole electrons transferred per mole reactant consumed. Thus, $n = 2$ for hydrogen (H_2) and

$n = 4$ for oxygen (O_2). Substitution of Equation (4.77) into Equation (4.80) yields the relation between the current generated and the concentration of the reactant:

$$J = nF \left(\frac{1}{h_m} + \frac{L}{\mathcal{D}^{\text{eff}}} \right)^{-1} (C_0 - C_I) \qquad (4.81)$$

This indicates that the current density generated in the electrode is proportional to the concentration difference $C_0 - C_I$, and J increases with C_0, the concentration of the reactant in the flow channel next to the electrode. For a given electrode and the operating condition, C_0 is fixed, then J increases as C_I is lowered. However, the minimum concentration at the backing/catalyst layer interface is $C_I = 0$, the maximum possible current density, or the limiting current density, becomes

$$J_L = nF \left(\frac{1}{h_m} + \frac{L}{\mathcal{D}^{\text{eff}}} \right)^{-1} C_0 \qquad (4.82)$$

For this condition of $C_I = 0$, physically it means that all the reactant supplied is completely consumed by the electrochemical reaction right at the backing/catalyst layer interface. Any further attempt to increase the current drawn from the electrode cannot be supported by the limited rate of reactant transport, hence the limiting condition for the current density drawn has been reached.

From Equations (4.82) and (4.79), it is seen that the limiting current density is influenced by the operating conditions (such as temperature T, pressure P, electrode flooding), design conditions (such as porosity ϕ, electrode-backing layer thickness L) and the flow conditions of the reactant in the flow channel, that influences the mass transfer coefficient h_m. Therefore, in general, we have

$$J_L = f(\text{operating conditions, design conditions, flow conditions})$$

Although J_L is difficult to predict accurately, it is much easier to measure experimentally.

Combining Equations (4.81) and (4.82) results in

$$\frac{J}{J_L} = 1 - \frac{C_I}{C_0} \qquad (4.83)$$

Then the concentration at the backing/catalyst layer interface can be obtained as

$$\frac{C_0}{C_I} = \frac{J_L}{J_L - J} \qquad (4.84)$$

This suggests that C_I may be easily determined if both J and J_L can be measured under the identical operating and flow conditions for a given fuel cell.

Now recall the Nernst equation derived in Chapter 2:

$$E_r(T, P_i) = E_r(T, P) - \frac{RT}{nF} \ln \prod_{i=1}^{N} \left(\frac{P_i}{P} \right)^{(v_i'' - v_i')/v_F'} \tag{4.85}$$

expressing the reversible cell potential in terms of the partial pressures for each reactant. The Equation (4.85) was derived for ideal gas mixtures, and the second term on the right-hand side of the equation, called the Nernst loss, is due to the decreased reactant concentration in the mixture. This can be illustrated that since for ideal gas mixtures

$$\frac{P_i}{P} = \frac{C_i}{C}$$

The Nernst loss can be expressed in terms of the reactant concentration.

Consider the concentration variation in an operation electrode shown in Figure 4.4, let us assume that all the electrochemical reactions in the electrode occur at the backing/catalyst layer interface with the concentration C_I, equivalent to a flat electrode. Further, all the processes associated with the surface reaction have negligible impact on the voltage losses, including the adsorption and desorption of the reactant on the electrode surface, the diffusion of the adsorbed reactant on the surface, the surface reaction including the charge transfer reaction described in the previous chapter, and the desorption of the reaction product on the reaction surface, as well as the transport of the desorbed product away from the surface, and so on. With these assumptions, the voltage loss arising from the reduced concentration at the reaction site, C_I, instead of the concentration as supplied to the electrode, C_0, becomes equal to the Nernst loss, thus the concentration overpotential may be written as

$$\eta_{\text{conc}} = \frac{RT}{nF} \ln \prod_{i=1}^{N} \left(\frac{C_I}{C_0} \right)^{(v_i'' - v_i')/v_F'} = \frac{RT}{nF} \ln \prod_{i=1}^{N} \left(\frac{C_0}{C_I} \right)^{(v_i' - v_i'')/v_F'} \tag{4.86}$$

Substituting Equation (4.84) yields

$$\eta_{\text{conc}} = \frac{RT}{nF} \ln \prod_{i=1}^{N} \left(\frac{J_L}{J_L - J} \right)^{(v_i' - v_i'')/v_F'} \tag{4.87}$$

Although concentrations of the reactants at the reaction sites are not easy to measure in operating fuel cells, the resulting overpotential is relatively easy to determine if Equation (4.87) is valid. Since the limiting current density is a function of the operating, design and reactant flow conditions, the concentration overpotential is also influenced by these conditions. One important objective of the fuel cell RD&D is to determine an optimal combination of these conditions that yields the best electrode (cell) performance and efforts are continuing to determine such an optimal combination.

Note that if reactants are consumed completely at the reaction sites,

$$v_i'' = 0$$

further for fuel as the only contributing factor for η_{conc}, Equation (4.87) can be reduced to

$$\eta_{conc} = \frac{RT}{nF} \ln\left(\frac{J_L}{J_L - J}\right) \tag{4.88}$$

an expression commonly seen in fuel cell literature.

It should be emphasized that Equation (4.87) or Equation (4.88) was derived under two main assumptions. The first assumption of a confined thin region of reaction is reasonable at high current densities, especially for the cathode side. However, at low-current densities, reaction occurs throughout the catalyst layer, and this assumption breaks down. The second assumption essentially implies that the rate of surface reaction processes is much faster than the rate of reactant transport so that the slow reactant transport process becomes the rate-limiting process for the entire physico-chemical processes occurring at the electrode — this is only true for operation at very high-current densities. Clearly, these two assumptions are only valid for high-current density conditions where the mass transport becomes rate determining. Thus, the use of Equations (4.87) or (4.88) in estimating η_{conc} should be cautioned and it is not valid for the majority of the cell polarization curve.

4.6.2   Reactant Supply From Bulk Phase: Engineering Approach

In the preceding subsection, the classic analysis leading to the Nernst voltage loss essentially implies that the current density distribution over the electrode surface is uniform. This is only valid when the current density is small when compared with the limiting current density J_L. Whereas for $J \gtrsim 0.5 J_L$, the current density distribution is no longer uniform, instead it decreases in the reactant flow direction over the electrode. This implication is clearly in conflict with the two assumptions underlying the classic analysis described in the previous subsection. As a result, the classic analysis is, strictly speaking, not valid for operating fuel cells, and a more consistent analysis is outlined in this subsection, which we shall call the engineering approach, or engineering type of analysis, because engineering correlations for the convection mass transfer will be employed for the analysis.

Consider a semi-infinite bulk reactant flow in parallel with a flat-plate electrode as shown schematically in Figure 4.4. By convention, the x coordinate is taken along the electrode surface pointing in the direction of the bulk reactant flow (which is shown as upwards in the figure). The electrode is taken as flat electrode with height H in the flow direction x, and width W in the transverse direction. For small current densities (say $J \lesssim 0.5 J_L$), uniform current density at the electrode surface is a better approximation to the reality. Then the convective mass transfer rate at the electrode surface is, according to Equation (4.58)

$$\dot{N}_s'' = h_m(C_{i,\infty} - C_{i,s}) \tag{4.89}$$

where $C_{i,s}$ and $C_{i,\infty}$ are the concentration of the reactant at the electrode surface and in the bulk reactant, respectively. From the physical problem, $C_{i,\infty} > C_{i,s}$ and the direction of the mass transfer, $\dot{N}_{i,s}''$ is from the bulk reactant to the electrode surface. This is opposite to the situation described in Section 4.5.3 where $C_{i,s} > C_{i,\infty}$ and the direction of the mass transfer is from the surface into the flow; hence, the difference between Equations (4.89) and (4.58) since it is known that the direction of mass transfer by diffusion is from the region of high concentration to the region of low concentration.

Substituting the mass transfer coefficient h_m from Equation (4.74) and considering the Faraday's law, Equation (4.80), the concentration difference can be written as

$$C_{i,\infty} - C_{i,s}(x) = 2.21 \frac{Jx}{nF\mathcal{D}_{i,j}} \text{Re}^{-1/2}\text{Sc}^{-1/3}; \quad \text{Sc} \gtrsim 0.6 \tag{4.90}$$

Clearly, the concentration difference increases as the function of $x^{1/2}$, so that the average concentration difference from the inlet $(x=0)$ to the outlet $(x=H)$ can be determined from, for the rectangular electrode of the area $A = WH$

$$\overline{C_{i,\infty} - C_{i,s}(x)} \equiv \frac{1}{A} \int_0^A (C_{i,\infty} - C_{i,s})dA = \frac{1}{wH} \int_0^H (C_{i,\infty} - C_{i,s})(Wdx)$$

$$= \frac{1}{H} \int_0^H (C_{i,\infty} - C_{i,s})dx \tag{4.91}$$

From Equation (4.90), we obtain the corresponding average concentration difference as

$$\overline{C_{i,\infty} - C_{i,s}} = \frac{2}{3}(C_{i,\infty} - C_{i,s})\big|_{x=H} \tag{4.92}$$

That is, the concentration difference averaged over the electrode surface is equal to two-thirds of the difference at the outlet $(x = H)$.

On the other hand, when the current density is large (say, $J > 0.5J_L$), $C_{i,s} =$ constant is a better surface condition compared to $J =$ constant assumed earlier for low-current density situation. For $C_{i,s} =$ constant boundary condition, Equation (4.72) provides the convective mass transfer coefficient at the electrode surface for steady laminar flow of the reactant, which is almost always valid under fuel cell operating conditions. Substitution into Equation (4.89) and combining with Equation (4.80), an expression for the distribution of the current density results:

$$J(x) = 0.332 \frac{nF\mathcal{D}_{i,j}}{x}(C_{i,\infty} - C_{i,s})\text{Re}^{1/2}\text{Sc}^{1/3}; \quad \text{Sc} \gtrsim 0.6 \tag{4.93}$$

The Equation (4.93) reveals that the current density decreases with x as the function of $x^{1/2}$ for large values of J (i.e., $J \longrightarrow J_L$). The average current density over the entire electrode surface can be calculated as

$$\bar{J} \equiv \frac{1}{A} \int_0^A J dA = \frac{1}{WH} \int_0^H J(x)(Wdx)$$

$$= \frac{1}{H} \int_0^H J(x) dx \tag{4.94}$$

or

$$\bar{J} = 2 J(x)|_{x=H} = 0.664 \frac{nF\mathcal{D}_{i,j}}{H} (C_{i,\infty} - C_{i,s}) \text{Re}_H^{1/2} \text{Sc}^{1/3}; \quad \text{Sc} \gtrsim 0.6 \tag{4.95}$$

where $\text{Re}_H = \rho v_\infty H / \mu$. That is, the average current density for this case is twice the current density at the outlet ($x = H$). The limiting current density J_L due to the resistance to convective mass transfer may be defined as the one corresponding to $C_{i,s} = 0$, from Equation (4.93), we have then

$$J_L(x) = 0.332 \frac{nF\mathcal{D}_{i,j}}{x} C_{i,\infty} \text{Re}_H^{1/2} \text{Sc}^{1/3}; \quad \text{Sc} \gtrsim 0.6 \tag{4.96}$$

Similarly, the average limiting current density can be evaluated according to Equation (4.94), and it is

$$\bar{J}_L = 2 J_L(x)|_{x=H} = 0.664 \frac{nF\mathcal{D}_{i,j}}{H} C_{i,\infty} \text{Re}_H^{1/2} \text{Sc}^{1/3}; \quad \text{Sc} \gtrsim 0.6 \tag{4.97}$$

It should be emphasized that the limiting current density derived here is due to the limitation in the convective mass transfer over the electrode surface. The limiting current density owing to the limitation in the mass transport through the porous electrode can be derived similarly. However, the latter is more complicated, arising from the porous nature of the electrode and the flooding of liquid water that almost always occurs for low-temperature fuel cells. Therefore, it will be considered in a separate subsection later.

Example 4.3 Hydrogen at 25 °C and 1 atm is saturated with water vapor, and the hydrogen–water vapor mixture flows over an anode at the velocity $v_\infty = 0.3$ m/s. The anode is rectangular and the length in the flow direction is $H = 25$ cm.

1. Determine the distribution of the limiting current density due to the limitation of the convective mass transfer $J_L(x)$ and the corresponding average value $\bar{J}_L$;
2. If the current density $J = 0.5$ A/cm^2 for a certain operating condition, then determine the distribution of the hydrogen concentration at the electrode surface, $C_{\text{H}_2,s}(x)$ and the average surface hydrogen concentration, $\bar{C}_{\text{H}_2,s}$.

The binary diffusivity for the pair hydrogen–water vapor at the given condition can be taken as $\mathcal{D}_{\text{H}_2-\text{H}_2\text{O}}(g) = 0.807 \times 10^{-4} \text{m}^2/\text{s}$.

SOLUTION

First we assume that the hydrogen–water vapor mixture behaves like an ideal gas mixture. From Appendix A.1, the saturation pressure of water at 25 °C is

$$P_{\text{sat}}(25\,°C) = 3.169\,\text{kPa}$$

Then, the mole fraction for the water vapor in the mixture is determined

$$X_{H_2O(g)} = \frac{P_{\text{sat}}(25\,°C)}{P} = \frac{3.169 \times 10^3\,\text{Pa}}{101{,}325\,\text{Pa}} = 0.0313$$

and the mole fraction for hydrogen in the mixture is obtained as

$$X_{H_2} = 1 - X_{H_2O(g)} = 0.9687$$

The total concentration in the mixture can be calculated as

$$C = \frac{P}{RT} = \frac{101{,}325\,\text{Pa}}{8.314\,\text{J}/(\text{mol} \cdot \text{K}) \times (273 + 25)\,\text{K}} = 40.90\,\text{mol/m}^3$$

Therefore, the concentration of hydrogen in the bulk gas phase, far away from the electrode surface becomes

$$C_{H_2,\infty} = X_{H_2}C = 0.9687 \times 40.90 = 39.62\,\text{mol/m}^3$$

Since water vapor concentration is very small, $x_{H_2O(g)} = 0.0313$, as determined earlier, the thermophysical properties of the mixture may be approximated as the properties of hydrogen. From Appendix B, we have, at 300 K and 1 atm

$$\rho_\infty = 0.08078\,\text{kg/m}^3,\ \mu = 89.6 \times 10^{-7}\,\text{N} \cdot \text{s/m}^2,\ \nu = 111 \times 10^{-6}\,\text{m}^2/\text{s}$$

Then,

$$Sc = \frac{\nu}{D_{H_2-H_2O(g)}} = \frac{111 \times 10^{-6}\,\text{m}^2/\text{s}}{0.807 \times 10^{-4}\,\text{m}^2/\text{s}} = 1.375$$

$$Re = \frac{\nu_\infty x}{\nu} = \frac{0.3(\text{m/s})x}{111 \times 10^{-6}\,\text{m}^2/\text{s}} = 2703x \quad (x\ \text{in m})$$

$$Re_H = \frac{\nu_\infty H}{\nu} = \frac{0.3\,\text{m/s} \times 0.25\,\text{m}}{111 \times 10^{-6}\,\text{m}^2/\text{s}} = 675.7$$

(1) Since $n = 2$ for hydrogen, Equation (4.96) yields

$$J_L(x) = 0.332 \frac{nF\mathcal{D}_{H_2-H_2O(g)}}{x} C_{H_2,\infty} Re^{1/2} Sc^{1/3}$$

$$= 0.332 \times \frac{2 \times 96,487 \times 0.807 \times 10^{-4}}{x}$$

$$\times 39.62 \times (2703\, x)^{1/2} \times 1.375^{1/3}$$

or

$$J_L(x) = 11,840 x^{-1/2}\, A/m^2 \quad (x \text{ in m})$$

Hence, $J_L = 23,680\, A/m^2 = 2.368\, A/cm^2$ at $x = H = 0.25$ m.
From Equation (4.97), the average limiting current density is

$$\bar{J}_L = 2 J_L(x)|_{x=H} = 2 \times 11,840 \times 0.25^{-1/2} = 47,360\, A/m^2 = 4.736\, A/cm^2$$

(2) Since $J = 0.5\, A/cm^2 < 0.5 J_L$, Equation (4.90) may be used to determine the surface concentration of hydrogen:

$$C_{H_2,s}(x) = C_{H_2,\infty} - 2.21 \frac{Jx}{nF\mathcal{D}_{H_2-H_2O(g)}} Re^{-1/2} Sc^{-1/3}$$

$$= 39.62 - 2.21 \frac{0.5 \times 10^4 x}{2 \times 96,487 \times 0.807 \times 10^{-4}}$$

$$\times (2703\, x)^{-1/2} \times 1.375^{-1/3}$$

$$= 39.62 - 12.27\, x^{1/2}\, mol/m^3 \quad (x \text{ in m})$$

Hence, $C_{H_2,s}(H = 0.25\, m) = 33.49\, mol/m^3$.
The average hydrogen concentration at the electrode surface can be calculated, similar to Equation (4.91)

$$\bar{C}_{H_2,s} = \frac{1}{H} \int_0^H C_{H_2,s}(x)dx; \quad H = 0.25\, m$$

$$= 39.62 - 12.27 \times \frac{2}{3} H^{1/2}$$

$$= 39.62 - 12.27 \times \frac{2}{3} \times 0.25^{1/2}$$

$$= 35.53\, mol/m^3$$

COMMENT

1. The values of the limiting current density J_L determined are very large, suggesting that the rate of mass transfer in the bulk gas phase is sufficiently fast and is not

significant in limiting the current density obtainable from an electrode, as long as very moderate forced convection is provided. However, mass transfer within the porous electrode may well be dominant factor in limiting the maximum current density attainable.

2. The hydrogen concentration at the electrode surface is only slightly smaller than the concentration value in the bulk gas phase. Also, $C_{H_2,s}$ reduces slightly from $x = 0$ to $x = H$ even though the current density drawn from the electrode is appreciable ($J = 0.5$ A/cm^2).

3. The readers are urged to plot $J_L(x)$ and $\bar{J}_L$ on the same diagram to observe visually the variation and magnitude of the current density. Similarly, plot $C_{H_2,s}$ and $\bar{C}_{H_2,s}$ on the same diagram.

4. This example illustrates the technique of estimating the forced convection needed to ensure the best performance an electrode is capable of. In fact, optimal fuel-cell performance may be determined as a balance between the pumping power associated with the reactant flow (power loss) and the power (current density) output from the cell (performance gain).

5. Note that in this example, we have evaluated, for simplicity, the thermophysical properties in the bulk gas phase as a reasonable approximation, instead of at the film condition.

4.6.3   Reactant Supply From Parallel Channel Flow: Engineering Approach

Since each cell potential is only about 0.7 V~0.8 V under practical operating conditions, many individual cells are often stacked together to increase the output potential and power density. Then it is imperative to reduce the spacings of the reactant supply compartment (often called the **reactant supply flow channel**) to minimize the size and weight of fuel cell stacks. The typical spacing of the flow channel b is on the order of mm or smaller. As a result, the boundary layers grow from each surface of the channel at the inlet and merge quickly at some distance downstream, typically on the order of 1 cm or less for laminar flow commonly encountered in fuel cells. Therefore, this short entrance length may be neglected and the entire flow in the channel may be considered fully developed. The reactant is transferred into the electrode surface for electrode reaction, while the other channel surface is mass impermeable. This situation is equivalent to one surface insulated and the other surface having heat transfer with the flow in the channel. For such a heat transfer problem, it has been shown that the Nusselt number for steady laminar flow is.[6]

$$\text{Nu} \equiv \frac{h_H D_h}{k} = \begin{cases} 5.39 & \text{For Uniform Surface Heat Flux, } q_s'' = \text{const.} \\ 4.86 & \text{For Uniform Surface Temperature, } T_s = \text{const.} \end{cases} \quad (4.98)$$

With the modified Reynolds analogy, the mass transfer coefficient may be obtained over the electrode surface, in analogy with the heat transfer

$$\text{Sh} \equiv \frac{h_m D_h}{\mathcal{D}} = \begin{cases} 5.39 & \text{For Uniform Surface Mass Flux, } \dot{N}_s'' = \text{const.} \\ 4.86 & \text{For Uniform Surface Concentration, } C_s = \text{const.} \end{cases} \quad (4.99)$$

where the hydraulic diameter for the channel flow of the spacing b is

$$D_h \cong 2b \tag{4.100}$$

because the channel depth, W, is much larger than the channel spacing, b.

In contrast, momentum loss due to the frictional viscous effect occurs at the both walls of the channel. For such internal flows, the momentum loss is expressed in terms of pressure loss, defined as

$$\Delta P = f \frac{H}{D_h} \frac{\rho v_m^2}{2} \tag{4.101}$$

where f is called friction factor (a dimensionless parameter), and v_m is the mean flow velocity in the channel, averaged over the channel cross section. From the definition of the friction coefficient, Equation (4.50), it can be shown that

$$C_f = \frac{f}{4} \tag{4.102}$$

For a steady laminar flow in a channel, the friction factor is [4]

$$f = \frac{96}{\text{Re}}; \quad \text{Re} = \frac{\rho v_m D_h}{\mu} \tag{4.103}$$

Clearly, the friction factor, representing the momentum change (actually pressure loss in this case), as well as the heat and mass transfer coefficients are constant, characteristic of fully developed flows inside conduits. Developing flows have larger values of the heat and mass transfer coefficients than the fully developed flows given above and the developing flow is always present, preceding the establishment of developed flow. Therefore, Equation (4.99) will be conservative in estimating the rate of mass transfer. Note that a steady laminar flow in the channel exists if Re $\lesssim 2000$.

The rate of convective mass transfer at the electrode surface is calculated, similar to Equation (4.89),

$$\dot{N}_{i,s}'' = h_m (C_{i,m} - C_{i,s}) \tag{4.104}$$

where $C_{i,m}$ is the mean concentration of the reactant species i in the flow channel, averaged over the channel cross section and it typically decreases along the flow direction (x) due to the reactant transfer into the electrode.

The variation of the mean concentration, $C_{i,m}(x)$, along the flow direction may be determined by the species conservation principle. Consider the channel flow shown in Figure 4.5, the mixture of the reactant species i and inerts moves at the total molar flow rate $C_m v_m A_c$ at the location x, a total molar flow carried by the bulk mean motion. Where A_c denotes the channel cross-sectional area.

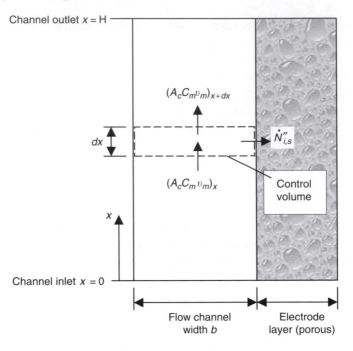

Figure 4.5 Control volume for channel flow and mass transfer into an electrode surface.

Applying the conservation of mass principle to the differential control volume shown, we have

$$(A_c C_m v_m)_x - (A_c C_m v_m)_{x+dx} - \dot{N}''_{i,s} W dx = 0$$

Expressing the second term, $(A_c C_m v_m)_{x+dx}$, in Taylor series and keeping up to the first-order term, results in

$$\frac{d}{dx}(A_c C_m v_m) = -\dot{N}''_{i,s} W$$

where W is the width of the electrode surface ($W \gg b$), over which mass transfer, $\dot{N}''_{i,s}$, occurs and hence $A_c = Wb$. Further for steady channel flow, $v_m =$ constant. Then Equation (4.105) becomes for the variation of the total concentration in the channel

$$\frac{dC_m}{dx} = -\frac{\dot{N}''_{i,s}}{b v_m}$$

Because the mixture consists of the reactant species i and inert species whose concentration remains constant, we finally have

$$\frac{dC_{i,m}}{dx} = -\frac{\dot{N}''_{i,s}}{b v_m} \tag{4.105}$$

Now consider the situation of small current density drawn form the electrode, say $J < 0.5J_L$, a better boundary condition at the electrode surface is approximated by $J = $ constant. From Faraday's law of electrochemistry, $J/(nF) = \dot{N}''_{i,s}$, we have $\dot{N}''_{i,s} = $ constant. Then Equation (4.105) yields upon integrating from $x = 0$

$$C_{i,m}(x) = C_{i,m,in} - \frac{J/(nF)}{bv_m}x \qquad (4.106)$$

where the subscript "in" denotes the quantity at the flow channel inlet. Accordingly, the mean concentration for the reactant decreases linearly along the channel. Also, Equation (4.104) indicates that the concentration difference $(C_{i,m} - C_{i,s})$ remains constant, implying that the surface concentration $C_{i,s}$ also decreases linearly, as $C_{i,m}$ does because h_m is a constant for fully developed flows. However, near the channel inlet, h_m is large due to the thinner boundary layers, thus $(C_{i,m} - C_{i,s})$ is smaller there and $C_{i,s}$ decreases more slowly in the region near the channel inlet as well.

Because of the linear distribution, the average concentration $C_{i,s}$ and the average concentration difference $(C_{i,m} - C_{i,s})$ are equal to the arithmetic average of their respective inlet and outlet values.

On the other hand, when the current density drawn is large, such as $J > 0.5J_L$, a closer approximation for the condition at the electrode surface is $C_{i,s} = $ constant. Then defining $\Delta C = C_{i,m} - C_{i,s}$, Equation (4.105) may be written as, with the help of Equation (4.104)

$$\frac{d(\Delta C)}{dx} = -\frac{h_m}{bv_m}(\Delta C)$$

Separating variables and integrating from the channel inlet to location x in the channel along the flow direction

$$\int_{\Delta C_{in}}^{\Delta C} \frac{d(\Delta C)}{\Delta C} = -\frac{h_m}{bv_m} \int_0^x dx$$

or

$$\frac{\Delta C}{\Delta C_{in}} \equiv \frac{C_{i,m}(x) - C_{i,s}}{(C_{i,m} - C_{i,s})_{in}} = \exp\left(-\frac{h_m x}{bv_m}\right) \qquad (4.107)$$

This result indicates that the mean concentration of the reactant decreases exponentially with the distance along the channel. Combining with the result in Equation (4.106), we now have a whole picture of the mean concentration change: $C_{i,m}$ changes from a linear decrease at low current densities to exponential decay as the current density is increased towards the limiting current density $(J \longrightarrow J_L)$. At the channel outlet, $x = H$, Equation (4.107) becomes

$$\frac{\Delta C_{out}}{\Delta C_{in}} \equiv \frac{C_{i,m,out} - C_{i,s}}{C_{i,m,in} - C_{i,s}} = \exp\left(-\frac{h_m H}{bv_m}\right) \qquad (4.108)$$

where the subscript "out" represents the quantity at the channel outlet.

Although determination of the total amount of mass transfer over the electrode surface, $\dot{N}_{i,s}$, is complicated by the exponential decay of the mean reactant concentration, a simple expression may be derived as follows. Consider the entire channel as the control volume, then the conservation of species yields

$$\dot{N}_{i,s} = (Wb)v_m(C_{i,m,\text{in}} - C_{i,m,\text{out}})$$
$$= (Wb)v_m \left[(C_{i,m} - C_{i,s})_{\text{in}} - (C_{i,m} - C_{i,s})_{\text{out}}\right]$$
$$= W(bv_m)(\Delta C_{\text{in}} - \Delta C_{\text{out}})$$

Substituting for (bv_m) from Equation (4.108), we have

$$\dot{N}_{i,s} = Ah_m\Delta C_{\text{lm}}; \quad \text{for } C_{i,s} = \text{constant} \tag{4.109}$$

where $A = WH$ is the electrode surface area over which mass transfer occurs, and ΔC_{lm} is the logarithmic mean difference for the concentration in the channel and at the electrode surface, defined as

$$\Delta C_{\text{lm}} = \frac{\Delta C_{\text{in}} - \Delta C_{\text{out}}}{\ln\left(\frac{\Delta C_{\text{in}}}{\Delta C_{\text{out}}}\right)} \tag{4.110}$$

and $\Delta C_{\text{in}} = C_{i,m,\text{in}} - C_{i,s}$, $\Delta C_{\text{out}} = C_{i,m,\text{out}} - C_{i,s}$. Note that ΔC_{lm} represents the average value of the concentration difference over the entire channel length and this logarithmic nature of the average concentration difference arises from the exponential nature of the concentration decay in the channel. It can also be shown that $\Delta C_{\text{lm}} = \Delta C_{\text{in}} = \Delta C_{\text{out}}$ if the concentration difference ΔC remains constant along the entire channel. Therefore, Equation (4.109) can be regarded as a general expression for the total rate of mass transfer over the entire electrode surface for both constant surface flux and constant surface concentration boundary conditions.

Substituting Equation (4.107) into Equation (4.104), we obtain the local current density corresponding to the rate of mass transfer is

$$J(x) = nFh_m \left(C_{i,m} - C_{i,s}\right)_{\text{in}} \exp\left(-\frac{h_m x}{bv_m}\right) \tag{4.111}$$

Similarly, the current density averaged over the entire electrode surface is

$$\bar{J} = nFh_m\Delta C_{\text{lm}} \tag{4.112}$$

The limiting current density is approached when $C_{i,s} \longrightarrow 0$, and thus

$$J_L(x) = nFh_m C_{i,m,\text{in}} \exp\left(-\frac{h_m x}{bv_m}\right) \tag{4.113}$$

$$\bar{J}_L = nFh_m \left[\frac{C_{i,m,\text{in}} - C_{i,m,\text{out}}}{\ln\left(\frac{C_{i,m,\text{in}}}{C_{i,m,\text{out}}}\right)}\right] \tag{4.114}$$

Thus, both the current density $J(x)$ and the limiting current density $J_L(x)$ decrease exponentially along the channel length and their respective average values, $\bar{J}$ and $\bar{J}_L$, are obtained with the logarithmic mean concentration difference.

Example 4.4 Consider a channel flow is formed by an impermeable wall and the flat electrode surface at a distance $b = 2$ mm apart. Air at $25\,^\circ$C and 1 atm is fed to the cathode electrode of $H = 30$ cm and the air flow is parallel to the channel walls with a steady velocity of $v_m = 3$ m/s. Determine the distribution of the limiting current density due to the limitation of the convective mass transfer $J_L(x)$ and the corresponding average value $\bar{J}_L$.

SOLUTION

First, we assume that air is composed of 21% O_2 and 79% N_2 by mole and it behaves like an ideal gas at the given temperature and pressure. These assumptions are realistic for the conditions stated in the problem.

The thermophysical properties are, at $T = 25\,^\circ$C $= 298$ K and pressure of 1 atm

$$\rho = 1.1614\,\text{kg/m}^3, \quad \mu = 184.6 \times 10^{-7}\,\text{N} \cdot \text{s/m}^2, \quad \nu = 15.89 \times 10^{-6}\,\text{m}^2/\text{s}$$

$$k = 26.3 \times 10^{-3}\,\text{W/(m} \cdot \text{K)}, \quad \mathcal{D}_{O_2 - N_2} = 0.18 \times 10^{-4}\,\text{m}^2/\text{s (at } T = 273\,\text{K)}$$

The binary diffusion coefficient at $T = 298$ K ($25\,^\circ$C) and 1 atm can be estimated as

$$\mathcal{D}_{O_2 - N_2}(298\,\text{K}) = \mathcal{D}_{O_2 - N_2}(273\,\text{K}) \left(\frac{298\,\text{K}}{273\,\text{K}}\right)^{3/2} \cong 0.21 \times 10^{-4}\,\text{m}^2/\text{s}$$

The hydraulic diameter for the channel flow is $D_h = 2b$. Hence, the Reynolds number is

$$\text{Re} \equiv \frac{\rho v_m D_h}{\mu} = \frac{v_m D_h}{\nu} = \frac{3\,\text{m/s} \times 2 \times 2 \times 10^{-3}\,\text{m}}{15.89 \times 10^{-6}\,\text{m}^2/\text{s}} = 755.2 < 2000$$

Therefore the flow is laminar. Assume the flow is fully developed, the convective mass transfer coefficient can be obtained from Equation (4.99), for the determination of the limiting current density which occurs at $C_{O_2, s} = 0$ (i.e., $C_s = $ constant)

$$h_m = \text{Sh} \frac{\mathcal{D}_{O_2 - N_2}}{D_h} = 4.86 \times \frac{0.21 \times 10^{-4}\,\text{m}^2/\text{s}}{2 \times 2 \times 10^{-3}\,\text{m}} = 0.02552\,\text{m/s}$$

Because the mole fraction of O_2 in the air is $X_{O_2} = 0.21$, the concentration of O_2 at the channel inlet can be calculated

$$C_{O_2, m, \text{in}} = X_{O_2}\left(\frac{P}{RT}\right) = 0.21 \times \frac{101{,}325\,\text{Pa}}{8.314\,\text{J/(mol} \cdot \text{K)} \times 298\,\text{K}} = 8.588\,\text{mol/m}^3$$

We can then obtain the limiting current density J_L corresponding to the rate of oxygen transfer from Equation (4.113)

$$J_L(x) = (nF)h_m C_{O_2,m,\text{in}} \exp\left(-\frac{h_m x}{bv_m}\right)$$

$$= (4 \times 96{,}487)\frac{C}{\text{mol } O_2} \times 0.02552 \,\text{m/s} \times 8.588\frac{\text{mol } O_2}{\text{m}^3}$$

$$\times \exp\left(-\frac{0.02552 \,\text{m/s} \times x}{2 \times 10^{-3}\,\text{m} \times 3\,\text{m/s}}\right)$$

$$= 8.459 \times 10^4 \exp(-4.253\,x); \quad \text{A/m}^2 \quad (x \text{ in m}).$$

or $J_L(x) = 8.459 \exp(-0.04253\,x);$ A/cm^2 (x in cm).

Hence, $J_L = 8.459\,\text{A/cm}^2$ at the channel inlet ($x=0$) and 2.362 A/cm^2 at the channel exit ($x = H = 30$ cm).

To calculate the average limiting current density $\bar{J}_L$ according to Equation (4.114), we need to determine the outlet average concentration of the oxygen. From Equation (4.108) with $C_{O_2,s} = 0$, we have

$$C_{i,m,\text{out}} = C_{i,m,\text{in}} \exp\left(-\frac{h_m H}{bv_m}\right)$$

$$= 8.588 \,\text{mol/m}^3 \times \exp\left(-\frac{0.02552 \,\text{m/s} \times 0.3\,\text{m}}{2 \times 10^{-3}\,\text{m} \times 3\,\text{m/s}}\right)$$

$$= 2.397 \,\text{mol/m}^3$$

Then

$$\bar{J}_L = (nF)h_m \left[\frac{C_{O_2,m,\text{in}} - C_{O_2,m,\text{out}}}{\ln\left(\frac{C_{O_2,m,\text{in}}}{C_{O_2,m,\text{out}}}\right)}\right]$$

$$= (4 \times 96{,}487)\,\text{C/mol } O_2 \times 0.02552\,\text{m/s} \times \left[\frac{(8.588 - 2.397)\,\text{mol } O_2/\text{m}^3}{\ln\left(\frac{8.588}{2.397}\right)}\right]$$

$$= 4.778 \times 10^4 \,\text{A/m}^2$$

$$= 4.778 \,\text{A/cm}^2$$

In fact, $\bar{J}_L$ can be easily obtained for this problem by integrating $J_L(x)$, and the same result is calculated by

$$\bar{J}_L = \frac{1}{H}\int_0^H J_L(x)dx$$

since $J_L(x)$ is already known.

COMMENT

1. The limiting current density $J_L(x)$, the average value $\bar{J}_L$ and the mean concentration of the oxygen in the flow channel have been computed for $0 \leq x \leq 0.3$ m, and the results are plotted in Figure 4.6. Clearly the marked decrease in $J_L(x)$ is due to the same marked decrease in the concentration of the reactant O_2. One important objective of the practical flow-channel design is to make the average reactant concentration over the electrode surface as high as possible for a given inlet condition.

2. One can verify that as the bulk flow velocity v_m is increased, both $J_L(x)$ and $C_{O_2,m}(x)$ will decrease more slowly over the channel length. Especially when $v_m \longrightarrow \infty$, both J_L and $C_{O_2,m}$ will approach to a constant equal to their respective inlet value. However, the pressure loss associated with the cathode air flow increases quickly with v_m and the parasitic power loss needed for air compression dictates that v_m should be as low as possible. Pressure loss for reactant gas flow is a significant issue in practical fuel cells where the flow channels typically have small cross sections.

3. The calculated values for the limiting current density $J_L(x)$ are obtained when the resistance to convection mass transfer in the bulk gas phase is the determining step in limiting the current density. In reality, practical fuel cells operating on atmospheric air at $25\,^\circ$C are not likely to be higher than $1\,A/cm^2$, implying that the processes in the electrode structure and in the electrolyte may impose more severe limitations on the cell performance. Thus, the primary purpose of the present calculations is two-fold: (i) to show that forced convection is required to obtain a good fuel cell performance; (ii) to show that channel flow with air is not the factor limiting the current density (or fuel cell performance) in

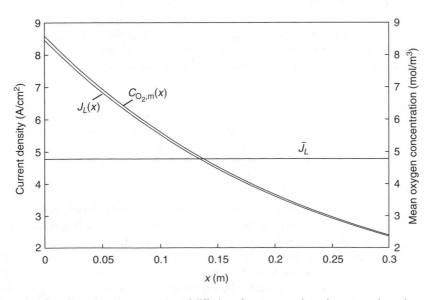

Figure 4.6 One-dimentional representation of diffusion of reactant gas through a porous electrode.

practice as long as moderate forced convection is used. Therefore, the practical significance of the present calculations is that they provide a method in estimating the amount of forced convection needed to ensure that channel flow does not become the dominant factor in limiting the cell current density and hence the cell performance.

4. The entrance length over which the convection mass transfer coefficient is higher can be estimated by

$$H_{entrance} \approx 0.05 \, ReSc \, D_h = 0.05 \, Re \left(\frac{\nu}{\mathcal{D}_{O_2-N_2}} \right) D_h$$

$$= 0.05 \times 755.2 \times \frac{15.89 \times 10^{-6}}{0.21 \times 10^{-4}} \times (2 \times 2 \times 10^{-3} \, \text{m}) = 0.114 \, \text{m}$$

or

$$\frac{H_{entrance}}{H} \approx 38\%$$

thus the entrance length may not be neglected and the actual rate of mass transfer and the limiting current density would be higher than what is calculated in this example. Thus the estimation on needed forced convection for good fuel cell performance is conservative when the entrance effect is neglected. However, for practical fuel cells the flow channel would be much longer than 30 cm given in this example, typically in the order of several meters to tens-of-meters long, for example, in PEM fuel cells. Hence, the entrance effect would be indeed minimal.

5. Note that in the present calculations, the thermophysical properties have been approximated as those of the reactant mixture at the channel inlet instead of at the bulk mean condition (i.e., the arithmetic mean of the inlet and outlet values) — that is, accounting for the changes in the gas composition along the channel length. Further, the effect of mass transfer rate on the transport processes has been neglected, which can be corrected as pointed out earlier.

Before concluding this section, it might be important to point out that the pressure loss ΔP associated with the reactant flow through the channel can be determined from Equations (4.101) and (4.103) and the power required to overcome this pressure loss (pumping power) is

$$\dot{P} = \Delta P \cdot \dot{Q} \tag{4.115}$$

where $\dot{Q}$ is the volumetric flow rate for the channel flow. The pumping power represents a form of energy loss for the fuel cell systems and since it is often derived from the power produced by the fuel cell in practice, it is one form of the parasitic energy loss for practical fuel cell systems. For laminar flows, the pressure loss (and hence the pumping power) increases with the flow velocity through the channel as well as the channel spacing b. In reality, an optimal design can be achieved by balancing between the fuel cell performance improvement and the pumping power requirement.

Example 4.5   Determine the pressure loss for the channel flow given in the previous example and the corresponding pumping power, assuming the width of the electrode $W = H = 30$ cm.

SOLUTION

Both the thermophysical properties and the Reynolds number are available from the previous example. From Equations (4.101) and (4.102), we have

$$\Delta P = f \frac{H}{D_h} \frac{\rho v_m^2}{2} = \frac{96}{\text{Re}} \frac{H}{2b} \frac{\rho v_m^2}{2}$$

$$= \frac{96}{755.2} \times \frac{0.3 \text{ m}}{2 \times 2 \times 10^{-3} \text{ m}} \times \frac{1.1614 \text{ kg/m}^3 \times (3 \text{ m/s})^2}{2}$$

$$= 49.83 \text{ Pa}$$

Then the pumping power required is

$$\dot{P} = \Delta P \cdot \dot{Q} = \Delta P \cdot (bW) v_m = 49.83 \text{ Pa} \times (2 \times 10^{-3} \text{ m} \times 0.3 \text{ m}) \times 3 \text{ m/s}$$

$$= 0.08969 \text{ W}$$

COMMENT

It seems that the pumping power is very small for this example because we have operated the cathode flow at the atmospheric pressure and temperature and the relatively large channel spacing. For example, if the channel spacing is reduced to $b/2 = 1$ mm while the flow velocity is increased to $2v_m = 6$ m/s so that the total flow rate through the channel remains the same, it can be shown that both $J_L(x)$ and $\bar{J}_L$ remain the same, while the pressure loss (here the pumping power) is increased by eight times! Typically, the pumping power for the air flow can be as much as over 30% of the power produced by a fuel cell stack.

4.6.4   Reactant Supply Through Porous Electrode Structure: Engineering Approach

The preceding sections have shown that the processes occurring in the bulk reactant phase flowing in the flow channel may not impose significant limitation on the fuel cell performance as long as appropriate design is employed. In this section, we carry out similar analysis for the reactant transport through the porous structure of an electrode in order to determine whether this process incurs limitations on the cell performance.

   The reactant supplied to a fuel cell usually flows in flow channels parallel to the electrode surface, and diffusion is the dominant mechanism for the cross-stream mass transport for the reactant through the porous electrode structure, as shown in Figure 4.7. Because we are interested in the impact of the diffusion process on the fuel cell performance (how it limits the current density achievable), we neglect the resistance to the convection mass transfer over the electrode surface, so that the reactant concentration at the electrode surface, $C_{i,1}$, is identical to the mean concentration $C_{i,m}(x)$ next in the flow channel, as determined in the previous section. Further, we

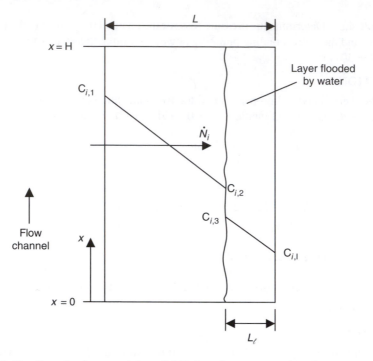

Figure 4.7 One-dimensional representation of diffusion of reactant gas through a porous electrode partially flooded by liquid water (or liquid electrolyte).

assume a portion of the electrode pore region is flooded by liquid water (or liquid electrolyte), as possible in low temperature H_2–O_2 fuel cells. In the present context of one-dimensional analysis, the thickness of the water flooded layer is denoted as L_ℓ and the thickness of the entire electrode-backing layer is L. Figure 4.7 illustrates the physical problem under consideration.

Using Fick's law of diffusion for the reactant in both the gas and liquid region, we obtain for steady process

$$\dot{N}_i'' = \mathcal{D}_{i-g}^{\text{eff}} \frac{C_{i,m} - C_{i,2}}{L_g} \tag{4.116}$$

and

$$\dot{N}_i'' = \mathcal{D}_{i-\ell}^{\text{eff}} \frac{C_{i,3} - C_{i,I}}{L_\ell} \tag{4.117}$$

where $\mathcal{D}_{i-g}^{\text{eff}}$ and $\mathcal{D}_{i-\ell}^{\text{eff}}$ are the effective diffusion coefficient corrected with porosity effect shown in Equation (4.79), and $L_g = L - L_\ell$.

Since the reactant gas is only weakly soluble in liquid water under the typical conditions encountered in fuel cell environment, Henry's law may be used to relate the reactant concentration in the liquid, $C_{i,3}$, to the concentration in the gas, $C_{i,2}$, at

Table 4.1   Henry's Constant for the
Dissolution of Oxygen and Hydrogen Gases
in Liquid Water at Moderate Pressures[7]

T (K)	$\mathcal{H}_{O_2}$ Pa · m³/mol	$\mathcal{H}_{H_2}$ Pa · m³/mol
273	45900	104000
280	54900	111000
290	68500	120000
300	82600	129000
310	93400	138000
320	103000	143000
323	106000	144000
333	115000	143000
343	123000	142000
353	128000	142000
363	132000	142000

the liquid–gas interface:

$$C_{i,3} = \frac{RT}{\mathcal{H}_i} C_{i,2} = \frac{P_{i,2}}{\mathcal{H}_i} \tag{4.118}$$

where $\mathcal{H}_i$ is Henry's constant for reactant gas dissolution in liquid water; it depends on the temperature at which the dissolution occurs and the pressure dependence can usually be neglected for the low pressures commonly encountered in fuel cell environment. For example, Table 4.1 shows the Henry's constant for the dissolution of oxygen and hydrogen in liquid water, respectively, at the low to moderate pressures.[7] Henry's constant for oxygen dissolution in fully hydrated Nafion membrane, as applicable in PEM fuel cells, may be determined from the following correlation[8]

$$\mathcal{H}_{O_2-m} = 0.1013 \exp\left(14.1 - \frac{666}{T} \right) \tag{4.119}$$

where the temperature T is in K, and $\mathcal{H}_{O_2-m}$ is in the unit of Pa · m³/mol or J/mol.

Combining Equations (4.116)–(4.118), we obtain the molar flux through the porous electrode structure

$$\dot{N}_i'' = \frac{\frac{RT}{\mathcal{H}_i} C_{i,m} - C_{i,I}}{\frac{RT}{\mathcal{H}_i} \frac{(1-\ell_\ell)L}{\mathcal{D}_{i-g}^{\text{eff}}} + \frac{\ell_\ell L}{\mathcal{D}_{i-\ell}^{\text{eff}}}} \tag{4.120}$$

Therefore current density obtainable for this molar flux is

$$J(x) = nF\dot{N}_i'' = \frac{nF\left(\frac{RT}{\mathcal{H}_i}C_{i,m} - C_{i,I}\right)}{\frac{RT}{\mathcal{H}_i}\frac{(1-\ell_\ell)L}{\mathcal{D}_{i-g}^{\text{eff}}} + \frac{\ell_\ell L}{\mathcal{D}_{i-\ell}^{\text{eff}}}} \tag{4.121}$$

where $\ell_\ell = (L_\ell A)/(LA)$ represents the fraction of the electrode structure that is flooded by liquid water. The two terms in the denominator are the mass transfer resistance for the unflooded and flooded region of the electrode, respectively. This result suggests that once flooding occurs, the rate of reactant transport, hence the corresponding current density obtainable can be calculated by treating the entire electrode as flooded and correcting the resistance to mass transfer for the unflooded region as well.

At high current densities (i.e., $J \longrightarrow J_L$), the reactant concentration $C_{i,I} \longrightarrow 0$. Thus, we derive the limiting current density as

$$J_L(x) = \frac{nF\left(\frac{RT}{\mathcal{H}_i}C_{i,m}\right)}{\frac{RT}{\mathcal{H}_i}\frac{(1-\ell_\ell)L}{\mathcal{D}_{i-g}^{\text{eff}}} + \frac{\ell_\ell L}{\mathcal{D}_{i-\ell}^{\text{eff}}}} \tag{4.122}$$

Following the similar derivation, we can easily obtain for unflooded electrode

$$J(x) = \frac{nF(C_{i,m} - C_{i,I})}{\frac{L}{\mathcal{D}_{i-g}^{\text{eff}}}} \tag{4.123}$$

and

$$J_L(x) = \frac{nFC_{i,m}}{\frac{L}{\mathcal{D}_{i-g}^{\text{eff}}}} \tag{4.124}$$

Comparing the results for flooded and unflooded situation reveals that flooding reduces significantly the reactant concentration, hence the current density J and the limiting current density J_L. This implies that flooding should be avoided for practical fuel cells.

Example 4.6   An air cathode operates on atmospheric air at $25\,°C$. The cathode electrode is $230\,\mu m$ thick with the porosity of $\phi = 40\%$. The spacing b and length H of the flow channel as well as the flow velocity in the channel v_m are the same as in Example 4.4. Determine the maximum current density attainable in accordance with the air transport through the cathode when (a) the cathode is not flooded by liquid water, and (b) 1% of the electrode is flooded by liquid water.

SOLUTION

Air at the given conditions behaves like an ideal gas and its composition can be approximated as 21% O_2 and 79% N_2 by mole. From Example 4.4, the bulk diffusion coefficient for O_2–N_2 pair is

$$\mathcal{D}_{O_2-N_2} \cong 0.21 \times 10^{-4}\, m^2/s \quad \text{at } T = 298\, K \text{ and } 1\, atm$$

Appendix B gives the bulk diffusion coefficient for O_2 in liquid water as

$$\mathcal{D}_{O_2-\ell} = 0.24 \times 10^{-8}\, m^2/s \quad \text{at } T = 298\, K \text{ and } 1\, atm$$

According to the Bruggeman-type correction, the effective diffusion coefficients for diffusion through the porous electrode structure are calculated as follows

$$\mathcal{D}_{O_2-N_2}^{eff} = \mathcal{D}_{O_2-N_2}\phi^{3/2} = 0.21 \times 10^{-4} \times 0.4^{3/2} = 0.5313 \times 10^{-5}\, m^2/s$$

$$\mathcal{D}_{O_2-\ell}^{eff} = \mathcal{D}_{O_2-\ell}\phi^{3/2} = 0.24 \times 10^{-8} \times 0.4^{3/2} = 0.6072 \times 10^{-9}\, m^2/s$$

At large current densities ($J \longrightarrow J_L$), the mean concentration of oxygen in the flow channel may be estimated by

$$C_{O_2,m}(x) = C_{O_2,m,in} \exp\left(-\frac{h_m x}{b v_m}\right)$$

with $b = 2 \times 10^{-3}$ m, $v_m = 3$ m/s, $h_m = 0.02552$ m/s from Example 4.4, we obtain

$$C_{O_2,m}(x) = 8.588 \exp(-4.253x)\, mol/m^3, \quad (x \text{ in m}).$$

(a) The limiting current density $J_L(x)$ for unflooded cathode is

$$
\begin{aligned}
J_L(x) &= \frac{n F \mathcal{D}_{O_2-N_2}^{eff} C_{i,m}(x)}{L} \\
&= \frac{\begin{array}{c}(4 \times 96{,}487)\, C/mol\, O_2 \times 0.5313 \times 10^{-5}\, m^2/s \\ \times 8.588 \exp(-4.253x)\, mol\, O_2/m^3\end{array}}{230 \times 10^{-6}\, m} \\
&= 7.657 \times 10^4 \exp(-4.253x)\, A/m^2, \quad (x \text{ in m}).
\end{aligned}
$$

Therefore, at the channel inlet ($x = 0$), $J_L = 7.657\, A/cm^2$; whereas at the channel outlet ($x = H = 0.3$ m), $J_L = 2.138\, A/cm^2$. The limiting current density averaged over the entire cathode surface is

$$
\begin{aligned}
\bar{J}_L &= \frac{1}{H}\int_0^H J_L(x)dx = \frac{7.657 \times 10^4}{-4.253\, H}\left[\exp(-4.253\, H) - 1\right] \\
&= \frac{7.657 \times 10^4}{4.253 \times 0.3}\left[1 - \exp(-4.253 \times 0.3)\right] \\
&= 4.326 \times 10^4\, A/m^2 = 4.326\, A/cm^2
\end{aligned}
$$

(b) From Table 4.1, Henry's constant for O_2 dissolution in liquid water is, at $T = 298$ K and after interpolation

$$\mathcal{H}_{O_2} = 79,780 \, \text{J/mol}$$

Then the limiting current density $J_L(x)$ for cathode with 1% water flooding is

$$J_L(x) = \frac{nF \frac{RT}{\mathcal{H}_{O_2}} C_{O_2,m}}{\frac{RT}{\mathcal{H}_{O_2}} \frac{(1-\ell_\ell)L}{\mathcal{D}_{O_2-N_2}^{\text{eff}}} + \frac{\ell_\ell L}{\mathcal{D}_{i-\ell}^{\text{eff}}}}$$

$$= \frac{(4 \times 96,487) \, \text{C/mol} \, O_2 \times \frac{8.314 \, \text{J/(mol·K)} \times 298 \, \text{K}}{79,780 \, \text{J/mol}}}{\frac{8.314 \, \text{J/(mol·K)} \times 298 \, \text{K}}{79,780 \, \text{J/mol}} \times \frac{(1-0.1) \times 230 \times 10^{-6} \, \text{m}}{0.5313 \times 10^{-5} \, \text{m}^2/\text{s}} + \frac{0.1 \times 230 \times 10^{-6} \, \text{m}}{0.6072 \times 10^{-9} \, \text{m}^2/\text{s}}}$$

$$= \frac{1.029 \times 10^5 \exp(-4.253x) \, \text{C/m}^3}{1.210 \, \text{s/m} + 37,880 \, \text{s/m}}$$

$$= 2.716 \exp(-4.253x) \, \text{A/m}^2, \quad (x \text{ in m}).$$

Thus,

$$J_L = 2.716 \, \text{A/m}^2 = 2.716 \times 10^{-4} \, \text{A/cm}^2 \quad \text{at } x = 0$$

and

$$J_L = 0.7583 \, \text{A/m}^2 = 0.7583 \times 10^{-4} \, \text{A/cm}^2 \quad \text{at } x = H = 0.3 \, \text{m}.$$

The corresponding average limiting current density is determined as

$$\bar{J}_L = 1.534 \, \text{A/m}^2 = 1.534 \times 10^{-4} \, \text{A/cm}^2$$

COMMENT

1. The limiting current density values for unflooded cathode are very large, suggesting that the mass transfer through an unflooded cathode does not impose significant limitations on practical fuel cells.
2. However, even a very small amount of flooding can have dramatic effect on the mass transfer though the liquid phase and hence the cell performance. It should be emphasized that actual electrode contains some small- and some large-sized pores and it is believed that small pores are easily flooded by liquid water due to the capillary effect while large pores would remain open for gas phase transport. Therefore, mass transport through a partially flooded electrode is inherently three dimensional, multiphase flow. Nevertheless, the present one-dimensional model of a flat layer of the electrode being flooded illustrates the importance of flooding in impairing the reactant supply to the reaction sites through the electrode and consequently in limiting the performance of fuel cells.

3. Comparing with the results obtained in Example 4.4, it reveals that the effect of convection mass transfer in the flow channel is comparable with the mass transport through an unflooded electrode so that convection mass transfer process in the flow channel may not be neglected in determining the performance of fuel cells.

4.7   TRANSPORT OF ELECTRICITY: OHMIC POLARIZATION η_{ohm}

Electricity in fuel cells is transported through the conducting solid-cell components through the flow of electrons and through the electrolyte through the flow of ions (charged species). The ohmic polarization arises due to electrical resistance in the cell components, including

- Resistance to the flow of ions in the electrolyte (ionic resistance),
- Resistance to the flow of electrons and ions in the catalyst layer (ionic and electronic resistance),
- Resistance to the flow of electrons through the electrode-backing layer, or gas-diffusion layer (electronic resistance), and
- Resistance to the flow of electrons through the interface contact and the terminal connections (electronic resistance).

The transport of electrons through the cell components and ions through the electrolyte is often expressed in terms of Ohm's law, so that the ohmic polarization can be determined as follows

$$\eta_{ohm} = I\mathcal{R} \tag{4.125}$$

where the total cell resistance $\mathcal{R}$ is the sum of electronic, ionic, and contact resistance and thus η_{ohm} is also commonly referred to as the resistance polarization, ohmic (or resistance) loss, and so on. For well-designed practical fuel cells, η_{ohm} is mainly caused by ionic resistance in the electrolyte.

In fuel cell literature, the use of the electrical resistance of the cell components, including electrolytes, is often avoided and reciprocal resistance or conductance $\mathcal{L}$ is commonly used instead. The conductance is simply the inverse of resistance $\mathcal{R}$

$$\mathcal{L} = \frac{1}{\mathcal{R}} \tag{4.126}$$

The resistance $\mathcal{R}$ depends on the material property as well as the geometry of the conductor as follows

$$\mathcal{R} = \rho\frac{\delta}{A} \tag{4.127}$$

where δ is the length of the conducting path, A is the cross-sectional area of the conductor normal to the conducting path and the electric field has been assumed uniform in arriving at the Equation (4.127). The specific resistance (or resistivity) ρ is a material property, representing the capability of the material in the transport of electricity, and it has the unit of $\Omega \cdot m$. The area specific resistance is also sometimes used in fuel cell literature and it is defined as $A\mathcal{R}$ with the unit of $\Omega \cdot m^2$. Hence, we have from Equation (4.127)

$$A\mathcal{R} = \rho\delta \tag{4.128}$$

Similarly, specific conductance (or conductivity) κ can be defined, and

$$\kappa = \frac{1}{\rho} \tag{4.129}$$

Conductivity is a material property having the unit of $(\Omega \cdot m)^{-1}$ or S/m; where S stands for Siemens. The area specific conductivity is also sometimes used, and it is defined as $\mathcal{L}/A$, and from Equations (4.126), (4.127), and (4.129) we have

$$\frac{\mathcal{L}}{A} = \frac{\kappa}{\delta} \tag{4.130}$$

Then Equation (4.127) can be expressed as

$$\mathcal{R} = \frac{\delta}{\kappa A} \tag{4.131}$$

This expression for the resistance to the transport of electricity is very similar to those for the resistance to the conduction of heat and to the transfer of mass by diffusion.

Therefore, we can write the ohmic overpotential for a cell, mainly due to the ionic resistance in the electrolyte, as

$$\eta = I\mathcal{R} = (JA)\left(\frac{\delta}{\kappa A}\right)$$

or simply

$$\eta = \frac{J\delta}{\kappa} \tag{4.132}$$

where A is interpreted as the active area of the cell and δ the thickness of the electrolyte layer. Clearly, the ohmic overpotential can be reduced by using thinner electrolyte layer (i.e., decreasing the electrode separation) and electrolyte of higher ionic conductivity. Similarly, using electrodes of higher electronic conductivity reduces cell ohmic overpotential. However, a thinner electrode may not necessarily yield a smaller cell overpotential, rather an optimal thickness exists for electrodes that gives rise to a smallest cell overpotential as a balance among the transport of electricity, reactant

Table 4.2 Conductivity Values for Some Typical Electrolytes[9]

Electrolyte	Temperature (°C)	Concentration (mol/ℓ)	Conductivity (S/cm)
H_2SO_4 in water	18	5	1.35
		10	1.41
		30	0.190
NaOH in water	18	5	0.345
		10	0.205
		15	0.110
KOH in water	18	5	0.528
		10	0.393
NaOH in water	50	5	0.670
		10	0.575
		15	0.440
	100	5	1.24
		10	1.41
		15	1.33
H_3PO_4 in water	18	6	0.625
		11	0.151
NaCl (fused)	750	—	3.40
Na_2CO_3 (fused)	850	—	2.92
KCl (fused)	900	—	2.76

supply to and product removal from the reaction sites, all of which are accomplished through the appropriate porous electrode structures. Furthermore, this delicate balance depends on the cell operating conditions, the flow-channel design and electrode structures which can be altered by the assemblying process of the cells. We discuss this issue further in later chapters.

It should be emphasized that the resistance R or more commonly the conductivity κ of the electrolyte is measured when no concentration gradient of the charged species (ions) exists in the electrolyte, otherwise mass transport of ions arises, which would contribute to the transport of electricity. This issue will be dealt with in the next section.

Typical values of conductivity are given in Table 4.2 for some electrolyte.[9]

Example 4.7 For alkaline fuel cells with electrolyte circulation (the so-called **mobile electrolyte AFCs**), the electrolyte conductivity at the cell operating temperature may be taken as 0.4 S/cm, and the typical operating current density at the design point is 200 mA/cm^2 when operating on H_2 and air. The electrolyte compartment thickness is about 2 mm, a minimum channel width required for liquid electrolyte flow without excessive pumping power due to the viscous resistance at the walls. Determine the ohmic overpotential caused by the alkaline electrolyte.

SOLUTION

The conductivity of electrolyte is given as

$$\kappa = 0.4\,\text{S/cm} = 0.4 \times 10^2\,\text{S/m} = 40\,\text{S/m}$$

The operating current density is

$$J = 200\,\text{mA/cm}^2 = 200 \times 10^{-3}\,\text{A}/(10^{-4}\,\text{m}^2) = 2000\,\text{A/m}^2$$

The electrolyte layer thickness is given as

$$\delta \cong 2\,\text{mm} = 2 \times 10^{-3}\,\text{m}$$

Therefore the ohmic overpotential caused by the electrolyte is determined as

$$\eta_{\text{ohm}} = \frac{\delta}{\kappa} J \cong \frac{2 \times 10^{-3}\,\text{m}}{40\,\text{S/m}} \times 2000\,\text{A/m}^2$$
$$= 0.1\,\text{V}$$

COMMENT

1. Since the potential of most alkaline fuel cells is about less than 1 V, the ohmic overpotential of 0.1 V is a significant factor in the reduction of such mobile electrolyte alkaline fuel cells. Note that the parameter values of κ, J, and δ are quite representative for this type of fuel cells.

2. Most practical fuel cells operate in the ohmic polarization region on the cell polarization curve, and Ohmic overpotential is directly proportional to the current density ($\eta_{\text{ohm}} \sim J$), therefore η_{ohm} is practically the limiting factor for the cell power density, cell potential and cell energy-conversion efficiency achievable.

3. Because of the above reasons, alkaline fuel cells with immobile electrolytes have been developed and the liquid alkaline solution is fixed in a thin porous matrix structure by the capillary effect. For this technology, the electrolyte layer can be as thin as $(0.1\sim0.2)$ mm. Then we can easily determine that

$$\eta_{\text{ohm}} = (0.005\sim0.01)\,\text{V}$$

for the same operating current density of 200 mA/cm^2. This value of η_{ohm} is quite small and negligible. In reality, immobile electrolyte design allows much thinner electrolyte layer, thus much higher current density with still negligible η_{ohm}, leading to much higher output power density, important for practical applications.

4. For PEMFCs, the proton-conducting solid polymer membrane is used as the electrolyte (immobile), the electrolyte can be as thin as 50 μm, and the corresponding operating-cell current density can be as high as $(700\sim800)$ mA/cm^2 or even higher still with acceptable ohmic overpotential. This is the part of reasons that PEMFCs have the highest power density compared with other types of fuel cells.

4.8   TRANSPORT OF MASS AND ELECTRICITY IN ELECTROLYTE

The analysis for mass transfer in the electrolyte is very similar to the reactant transfer to the electrode described in the preceding sections. The only exception is that mass transfer can arise in the electrolyte due to the motion of mobile ionic species in the electric field set up between the anode and cathode, or the migration of the ions. We consider both situations in this section, with and without the ion migration effect.

4.8.1   Mass Transfer by Diffusion of Ions Only

Consider the situation of a mobile ionic species i that is being discharged at an electrode surface. The concentration of the mobile ionic species i is denoted as $C_{i,\infty}$ in the bulk electrolyte (sufficiently far away from the electrode surface, or realistically outside of the concentration boundary layer), and $C_{i,s}$ at the electrode surface. The transfer process for the ionic species is thus caused by the concentration difference (diffusion mechanism), as shown in Figure 4.8, where the coordinate x is pointing away from the electrode surface and into the bulk electrolyte.

The rate of diffusion towards the electrode surface for the ionic species i is, in terms of the molar flux and according to the Fick's law

$$\dot{N}_i'' = -\mathcal{D}_i \frac{dC_i}{dx} \tag{4.133}$$

where $\mathcal{D}_i$ is the diffusion coefficient of the ionic species i with respect to the electrolyte solution and it depends on the nature and molecular size of the solute species, on the viscosity of the electrolyte solution μ, and on the temperature T. A typical value of $\mathcal{D}_i$ in common aqueous solution based electrolyte at the room temperature is on the order of 10^{-9} m²/s, a much smaller value than the corresponding diffusion coefficient in the gas medium. In general, the diffusion coefficient $\mathcal{D}_i$ can be increased by lowering the electrolyte viscosity μ and increasing the medium temperature.

The current density corresponding to the rate of ion transport given in Equation (4.133) is

$$J_i = n_i F \dot{N}_i'' = -n_i F \mathcal{D}_i \frac{dC_i}{dx} \cong n_i F \mathcal{D}_i \frac{C_{i,\infty} - C_{i,s}}{\delta_i} \tag{4.134}$$

where n_i is the charge number of the ion i and δ_i here represents the thickness of the diffusion layer adjacent to the electrode surface, and a typical value for δ_i is about $300\,\mu$m for unstirred (initially quiescent) solutions and can be as thin as $1\,\mu$m for vigorously stirred (strongly convective) solutions.

The limiting current density corresponds to the maximum rate of ion transport to the electrode surface, which occurs when the ion concentration at the electrode surface vanishes, or $C_{i,s} = 0$. Then

$$J_{L,i} = n_i F \mathcal{D}_i \frac{C_{i,\infty}}{\delta_{L,i}} \tag{4.135}$$

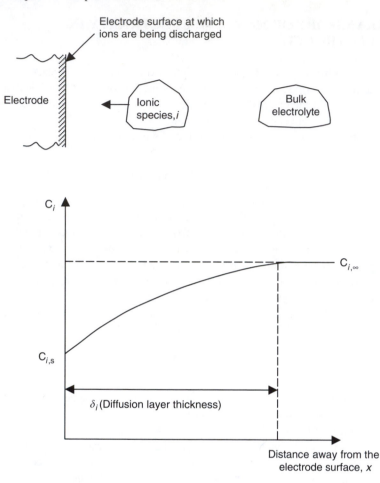

Figure 4.8 Illustration of the concentration distribution of ionic species which is discharged at the electrode and whose transport is set up by a diffusion mechanism.

where $\delta_{L,i}$ is the diffusion layer thickness at the limiting current density for the ionic species i. If the diffusion layer thickness could be assumed almost invariant (which may not be necessarily true; see comments at the end of this section), then $\delta_i \cong \delta_{L,i}$. Combining Equations (4.134) and (4.135) yields

$$\frac{C_{i,\infty}}{C_{i,s}} \cong \frac{J_{L,i}}{J_{L,i} - J_i} \tag{4.136}$$

Thus the concentration of the discharging ions at the electrode surface, $C_{i,s}$, is smaller than its concentration in the bulk electrolyte, $C_{i,\infty}$, which is far away from the electrode surface and is supplied externally for the fuel cell operation. The lowering of the concentration across the diffusion layer is due to the limited finite rate of mass transfer and it translates to a voltage loss, $\eta_{conc.}$.

Assuming all other processes at the electrode are reversible, this voltage loss due to the lowering of the ion concentration at the electrode surface can be estimated from the Nernst equation as described previously. Specifically,

$$\eta_{conc.} = \frac{RT}{n_i F} \ln \left(\frac{J_{L,i}}{J_{L,i} - J_i} \right)^{(v_i'' - v_i')/v_F'} \tag{4.137}$$

Again it must be pointed out that the diffusion layer introduced here has also been called the Nernst diffusion layer in honor of the contribution made by Nernst in 1904 when the flow of liquids over solid surfaces was not well-understood. In view of the boundary layer concept introduced some two decades later (described earlier in this chapter), the Nernst diffusion layer does not exist in reality; thus this analysis needs to be cautioned despite its popularity in the conventional electrochemistry.

4.8.2   Mass Transfer by Migration of Ions Only

Electrolyte contains positively and negatively charged species called ions, as well as other neutral species such as water for aqueous electrolyte solutions. These ions are constantly in random thermal motion at the microscopical level, as shown in Figure 4.9. However, they start to accelerate in the direction of the electric field once

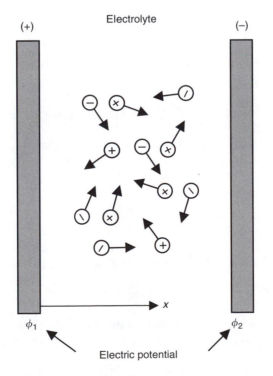

Figure 4.9 Schematic of ion motion in the electrolyte.

an external electric field is set up across the electrolyte layer. The acceleration of ions through the electrolyte is retarded by viscous drag and electrical forces arising mainly from ionic interactions. Eventually the driving electrical forces are balanced by the retarding forces and the velocity of the migration of an ion in the electrolyte reaches a constant value, which may be called the terminal velocity of the ion. It might be expected that the terminal velocity for the ionic species i, v_i, is proportional to the driving electrical force due to the charge on the ion in the external electric field since the retarding forces are passive and reactive to the driving forces. Hence we may write

$$v_i = -u_i(n_i F)\frac{d\Phi}{dx}, \quad \text{m/s} \tag{4.138}$$

where the charge number of the ion i, n_i, is essentially the number of the mole electron equivalent charge per mole of the ion, F is the Faraday constant ($= 96,487$ C/mole); thus $n_i F$ represents the electrical charge per mole ion. The electric field is represented by the gradient of the electrical potential Φ in the x direction, which is defined in Figure 4.9. The proportionality factor u_i is referred to as the mobility of the ion i and its SI unit is $m^2 \cdot mol/(J \cdot s)$. The negative sign in Equation (4.138) reflects the fact that positively charged ions ($n_i > 0$) move in the direction of decreasing potential Φ and the negatively charged ions ($n_i < 0$) migrate towards the higher potential.

The molar flux of the ion i arising from the terminal velocity v_i driven by the electric field effect is

$$\dot{N}_i'' = C_i v_i = -C_i u_i n_i F \frac{d\Phi}{dx}, \quad \text{mol/(m}^2 \cdot \text{s)} \tag{4.139}$$

A current density J_i arises from the migration of the ion i which can be determined by the following expression

$$J_i = (n_i F)\dot{N}_i'' = n_i F C_i v_i, \quad \text{A/m}^2 \tag{4.140}$$

where C_i is the molar concentration of the ion i, thus $C_i v_i$ represents the molar flux of the ion i with the unit of mol/(m$^2 \cdot$ s) due to its migration in the electrical field $d\Phi/dx$ at the terminal velocity v_i. Substituting Equation (4.138), Equation (4.140) becomes

$$J_i = -u_i(n_i F)^2 C_i \frac{d\Phi}{dx}; \quad \text{A/m}^2 \tag{4.141}$$

The coefficient in Equation (4.141) between the current density J_i and the gradient of the potential is known as the conductivity of the ion i, such that

$$\kappa_i = u_i(n_i F)^2 C_i \tag{4.142}$$

The conductivity κ_i is a transport property of the ion i in the electrolyte and it has the unit of $(\Omega \cdot m)^{-1} = $ S/m. Then, Equation (4.141) can be expressed as

$$J_i = -\kappa_i \frac{d\Phi}{dx} \tag{4.143}$$

which is simply an expression of Ohm's law. It states that the current density arising from the migration of the ion i is proportional to the electrical field strength, $d\Phi/dx$, with the proportionality constant equal to the conductivity of the ion i in the electrolyte. It should be emphasized that this equation is valid when the concentration C_i is uniform throughout the electrolyte; otherwise, it is not valid in general as described in Section 4.8.3.

The total current density J resulting from the migration of all the ions in the electrolyte is equal to the sum of the contribution made by the transport of each of the ions, or

$$J = \sum_i J_i = - \left(\sum_i \kappa_i \right) \frac{d\Phi}{dx} \tag{4.144}$$

Therefore, the conductivity of the electrolyte may be written as

$$\kappa = \sum_i \kappa_i \tag{4.145}$$

and Ohm's law is valid for the total current density contributed by the migration of all the ions in the electrolyte. Again this equation is valid only when all the concentrations in the electrolyte is uniform so that diffusional ion transport is absent.

Although the migration of each ion in the electrolyte contributes to the total current density carried by the electrolyte, the contribution is not equal. To quantify the relative contributions, the transference number of the ion i is defined as the fraction of the total current density that is carried by the given ion i, or

$$t_i = \frac{J_i}{J} \tag{4.146}$$

Substituting Equations (4.143) and (4.144) into the Equation (4.146), we obtain

$$t_i = \frac{\kappa_i}{\sum_i \kappa_i} = \frac{\kappa_i}{\kappa} \tag{4.147}$$

Since both κ_i and κ are the transport property, t_i is also a transport property of the electrolyte and the ion i, and can be determined by the Equation (4.147). Transference numbers, once determined, could be very useful in fuel cells to express the fraction of the current density carried by each individual ion; that is, $J_i = t_i J$ is the current density contributed by the migration of the ionic species i.

Substituting this expression into the first equality in Equation (4.144), we have

$$\sum_i t_i = 1 \tag{4.148}$$

which is valid for electrolytes of all types.

Now combining Equations (4.144) and (4.145) together and integrating $d\Phi/dx$ across the electrolyte layer thickness, we obtain the ohmic overpotential for ionic transport through the electrolyte layer by migration

$$\eta_{\text{ohm}} = \Phi_1 - \Phi_2 = \int_0^{\delta} \frac{J}{\kappa} dx = \frac{J\delta}{\kappa} \tag{4.149}$$

This relation is identical to Equation (4.132) and is valid only if the concentration is uniform in the electrolyte.

4.8.3   Mass Transfer by Simultaneous Diffusion and Migration of Ions

In electrolytes, mass transfer occurs as a result of the ionic transport due to the motion of the ions under the influence of the external electric field — which is often called the migration of ions. Ion migration causes the accumulation of a certain ion in one region and the depletion of the same ion in another region of the electrolyte, leading to the establishment of the concentration gradient for the ion. Then mass transfer occurs due to the diffusion mechanism driven by the concentration gradient. Therefore, in any fuel cell, electrolyte mass transfer may arise because of simultaneous mass diffusion and ionic migration.

From the mass transfer rate due to diffusion given by Equation (4.133) and due to ionic migration given by Equation (4.139), we have the total molar flux for the ion i under the combined effect of the diffusion and ion migration

$$\dot{N}_i'' = -\mathcal{D}_i \frac{dC_i}{dx} - C_i u_i n_i F \frac{d\Phi}{dx} \tag{4.150}$$

The corresponding current density arising from this transfer of the ion i is $J_i = (n_i F) \dot{N}_i''$, or

$$J_i = -(n_i F) \mathcal{D}_i \frac{dC_i}{dx} - C_i u_i (n_i F)^2 \frac{d\Phi}{dx} \tag{4.151}$$

When a fuel cell has reached a steady state, all the current is carried by the transport of ions that participate in the electrode reactions. For all other ions present in the electrolyte, their flux due to the electric field effect (migration) is counterbalanced by the flux due to the concentration gradient (diffusion) and no net transport of electricity arises from the motion of these ions.

At this point we may emphasize that both the diffusion coefficient $\mathcal{D}_i$ and the ion mobility u_i are the transport property of the ion i through the electrolyte solution and they may be related to each other. Indeed, this is the case as we can show in the following text.

Consider a situation where there is no net current flow ($J = 0$) through the electrolyte, that is, the current arising from the ion diffusion due to concentration gradient and from the ion migration in an applied electric field reaches the same magnitude, but in the opposite direction. Thus, thermodynamic equilibrium is attained and the concentration of (or the number density for) the ion i with the charge number n_i at the location x in the electrolyte and under the influence of electric field with the local potential Φ is given by the thermodynamic Boltzmann distribution

$$C_i = C_{i,\infty} \exp\left(-\frac{n_i F \Phi}{RT}\right) \tag{4.152}$$

where $C_{i,\infty}$ is the concentration of the ion i corresponding to $\Phi = 0$.

On the other hand, setting $J = 0$ and integrating Equation (4.151) from $C_i = C_{i,\infty}$ at $\Phi = 0$ to C_i at Φ, we obtain

$$C_i = C_{i,\infty} \exp\left(-\frac{u_i n_i F \Phi}{\mathcal{D}_i}\right) \tag{4.153}$$

Comparing Equation (4.152) with Equation (4.153) indicates that

$$u_i = \frac{\mathcal{D}_i}{RT} \tag{4.154}$$

The preceding expression is known as the Nernst–Einstein relation in honor of the contributions made by Nernst (1888) and Einstein (1905). It relates the transport processes of ion diffusion and migration in an electrolyte and provides an important link between mass diffusion and electrical conductance. Substituting Equation (4.154) into Equations (4.142) and (4.145), respectively, we obtain the conductivity of the ion i and the electrolyte as follows

$$\kappa_i = \frac{(n_i F)^2 C_i \mathcal{D}_i}{RT} \tag{4.155}$$

$$\kappa = \frac{F^2}{RT} \sum_i n_i^2 C_i \mathcal{D}_i \tag{4.156}$$

It should be pointed out that the Nernst–Einstein relation is valid even when the total current through the electrolyte does not vanish, although it was derived when $J = 0$. This is because both u_i and $\mathcal{D}_i$ are the (transport) property of the ion i in the electrolyte.

Considering Equations (4.142) and (4.154), Equation (4.151) can be rearranged as follows:

$$-\frac{d\Phi}{dx} = \frac{J_i}{\kappa_i} + \frac{RT}{n_i F C_i} \frac{dC_i}{dx}$$

Then integration of the preceding equation from the electrode surface where $x = 0$, $\Phi = \Phi_1$ and $C_i = C_{i,s}$ to the bulk electrolyte sufficiently away from the electrode surface where $\Phi = \Phi_2$ and $C_i = C_{i,\infty}$ results in

$$\Phi_1 - \Phi_2 = \int_0^x \frac{J_i}{\kappa_i} dx + \frac{RT}{n_i F} \ln\left(\frac{C_{i,\infty}}{C_{i,s}}\right) \tag{4.157}$$

Since for most electrolytes in fuel cells, a single ion carries the current, such as OH^- for alkaline fuel cells and H^+ for acid-electrolyte fuel cells, the first term in the right-hand side of Equation (4.157) represents ohmic overpotential, while the second term is due to the concentration variation. Also notice that κ_i depends on the concentration C_i, which varies, the conductivity κ_i must be included under the integral sign.

Therefore, the potential drop from an electrode to the bulk electrolyte may be considered as composed of an ohmic overpotential η_{ohm} and a concentration overpotential $\eta_{conc.}$. It is known that η_{ohm} is directly proportional to the current and vanishes immediately once the current is interrupted (equivalent to open circuit, or no current is drawn from the electrode or cell). Then, the total potential difference $\Phi_1 - \Phi_2$ is measured when the current is drawn and the concentration overpotential $\eta_{conc.}$ is the value measured right after the current is interrupted but before the concentration distribution can change by diffusion or convection. We can then determine the ohmic overpotential from

$$\eta_{ohm} = (\Phi_1 - \Phi_2) - \eta_{conc.} \tag{4.158}$$

It should be noted that right after the external current is interrupted, the local current density may not vanish everywhere in the electrolyte. It must be cautioned then when the current interruption method is used to measure the ohmic overpotential.

4.9   CELL PERFORMANCE AND CELL CONNECTION (STACKING) ARRANGEMENTS

As described in Chapter 2, cell performance can be represented entirely by cell potential variations as a function of the current density drawn from the cell, or commonly referred to as the **cell polarization curve**; and the actual cell potential may be determined by subtracting all forms of voltage drop from the reversible cell potential. Therefore, we have

$$E = E_r - \eta_{act} - \eta_{ohm} - \eta_{conc.} \tag{4.159}$$

The activation overpotential may be further subdivided into the overpotential associated with the anodic and cathodic reactions, and they may be approximated by the Tafel equation when reasonably current density is drawn from the cell, as described in Chapter 3. Substituting the forms of the ohmic and concentration overpotentials developed earlier in this chapter, we may write the actual cell potential as follows:

$$E = a_1 \ln\left(\frac{J}{J_L}\right) + a_2 J - a_3 \ln\left(\frac{J_L - J}{J_L}\right) + a_4 \tag{4.160}$$

where a_1 represents the Tafel slope; a_2 is essentially the equivalent cell conductivity when all cell components are taken into account, including the electrolyte, electrodes, cell connections, and surface contacts; a_3 is proportional to RT/F and a_4 is related to (but not equal to) E_r. Although these various coefficients, a_i's, may be determined from an analysis of the overpotentials involved, as presented earlier in this chapter, they are best determined experimentally for a given cell to account for the variability of the cell performance due to the process of manufacturing and the associated variability in the microscopic structure of the electrodes.

The potential of a working cell is typically around 0.7~0.8 V and it is normally too small for practical applications, also because of the limited power available from a single cell. Therefore, many individual cells are connected (or stacked) together to form a fuel cell stack. Although many stacking configurations are possible, the overpotential associated with the transport processes discussed in this chapter imposes limitations and technical difficulties, making cell stacking one of the significant technical challenges in the drive for the fuel cell commercialization. We briefly discuss the stacking options in the following text and the associated transport-related issues.

The electrical connection among the individual cells may be arranged in parallel or in series, as shown in Figure 4.10. The parallel connection still provides a low-voltage output from the stack, but a very high-current output since the stack current

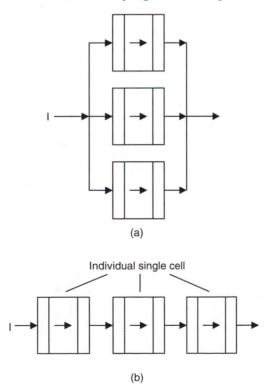

(a)

Individual single cell

(b)

Figure 4.10 Cell connection (stacking) configurations. (a) Parallel connection; (b) series connection. The arrows in the diagram represent the direction of the current flow.

is the sum of the current produced in each cell. Such an extremely large current flow causes an excessively large ohmic voltage loss in the stack components and at the surface contacts among the components. Thus parallel connection is typically avoided unless for small current or power applications.

Series connection can have two typical arrangements: unipolar and bipolar, as shown in Figure 4.11. Unipolar design has one fuel stream supplying fuel to two anode electrodes for the two adjacent cells, and one oxidant stream delivering oxidant to two cathode electrodes for the two adjacent cells. This arrangement of one reactant stream serving for two adjacent electrodes simplifies the reactant flow channel design. However, it forces the electrical current generated in each cell to be collected at the edge of the electrodes. Since electrode is very thin (<1 mm), while the other electrode dimensions (in the direction of the current flow) are at least on the order of centimeters or larger, the ohmic resistance tends to be very large, according to Equation (4.131). Thus edge collection of current, although used in early fuel cell stack designs, is generally avoided in recent fuel cell stack development, primarily due to the excessively large ohmic voltage losses.

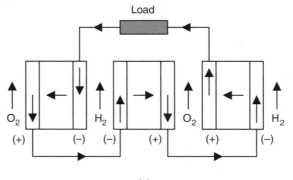

(a)

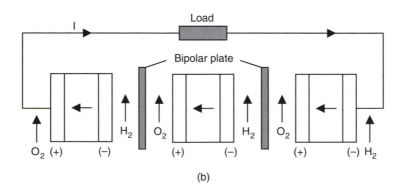

(b)

Figure 4.11 Series cell connection (stacking) configurations. (a) Unipolar arrangement with the edge collection of the current generated in each cell; (b) Bipolar arrangement with the endplate collection of the current.

The bipolar arrangement has the current flow normal to the electrode surface, instead of along the electrode surface as in the unipolar arrangement, thus the current flow path is very short, while the cross-sectional area available for the current flow is very large. By Equation (4.131) the ohmic voltage loss for this case is very small in comparison and bipolar design is favored in recent fuel cell stack technology. However, this endplate collection of current results in the complex design for the reactant flow channels and complex organization for the reactant stream and the bipolar plate has to fulfill several functions simultaneously in order to obtain a good overall stack performance: Bipolar plates must serve as current collectors; reactants must be delivered to the electrode surface; cell reaction product (i.e., water) must be removed; and integrity of the cell/stack must be maintained. These functions for the bipolar plates may be contradictory to each other and optimal design for the bipolar plates represent one of the significant technical challenges for practical fuel cells.

Typical configuration of a bipolar plate is shown in Figure 4.12. The reactant flow follows the flow channels made on the bipolar plate, thus distributing the reactant over the electrode surface. On the other hand, the land between the adjacent flow channels serves as the passage for the current flow from one cell to the next. Therefore, a wide flow channel is beneficial for the reactant distribution over the electrode surface and for the reaction product removal; while a wide land is beneficial for the electron flow and for the mechanical integrity of the cells and the entire stack. Normally the same cell unit as shown in Figure 4.12c is repeated to form a stack. The plate at the end of the stack only has flow channels on one side of its surface.

From the cell repeating unit in a stack, it is clear that the reactant concentration decreases along the flow direction following the flow channel design and into the electrode due to convectional and diffusional mass transfer. Thus, the concentration field is three dimensional, so is the electrical field due to the flow channels made on the bipolar plate surfaces, the nonuniform rate of electrochemical reactions in the catalyst layers, and the three-dimensional distribution of the reactant concentrations and the temperature. Whereas the cell performance in terms of the polarization curve given in Equation (4.159) or Equation (4.160) is the result of essentially assuming one-dimensional concentration and electrical field. Therefore, an accurate prediction of the cell performance requires a three-dimensional analysis based on solving the conservation equations governing the transport phenomena of the reactant flow, species concentration, temperature and electric fields, incorporating the in-cell electrochemical reaction processes — such a comprehensive approach has been attempted in the literature. However, it is deemed to be too advanced to be covered in this introductory text. Readers may wish to consult additional references for further discussion and description of the transport phenomena and their impact on fuel cell performance.[10,11]

4.10 SUMMARY

The primary objective of this chapter is to describe the transport phenomena of mass, momentum, energy, and electricity and their impact on the performance and design of practical fuel cells. We first present the various descriptions for multicomponent

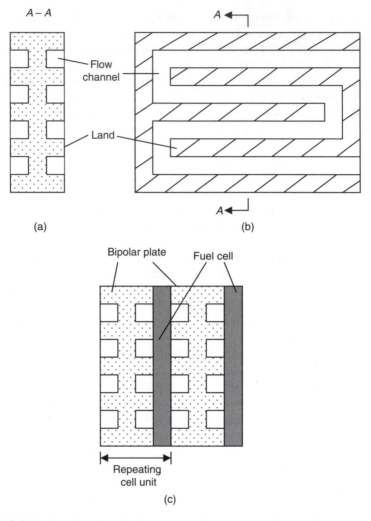

Figure 4.12 Typical configuration of a bipolar plate. (a) Cross-sectional view; (b) face view; (c) repeating cell unit in a stack.

mixtures, and then develop the rate of their transport due to diffusional and convectional mechanisms. The diffusional transport is related to the gradient of velocity, temperature, and concentration for the momentum, thermal energy and species diffusion, and the concept of velocity, thermal and concentration boundary layer is described for the transport of momentum, thermal energy and species near a solid surface (i.e., electrode surface). The similarity and difference for these transport processes are then examined in terms of the dimensionless parameters like the Prandtl number, Lewis number, and Schmidt number. The rate of convectional transfer is expressed in terms of the dimensionless friction coefficient, Nusselt, and Sherwood number for the transport of momentum, thermal energy, and species, respectively. Then, the effect of the transport processes on fuel cell performance is developed in terms of concentration and ohmic polarization as well as the parasitic load arising from the pumping power required to force the reactant or electrolyte flow under a specified condition. This chapter illustrates how to estimate whether a certain transport process plays the dominant role in limiting fuel cell performance. Finally, the implication of the transport phenomena on the current collection from fuel cells and the design of fuel cell stacks is described.

BIBLIOGRAPHY

1. Bird, R. B., W. E. Stewart and E. N. Lightfoot. 2000. *Transport Phenomena – 2*. New York: Wiley.
2. Blasius, H. and Z. Math. 1908. *Phys.*, 56: 1.
3. Schlichting, H. 1979. *Boundary Layer Theory – 7*. New York: McGraw-Hill.
4. White, F. M. 1991. *Viscous Fluid Flows – 2*. New York: McGraw-Hill.
5. Incropera, F. P. and D. P. DeWitt. 2002. *Fundamentals of Heat and Mass Transfer – 5*. Wiley.
6. Kays, W. M. and M. E. Crawford. 1980. *Convective Heat and Mass Transfer*. New York: McGraw-Hill.
7. Spalding, D. B. 1963. *Convective Mass Transfer*. New York: McGraw-Hill.
8. Bernard, D. and M. Verbrugge. 1992. *J. Electrochem. Soc.*, 139(9): 2477–2491.
9. McDougall, A. 1976. *Fuel Cells*. London: MacMillan.
10. Liebhafsky, H. A. and E. J. Cairns. 1968. *Fuel Cells and Fuel Batteries*. New York: Wiley.
11. Newman, J. S. 1991. *Electrochemical Systems – 2*. Englewood Cliffs, NJ: Prentice Hall.

PROBLEMS

4.1. Consider air is composed of 79% N_2 and 21% O_2 by volume. If the velocity of N_2 and O_2 is in the same direction for a certain flow situation, and the velocity magnitude is $v_{N_2} = 1$ m/s and $v_{O_2} = 2$ m/s, determine the following:
   (a) mass- and molar-average velocity;
   (b) mass- and molar-diffusion velocity for N_2 and O_2;
   (c) total mass and molar flux for N_2 and O_2, respectively, if the mixture is at 1 atm and 80 °C;
   (d) diffusional mass and molar flux for both N_2 and O_2 at 1 atm and 80 °C;
   (e) the respective sum of the diffusional mass and molar flux as calculated in (d).

4.2. Starting from Equation (4.26), show that the diffusional molar flux of species i can be expressed as

$$\dot{N}_{i,d}^{''} = -CX_iX_j \left(\mathbf{V}_j - \mathbf{V}_i \right)$$

in a binary mixture consisting of species i and j only. Where C is the total concentration of the mixture, X_i and X_j are the molar fraction of the species i and j, and $\mathbf{V}_i$ and $\mathbf{V}_j$ are the mass-diffusion velocity defined in Equation (4.11).

4.3. A fuel stream consists of hydrogen gas, fully saturated with water vapor, at the temperature of 80 °C and 1 atm. The fuel stream flows over, and parallel to, an anode electrode surface at the velocity of $v_\infty = 1$ m/s. If the anode is rectangular and the length along the flow direction is $H = 10$ cm, determine

    (a) the limiting current density profile corresponding to the rate of convective mass transfer, $J_L(x)$, as well as its average value $\bar{J}_L$;

    (b) the hydrogen concentration variation at the electrode surface, $C_{H_2,s}(x)$, and its average value, $\bar{C}_{H_2,s}$, if the current density drawn from the electrode is $J = 0.5\,\text{A/cm}^2$.

4.4. Repeat Problem 4.3 if the fuel stream is at $80\,°\text{C}$ and 3 atm.

4.5. Consider an air flow over a rectangular cathode-electrode surface with the velocity of $v_\infty = 3\,\text{m/s}$, and the electrode length in the flow direction is $H = 10\,\text{cm}$. Determine the distribution of the limiting current density, $J_L(x)$, that corresponds to the limitation of the convective mass transfer over the electrode surface, and its average value, $\bar{J}_L$, for the following conditions of the temperature and pressure of the air stream:

    (a) $T = 25\,°\text{C}$ and $P = 1$ atm;
    (b) $T = 25\,°\text{C}$ and $P = 3$ atm;
    (c) $T = 80\,°\text{C}$ and $P = 1$ atm;
    (d) $T = 80\,°\text{C}$ and $P = 3$ atm.

State any observation or conclusion that may be made from these calculations.

4.6. Repeat Problem 4.5 if the flow is over the anode electrode and the fluid is the fully humidified hydrogen gas.

4.7. Estimate the pumping power required for the conditions given in Problem 4.5 and assume that the electrode surface is square (i.e., the electrode width and height are identical, $W = H$). Observe the trend of variation for the electrode-performance and pumping-power requirement.

4.8. Estimate the pumping power required for the conditions given in Problem 4.6 and assume square electrode surface. Observe the trend of variation for the electrode-performance and pumping-power requirement.

4.9. Consider a fully humidified hydrogen stream flowing over an anode electrode of $200\,\mu\text{m}$ thickness and 40% porosity. The anode flow channel has a spacing of $b = 1\,\text{mm}$ and length of $H = 10\,\text{cm}$. The hydrogen stream is at $25\,°\text{C}$ and 1 atm and flows at a mean velocity of $v_m = 1\,\text{m/s}$. Determine the maximum current density attainable for the hydrogen mass transfer through the anode electrode when

    (a) the anode is not flooded by liquid water;
    (b) 1% of the electrode is flooded by liquid water.

4.10. Atmospheric air at $25\,°\text{C}$ flows over a cathode electrode at a mean velocity of $v_m = 5\,\text{m/s}$. The cathode electrode has a thickness of $200\,\mu\text{m}$ and porosity of 40%. The flow channel next to the electrode surface has a spacing of $b = 1\,\text{mm}$ and length of $H = 10\,\text{cm}$ in the flow direction. Determine the limiting current density corresponding to the limitation of oxygen mass transfer through the porous cathode electrode when

    (a) the cathode is not flooded by liquid water;
    (b) 0.1% of the cathode electrode is flooded by liquid water;
    (c) 0.1% of the cathode electrode is flooded by fully hydrated Nafion membrane.

4.11. Determine the ohmic overpotential arising from the ion migration through the electrolyte at the temperature of $18\,°C$ and the cell current density of $500\,mA/cm^2$ for the following electrolytes:

(a) 10 mol/ℓ H_2SO_4 in water;

(b) 10 mol/ℓ NaOH in water;

(c) 10 mol/ℓ KOH in water;

(d) 10 mol/ℓ H_3PO_4 in water.

ALKALINE FUEL CELLS (AFCs)

5.1 INTRODUCTION

Alkaline fuel cells (AFCs) were among the first fuel cells to have been studied and developed for practical applications, and in fact, AFCs are the first and the only type of cells to have reached successful routine applications, mainly in space explorations such as space shuttle missions in the United States. Because of this splendid space application, the largest number of fuel cell development programs in the world, particularly in Europe, once had been initiated to bring the AFCs from space application to terrestrial commercial applications and unfortunately almost all the AFC development programs have ended, mainly due to the so-called **CO$_2$ syndrome** described later in this chapter. Current R&D activities in AFCs are fairly limited.

The attraction of AFC-based power system lies in that it has, among all type of fuel cells under development, the best performance when pure hydrogen and oxygen are used as reactants. Figure 5.1 shows the oxygen reduction reaction in several acid and alkaline electrolytes. It is clear that oxygen reduction has the smallest activation polarization in 30% KOH electrolyte operating at 70 °C, while the voltage loss for 96% phosphoric acid as the electrolyte is much larger, even at a much higher operating temperature of 165 °C. The AFC power system can reach as high as about 60% chemical–electrical energy conversion efficiency for low-temperature operation (<100 °C), and it can achieve over 70% energy efficiency at elevated temperatures, such as the AFC units used in the U.S. space shuttle programs.

Another major advantage is that alkaline fuel cells can use a wider range of possible catalysts, at least thermodynamically. For example, AFCs can possibly use non-Noble catalysts such as nickel which is less costly than the noble metals required for other low-temperature acid electrolyte fuel cells currently under development.

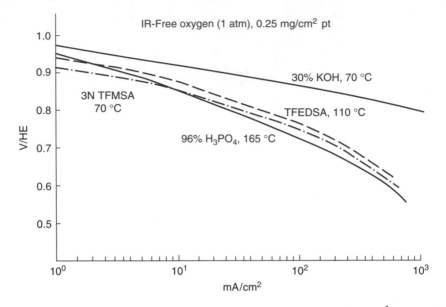

Figure 5.1 Comparison of oxygen reduction reaction (pure O_2 at 1 atm) on 0.25 mg/cm^2 supported Pt electrodes (10 wt% on carbon) in aqueous acid and alkaline electrolytes[2]. TFEDSA = Tetrafluoroethane-1,2-disulfonic acid; TFMSA = Trifluoromethylsulfonic acid.

In practice, however, low temperature, as well as high-performance high temperature AFCs use significant quantities of Noble metal catalysts to guarantee good performance and long lifetime. In U.S. space shuttle fuel cells, the anode consists of 10 mg/cm^2 of pure Noble metal blacks (80% platinum and 20% Pd), the cathode contains 20 mg/cm^2 of a mixture of 90% gold and 10% platinum. Operating at 0.86 V per cell at 470 mA/cm^2 on hydrogen and oxygen, this represents a Noble metal catalyst requirement of 74 mg/kW! The Elenco's H_2–air AFCs, developed as power source for buses, use supported platinum at a total loading of 0.7 mg/cm^2 for the anode and cathode. With performance of 0.7 V per cell at 100 mA/cm^2, the Elenco cell requires 10 mg of platinum per kilowatt. Platinum loadings of 0.15 mg/cm^2 have been suggested as being possible.

Although interest in alkaline fuel cells has declined, primarily because of the CO_2 syndrome, its shining return may be expected[1] if hydrogen economy is realized. Pure hydrogen is expected to be available from the electrolysis of water and the aforementioned two advantages will win the competition for the alkaline fuel cell over other fuel cell types. This chapter is devoted to the alkaline fuel cell because of this reason, as well as the historical significance of its success in space applications that had inspired and maintained the momentum for fuel cell R&D activities.

AFC's major disadvantage is that alkaline electrolytes (i.e., KOH and NaOH) do not reject carbon dioxide, even the 300~350 ppm of carbon dioxide in the atmospheric air is not tolerated. Consequently, AFCs are currently restricted to specialized applications where pure hydrogen and oxygen are utilized, such as the Apollo space program and the Space Shuttle Orbiter in the United States. For terrestrial commercial applications, it is necessary to use air as oxidant and either pure hydrogen or reformed hydrocarbon as fuel. Then, it is mandatory to scrub carbon dioxide from air and possibly from the fuel stream if nonpristine hydrogen is used. Carbon dioxide removal contributes significantly to a lower system efficiency and output power density (both volume and weight basis) and a higher capital and maintenance cost.

Because of low-operating temperature for current AFCs, the overall efficiency for large-size power plants are lower than the high-temperature fuel cell systems. Hence, AFC systems may be suitable for small-size mobile applications if pure hydrogen is available at a reasonable cost. Perceived practical applications include transportation, remote site use, military, and space applications such as the submarine and the space shuttle where cost is less a concern in comparison with commercial applications.

5.2   STATUS OF AFC DEVELOPMENT

Although the fuel cell concept was demonstrated in the nineteenth century, via the reversal of H_2/O_2 from water electrolysis, the practical alkaline fuel cell was initially developed, almost single-handed by Bacon from 1932 to 1955. Bacon's H_2–O_2 cells were operated at 200 °C~240 °C and 40~55 atm with 45% KOH. High pressure was used to prevent the electrolyte from boiling. The anode was sintered nickel with dual porosity (16-μm maximum pore diameter on the electrolyte side and 30-μm pore diameter on the gas side); the cathode was lithiated NiO; and the electrolyte was circulated. Bacon's work on alkaline fuel cells culminated in a 5-kW system demonstrated in 1959.

From late 1950s to early 1960s, many AFC development programs were initiated by industry, such as United Technology Corp. (UTC), Union Carbide, Allis Chalmers, and so. In 1960s, AFCs developed by UTC reached practical use for Apollo space missions. The Apollo missions included three 1.42-kW AFC power units, operating on pure hydrogen and oxygen with concentrated electrolyte (85% KOH). This concentrated KOH was needed to permit cell operation at a lower pressure (4.2 bars) without electrolyte boiling at the cell temperature of 260 °C. Each fuel cell module was about 5 cm diameter, 112 cm high, and weighed about 110 kg. The cell operating condition was about 0.85 V at 150 mA/cm². The electrolyte was immobile with passive heat and water removal. The AFC power units had a design lifetime of 500 hours. The AFC system was modified so that in later version, each cell was operated at 0.8 V at 400 mA/cm². The anode was porous nickel and the cathode was Li-doped NiO. Carbon dioxide-scrubbed air was used as oxidant.

From 1970s to the present, AFCs have been used for the Space Shuttle Orbiter applications. It includes three 12-kW AFC power modules, each module is operated at 0.85 V per cell at 470 mA/cm² with the power density of about 4 kW/m². The operating pressure is about 4~4.4 bars and the cell temperature is about 80 °C with a design lifetime of 2000 hours. Each fuel cell module is 38 cm wide, 147 cm high and 101 cm long. It weighs 91 kg with a power output of 12 kW and the total output voltage of 27.5 V. The electrolyte is 35%~45% KOH held in asbestos matrix (immobilized).

The anode is silver (Ag)-plated nickel screen with 10 mg (80% Pt and 20% Pd)/cm^2, and the cathode is silver (Ag)-plated nickel screen with 20 mg (90% Au and 10% Pt)/cm^2. The UTC lightweight alkaline cells have also achieved a power density of 21 kW/m^2 and 4.2 kW/kg.

In 1980s, UTC was developing advanced alkaline cells for the Strategic Defense Initiative (SDI) more commonly known as the Star Wars program, aiming for 27 kW/kg in MW-size stacks.

In Europe, Siemens (Germany) and CGE/IFP (France) developed in 1960s and 1970s a few kW units for military use. Alsthom/Occidental and Royal Institute of Technology (Sweden) also pursued a low-cost version of AFC power systems. From early 1970s to mid-1990s, Elenco (Belgium) developed and marketed small size (a few kW to 10-kW) modules for mobile applications, and Elenco's technology was transferred to, and continued by Zevco!

5.3   BASIC PRINCIPLES AND OPERATIONS

The composition and function of a typical alkaline fuel cell has been described in Chapter 3 along with reaction kinetics. As mentioned earlier, the electrolyte in alkaline fuel cell is an aqueous solution and can be either mobile (circulated) through, or held immobile in, the electrolyte compartment. Figure 5.2 shows schematically the cell structure for mobile electrolyte, illustrating the principle of its operation. Hydrogen oxidation occurs at the anode where hydrogen and hydroxyl ions combine to produce water and electrons. Electrons migrate through the external electric circuit, generating useful electric energy output; while a part of water generated is transferred to the

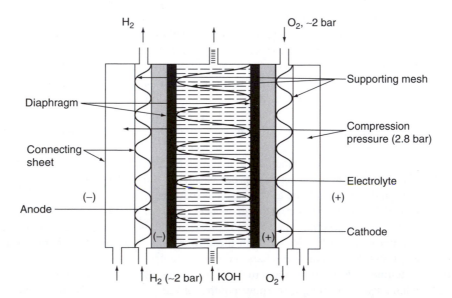

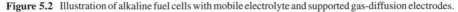

Figure 5.2 Illustration of alkaline fuel cells with mobile electrolyte and supported gas-diffusion electrodes.

cathode where molecular oxygen is reduced to form hydroxyl ions. The ions are transported to the anode for hydrogen oxidation reaction, thus carrying electric current through the electrolyte — a complete electric circuit is therefore formed.

For both mobile and immobile electrolyte alkaline fuel cells, the overall electrochemical reactions are

$$H_2 + 2OH^- \longrightarrow 2H_2O + 2e^- \qquad (5.1)$$

at the anode electrode,

$$\frac{1}{2}O_2 + H_2O + 2e^- \longrightarrow 2OH^- \qquad (5.2)$$

at the cathode electrode and the net cell reaction is

$$H_2 + \frac{1}{2}O_2 \longrightarrow H_2O + \text{Electric energy} + \text{Heat} \qquad (5.3)$$

Thus the reaction by products, water and heat, have to be removed continuously from the cell in order to maintain steady operation. From a system point of view, the heat and water management are critical for the effective and efficient operation of alkaline fuel cells. This is because water formed is in the liquid state if the cell operation temperature is below its saturation temperature (e.g., 100 °C for atmospheric operation) and it may cause the dilution of the electrolyte solution as well as the water flooding of the electrodes. On the other hand, water would be in the vapor state if the operation temperature is above its saturation temperature, and the water vapor bubbles coming out of the liquid electrolyte can carry some of the electrolyte away, accelerating the electrolyte depletion, thus limiting the cell operation lifetime.

Typically, the alkaline fuel cell of today operates at the temperature of around 60 °C~80 °C and the pressure of one to a few atmospheric pressures. Because carbon dioxide is not tolerated in alkaline fuel cells, whether it is in the fuel or oxidant stream, pure hydrogen and oxygen are required as the reactants, thus restricting its application.

Potassium hydroxide (KOH), which is the most conducting of all alkaline hydroxides, has almost always been the electrolyte of choice. The hydroxyl ions are the conducting species in the electrolyte. Water is produced at the anode in this cell, but some water migrates to the cathode due to concentration gradient. Therefore, product water leaves the cell from both the anode (about two thirds) and the cathode (about one third). Because of this water transport through the electrolyte and removal from both the anode and cathode streams, electrolyte concentration would be varying from location to location if rigorous mixing (or stirring) is not provided. Electrolyte circulation provides one means of electrolyte management, but it brings about a number of undesirable effects, including higher ohmic polarization, additional pumping, and pipe networks — all leading to a poorer performance.

5.4   CELL COMPONENTS AND CONFIGURATIONS

5.4.1   Electrolytes

The most common alkaline electrolyte is a potassium hydroxide (KOH) water solution, with normalities of 6 to 8. The KOH used must be sufficiently pure so that no impurities can cause catalyst poisoning.

The electrolyte can be arranged in the cell either in the form of a mobile or an immobile manner during cell operation. A mobile electrolyte, as in the design of Elenco and Siemens, is pumped through the cells and the moving electrolyte takes both the reaction product water and waste heat out of the cell. The water and heat are removed from the electrolyte solution and the resulting electrolyte is then circulated back to the cell. Therefore, this type of the arrangement is also called a **circulating electrolyte**. The electrolyte compartment in the cell typically has a large flow channel width, about 2 to 3 mm, to minimize the flow resistance, and hence pumping power required. Such a large thickness for the electrolyte layer results in a large value of ohmic polarization. Hence, electrolyte resistance is often a controlling factor for the performance of AFCs for the mobile electrolyte design.

For the design of immobile electrolyte adopted by IFC and used in the U.S. Space Shuttle program, the electrolyte is held in the asbestos matrix and an electrolyte reservoir plate acts as an electrolyte buffer. For this arrangement, the electrolyte layer can be as thin as 0.05 mm; thus the ohmic polarization associated with the electrolyte layer is very small in comparison, in addition to saving the pumping power for the electrolyte circulation. Hence, this design offers high cell performance and high energy-conversion efficiency. However, this design requires high purity for the reactants, as a stationary electrolyte does not offer a mechanism for the removal and regeneration of the electrolyte.

The electrolyte used is typically 30% (or $6\,mol/\ell$) potassium hydroxide water solution, which yields the optimal value for ion conductivity, when the cells are operated at low-temperatures of $60\,°C$–$80\,°C$. Because highly concentrated potassium hydroxide in water has an extremely low vapor pressure, molten potassium hydroxide (80% by mass or $14.3\,mol/\ell$ KOH) has also been used, especially in space applications where cells operate at high pressure and temperature for enhanced cell performance in terms of high energy conversion efficiency and high power density, although high operating temperature and pressure result in expensive materials required for cell construction, leading to high capital and maintenance costs.

The effect of potassium hydroxide concentration at $25\,°C$ on the cell reversible potential is shown in Table 5.1[3] and Figure 5.3, respectively.[4] It is clear that increasing KOH concentration is beneficial for cell performance. However, it is not practical and feasible to use very high concentrations of KOH in water due to the nonuniformity of KOH concentrations in operating cells due to the water consumption at the cathode electrode as shown in Equation (5.2). The oxygen reduction reaction may reduce the water concentration near the cathode electrode to such low levels that the electrolyte solution may solidify and plug the electrode pore regions which severely impaires reactant transport.

Table 5.1 Effect of KOH Concentration on Reversible
Cell Potential for a Hydrogen–Oxygen Cell Operating
at Standard Temperature and Pressure[3]

KOH Concentration (mol/kg)	E_r^0 (V)
0.18	1.229
1.8	1.230
3.6	1.232
5.4	1.235
7.2	1.243
8.9	1.251

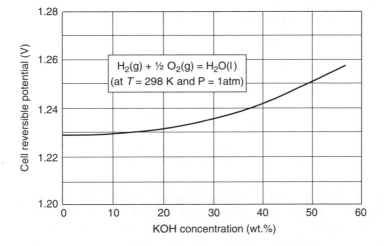

Figure 5.3 Effect of KOH concentration on the reversible cell potential for a hydrogen–oxygen cell operating at standard temperature and pressure.[4]

Sodium hydroxide (NaOH) has been envisaged as an alternative for potassium hydroxide as the electrolyte of choice. However, the performance characteristics of NaOH are not very good compared to KOH. For example, the conductivity of KOH is larger than that of NaOH and increases with temperature faster than NaOH; while NaOH is more sensitive to the presence of carbon dioxide for the formation of sodium bicarbonate, which is not very soluble in concentrated (necessary for sufficiently high conductivity) NaOH solutions and therefore can block the pores of, or even damage, the electrode system, preventing proper reactant supply for the cell electrochemical reactions, and reducing both the performance and lifetime of the cells.

5.4.2   Matrix

For alkaline fuel cells with immobilized electrolyte, the liquid electrolyte solution is held in asbestos matrix by the capillary effect, such as the AFCs used in U.S. space shuttle orbiters. The porous matrix has pore sizes less than that of both anode and cathode electrodes. However, asbestos poses a health hazard and is an undesirable material that has been banned for use in some countries. Alternative materials are needed for replacement.

5.4.3   Electrode and Catalyst

Significant amounts of research and development have been conducted over the past 50 years for the proper choice, design, and fabrication of electrodes and catalysts for alkaline fuel cells. The type and manufacturing of electrodes are linked to the kind of catalysts used.

One of the advantages of the alkaline fuel cell is that it can use both precious and nonprecious metal catalysts. Precious metal catalysts include platinum or platinum alloys. Catalyst loading has been reduced from about 10 mg Pt/cm^2 to 0.5 mg Pt/cm^2 per electrode and it can be as low as about 0.25 mg Pt/cm^2 for each electrode. Precious metals used as a catalyst typically in powder form, deposited on carbon supports, or manufactured as a part of metallic electrodes which are generally based on nickel substrates.

The most commonly used nonprecious metal catalysts include Raney nickel powders for the anodes (about 120 mg nickel/cm^2 along with small amount of titanium and aluminium for the control of sintering) and silver-based powders for the cathodes, such as Ag-supported on carbon (typical loading of 1.5~2 mg Ag/cm^2).

Electrodes may be hydrophobic or hydrophilic. Hydrophobic electrodes are carbon-based with PTFE (polytetrafluoroethylene) and are only partially wetted by the liquid electrolyte. Hydrophilic electrodes are usually made of metallic materials such as nickel and nickel-based alloys. Electrodes usually consist of several layers and each layer has different porosities in order to organize within and through the electrodes the respective flows of liquid electrolyte, gaseous fuel, and oxidant. In general, electrodes with layers of various void fractions are made of powders by the following techniques: mixing powders together, pressing or calendaring into layers, sedimentation, spraying technique, and high-temperature sintering to ensure good mechanical stability.

5.4.4   Stacks

Alkaline fuel cells are connected either with bipolar arrangement, as in Siemens' design, or unipolar electrodes with edge collection of electric current as adopted by Elenco (Zevco). The AFCs power units used in the U.S. shuttle program use a combination of bipolar and edge current collection, which is manufactured by International Fuel Cells.

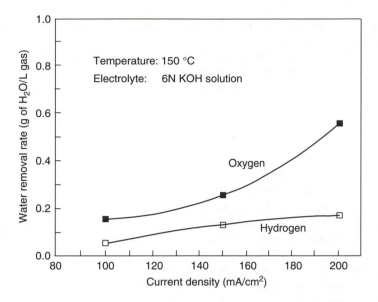

Figure 5.4 Water removal by H_2 and O_2 gas stream in a high temperature alkaline fuel cell as a function of cell current density.[4] $T = 150\,°C$; H_2- and O_2-circulated; KOH concentration: $\sim 16\,N$.

5.4.5 Systems

The stacks are the core of a fuel cell system. In addition to stacks, an alkaline fuel-cell system consists of cooling, waste removal, as well as fuel- and oxidant-supply subsystems. The cooling subsystem is needed to remove excess heat from the AFC stacks. When the electrolyte is circulated, excess heat is taken out of the stacks by the electrolyte and is then removed in an external heat exchanger. For immobile electrolyte design, thermal management can be accomplished by circulating a coolant liquid or processing air through the stacks by recirculating either fuel and/or oxidant stream such as the AFC units used in Apollo space program, or by using cooling plates (with dielectric liquid circulation) such as the case for the AFC units used in the U.S. space shuttle flights.

The reaction product water can be removed from the stacks by the same technique as for heat removal as discussed earlier. For example, circulating electrolyte can carry water out of the stacks, then water can be removed in an external evaporator as used by Elenco (Zevco). Circulating fuel stream (hydrogen) can take water out of the stack, then water is removed in an external condenser as used in Apollo missions. Figure 5.4 shows the amount of water removal by the reactant gas stream as a function of the cell-operating current density for a high-temperature alkaline fuel cell.

Because AFCs cannot tolerate the presence of carbon dioxide in both fuel and oxidant streams, the reactant supply subsystem is very much dependent on the type of reactants chosen. It is the simplest if pure hydrogen is used as fuel and pure oxygen is used as the oxidant, as used in the space explorations. Otherwise, carbon dioxide has to be removed in both the fuel and oxidant streams to below approximately 10 ppm by volume and such a stringent requirement on carbon-dioxide removal represents the biggest obstacle to the commercialization of the alkaline fuel cells for terrestrial commercial applications.

In addition, the DC electric power from fuel cell stacks may need conditioning, or conversion to AC power before connection to the electric loading. Further a control subsystem is required to coordinate the operation of the various system components.

5.5   MATERIALS AND MANUFACTURING

For alkaline fuel cells, the most important components are the electrode and catalyst. The materials of choice determine the design and fabrication techniques adopted. Although both precious and nonprecious metals can be used as catalysts for AFCs, as pointed out earlier, as a thumb of rule, nonprecious metals are used for high-temperature AFCs and precious metals are used for low-temperature AFCs ($< 100\,°C$). This is because additional electrocatalytic activity is needed from the precious metals in order to keep the activation overpotential at the acceptable levels for low-temperature operations. A number of metals have sufficiently high electrocatalytic activity for applications in AFCs as the catalyst, such as platinum, palladium, nickel, and silver, and some intermetallic compounds and alloys. Of course, electrodes constructed from these materials do not need a separate catalyst layer as well. Further detail on AFC electrode development may be found in References 5–8.

5.5.1   Raney Metal Electrodes

Raney metals are highly porous (large surface area for a given mass) and very active, thus highly desirable for electrodes. They are fabricated by mixing the active metals (e.g., nickel or silver) with an inactive metal, usually aluminium. This well-dispersed mixture of metals is not a true alloy and the inactive metal, aluminium, is removed by dissolution in strong alkali, leading to the porous structure for the remaining active metals — a good structure for porous electrodes.

Raney nickel electrodes can be made with this method and aluminium removal is accomplished by treating with an aqueous potassium hydroxide solution. The resulting electrode is highly active after treatment with hydrogen gas and will catch fire when it is dry and exposed to air; it has to be stored in water after being made. Therefore, Raney nickel electrode is usually used for the anode and Raney silver electrode for the cathode.

Raney nickel is attractive as catalytical material for alkaline fuel cells due to its low cost and high activity. The performance of such electrodes depends mainly on the structure of the electrodes and can be improved by doping with several transition metals such as titanium, iron, molybdenum, and chromium. For example, hydrogen electrodes made of titanium-doped Raney nickel may reduce the ohmic polarization by a factor of 2.4 at the minimum, doping with chromium may reduce the overall polarization dramatically when Cr concentration exceeds 0.2%, and the ohmic polarization may be reduced by a factor of 4. Raney nickel electrodes doped with Cr are more active than those doped with Ti, and provide high activity rates over a wide range of dopant content. However, the resulting alloy powder becomes excessively soluble in KOH solution if Cr content exceeds 1%.

5.5.2   Palladium Electrodes

Palladium has several unique properties that make it an advantageous material for a hydrogen electrode. Hydrogen can diffuse through its lattice structure; in fact, hydrogen is the only gas that can diffuse through it. Therefore, a hydrogen electrode made of palladium can tolerate impurities in the hydrogen gas stream since only hydrogen can diffuse through to reach the three-phase boundary for reaction to proceed, and all impurities are blocked from reaching the active sites. However, hydrogen concentration at reaction sites is reduced after diffusion through palladium, thus the electrode performance is lowered. This also suggests the use of a thin palladium membrane as the electrode. In practice, to have better mechanical strength for the resulting electrode, a thin membrane may be made with palladium-silver alloy as the substrate in the middle and a fine layer of palladium black on each of the surfaces. Such membranes can also be used for hydrogen gas purification as well.

Other advantages for a palladium electrode include its high electrocatalytic activity for hydrogen oxidation reaction and its ability to resist the flooding of its pore regions by the liquid electrolyte. However, such electrode is not feasible for commercial fuel cells because of the high cost of palladium and lowered performance resulting from the reduced hydrogen concentration after crossing the thin electrode.

5.5.3   Carbon-Based Electrodes

Porous carbon electrodes are attractive due to their low cost. Carbon used for this purpose is often made by treating suitable coals at high temperature in an air-free environment. Such carbon itself is catalytically active for an oxygen-reduction reaction in alkalyte electrolytes, but in practice, a small amount of active metals is added to enhance its activity significantly. On the other hand, carbon is not active for hydrogen-oxidation reaction, thus requires active metals like platinum and palladium as catalyst. Porous carbon structures can be easily flooded by the liquid electrolyte, thus requiring hydrophobic material, often PTFE, in the electrode structure. The PTFE also acts as a binder to keep the porous structure together, and sometimes carbon fibre may also be used to increase the strength and conductivity of the electrodes. Carbon electrodes may be prepared with either wet or dry method as shown in Figure 5.5.

To further improve the mechanical strength and conductivity, high porosity carbon, mixed with PTFE, can be rolled onto a metal mesh in the shape of the electrode needed, and the metal mesh, often made of nickel, acts as the current collector as well. Due to the high conductivity of the metal mesh, edge collection of current is viable with acceptable ohmic losses.

It is also noted that carbon-electrode performance reduces considerably when the cathode gas is changed from pure oxygen to air because of nitrogen dilution and small oxygen diffusion coefficient (or small rate of oxygen transport). Furthermore, carbon dioxide in the air can impose a significant adverse effect on the cathode-electrode

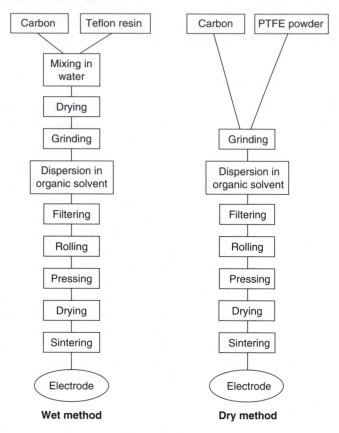

Figure 5.5 Flow diagram for the preparation of carbon electrodes.

performance and lifetime, and the removal of carbon dioxide from the air becomes necessary.

5.5.4   Bimetallic Catalysts

As mentioned earlier, carbon-based electrodes require the use of active metal catalysts for both anode and cathode. Also for immobile electrolyte-alkaline fuel cells operating at low-temperatures, Raney nickel electrodes do not provide sufficient catalytic activity and a precious metal such as platinum must be used. Platinum is a widely used catalyst in low- and medium-temperature oxidation of oxygen, as pointed out in Chapter 3.

However, platinum catalyst experiences corrosion and agglomeration over time. This causes a reduction in the reactive surface area, decreases electrode performance, and eventually limits the cell lifetime. Therefore, bimetallic catalysts are often used. Combining platinum with a transition metal such as Ti, Cr and Co, and so on may have a future as catalysts for oxygen reduction in alkaline fuel cells because they have shown

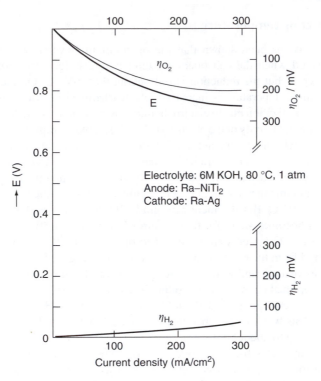

Figure 5.6 A typical performance of an alkaline fuel cell showing the polarization characteristics of both anode and cathode.[3]

enhanced catalytic activity and stable chemical properties. Also, their cost is attractive when compared with precious bimetallic catalysts such as platinum/palladium alloys.

5.6 PERFORMANCE

A typical performance of alkaline fuel cell is shown in Figure 5.6 at the operating condition of 80 °C and 1 atm. The electrolyte is a 6-M KOH solution, the anode is the Raney nickel doped with titanium and Raney silver serves as the cathode electrode. Figure 5.6 also shows the polarization characteristics of both anode and cathode, respectively. It is clear that the anode overpotential in alkaline fuel cell is smaller than the cathode counterpart, but it is not negligible, about 20% of the total polarization at high current densities. This appreciable anode overpotential indicates that the reduction in the anode overpotential is still important in the alkaline fuel cells. As expected from the previous three chapters, the performance of alkaline fuel cell is influenced by the operating temperature, pressure, reactant concentration, and utilization, and so on. Also impurities in the reactant streams can have adverse effect on cell performance as discussed briefly in the following text.

5.6.1   Effect of Temperature

In Chapter 2, it has been shown that the open-circuit (reversible) cell voltage is actually lowered when the cell operating temperature is increased for hydrogen–oxygen fuel cells, but the reduction is only about 0.85 mV/°C. On the other hand, as cell operating temperature is increased, electrochemical reactions proceed more easily, so that polarization due to electrochemical overpotential, η_{act}, decreases; at the same time, the conductivity of the electrolyte (KOH) increases rapidly such that ohmic losses decrease with the temperature. Also, diffusion coefficients for both hydrogen and oxygen increase with temperature so that the rate of mass transport increases as well. As a result, the actual cell potential E is increased when the cell temperature is increased; and this increase in E is very significant at low temperatures ($<60\,°C$), as much as 4 mV/°C. But the increment gradually decreases as the temperature is increased and becomes small for temperatures beyond about 80 °C. Therefore, the state-of-the-art AFC systems are designed to operate at 60 °C~80 °C. This can be clearly inferred from the experimental results shown in Figure 5.7.

For air as oxidant and at the operating pressure of 1 atm, experience [3] indicates that the optimal operating temperature is about 70 °C~80 °C for electrolytes of 6 ~ 7 N KOH concentrations and about 90 °C for 8 ~ 9 N KOH concentrations. The output power density at room temperature is about half of that at 70 °C and it increases almost linearly up to about 50 °C~60 °C.

For space applications, higher operating temperature is also used, for example, the one used for the Apollo space flights which was based on the historical Bacon cells, operated at 200 °C, but in order to prevent electrolyte from boiling the operating pressure had to be at least 4 bars.

Other implications beyond the cell performance need consideration as well. For example, high-temperature AFCs can utilize less expensive catalysts, but are more prone to corrosion and catalyst recrystallization than the low-temperature version. The high-temperature version also requires some time to start up from a cold start. The low-temperature version can be started up easily, usually in a matter of minutes.

5.6.2   Effect of Pressure

As described in Chapter 2, the reversible (open-circuit) cell potential E_r increases with operating pressure logarithmically as follows:

$$\Delta E_r = -\frac{(2.3\Delta N)RT}{2F}\log\left(\frac{P_2}{P_1}\right) \qquad (5.4)$$

where ΔE_r represents the reversible-cell potential increase due to the increase in operating pressure from P_1 to P_2 and $\Delta N = -1.5$ and -0.5, respectively, for reaction product water in a liquid and vapor state. Clearly, the increase in E_r depends on the cell temperature as well, and is about 50 mV/decade of the pressure increase at cell

(a)

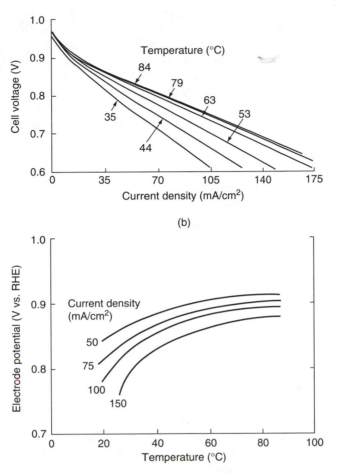

(b)

Figure 5.7 Effect of cell operating temperature on AFC performance. (a) Cell polarization curve obtained for a cell of 289 cm^2 active area, carbon-based Pd anode and Pt cathode and 50% KOH electrolyte[9]; (b) Cathode potential for O$_2$ (air) reduction in 12-N KOH electrolyte and carbon-based porous electrodes with 0.5 mg Pt/cm^2 cathode.[10]

temperature of 60 °C. It is also evident that as pressure is increased, the increase in E_r becomes less and less significant.

On the other hand, as the cell operating pressure is increased, the concentration of the reactants is increased as well and more reactants become available to participate in the electrochemical reactions. Enhanced electrochemical kinetics and mass transport process result in a much more significant increase in the actual cell potential than what Equation (5.4) would predict. This is true for almost all types of fuel cells operating on hydrogen and oxygen. Consequently, it is advantageous to have operating pressures higher than atmospheric pressure (mostly about a few bars).

However, increasing operating pressure much higher than the atmospheric pressure would only have minor improvements from the thermodynamic and electrochemical point of view, but the weight increases associated with the higher mechanical strength needed due to higher pressure become excessive and overshadow the benefit gained. A major concern with reactant pressure increase is that the pressure difference between reactant streams and between reactant and the electrolyte must at all times be maintained within given limits, dictated by the nature of the electrodes and their composition, porosity, and so on. Otherwise, reactants may bubble in the electrolyte, or the liquid electrolyte may leak into the reactant streams, leading to fire hazards and safety concerns. This careful pressure control and balance is difficult during the start-up and shut-down process and elaborate and expensive sensors and control mechanisms must be employed, which become quite costly as well.[11] It should be emphasized that higher pressures are required to prevent the electrolyte from boiling if the operating temperature higher than 100 °C is used. In space application, pressures up to 50 atm at 200 °C have been used together with pristine hydrogen as the fuel.

At this point, it must be emphasized that the selection of the operating conditions, such as temperature and pressure, is directly related to the selection of the materials for cell construction, including the concentration of KOH electrolyte, the electrode, and the catalyst. Thus, for a given cell construction, the optimal operation conditions are fixed as well.

5.6.3   Effect of Reactant Gas Concentration and Utilization

In Chapter 2, it is shown that for a given operating temperature and pressure, the cell reversible potential E_r increases with the concentration or partial pressure of the reactants and reduces with the concentration of the products. This dependence on the composition can be expressed similarly to Equation (5.4) if the total pressure is replaced by the partial pressures of the relevant species involved. Due to the reduction in the irreversible losses, the increase in the actual cell potential is much more than the increase in E_r, similar to the case for total pressure change discussed in the previous subsection. Figure 5.8 shows the effect of oxidant and fuel concentrations on the respective performance of the cathode and anode. As pointed out earlier, the potential loss occurs mainly at the cathode, yet the loss at the anode is still significant, especially at lower oxidant and fuel concentrations.

It should be pointed out that the AFC stacks developed and optimized for pure oxygen as oxidant cannot operate on air satisfactorily, because the porous electrodes cannot cope with the large amount of inert gas (nitrogen), whereas stacks developed for air can use pure oxygen or any form of oxygen-enriched air with enhanced performance.

As discussed in Chapter 2, reactants are almost always supplied more than necessary for current output requirements, in order to have reasonably good fuel cell performance. For hydrogen-air fuel cell, air flow stoichiometric ratio of 2.5 is usually used and the excessive air flow can also be used for product water removal. For hydrogen–oxygen fuel cell, water removal at the hydrogen side can be achieved with hydrogen stoichiometric ratio in the range of 4 to 6.

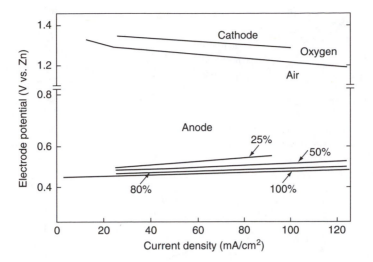

Figure 5.8 Effect of oxygen and hydrogen concentrations on the anode and cathode performance.[12] Carbon-based porous electrode with Noble metal loading of 0.5 to 2.0 mg/cm^2; 9-N KOH electrolyte; at 55 °C to 60 °C.

5.6.4   Effect of Impurities

The main, and perhaps the only, disadvantage of AFCs is its drastic performance loss with fuels containing carbon dioxide from reformed fuels and from the presence of carbon dioxide in air. Therefore, carbon dioxide must be removed, from both in the fuel and oxidant stream, down to about 10 ppm by volume or less, in order to prevent the hydroxide (OH^-) from becoming carbonate ($CO_3^=$).

The negative impact of carbon dioxide arises from its reaction with hydroxide as follows:

$$CO_2 + 2OH^- \longrightarrow CO_3^= + H_2O \tag{5.5}$$

The formation of carbonate $CO_3^=$ produces the following effects: it reduces OH^- concentration and interferes with electrochemical kinetics; it increases electrolyte viscosity, lowers diffusion coefficients and limits current output (i.e., increases concentration polarization); it causes eventual precipitation of carbonate salts in the pores of the porous electrodes, thus blocking the reactants from reaching the reaction sites; it reduces the oxygen solubility in the electrolyte; and it reduces the electrolyte conductivity, hence increases the ohmic polarization. The effect of carbon dioxide on the performance of cathode electrode is shown in Figure 5.9. Performance degradation is clearly shown for both carbon dioxide-free air and carbon dioxide-containing air. However, with CO_2-containing air, the performance decreases much faster.

Higher concentrations of KOH are detrimental to the life of oxygen electrodes operating with CO_2-containing air, but operating the electrode at higher temperature is beneficial because it increases the solubility of $CO_3^=$ in the electrolyte.

Sulfur dioxide (SO_2) and hydrogen sulfide (H_2S) also have deleterious effects on catalysis and must be removed from both fuel and oxidant streams. Other impurities

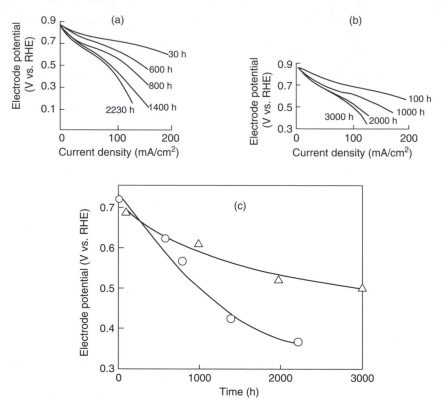

Figure 5.9 Effect of carbon dioxide on air electrode performance (0.2 mg Pt/cm² supported on carbon black, 6-N KOH electrolyte, at 50 °C). (a) For CO_2-containing air; (b) for CO_2-free air; both operated at 32 mA/cm²; and (c) at 100 mA/cm² with CO_2-containing (○) and CO_2-free (△) air.[13]

have to be avoided as well, although they may cause quite different effects. For example, the presence of mercury (Hg) in hydrogen originating from some chlorine plants may lead to accumulation effects in the anodes, with irreversible and continuously aggravating degradation.

Some amount of carbon monoxide in the hydrogen somewhat decreases the power level, but it will remain stable with time and the decrease is reversible. When hydrogen and oxygen are used, they are normally circulated in closed loops. In this case a regular purge must be performed in order to avoid accumulation of inert gases and impurities in the stacks.

5.6.5 Lifetime of AFC Operations

At present both H_2–O_2 and H_2–air AFCs have stacks sufficiently stable over at least 5000 hours, with performance degradation rates of 20 μV/h or less, for example, Siemens has achieved more than 8000 h of operation for its stacks. American space shuttle fuel cells have, on average, accumulated some 2000 operating hours each.

However, for electric power plants (utility) application, lifetimes of at least 40,000 hours with acceptable performance decays are required. For transportation application, a multiple of 5000 or even 10,000 hours of fuel cell stacks are needed. Clearly, much work needs to be done for these two applications. Further, the high weight or size of AFCs (or lower power density on a volume or weight basis) presents another problem for transportation application. At present, the AFC power system is also too expensive comparing with the conventional heat engine-based power plants.

5.6.6   Some Examples of Performance Data

Siemens: 6~7 kW size, pressurized H_2–O_2 AFC systems, 85 kg in weight. This is equivalent to about 12 kg/kW for the system; and 7 ± 1 kg/kW for the stack. The cell is operated at 420 mA/cm^2 and 0.78 V.

Elenco (Zevco): H_2–air system operated at atmospheric conditions. The specific weight is about 12 kg/kW at 110 mA/cm^2. It is said it can potentially reached 6.5 kg/kW at the cell operating conditions of 200 mA/cm^2 and 0.88 V and 3.5 kg/kW at 400 mA/cm^2 and 0.78 V.

For special purposes: U.S. space defence projects, requiring only extremely short operating times of fractions of minutes, possible performance figures for alkaline cells of 0.15~0.25 kg/kW have been quoted.

5.7   METHODS FOR CARBON DIOXIDE REMOVAL

Since the adverse effect of carbon dioxide is the main bottleneck for the terrestrial commercial application of the alkaline fuel cell, it might be worth mentioning techniques for carbon dioxide removal from the reactant streams. A number of methods exist to remove carbon dioxide from the reactant gas mixtures. Liquid-adsorption towers can be used for high-concentration carbon dioxide removal. Basically a liquid solution containing 15% KOH and 10% amine borate is sprayed into the gas stream which contains carbon dioxide. One pass through the tower can reduce carbon dioxide concentration from 25% to about 0.5%; three consecutive passes can reduce CO_2 to about 100 ppm. If followed by thermally regenerative adsorption with ethanolamine:

$$2RNH_2 + CO_2 \underset{115\,°C}{\overset{27\,°C}{\rightleftarrows}} RNHCONHR + H_2O \qquad (5.6)$$

carbon dioxide concentration can be further reduced to about 50 ppm for just one pass.

In physical adsorption, the selexol process with dimethyl ether polyethylene glycol can reduce carbon dioxide concentration by 250 times for one pass. This technique is also called **pressure swing adsorption** where adsorption occurs at high pressure (about 100 psi) and regeneration occurs at low pressure.

For solid beds, one pass through molecular sieves can reduce CO_2 concentration to about 1 ppm with temperature regeneration. Silicon rubber membrane can also be used to separate carbon dioxide from the gas mixture. Electrochemical processes can also be used for the removal of carbon dioxide (i.e., H_2 pump operating at 0.1 V and 600 mA/cm^2).

5.8   FUTURE R&D

The future and the revival of the alkaline fuel cells for commercial applications perhaps lies considerably, if not entirely, in the invention of an economic, efficient and compact carbon-dioxide removal device. The prevention of cell components from deterioration will be essential to reach the cell lifetime targets; this includes improved methods of the cell material processing and fabrication, and possibly the development of new materials which can withstand the hostile reducing and/or oxidizing environment in fuel cells. A successful new material for cell components should have advantage in terms of weight, size and cost. A possible alternative under consideration is conducting plastics.

If hydrocarbons are used as the primary fuel, the economics and efficiency of fuel reformer (or processing) and heating required for the reformation process become another significant area for further research and development. Techniques for electrolyte circulation and regeneration should be investigated if hydrogen is not pristine and air is used as oxidant. From the cost point of view, much work is needed to find acceptable non-Noble metal catalysts for both anode and cathode, particularly transition metal oxide and pyrolyzed "macrocycles," with energy conversion efficiency of 60%.

The design and fabrication of high-access porous electrodes are needed for high performance AFCs. The target is to achieve high current densities on the order of several A/cm^2 with reasonable cell voltage and energy efficiency and high output power density (on the order of kW/kg).

5.9   SUMMARY

In this chapter, we provide brief coverage on the alkaline fuel cell. The advantages and disadvantages are described, and it is shown that the alkaline fuel cell provides the best performance among all fuel cells if pure hydrogen and oxygen are available as reactants. An understanding of the developmental status for this type of fuel cell was developed and a description of the operational principle including the half-cell and entire-cell reactions was given. Cell components and their geometrical configurations are described along with the typical materials used and fabrication processes involved. The performance of the alkaline fuel cell as well as the various influence parameters such as the operating temperature, pressure, reactant concentrations and utilizations, current density, and so on, were outlined. In particular, the significance of the carbon dioxide poisoning in hampering the commercialization of this type of fuel cell was emphasized. This chapter ends with a brief discussion on the method for carbon dioxide removal and other technical issues for the alkaline fuel cell.

BIBLIOGRAPHY

1. Appleby, A. J. 1986. *Energy*, 12: 13.
2. Appleby, A. J. and F. R. Foulkes, 1988. *Fuel Cell Handbook*. New York: Van Nostrand Reinhold.
3. Blomen, L. J. M. J. and M. N. Mugerwa. 1993. *Fuel Cell Systems*. New York: Plenum Press.
4. Kivisari, J. 1995. Fuel Cell lecture notes.
5. Kenjo, T. 1985. *J. Electrochem. Soc.*, 132: 383–386.
6. Tomida, T. and I. Nakabayashi. 1989. *J. Electrochem. Soc.*, 136: 3296–3298.
7. Stashewski, D. 1992. *Intl. J. Hydrogen Energy*. August: 643–649.
8. Kiros, Y. 1996. *J. Electrochem. Soc.*, 143: 2154–2157.
9. McBreen, J., G. Kissel, K. V. Kordesch, F. Kulesa, E. J. Taylor, E. Gannon and S. Srinivasan. 1980. *Proc. of the 15th Intersociety Energy Conversion Engineering Conf.*, 2, New York: AIAA.
10. Tomantschger, K., F. McClusky, L. Oporto, A. Reid and K. Kordesch. 1986. *J. Pow. Sour.*, 18: 317.
11. Warshay, M., P. R. Prokopius, M. Le and G. Voecks. 1996. NASA fuel cell upgrade program for the space shuttle orbiter. *Proc. of the Intersociety Energy Conversion Engineering Conf.*, 1: 1717–1723.
12. Clark, M. B., W. G. Darland and K. V. Kordesch. 1965. *Electro-Chem. Tech.*, 3: 166.
13. Kordesch, K., J. Gsellmann and B. Kraetschmer. 1983. In Power Sources 9, ed. J. Thompson. New York: Academic Press.

PROBLEMS

5.1   Describe briefly the advantage and disadvantage as well as the areas of applications for alkaline fuel cells.

5.2   Describe the operation principle such as half-cell and whole-cell reaction, the primary fuels expected to be used for alkaline fuel cells.

5.3   Discuss briefly the typical (or target) operating conditions such as cell voltage, cell current density, temperature, pressure, fuel and oxidant utilization, and chemical–electrical energy conversion efficiency, and so on.

5.4   Describe the effect of operating conditions on the cell performance.

5.5   Describe the geometrical configuration of cell and stack, typical materials used for manufacture, and the thickness as well as other dimensions of the cell components.

5.6   Describe the components of the alkaline fuel cell system such as cooling (thermal management), product removal, electrolyte management, fuel processing, control, and so on.

5.7   Describe the factors affecting the short- and long-term performance of alkaline fuel cells.

5.8   Describe the critical technical barriers to be overcome for the commercialization of alkaline fuel cells, the possible solutions, and their pros and cons.

PHOSPHORIC ACID FUEL CELLS (PAFCs)

6.1 INTRODUCTION

Phosphoric acid fuel cells (PAFCs) are the most advanced type of fuel cells, and are considered to be technically mature since commercial PAFC units are already available, commonly in 200-kW size range, after nearly 30 years of development. Therefore, PAFCs have been referred to as the first-generation fuel cell technology. Unlike the alkaline fuel cell which was primarily developed for space applications, the PAFC was targeted initially for terrestrial commercial applications with the CO_2-containing air as the oxidant gas and hydrocarbons (natural gas in particular) as the primary fuel for power generation. Hydrogen is still the fuel feeding the fuel cell operation and it is produced through the steam reforming of hydrocarbon fuels.

The PAFC was the result of a compromise caused by the CO_2 syndrome. Although the AFCs have achieved routine application in space exploration, where pure hydrogen and pure oxygen are employed, they cannot tolerate any carbon dioxide in both cathode and anode gases (CO_2 concentration must be less than 10 ppm by volume, as discussed in the previous chapter). For terrestrial commercial applications, fuel cells would have to use air as the oxidant and reformed hydrocarbon gases as the fuel. This is because cheap pure hydrogen is not available and hydrocarbon reformation provides the cheapest source of hydrogen gas, albeit in the mixture form containing a significant amount of carbon dioxide (as much as more than 20%) and some carbon monoxide (typically around 1%~2%). Because acid electrolytes do not react with the carbon dioxide in the reformed fuel gas and the air stream, unlike the alkaline electrolyte shown in Equation (5.5), acid electrolytes were sought after for fuel cell applications and the phosphoric acid fuel cell was recognized in 1960s as the only choice of technically acceptable acid fuel cell, especially for small kW commercial units.

Despite its tolerance to almost any CO_2 concentrations in the reactant gases, the PAFC cannot tolerate the presence of carbon monoxide, which poisons the catalyst platinum required for acid-electrolyte fuel cells, as pointed out in Chapter 3. Further, the PAFC has poorer performance than the corresponding AFC when operating on the same pure hydrogen and oxygen as reactants, as shown in Figure 5.1, even when the PAFC is operated at much higher temperature. This lower performance of the PAFC is primarily due to the slow oxygen reaction rate at the cathode. Therefore, the PAFC is typically operated at higher temperature (near 200 °C) for better tolerance to carbon monoxide in the fuel gas and better electrochemical reactivity and for smaller internal resistance, which is mainly due to the phosphoric acid electrolyte. However, reforming of hydrocarbons, such as methane (natural gas), occurs at higher temperatures, commonly more than 500 °C; the waste heat from the PAFC at around 200 °C can only be used for the preheating of the fuel and air, and additional heating and energy for the endothermic reforming reaction have to be from the burning of fresh fuel. This affects adversely the efficiency of the overall PAFC system, which is limited to around 40%–45% based on HHV. As a result, the PAFC exhibits the problems of both high- and low-temperature fuel cells, but possibly none of the advantages of either options.

Despite the initial belief that, among the low-temperature fuel cells, PAFCs were the only fuel cells which showed relative tolerance for reformed hydrocarbon fuels and thus could have widespread applicability in the near term, recent advances in polymer electrolyte membrane fuel cell have also called that choice of PAFC into question. But PAFC is still being evaluated for stationary applications, including onsite co-generation and dispersed electricity generation, transportation applications as vehicular power generators for buses and forklift trucks, and small-capacity trans-portable systems, mainly for military applications. The state-of-the-art PAFC units used for onsite co-generation have a chemical–electrical energy efficiency of 40% and thermal energy efficiency of 50%, giving an unprecedented 90% combined efficiency.

6.2   BASIC PRINCIPLES AND OPERATIONS

The phosphoric acid fuel cell is just one type of the acid-electrolyte fuel cells, there-fore, its basic cell structure as well as the half-cell and total-cell overall reaction has already been described in Sections 1.3.2 and 3.2. As the name implies, the PAFC uses the phosphoric acid (H_3PO_4) as its electrolyte. The phosphoric acid is usually in highly concentrated form (95% or higher, hence in the form of pyrophosphoric acid, or $H_4P_2O_7$), with a high ionic conductivity of more than 0.6 S/cm. The electrolyte is often immobilized in a porous silicon carbide (SiC) matrix by capillary action. Pure hydrogen or hydrogen-rich gases can be used as fuel and air is almost invariably used as oxidant. Because the acid electrolyte is tolerant to the presence of CO_2 in the reactant gas streams, as mentioned earlier, hydrogen produced by steam reforming of organic fuels, such as hydrocarbons (typically natural gas or methane) and alcohols (mainly methanol or ethanol), is often used as the anodic reactant.

The steam reformation of hydrocarbons is very similar in principle, and may be illustrated with that of natural gas (mostly methane)

$$CH_4 + H_2O \longrightarrow 3H_2 + CO \tag{6.1}$$

This reaction is endothermic and usually occurs over a supported catalyst (i.e. nickel) to accelerate reaction kinetics. The carbon monoxide, produced in the above steam reforming process, can severely poison the Noble metal catalyst, as pointed out earlier. Thus, CO removal or cleanup is necessary after the steam reforming of the hydrocarbon fuels and before the hydrogen-rich reformate gas mixture is supplied to the phosphoric acid fuel cell. Carbon monoxide removal is usually achieved over a supported catalyst (i.e., Fe–Cr or Cu–Zn) at elevated temperatures via the water–gas shift reaction:

$$CO + H_2O \longrightarrow CO_2 + H_2 \tag{6.2}$$

Thus, CO concentration is reduced by reaction with water to form CO_2, which is only an inert diluent for PAFCs. Normally, excess water is provided in the reforming process so that a part of the CO from the reforming process is converted to CO_2 before the CO removal stage. Despite this, water–gas shift reaction needs to be carried out in two stages at different temperatures in order to reduce CO concentration to the levels acceptable for PAFC operation. Therefore, the overall reaction for the fuel processing may be written as

$$CH_4 + 2H_2O \longrightarrow 4H_2 + CO_2 \tag{6.3}$$

which is simply the result of combining both Equations (6.1) and (6.2) together. Clearly, hydrocarbon fuel processing involves a multistep process and in practice occurs in different devices at different temperatures and even possibly different pressures.

A phosphoric acid fuel cell normally operates at the temperature of around 170 °C~210 °C and 1~8 atm, both water and the waste heat are the byproduct of the electrochemical reactions occurring in the cell (on the cathode side), and they need to be removed in order to ensure continuous and stable operation of the fuel cell. Water removal is relatively easy in PAFCs because the product water is in the vapor form at the typical PAFC operating temperatures and can be carried out of the cell via the excess oxidant stream. Thermal management is more involved and various cooling systems have been used for the waste heat removal, employing the convective heat transfer of gas (typically air) or liquid coolants (typically water or dielectric liquid oil) and the convective–evaporative cooling of liquid coolants (typically water).

Example 6.1 Normally the steam reforming reaction of hydrocarbon fuels is endothermic, thus, heat addition is necessary to produce hydrogen gas. Calculate the enthalpy of reaction for reactions (6.1), (6.2), and (6.3) at 25 °C and 1 atm to determine the amount of heating required for the natural gas reforming process.

SOLUTION

The enthalpy of reaction can be calculated from Equation (2.4)

$$\Delta h_{reaction} = h_P - h_R$$

Because the reaction occurs at the standard temperature and pressure, the sensible enthalpy for each species vanishes, we have for reaction (6.1), with the standard enthalpy of formation given in Appendix A.2

$$\Delta h_{reaction} = 3h_{f,H_2} + h_{f,CO} - [h_{f,CH_4} + h_{f,H_2O(g)}]$$
$$= 3 \times 0 + (-110,530) - [(-74,850) + (-241,826)]$$
$$= +206,100 \, J/mol = +206.1 \, kJ/mol$$

where the positive sign before the numerical number indicates that heat needs to be added to the system in order to maintain the reaction going forward at the given temperature and pressure, or the reaction is endothermic. Similarly, we can determine for reaction (6.2) and (6.3), respectively

$$\Delta h_{reaction} = h_{f,CO_2} + h_{f,H_2} - [h_{f,CO} + h_{f,H_2O(g)}]$$
$$= (-393,522) + 0 - [(-110,530) + (-241,826)]$$
$$= -41,170 \, J/mol = -41.17 \, kJ/mol$$
$$\Delta h_{reaction} = 4h_{f,H_2} + h_{f,CO_2} - [h_{f,CH_4} + 2 \times h_{f,H_2O(g)}]$$
$$= 4 \times 0 + (-393,522) - [(-74,850) + 2 \times (-241,826)]$$
$$= +165,000 \, J/mol = +165.0 \, kJ/mol$$

Therefore, the water–gas shift reaction is exothermic with 41.17 kJ/mol of the heat released during the reaction; while the overall steam reformation of methane is endothermic, requiring 165.0 kJ of heat addition per mole of methane reformed.

COMMENT

1. Reactions (6.1)–(6.3) can also be written in the following form with the enthalpy of reaction stated:

$$CH_4 + H_2O(g) \longrightarrow 3H_2 + CO + 206.1 \, kJ/mol$$
$$CO + H_2O(g) \longrightarrow CO_2 + H_2 - 41.17 \, kJ/mol$$
$$CH_4 + 2H_2O(g) \longrightarrow 4H_2 + CO_2 + 165.0 \, kJ/mol$$

2. Steam reforming of methane is highly endothermic and the design of heat supply for this reaction often complicates the overall fuel cell power system design with

a multitude of heat exchangers employed. Thus, thermal integration and system optimization become important issues.

3. Although the overall steam reforming of methane is endothermic, the water–gas shift reaction is exothermic, thus CO removal via this reaction is better carried out at low temperatures, at least lower than the temperature at which reforming reaction occurs.

Example 6.2   For the steam reforming of methane

$$CH_4 + 2H_2O(g) \longrightarrow 4H_2 + CO_2$$

Take the water (steam) to methane mole ratio of 2 in the reactant mixture. Determine the equilibrium composition for the reforming product mixture considering the water–gas shift reaction and the incomplete reforming process. Assume the reforming reaction occurs at the various temperatures (i.e., around 1000 K) and 1 atm.

SOLUTION

To determine the equilibrium composition for the reforming product mixture, we need first to determine what species is present in the product mixture. Obviously, H_2 and CO_2 are present as they are the final product if the reforming reaction proceeds to completion. In reality, the reforming reaction, as any other reaction, is incomplete due to thermodynamic limitations, thus some CH_4 and H_2O are left in the mixture. Further, CO is present because it is the intermediate species formed during the reforming process.

Solid carbon particle formation is real and possible in the reforming process of hydrocarbon fuels and is one of the most critical issues in the fuel cell systems with hydrocarbons as the primary fuel. This is because carbon particles can deactivate the catalyst used to promote the reforming process and may even plug the reformer, leading to reduced H_2 production. Carbon formation begins to occur for steam/carbon ratio much less than the stoichiometric value. Steam/carbon (S/C) ratio is defined as the mole ratio of the amount of steam to carbon in the reactant mixture before the reforming reaction. For example, carbon begins to form if S/C ratio becomes less than 1.5. To avoid carbon formation, S/C ratio of 2 to 3 is usually used in practice in the steam reforming process. For the present situation of S/C $= 2$, solid carbon would be avoided. The formation of solid carbon particles is taken up again in more detail in Chapter 8.

Therefore, the reforming product mixture for the present problem can be considered to include the following species

$$H_2, CO_2, CO, H_2O, \text{ and } CH_4$$

and the reforming reaction can be expressed as

$$CH_4 + 2H_2O(g) \longrightarrow n_1H_2 + n_2CO_2 + n_3CO + n_4H_2O + n_5CH_4$$

where n_i is the number of moles for the species i and $i = 1, 2, 3, 4, 5$ corresponds to the species H_2, CO_2, C, H_2O and CH_4, respectively.

To solve for these five unknown n_i's, five independent equations must be formulated. Three of them can be established based on the conservation of atomic species:

$$C: \quad n_2 + n_3 + n_5 = 1$$
$$H: 2n_1 + 2n_4 + 4n_5 = 4 + 4$$
$$O: \quad 2n_2 + n_3 + n_4 = 2$$

The other two equations can be obtained from the equilibrium of the water–gas shift reaction and the methane reforming:

$$CO + H_2O(g) \rightleftharpoons CO_2 + H_2$$
$$CH_4 + H_2O(g) \rightleftharpoons 3H_2 + CO$$

and the corresponding equilibrium constant for partial pressure becomes

$$K_{P,1} = \frac{P_{CO_2} P_{H_2}}{P_{CO} P_{H_2O}}$$

$$K_{P,2} = \frac{\left(P_{H_2}/P_{ref}\right)^3 \left(P_{CO}/P_{ref}\right)}{\left(P_{CH_4}/P_{ref}\right)\left(P_{H_2O}/P_{ref}\right)} = \frac{P_{H_2}^3 P_{CO}}{P_{CH_4} P_{H_2O}} \frac{1}{P_{ref}^2}$$

Assume ideal gas behavior for the gas mixture, we have

$$P_i = \frac{n_i}{n_T} P_T$$

where $n_T = n_1 + n_2 + n_3 + n_4 + n_5$ represents the total number of moles in the reforming product mixture and P_T is the total pressure of the mixture. Substituting the expressions for the partial pressures, we have for the equilibrium constants as follows

$$K_{P,1} = \frac{n_1 n_2}{n_3 n_4}$$

$$K_{P,2} = \frac{n_1^3 n_3}{n_4 n_5} \frac{P_T^2}{(n_T P_{ref})^2}$$

where $P_{ref} = 1$ atm is the reference pressure; and $K_{P,1}$ and $K_{P,2}$ are functions of temperature only at the given P_{ref}, and are often tabulated or given in correlation. For example, their values can be obtained from Appendix C and given below:

	$K_{P,1}$	$K_{P,2}$
500 K	138.3	8.684×10^{-11}
1000 K	1.443	26.64

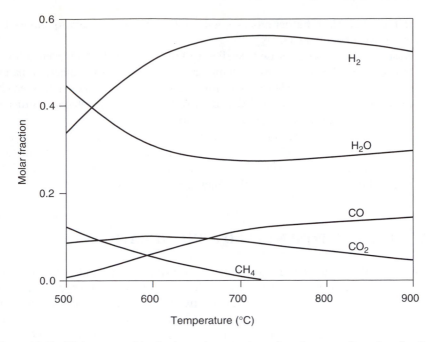

Figure 6.1 Equilibrium composition for the product gas mixture from the steam reformation of methane with steam-to-carbon ratio of 2 and at the total pressure of 1 atm.

It is clear that for the endothermic reaction of methane steam reforming, the equilibrium constant $K_{P,2}$ increases rapidly with temperature, so that the reaction product H_2 and CO also increases rapidly with temperature. On the other hand, for the exothermic water–gas shift reaction the equilibrium constant $K_{P,1}$ decreases with temperature, so that the reaction product CO_2 and H_2 also decreases, but the concentration for CO as the reactant increases. Therefore, it might be expected that CO concentration increases with temperature, but H_2 concentration peaks as temperature is increased. This is indeed the case, when the above five equations formulated are solved simultaneously with Newton–Raphson method and the results are plotted in Figure 6.1.

COMMENT

1. Based on the principle of Le Chatelier, which says that for any reaction initially in equilibrium, once its condition such as pressure and temperature is changed, its composition will change accordingly in such a manner as to minimize the effect of the change. Thus, the effect of temperature increase is as explained earlier for the variation of K_Ps with temperature. For an increase in pressure, it means that the reaction moves toward the direction of producing fewer total number of moles in the reaction mixture. For the methane steam reforming as given in Equation (6.1), a lower pressure will favor the production of more H_2, which accompanies with the formation of more CO; whereas the water–gas shift reaction is hardly affected by pressure change. This usually results in the CO concentration being too high in the reforming product mixture to be used directly

in fuel cells. Thus, further processing of fuel stream is necessary to remove the excessive amount of CO.

2. Based on the principle of Le Chatelier, CO removal is better achieved at lower temperatures than the reforming temperature because of its exothermic nature of the reaction. This is also evident from Figure 6.1. In practice, multistage CO removal may be implemented at successively lower temperatures in order to achieve successively lower CO concentrations.

3. Figure 6.1 suggests that the temperature for the steam reforming of methane should be at least 650 °C~700 °C in order to produce more hydrogen, although in practice the rate of hydrogen production will also depend on the catalytic effect of the catalyst used (i.e., the kinetics of the reaction).

6.3 CELL COMPONENTS AND CONFIGURATIONS

Basic fuel cell structure consists of an electrolyte (phosphoric acid), contained in a porous silicon carbide (SiC) matrix by capillary action and so immobilized in the PAFC cell structure and sandwiched between two electrodes, the anode and the cathode. Figure 6.2 illustrates the schematic of a phosphoric acid fuel cell unit, and many such units may be stacked together, as shown in Figure 4.11, to form a PAFC stack. Notice that the bipolar plate shown in Figure 4.11 is also called separator plate, or simply separator, in PAFCs. An enlarged view of the single PAFC is presented in Figure 6.3 showing clearly the structures of the catalyst layer as well as the electrolyte matrix. The liquid electrolyte penetrates into the catalyst layer, equilibrated with the reactant gas on both anode and cathode side, thus forming a three-phase zone in each of the catalyst layers. The physical and chemical processes occurring in three-phase zones have already been described in Chapter 1 in relation to Figure 1.7. Especially, the half-cell electrochemical reactions occur at the surface of the catalyst in the three-phase zones where the reactant gas (hydrogen in the anode and oxygen

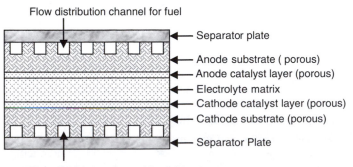

Figure 6.2 Schematic of a phosphoric-acid fuel cell unit. Notice in the figure that the flow distribution channels for the fuel and oxidant are made on the electrode substrates (or the so-called **ribbed substrate design**). It is also possible to make the flow channels on the separator plate (or the so-called **ribbed separator design**).

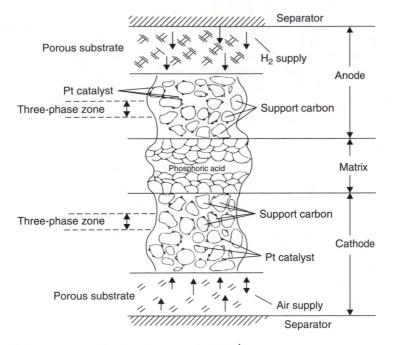

Figure 6.3 Enlarged view of the three-phase zones in PAFCs[1].

in the cathode) must be available at the catalyst surface, which must be wetted by the liquid electrolyte. As a result, the reactant supply to, and reaction product removal from, the catalyst surfaces (or reaction sites) must be carefully organized by appropriate design of the microscopic structures of the electrode backing layers and catalyst layers, as well as the electrolyte matrix for appropriate electrolyte penetration into the catalyst layer and for the wetting of the catalyst surfaces. This structural design must be optimized with corresponding operating conditions since the electrolyte penetration into the catalyst layer can be altered by the pressure of the anode and cathode gas streams.

6.3.1 Electrolytes

As pointed out earlier, phosphoric acid was chosen among all type of acids as the electrolyte for fuel cells because of the need to tolerate the presence of carbon dioxide in the reactant gas streams, especially the hydrogen-rich anode fuel stream arising from the reformation of hydrocarbon fuels. Phosphoric acid as electrolyte has the following advantageous characteristics when compared with other acids

- Tolerance to any amount of carbon dioxide, CO_2;
- Reasonable tolerance to carbon monoxide CO, up to 1%~2% at the cell operating temperature of 200 °C;
- Low vapor pressure, resulting in low rate of electrolyte loss due to evaporation and thus long periods of operation without the need of electrolyte replenishment;

- High oxygen solubility, thus fast cathode reactions since cathodic activation polarization represented one of the largest single energy loss mechanism in acid-electrolyte fuel cells as discussed in Chapter 1;
- Reasonably good ionic conductivity at relatively high temperatures, typically >0.6 S/cm at 200 °C) and no electronic conductivity;
- Relatively low corrosion rate at the high temperature of around 200 °C;
- Large contact angle (>90°), thus good wettability for the catalyst surface, which is needed for the formation of three-phase zones.

Phosphoric acid is a colorless, viscous, and hygroscopic liquid. At temperatures below 150 °C, oxygen reduction kinetics is very slow because the adsorption of the phosphoric acid molecule and/or anions of this acid inhibits oxygen adsorption on the catalyst surface, which is a necessary intermediate step in the oxygen-reduction reaction as described in Section 3.7.2. Above 150 °C, phosphoric acid is predominantly in the polymeric state as pyrophosphoric acid ($H_4P_2O_7$), which is strongly ionized, and probably because of the relatively large size of the anions ($H_3P_2O_7^-$) of this acid, anion adsorption is minimal and oxygen reduction reaction proceeds reasonably fast.

Generally speaking, 100% phosphoric acid (H_3PO_4) is used in PAFCs. It contains 72.43% phosphoric anhydride (P_2O_5), and has a density of 1.863 g/mL at 20 °C. Its density increases as P_2O_5 content is reduced. 100% phosphoric acid has a high solidification temperature of 42 °C, and its solidification results in a volume increase, which can damage the porous electrode and matrix structures, lower cell performance and shorten cell lifetime. Therefore, PAFC stacks must be maintained at temperatures above the solidification temperature, typically above 45 °C, even during off-load conditions. This is undesirable for practical applications.

The temperature at which the solidification of phosphoric acid occurs depends very much on the concentration of H_3PO_4, as shown in Table 6.1 and Figure 6.4. Clearly solidification temperature is the highest for concentrated phosphoric acid of around 100% (by weight), and decreases rapidly as H_3PO_4 concentration is reduced. Low concentration is generally used to avoid electrolyte freezing during transportation of the acid from factory to the site, then concentrated into strong phosphoric acid before entering the stack into operation. Once the stack is in operation, it must be

Table 6.1 Solidification (Freezing) Temperature of Phosphoric Acid as a Function of Its Concentration[2]

H_3PO_4 Concentration (% by weight)	Solidification Temperature (°C)
100	42.4
91.6	29.3
85	21
75	−20
62.5	−85

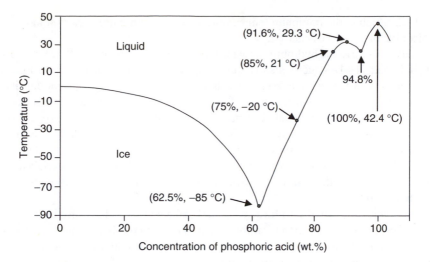

Figure 6.4 Phase diagram showing the solidification (freezing) temperature of phosphoric acid as a function of its concentration.[2]

kept warm continuously ($T \geq 45\,°C$) even during offload conditions. Hence the stack must be equipped with an appropriate heating device.

Although the vapor pressure of H_3PO_4 is very low, the acid loss due to evaporation becomes significant over long periods of operation (4~5 years) at the high-temperature cell operating conditions. The rate of the acid loss increases with reactant gas velocity and electric current density output. Also high current density produces a large amount of water vapor which entrains (or carries away) the acid in the form of small droplets. Because of phosphoric acid loss during operation, an acid replenishment system is necessary. Alternatively, a sufficient amount of H_3PO_4 should be stored within the cell for the total duration of operation.

At the operating temperature range of $170\,°C$~$210\,°C$, high-concentration phosphoric acid H_3PO_4 dissociates into $H_4P_2O_7$ (pyrophosphoric acid):

$$2H_3PO_4 \longrightarrow H_4P_2O_7 + H_2O \qquad (6.4)$$

and the catalytic effect for fuel cell reaction is reduced due to the acid dissociation. In this respect, fluorinated sulfonic acids such as trifluoromethane-sulfonic acid (CF_3SO_3H) is much better, even at low temperatures. However, fluorosulfonic acids lose water at temperatures higher than $110\,°C$ and then exist as hydrates with reduced conductivity (<0.1 S/cm) as temperature is further increased.

Other acids such as sulfuric, perchloric, and hydrofluoric acid are not suitable as acid electrolyte for fuel cell applications. This is because sulfuric acid is unstable at the operating potentials of the anode where it is reduced to sulfurous acid and to some extent even to H_2S or S. Perchloric acid is a strong oxidizing agent and can cause the fuel to explode. Further both sulfuric and perchloric acids have relatively high vapor pressure and lose water at temperature above about $100\,°C$. Therefore, these two acids cannot compete with phosphoric acid, particularly for fuel cells operating at over

150 °C, which is the minimum temperature essential to minimize carbon monoxide poisoning of the electrocatalyst.

Hydrochloric and hydrobromic acids are the ideal electrolytes for the regenerative H_2–Cl_2 and H_2–Br_2 fuel cells. The chlorine and bromine electrode reactions are considerably faster than the oxygen-electrode reaction (exchange current density of about 10^{-3} A/cm^2 as compared to 10^{-9} A/cm^2 for O_2 on smooth electrode surfaces). Further, there are no complicated changes in the chemistry of the surfaces (e.g., oxide formation), no different electrocatalysts required for forward and backward reactions; the performances of regenerative H_2–Cl_2 and H_2–Br_2 fuel cells are excellent in terms of efficiency and power density. The main problem is to find stable materials as fuel cell components in the highly corrosive environment such that the resulting fuel cells can have sufficient lifetime.

Suitable fluorinated sulfuric acids as fuel cell electrolytes have also been investigated because trifluoromethane sulfonic acid (TFMSA) has high ionic conductivity, good thermal stability, higher solubility of O_2 than in nonfluorinated acids and low degree of adsorption of the acid anion on the electrode surface. The latter properties result in oxygen-reduction kinetics on a platinum surface more than one order of magnitude faster than in other aqueous acids, leading to good cathodic performance as shown in Figure 5.1. The problem arises from its high wettability of Teflon, which causes electrode flooding, and acid concentration management at the operating temperature of ≤ 100 °C (i.e., product water dilution and removal).

To overcome the problems encountered with TFMSA, but still retaining the attractive features of fluorinated acids (the so-called "superacids"), similar acids have been developed with higher molecular weights for use in fuel cells at above 100 °C and with two or more acid groups, such as disulfonic acid. Instead of fluorinated sulfonic acids, fluorinated carboxylic, phosphoric, and antimonic acids as well as fluorinated disulfone imide acid have also been investigated as a possible fuel cell electrolyte; and all attempts to find an alternative acid to phosphoric acid have not been successful. But significant progress has been made in developing the solid polymer electrolyte fuel cell, which uses a perfluorinated sulfonic acid polymer as the electrolyte — this type of fuel cell is often called a polymer electrolyte membrane (PEM) fuel cell since fuel cells are commonly classified based on the type of electrolyte used. The PEM fuel cell are described in Chapter 7.

6.3.2   Matrix

The purpose of the electrolyte matrix is to hold phosphoric acid as an integral part of the cell structure and to prevent crossover of reactant gases into the opposite electrode compartments. The electrolyte matrix is ionically conductive but electrically nonconductive (insulating). Phosphoric acid, as discussed earlier, is contained in a so-called matrix structure made of a fine silicon carbide (SiC) powder bound with a small amount of PTFE. The matrix is porous and holds the acid by capillary action. The matrix should be as thin as possible to minimize internal resistance and is typically about 0.1∼0.2 mm thick. The matrix should also have high thermal conductivity for waste heat dissipation and sufficient chemical stability and mechanical strength at the high temperature cell operating conditions.

Silicon carbide matrix structures can satisfy the preceding requirements except for mechanical strength, since they may only withstand up to about one-tenth of atmospheric pressure difference between the anode and cathode stream. Therefore, the pressure differential must be controlled even under transient startup and shutdown operations.

6.3.3   Electrode and Catalyst

The electrode is composed of the catalyst layer and the substrate on which the thin catalyst layer is mechanically supported, as illustrated in Figure 6.3. Under suitable pressure balance, the three-phase zone is established within the catalyst layer and the electrochemical reactions take place in the three-phase zone.

Catalysts   The key components of the catalyst layer are carbon-supported, highly dispersed platinum catalyst, and hydrophobic agents such as PTFE. The thickness of the catalyst layer is about 0.1 mm or less. The hydrophobicity due to PTFE is optimized to control electrolyte wettability and gas diffusion, that is, to prevent the acid flooding of the electrode. To prevent overpenetration of reactant gases, the adjacent layer of the electrode to the matrix should have high capillary action (small pore size). The effect of hydrophobic agent content on electrode activity is shown in Figure 6.5 for the cathodic oxygen-reduction reaction. Typically the optimal PTFE content is about 20%~30%.

Platinum alloys in high-surface-area form are currently the preferred catalyst material and their catalytic activity depends on the kind of catalyst, crystallite size, and specific surface area. Smaller crystallite size and larger specific surface area yield higher catalyst activity. Crystallite sizes of down to the order of 2 nm and specific surface areas of up to $100 \, \text{m}^2/\text{g}$ or more of platinum have been achieved, as shown in Table 6.2. The typical platinum loading is about $0.1 \, \text{mg/cm}^2$ for the anode and $0.5 \, \text{mg/cm}^2$ for the cathode.

Carbon is preferred as supporting material to have high electric conductivity of the catalyst layer, thus to reduce ohmic losses in the electrode. The main function of the carbon support is to disperse the platinum catalyst, to provide numerous micropores in the electrode for gas diffusion and to increase the electrical conductivity of the catalyst layer. At present, two types of carbon black are used as support carbon: acetylene black and furnace black. Both are heat treated to achieve high electrical conductivity, corrosion resistivity, and specific surface area. The platinum supported on carbon, or Pt/C, is bonded by the PTFE to form the catalyst layer.

Electrode Substrate   It is a reinforced material adjacent to the catalyst layer and permits the flow of electrons and reactant gases, or it is also commonly called gas diffusion layer (GDL). At the operating temperature of the PAFCs, 100% phosphoric acid is very corrosive and graphitized carbon material is usually used in the form of carbon paper or carbon cloth. The substrate usually has about 60%~65% porosity and 20~40 μm diameter of pore size. For the ribbed substrate design, the thickness of the substrates is about 1~1.8 mm, which contains around 0.6~1.0 mm rib thickness

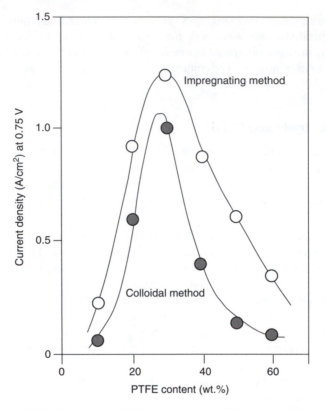

Figure 6.5 Effect of hydrophobic agent PTFE content on the activity of cathodic oxygen-reduction reaction.[3]

Table 6.2 Typical Platinum Catalyst Particle Sizes and Surface Areas[3]

Particle Size (nm)	% Pt Atom on Surface	BET Surface Area (m^2/g)
0.8	90	200
1.8	67	130
5.0	25	50

for each electrode. However, for the ribbed separator design, the electrode substrate layer can be much thinner, as thin as around 200 μm.

6.3.4 Stacks

PAFCs are arranged in bipolar configuration with separator plate between each cell. Typical design uses either a ribbed separator or a ribbed substrate. A ribbed separator design has ribs or flow channels made on the separator for the reactant gas supply;

whereas a ribbed substrate design has ribs or flow channels made on the substrate layer of the electrode. The ribbed substrate design is generally used in large stacks for better and uniform gas diffusion to the active reaction sites. Also a porous substrate may be used for phosphoric acid storage in order to have sufficient electrolyte in the cell during the design lifetime.

Other configuration has also been proposed such as the integrated electrode substrate. Basically, it consists of two electrodes with separator plate sandwiched in between and all these three components are fabricated into a single structure for better electrical and thermal conductivity and easy assembly of the stack components.

Separator (or Bipolar) Plate   Its purpose is to prevent the mixing of hydrogen-rich anode gas and air in the cell structure and to connect the two adjacent electrodes electrically. Thin glassy carbon plate (or vitreous carbon, polymer carbon) is generally used as separator and should be as thin as possible for low electric and thermal resistance. However, its thickness is limited by mechanical strength needed to withstand the pressure difference between the anode and cathode gas and the possible crossover of reactant gases, especially due to hydrogen diffusion. Typically it is less than 1-mm thick for the ribbed substrate design since the separator plate is flat for both of its surface. However, for ribbed separator design the separator plate would be much thicker, typically in the range between 6~8 mm, to accommodate the flow channels made on both of its surfaces and graphite plate is often used for this purpose.

Gas-Supplying Structure (Manifolds)   Stack manifolds are used to supply the reactant gas uniformly to each cell in order to obtain the same optimal performance for each cell in the stack. This is especially important for the hydrogen-rich reformed fuel gas stream because of density difference between hydrogen and carbon dioxide. The upper portion of a large stack has the tendency to become hydrogen-rich. If this happens, it would be advisable to have the stack divided into several smaller stack blocks for uniform gas supply through each manifold to the smaller stack blocks.

Two types of manifold structures are often used: internal and external manifolding, as shown in Figure 6.6. The former is in the form of holes through the stack itself, and for the latter, the manifold box is usually attached to the sides of the stack. For internal manifolding, the anode and cathode gas flow can be easily arranged in co-flow, counterflow, crossflow, or any other complex flow arrangements; whereas external manifolding only allows the crossflow arrangement for the anode and cathode gas stream.

6.3.5   Systems

In addition to fuel cell stacks, PAFC systems consist of product water removal, cooling subsystem for thermal management, fuel and oxidant supply and processing/conditioning subsystem, the inverter, and the control subsystem.

Product Water Removal   Since PAFCs are operated at about $190\,°C \sim 210\,°C$, product water produced at the cathode can be removed easily with proper electrode

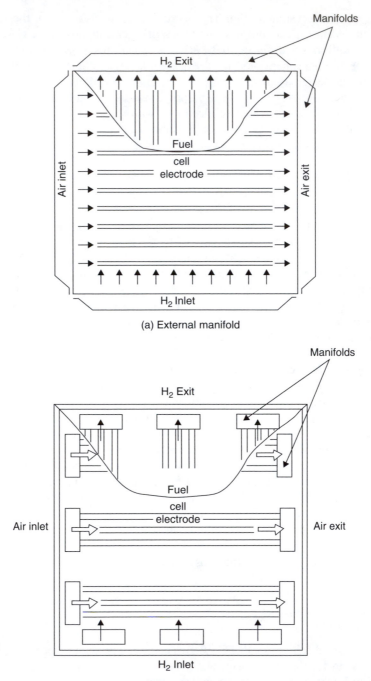

Figure 6.6 Stack design with internal and external manifolding for the reactant gas supply to each cell in the stack.[4]

structure. At the operating temperature of PAFCs, water can be removed by natural vaporization in the form of steam through the porous electrode and be taken away by the process air flow. Although the vapor pressure of water at the operating temperature is higher than that of the acid, some acid is inevitably entrained in the water vapor, particularly at high temperatures, and must be replenished. Therefore, for PAFCs water management is easy as compared to other low-temperature fuel cells such as alkaline and proton exchange membrane fuel cells.

It should be noted that at $T \leq 190\,°C$, product water would dissolve in the electrolyte, diluting the acid electrolyte and increasing its volume and consequently flooding the pores of the electrodes. At $T \geq 210\,°C$, phosphoric acid begins to decompose and corrosion becomes more serious; even the graphite components are subject to corrosion.

Cooling Methods   As for other types of fuel cells, PAFCs need to be cooled to dissipate the waste heat generated within the stacks. In general, three different cooling techniques can be used: air cooling, dielectric liquid cooling (synthetic oil cooling), and water cooling.

Water Cooling   It is most popular because of its high cooling capacity, efficient operation, and compatibility with system components; its performance is better than the other two methods; and is suited for large stacks and for co-generation. This cooling method has two possible variates:

1. Evaporative–convective cooling through phase change: Due to the large enthalpy of vaporization, this method yields almost uniform cell-temperature distribution, which results in higher fuel cell performance. Also a much smaller amount of water is required for cooling, leading to lower pumping power (parasitic load).
2. For convective cooling without the phase change (single-phase convection): Cooling water is normally pressurized by cooling pump for circulation and the temperature difference between the cooling inlet and outlet of the PAFC stack is higher than that by using evaporative–convective cooling, but still lower than that by using the other two techniques, primarily due to higher heat capacity of water.

    A comparison is shown in Figure 6.7 for the temperature increase of cooling water along the cooling flow path with and without the phase change. Clearly significant incell temperature difference may occur for the cooling without phase change, especially for large cell-active areas and that can have significant negative impact on the cell performance. Of course, the temperature difference along the cooling flow path can be reduced by using higher flow rate of cooling water, but that increases the pumping power required for the coolant circulation.

Water cooling is typically implemented through a separate cooling plate placed periodically (every few active cells) in the stack. Figure 6.8 shows the temperature distribution across five active cells sandwiched in between two cooling plates in

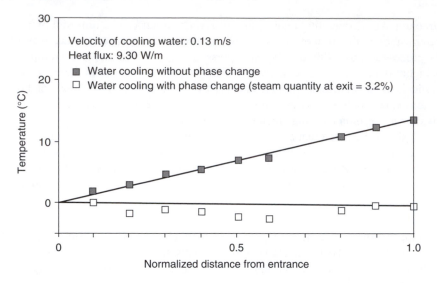

Figure 6.7 A comparison of the temperature increase for cooling water along the cooling flow path in the stack with and without phase change.[5]

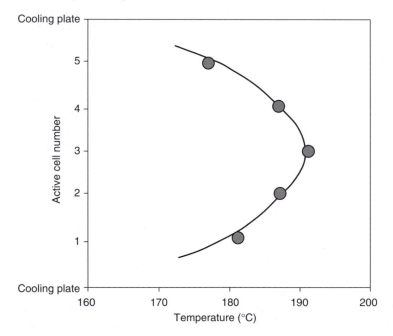

Figure 6.8 Temperature distribution along the stack and across the five active cells sandwiched between two cooling plates.[5]

a PAFC stack. As expected, the temperature across the cells exhibits a parabolic distribution due to the heat generation in each cell, and the peak temperature occurs for the middle cell. The temperature variation in a PAFC stack could become excessive with local temperature exceeding the ceiling operation temperature of 210 °C for

PAFCs, if this across-cell temperature increase is combined with the incell temperature increase described earlier, thus severely reducing the cell performance and lifetime. The across-cell temperature variation can be reduced easily by placing the cooling plate for every active cell, but this increases the stack volume and weight due to the increase in the number of cooling plates and also the cooling pumping power is increased due to the increased cooling flow path length within the stack. Clearly, an optimal design can be achieved by considering the increase in the cell performance for more uniform temperature distribution and high stack power density that results from less cooling plates and higher cell performance. It should be pointed out that both incell and across-cell temperature variations occur in every type of fuel cells and they are important design considerations for all fuel cell stacks.

It should be mentioned that tap water cannot be used as cooling water for PAFC because the cooling water should be of high quality, at least similar to the boiler feedwater for modern thermal power plants in order to avoid corrosion of cooling pipes and plates at high pressure and temperature conditions. Thus a water treatment (or purification) system is necessary, which adds cost in capital and maintenance as well. In summary, water cooling is most effective, but requires water purification and complex system design.

Air Cooling   It is the simplest of all cooling methods, but is less efficient due to the lower heat removal rate arising from the rather low thermal capacity of the air. As a result, a large amount of air is needed for the cooling purpose, and thus a large pumping power for air circulation. If the cooling air is also passed through cathode flow channels, this creates excessively high air-flow rate at the cathode, requiring very large cathode-flow channels and taking away higher electrolyte vapor, reducing the cell lifetime or shortening the intervals for electrolyte replenishment. Similarly, a separate cooling plate with cooling flow channels is needed. In practice, air cooling is usually preferred for small stacks.

Dielectric Liquid Cooling   This method uses a single-phase dielectric liquid to remove waste heat convectively. Its cooling performance and system complexity are between that of air and water cooling, and it is often used for relatively small to medium sized units such as vehicles, on-site electric generation and other special applications where compactness is necessary. However, coolant flow passage must be sealed well to prevent the coolant leakage which can cause problems. Therefore, a cooling tube (made of Teflon) is often imbedded in the cooling plate for dielectric liquid cooling, instead of machining flow channels in the cooling plates as commonly used for other two types of cooling methods.

Oxidant Supply   The PAFC operates on air without the purification, as required for alkaline fuel cells. For the pressurized system, air is compressed to 3~8 atm (so is the fuel) by a compressor driven by a motor or turbine. In the atmospheric pressure system, the air is supplied by a blower. The air is needed not only in the fuel cell stacks, but also in the reforming process, typically as oxidant for the burning of anode exhaust to provide heat for reforming purpose. The air should be conditioned to proper temperature and pressure before entering the fuel cell stacks, hence it may go through a series of heat exchangers in an optimized fuel cell power system.

Fuel Supply and Conditioning   One advantage of acid electrolyte fuel cells is that their performance is not degraded by carbon dioxide in the fuel gas, as is the case with alkaline fuel cells. The PAFC is also relatively insensitive to traces of carbon monoxide in the fuel stream. Thus, the PAFC can operate on reformed hydrocarbon fuels without significant purification of the reformed gas stream.

Natural gas (mainly methane), liquid petroleum gas (a mixture of propane and butane), methanol (CH_3OH) and naphtha (on average C_6H_{14}) presently constitute the principal primary fuels for PAFC plants. Since these primary fuels need to be transformed into hydrogen-rich gas before entering fuel cell stacks, the fuel supply system is also called a fuel conditioning (or processing) system.

In general, fuel processing system includes desulfurization, steam reforming and CO shift processes since most of hydrocarbon fuels, especially natural gas, contain a small amount of sulphur, which can poison the catalysts for both steam reforming and fuel cell reactions, and must be removed from the fuel stream before the reforming stage.[6] Desulfurization is often achieved at the temperature range of 200 °C–400 °C and by passing the hydrocarbon fuels over Co–Mo (or Ni–Mo) and then ZnO catalysts through the following chemical reaction

$$R\text{–}SH + H_2 \longrightarrow R\text{–}H + H_2S \tag{6.5}$$

$$H_2S + ZnO \longrightarrow ZnS + H_2O \tag{6.6}$$

where R designates for hydrocarbon fuels.

There are in general three techniques available for hydrocarbon reforming and production of hydrogen-rich fuel gas, including steam reforming, partial oxidation and autothermal reforming. Normally steam reforming is preferred for low carbon fuels such as natural gas. As illustrated in Example 6.2, efficient steam reforming process occurs at about 750 °C–850 °C over the nickel catalyst, and high steam-to-carbon ratios are used, ranging from 2 to 4, in order to avoid the formation of solid carbon particles. But as much as more than 10% carbon monoxide may be produced in the reforming process.

Since PAFCs can tolerate no more than about 1% carbon-monoxide concentration in the fuel stream, often two stages of CO shift reaction may be implemented in order to reduce the CO concentrations to acceptable levels. CO reduction is achieved via the water–gas shift reaction shown in Equation (6.2). The high-temperature CO shift reaction occurs at the temperature range of 320 °C–480°C, and low temperature shift at 180 °C–280 °C. This is because the thermodynamic equilibrium concentration for CO is reduced as temperature is lowered, as described in Examples 6.1 and 6.2. However, the rate of reaction is higher at higher temperatures, even though catalysts like Fe–Cr and Cu–Zn are almost always used to accelerate the shift reaction. Two stages of shift reaction for CO reduction add significant size and weight to the fuel cell power system.

The pressure for desulfurization, reforming and CO shift processes may range from atmospheric to about 10 atm in order to match the required pressure for fuel cell operation; however, the corresponding temperature for these processes differs significantly as described above. Therefore, a series of heat exchangers are used for the proper temperature control and flow paths must be optimized in order to utilize waste heat from the various processes. As a result, the actual PAFC power plant may be very complicated in the flow arrangement, sensing/monitoring and control. A typical example of PAFC power system is shown in Figure 6.9 with natural gas as the primary fuel. Clearly, fuel processing occupies a significant portion of the overall system layout. As a rule of thumb, fuel processing can be as much as over one-third of the total size (weight) and cost of the entire fuel cell system. Also important to mention is the stack thermal management near the lower right corner of the diagram, where a pump is shown to circulate the coolant through the fuel cell stack and steam separator to remove the coolant water vapor from entering the stack. Also shown inside the cooling loop is an auxiliary boiler, which is needed to provide the necessary heating to avoid electrolyte solidification during offload conditions, as described earlier in Section 6.3.1.

Power Conditioning   Power conditioning is necessary to regulate the electrical power output from the fuel cell stacks that matches the electrical load variations. Normally the power to the load is varied through current change while the voltage is maintained constant. Depending on whether DC or AC power is needed, DC–DC or DC–AC inverter can be used to convert the DC power generated by the fuel cell stacks to the appropriate form of power. An inverter has very high efficiency, normally in the upper 90%.

Sensing/Monitoring and Control   Close sensing/monitoring and control is essential for the proper operation and effective management of the integrated fuel cell power system because of the significantly different response times involved: Some components respond very quickly, almost instantaneously, such as the fuel cell and inverter (typically in the order of milliseconds); and some very slowly such as the reformer (in the order of several tens of seconds). Proper and precision monitoring and control is especially important during the startup and shutdown process when the transient effect can cause pressure surges in the reactant gas streams entering the fuel cells and pressure surges may affect the proper wetting of the electrolyte in the catalyst layers (formation of the three-phase zones) and in severe cases reactant gases may even penetrate through the liquid electrolyte layer. As discussed earlier in Section 6.3.2, the pressure differential between the anode and cathode streams must be limited to within a tenth of atmospheric pressure. To minimize the transient effects startup process may take several hours, gradually turning on the entire power system. Similarly normal shutdown process may take hours, however, immediate shutdown is required in emergency (within a second). Therefore, proper sensing/monitoring and control is a must for fuel cell power system.

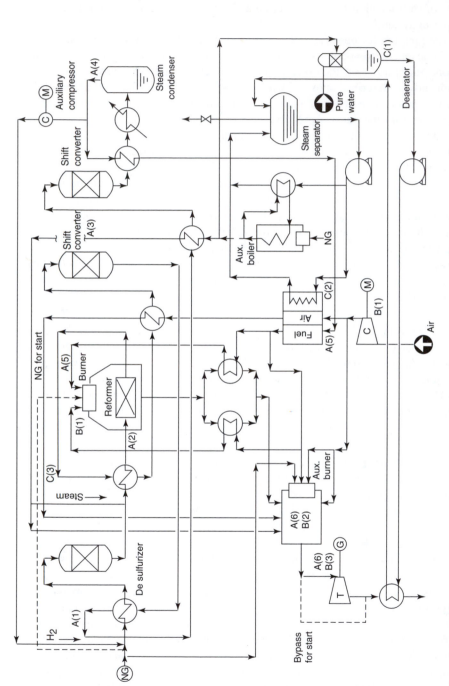

Figure 6.9 A typical PAFC power system with natural gas as the primary fuel, including fuel processing, oxidant supply, stack thermal management, and many heat exchangers for integrated temperature control for the various processes involved.[4]

6.4   MATERIALS AND MANUFACTURING

For PAFCs the electrode-backing layer, or gas-diffusion layer, is made of carbon paper or carbon cloth, as described in Section 5.5.3: the carbon-based electrodes for alkaline fuel cells, with essentially the same method for the electrode fabrication. The catalyst used is almost invariably platinum or platinum alloys (often called platinum black) supported on carbon particles, since PAFC belongs to the acid-electrolyte fuel cell, as described in Section 3.8. Often the same catalyst is used for both anode and cathode electrodes, and typical catalyst loading is about $0.5 \, mg/cm^2$ for each electrodes; although it can be lower for the anode (such as $0.1 \, mg/cm^2$). The separator (or bipolar) plate is made of graphite with machined flow channels for reactant gas distribution over the active cell area. Basically the materials and fabrication techniques used for these cell and stack components have been used in the polymer electrolyte membrane (PEM) fuel cell because PEM fuel cell also uses acid electrolyte and currently significant activities are in the area of PEM fuel cell. Therefore, the relevant topics are described in more detail in the next chapter on PEM fuel cell.

6.5   PERFORMANCE

The discussion of the PAFC performance is meant within the typical ranges of its operation conditions, which are stated briefly as:

Operating Temperature   In reality, the cell temperature varies along the direction of the fuel- and air-stream flow and is largely influenced by the design and operation of cooling subsystems, such as cooling channel configuration and coolant flow conditions. However, the average cell temperature is within the range of $180 \, °C$ to $210 \, °C$. The ceiling temperature is limited by the corrosion of cell components and hence the cell lifetime. On the other hand, the lower bound of the operating temperature is determined by the need of reasonable tolerance of carbon monoxide concentration in the reformed fuel gases and low internal resistance of the PAFCs. In general, the higher the operating temperature, the higher the energy conversion efficiency of the PAFCs. Therefore, it is extremely important to maintain cell temperature as uniform as possible.

Operating Pressure   The pressure of the reactant gases also changes along the direction of their flow. This is because pressure losses occur due to frictional effects and reactants are consumed by the electrochemical reactions for energy conversion with the accompanied product water vapor formation. Therefore, the operating pressure is normally referred to the pressure at the anode and cathode inlet. For PAFCs, the pressure ranges from one to several atmospheric pressures (maximum around 8 atm). In general, small capacity PAFCs operate at one atmospheric pressure and large capacity power plants use higher pressures. High operating pressures improve the energy conversion efficiency, but bring about other problems such as requiring more elaborate sealing, larger and heavier cell components for sufficient mechanical strength, leading to bulky and heavy fuel cell power plants.

Fuel Utilization Factor   Typically natural gas after steam reforming is used as fuel for PAFC power plants, and reformed fuel gas contains approximately 78% H_2, 20% CO_2, and small amounts of other trace species such as CH_4, CO, sulfur compounds, nitrogen compounds, etc. Also some water vapor remains in the mixture after the reforming process. Often reformed gas stream cleaning is required to reduce the concentration of impurities to an acceptable level. It is the general practice that hydrogen utilization is about 70%~85%, that is, 70%~85% of hydrogen in the anode fuel gas is consumed in the cell for electricity generation. The remaining amount of hydrogen in the anode exhaust gas is usually burned to provide the heat required for the endothermic steam reforming of natural gas.

Oxidant Utilization Factor   Air is usually supplied to the cathode for electrochemical reactions. The oxygen concentration in air is approximately 21% by mole (or by volume since air can be approximated as an ideal gas). Representative oxidant utilization in PAFCs is about 50%~60%, which means 50%~60% of the oxygen in air is consumed in the cells for electric power generation.

Cell Potential and Current Density   For PAFCs, cell potential over 0.8 V contributes to accelerated electrode corrosion, leading to short cell lifetime. As a result, PAFCs often operate at a voltage of 0.6~0.7 V per cell and a current density of 150~350 mA/cm^2. The stack power density can be as high as about 0.25 W/cm^2, although for single cells or small stacks it can be as high as over 0.3 W/cm^2.

Cell Efficiency   As shown in Chapter 2, the cell (or stack) free energy-conversion efficiency can be calculated from the expression

$$\eta = -\frac{nFE}{\Delta h}$$

where Δh is equal to the negative of the heating value of the fuel. For hydrogen, the higher heating value (HHV) is 285 kJ/mol H_2 when the product water is in liquid form and the lower heating value of 242 kJ/mol H_2 results if the product water is in vapor state (note that both HHV and LHV are measured for reaction occurring at 25 °C and 1 atm). Therefore, from the above expression we can have

$$\eta = 0.675\,E \quad \text{(HHV Base)} \tag{6.7}$$

$$\eta = 0.799\,E \quad \text{(LHV Base)} \tag{6.8}$$

where the cell potential E should be in Volts (= J/Coloumb). Hence, fuel cell efficiency is directly proportional to the cell potential only and the improvement of cell performance is critical for improving fuel cell plant efficiency. For the typical operating voltage of $E = 0.6$~0.7 V, the efficiency of the PAFC stack is $\eta = 40\%$~47% based on HHV and $\eta = 48\%$~56% based on LHV of hydrogen fuel.

It should be emphasized that this efficiency relation is valid when all the hydrogen fuel entering the fuel cell is consumed completely in the cell, or the fuel stream is recirculated for complete hydrogen consumption. If the fuel stream is not recirculated, then the fuel utilization factor is less than 100% (so that some portion of the fuel is not utilized), then the current efficiency $\eta_I < 1$, or the previous efficiency expression should be modified by multiplying the fuel utilization factor. Then fuel cell efficiency is lowered due to the fact that not all the fuel entering the cell is used for electric energy generation.

6.5.1   Effect of Temperature

The effect of temperature on cell performance can be separated into two parts: the effect on the cell reversible potential E_r and the various forms of overpotential, or the effect on the actual cell potential E. As shown in Chapter 2, the reversible cell potential E_r for hydrogen and oxygen fuel cells decreases as the temperature is increased, specifically

$$\left(\frac{\partial E_r}{\partial T}\right)_p = \frac{\Delta s}{nF} \sim -0.27\,\mathrm{mV/°C}$$

or the reduction in E_r is only about $0.27\,\mathrm{mV/°C}$ for H_2/O_2 at 1 atm and 25 °C, and for product water in vapor state.

However, as discussed in Chapter 3, higher cell operating temperature has a beneficial effect on cell performance due to higher reaction rates, lower cell resistance, enhanced mass transfer, and reduced voltage losses. The kinetics for oxygen reduction on platinum improves as the cell temperature is increased. However, the anode shows no significant performance improvement from 140 °C to 180 °C on pure hydrogen, but in the presence of carbon monoxide, increasing temperature results in a marked improvement in performance, as is discussed in Section 6.5.4.

Therefore, the actual cell potential usually increases with the operation temperature and empirical results for cell potential gain ΔE_T with increasing cell operating temperature may be written as

$$\Delta E_T = K_T(T - T_1)\ \mathrm{(mV)} \tag{6.9}$$

where T_1 is the reference temperature within the range of typical PAFC operation temperature and the temperature coefficient K_T depends on the operating conditions, such as current density J, pressure and the operating time of stacks, and so on. It may also depend, on the size, design and fabrication process of PAFCs involved. A value of $K_T = 1.15\,\mathrm{mV/°C}$ has been reported for relatively high operating pressures (3~4 atm) and at the beginning stage of the operation (new stack). For lower current densities J, K_T is lowered, as well, e.g., $K_T = 1.05\,\mathrm{mV/°C}$ for $J = 100$ mA/cm^2 has been measured. Also $K_T = 0.8\,\mathrm{mV/°C}$ has been cited at ambient room temperature. Other measurements have indicated that the value of K_T may be in the range of 0.55~0.75 mV/°C.

Although operating temperature has only minimal effect on the hydrogen oxidation reaction at the anode, it is important in terms of anode catalyst poisoning.

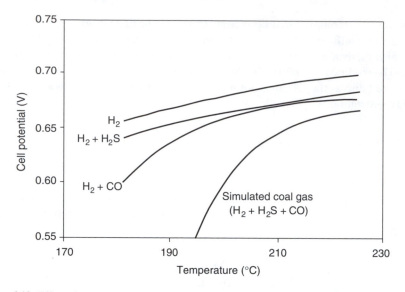

Figure 6.10 Effect of temperature on the tolerance to CO and H_2S poisoning at the current density of 200 mA/cm^2. Oxidant: air; Fuel: H_2, $H_2 + 200$ ppm H_2S, $H_2 + CO$, simulated coal gas, respectively.[7]

As shown in Figure 6.10, higher temperature results in increased anode tolerance to CO poisoning, due to reduced CO adsorption. A similar trend is observed for H_2S poisoning, and a strong temperature effect is also seen for simulated coal gas, which is assumed to contain both CO and H_2S gases. The cell potential drops significantly for $T < 200\,°C$. This figure also indicates that the effect of contaminants is not additive, implying that interaction between CO and H_2S exists.

In summary, higher operating temperature enhances cell performance, it also increases catalyst sintering, component corrosion, electrolyte degradation, evaporation, and concentration change. All these adversely affect stack lifetime. Hence, the highest allowable operating cell temperature is about 220 °C for peak and 210 °C for continuous operation. As a result, uniform temperature distribution in each cell is an important factor to increase cell performance and cell lifetime. This requires effective cooling, heat, and water management.

6.5.2 Effect of Pressure

The effect of pressure on cell performance can also be separated into two parts: the effect on the cell reversible potential E_r and the effect on the irreversible cell potential E. The reversible cell potential E_r increases with cell operating pressure logarithmically, as given in Equation (5.4). At $T = 190\,°C = 463$ K, we have

$$\Delta E_r = 23 \ (\text{mV}) \log\left(\frac{P_2}{P_1}\right) \tag{6.10}$$

Clearly, a ten-fold increase in the operation pressure only brings about 23 mV of the improvement in the reversible cell potential.

However, higher operating pressure increases electrochemical reactions, particularly at the cathode by increased O_2 contents (hence lower diffusion polarization) and water partial pressure (i.e., water contents). Higher water partial pressure lowers acid electrolyte concentration which increases ionic conductivity and results in higher exchange current density, hence leading to a reduction in ohmic losses. It has been reported that an increase in the pressure of a PAFC cell (100% H_3PO_4, 169 °C) from 1-to-4.4 atm produces a reduction in acid concentration to 97%, and a decrease of about 0.001 Ω in the resistance of a small six cell stack (350 cm^2 electrode area).[8] Therefore, higher cell operating pressure increases the performance of PAFCs and this increase becomes more significant at higher current densities due to significant improvement of electrochemical reactions at the cathode.

Empirical correlations are often developed in the form of

$$\Delta E_P = K_P \log\left(\frac{P_2}{P_1}\right) \tag{6.11}$$

where ΔE_P is the cell potential difference for different cell-operation pressure P_1 and P_2. The coefficient K_P is often much larger than the value of 23 mV/decade shown earlier for the reversible cell potential, and $K_P = 146$, 142 and 125 mV/decade have been reported in literature.[4,8]

It should be noted that the cell potential depends on many factors, such as cell structure, kind and amount of catalyst loading, current density, utilization factors of reactant gases, and many more (e.g., impurities, pure hydrogen or oxygen versus reformed fuel and air, etc.). Therefore, these correlations are just indicative, practical applications require correlations be developed for specific cells and stacks by specific manufacturer.

6.5.3   Effect of Reactant Gas Concentration and Utilization

Pure hydrogen and pure oxygen are not usually available for commercial applications, reformed gas from fossil fuels, such as natural gas, methane, liquid petroleum gas, and so on, and air are used as fuel and oxidant gases, respectively. Only in special circumstances such as in chloroalkaline industries, pure hydrogen is available as a byproduct. In general, increasing reactant gas utilization or decreasing inlet reactant gas concentration results in decreased cell performance due to higher concentration polarization and Nernst losses. These effects are related to the partial pressures of reactant gases, and are considered in the following text.

Fuel   Hydrogen for PAFCs is typically derived from the conversion of a wide variety of primary fuels, such as methane (e.g., natural gas), petroleum products (e.g., naphtha), coal liquids (e.g., methanol) or coal gases. Besides hydrogen, carbon monoxide and carbon dioxide are also produced during the conversion, and some unreacted hydrocarbons are also present. After steam reforming and shift conversion reactions, these reformed fuel gases contain also some low levels of carbon monoxide which cause anode poisoning in PAFCs. The CO_2 and unreacted hydrocarbons (e.g., CH_4) are electrochemically inert and act as diluents. Because the anode reaction is nearly reversible, the fuel composition and hydrogen utilization generally do not strongly influence cell performance, in contrast to the alkaline fuel cells described in the previous chapter.

The change of cell performance with respect to the partial pressure of hydrogen, which is affected by fuel concentration and utilization, can be described as follows

$$\Delta E_{H_2} = (55{\sim}77) \log\left(\frac{[\bar{P}_{H_2}]_2}{[\bar{P}_{H_2}]_1}\right) \text{ (mV)} \tag{6.12}$$

where $\bar{P}_{H_2}$ is the average partial pressure of the hydrogen gas in the anode stream, and ΔE_{H_2} represents the cell potential change from one H_2 partial pressure to another. Note that this correlation is valid for the same total pressure in the anode while hydrogen partial pressure varies from one value to another.

From this correlation, it can be estimated that the actual potential will reduce by about $10\,mV$ if pure hydrogen is replaced by reformate gas (80% H_2 and 20% CO_2) as the fuel. Similarly, the presence of 10% CO_2 in the fuel stream can cause a voltage loss of about $2\,mV$. Thus, diluents in low concentrations are not expected to have a major effect on electrode performance. However, relative to the total anode polarization for the case, which is approximately $3\,mV$ at $100\,mA/cm^2$, the effect is significantly large.

It has also been reported that with pure hydrogen as fuel, the cell potential E at $215\,mA/cm^2$ remains nearly constant at hydrogen utilizations up to 90%, and then it decreases sharply at higher utilization factors.

Oxidant   The oxidant composition and utilization affect the cathode performance. Air, which contains approximately 21% O_2 and 79% N_2 by volume (or mole), is usually the oxidant of choice for PAFCs. The use of air instead of pure oxygen leads to a decrease in the current density of about a factor of three at constant electrode potential. The polarization at the cathode increases with an increase in oxygen utilization. The voltage loss due to a change in oxidant composition and utilization may be correlated as follows

$$\Delta E_{O_2} = K_{O_2} \log\left(\frac{[\bar{P}_{O_2}]_2}{[\bar{P}_{O_2}]_1}\right) \tag{6.13}$$

where ΔE_{O_2} is the cell potential change between two cell-average oxygen partial pressure of $[\bar{P}_{O_2}]_1$ and $[\bar{P}_{O_2}]_2$; and the coefficient K_{O_2} may be taken as $103\,mV/decade$ of the oxygen partial-pressure change.

Then it can be estimated that the cell voltage decreases by $84\,mV$ if air, instead of pure oxygen, is used as the oxidant. The difference in cell voltage between running on pure oxygen and air is quite large and is much greater than in the case of changing anode feed gas from pure hydrogen to reformed gas. That is, the main voltage losses occur at the cathode due to reactant gas concentration and utilization.

More accurate correlation can often be obtained for actual fuel cell operation by having the coefficient K_{O_2} valid in a smaller concentration range. For example, the following two values have been proposed[8]

$$K_{O_2} = 148 \ (\text{mV}), \quad \text{for } 0.04 \leq \left(\frac{[\bar{P}_{O_2}]_2}{[\bar{P}_{O_2}]_1} \right) \leq 0.20 \qquad (6.14)$$

$$K_{O_2} = 96 \ (\text{mV}), \quad \text{for } 0.20 \leq \left(\frac{[\bar{P}_{O_2}]_2}{[\bar{P}_{O_2}]_1} \right) \leq 1.00 \qquad (6.15)$$

and Equation (6.14) is generally used for fuel cells using air as the oxidant (the low oxygen concentrations), whereas Equation (6.15) for fuel cells using an oxygen-enriched oxidant (high oxygen concentrations).

It might be pointed out that the effect of the hydrogen and oxygen partial pressures, as shown in Equations (6.12)–(6.15), is logarithmic. This functional dependence is similar to the effect on the cell reversible potential E_r given in Equation (2.20). The reversible cell potential E_r decreases according to Equation (2.20) if not pure hydrogen and oxygen are used in the cell. This decrease is often called the Nernst losses. Whereas at a finite current density, additional voltage losses occur due to increased concentration polarization because of the impedance of the inert gas components to the reactant transport. The correlations given in Equations (6.12)–(6.15) include both losses.

Low utilization factors, particularly oxygen utilization, yields high performance. However, low utilizations result in poor fuel usage and excessive oxidant flow rate. The latter not only incurs excessive parasitic power losses, but also leads to high electrolyte loss. Optimization of reactant utilization is required. State-of-the-art utilizations used are on the order of 85% and 50% for the fuel and oxidant, respectively.

6.5.4   Effect of Impurities

As explained earlier, reformed gas is composed of about 80% H_2 and 20% CO_2, and other impurities. A typical reformed gas composition acceptable for PAFCs is 78% H_2, 20% CO_2, <1% CO, <1 ppm H_2S, Cl, NH_3, and so on. The maximum levels of impurities permissible in reformed gases for the PAFC are given in Table 6.3.

As discussed earlier, for PAFCs CO_2 has no impact on cell performance other than the effect of Nernst losses through the changes in hydrogen partial pressure in the fuel gas mixture. Other impurities, however, have a significant impact on the cell performance in various ways, though their concentration entering the PAFCs is very low relative to that of diluents and reactant gases. Some impurities, such as, sulfur compounds, originate from the fuel gas entering the fuel processor (or reformer) and are carried into the fuel cell with the reformed fuel, whereas others, such as CO, are produced in the fuel processor.

Table 6.3   The Maximum Allowable Limits of Impurities in Reformed Gases for PAFCs[4]

Impurities	Allowable Concentrations
CO_2	Diluent
CH_4	Diluent
N_2	Diluent
H_2O	10%–20%
CO	<1% at 175 °C
	<1.5% at 190 °C
	<2% at 200 °C
H_2S, COS	<100 ppm
C_2^+	<100 ppm
Cl^-	<1 ppm
NH_3	<1 ppm
Metal ions (Fe, Cu, etc.)	Nil

Effect of Carbon Monoxide   The presence of carbon monoxide in a hydrogen-rich fuel has a significant effect on the anode performance due to CO poisoning of the electrocatalytic activity of platinum catalyst. The anode performance degradation becomes more significant as cell operating temperature is decreased. The cell voltage loss resulting from CO poisoning as a function of temperature may be correlated as[8]

$$\Delta E_{CO} = K_{CO}(T) \{[CO]_2 - [CO]_1\} \tag{6.16}$$

where carbon monoxide concentration, [CO], is expressed in terms of percent-CO content in the fuel gas, and $K_{CO}(T)$ is a constant independent of CO concentration, but is a strong function of cell operating temperature. The values of $K_{CO}(T)$ can range from -1.30 mV/% at $T = 218$ °C to -11.1 mV/% at $T = 163$ °C. It is obvious that CO tolerance depends on the amount of platinum catalyst loadings as well. Thus, other values of $K_{CO}(T)$ have also been reported.

Equation (6.16) suggests that the cell performance loss is proportional to CO content in the fuel gas and it can be estimated that for a given change in CO content ΔE_{CO} is about 8.5 times larger at the temperature of 163 °C than at 218 °C. Therefore, the permissible level of CO depends on the cell temperature, as shown in Table 6.3. For $T \geq 190$ °C, 1% CO in the fuel gas stream is acceptable without noticeable adverse effects on the cell performance. This CO concentration compares favorably with other low-temperature fuel cells. For alkaline and proton exchange membrane fuel cells, CO concentration must be limited to less than a few ppm level primarily because of their low operating temperatures (typically ≤ 80 °C).

CO poisoning is common for low-temperature acid electrolyte fuel cells because of the platinum used as catalysts. Since PEM fuel cells operate at much lower temperatures than PAFCs, CO poisoning becomes much more severe and restrictive for the PEM fuel cells. Therefore, the mechanism of CO poisoning is presented in the next chapter on PEM fuel cells. It suffices to say that the cell voltage loss due to CO poisoning is reversible and can be offset by increasing the cell operating temperature.

Effect of Sulphur Compounds   Commercial natural gas contains a small amount of sulphur for leak detection purpose. Anode gases from coal gasifiers may contain a total sulfur of 100~200 ppm, depending on the quality of coal used. Hydrogen sulfide (H_2S) and carbonyl sulfide (COS) are impurities in fuel gases from fuel processors and coal gasifiers in PAFC power plants.

The concentration levels of H_2S in an operating PAFC that can be tolerated by platinum anodes without suffering a destructive loss in cell performance are <50 ppm ($H_2 + COS$) or <20 ppm (H_2S). Rapid cell failure occurs with fuel gas containing more than 50 ppm H_2S. Therefore, raw fuel usually goes through a desulfurization process before entering the reformer. As shown in Figure 6.10, the presence of both CO and H_2S can intensify the poisoning effect. Similarly to CO poisoning effect, the performance loss due to H_2S poisoning is reversible and decreases for higher cell operating temperature.

The H_2S poisoning occurs when H_2S absorbs on platinum surface and blocks the active sites for hydrogen oxidation in a manner much the same as the CO poisoning.[9]

6.5.5   Effect of Internal Resistance

The cell potential is reduced by lower ionic flow in the electrolyte and electronic conduction in the electrodes, current collectors, and interfaces. The interfacial contact resistance may be substantial for improperly assembled cells; it decreases and approaches zero as the cell compression force is increased to appropriate levels. The cell potential loss due to internal resistance may be correlated for properly assembled cells as[8]

$$\Delta E_{IR} = -0.20J \text{ (mV)} \qquad (6.17)$$

where the current density J is in mA/cm^2. This potential loss mainly arises from the resistance to protonic migration through the phosphoric acid electrolyte.

6.5.6   Effect of Cell Operating Time

One of the primary areas of research and development activities for fuel cells is to increase the cell lifetime. The goal for PAFCs is to maintain the cell stack performance during a standard utility application, or approximately 40,000 h of operation (~4–5 years). As the PAFCs are operated over time, their performance degrades gradually as the operation time lengthens. A typical PAFC degradation over time may be represented by

$$\Delta E_{Time} = -3 \text{ mV}/1000 \text{ h of operation} \qquad (6.18)$$

More details on the PAFC lifetime and performance decay are given in the next section.

6.6   PERFORMANCE DECAY AND LIFETIME

The lifetime of a fuel cell is often defined as the time of operation after which the cell output voltage (or potential) at rated output current has decreased by 10% of the initial voltage, that is

$$\left| \frac{E_{final} - E_{initial}}{E_{initial}} \right| = 10\% \tag{6.19}$$

where $E_{initial}$, the initial voltage, is the cell output voltage after about 100 h of trial operation, which is needed for a PAFC cell to reach steady-state initial condition. The lifetime of PAFCs is estimated to be about 40,000 h (close to 5 years). That is, the cell performance is expected to decay from, approximately 0.7 V @ 200 mA/cm^2 at the initial stage to about 0.63 V at the same current density of 200 mA/cm^2 after 40,000 h of operation.

Cell life depends very much on operating conditions, such as operating pressure and temperature, cell voltage, and load variation (e.g., startup, shutdown, etc. or load cycling). The performance decay of a PAFC is mainly due to the sintering of platinum catalyst particles, corrosion of carbon support, progressing of electrolyte flooding of the electrodes, and the depletion of the electrolyte in the cell structure.

6.6.1   Sintering of Platinum Particles

Platinum particles have the tendency to migrate on the surface of the carbon support and to coalesce (agglomerate) into larger size particles, thus decreasing the active surface area and cell performance. The rate of sintering is proportional to the logarithm of time and depends mainly on the operating temperature of the cell. At temperatures above 150 °C, surface migration of platinum particles begins to occur and at higher temperatures crystallite coalescence is predominant.

6.6.2   Corrosion of Carbon Support

The corrosion of carbon support would eventually lead to the breakdown of the large carbon support particles into smaller and isolated carbon particles, thus a continuous path for electron migration (or current collection) is disrupted, rendering the platinum catalyst particles on the smaller isolated carbon support particles inactive. This is equivalent to a loss of platinum particles and also accelerates the wetting of the carbon surface by the liquid electrolyte due to increased pore size. These two phenomena decrease the active area of catalyst and prevent gas diffusion to the catalyst layer.

The rate of corrosion of the carbon support depends on cell potential, operating temperature and the type of carbon used (e.g., whether acetylene black or furnace black). Higher potential and temperature contribute to fast corrosion. Therefore, cell potential should never exceed 0.8 V. This implies that the cell should not be left in an open-circuit condition at temperatures above 180 °C. Carbon corrosion is drastically enhanced if water vapor partial pressure (or concentration) is approximately above

100 mm Hg. It might be emphasized that water vapor is unavoidable as it is formed at cathode due to electrochemical reactions. Therefore, heat treatment of the porous carbon (graphitization at 2700 °C) while maintaining a high surface area (200 m^2/g or more) is essential for long cell lifetime.

Experimental observations indicate that the corrosion rate at 180 °C with an open-circuit potential of up to 1.0 V is about eight times that at temperatures below 180 °C and current density of 100 mA/cm^2 operating condition. The corrosion is usually more severe for cathode than for anode because of the water vapor presence in the cathode. Therefore, during offload or plant shutdown, cell voltage should be kept below 0.8 V by either purging electrodes with nitrogen or short-circuiting the stack. Similar procedure should be adopted as well during startup process to extend the lifetime of PAFC stacks.

6.6.3   Electrode Flooding

It is caused mainly by the decay of electrode hydrophobicity and the gradual corrosion of the carbon support, as discussed previously. Normally, the hydrophobic electrode is made by adding appropriate amounts of waterproof material (usually PTFE powder) to the carbon-supported platinum powder, and possesses sufficient hydrophobic characteristics which deteriorate gradually over time.

6.6.4   Electrolyte Depletion

Electrolyte depletion occurs over the time of operation mainly due to the evaporation of the liquid electrolyte and the corrosion of electrode. Despite the low vapor pressure of phosphoric acid electrolyte, evaporation does occur and the electrolyte vapor is taken out of the cell structure by the fuel and oxidant stream; therefore, the rate of evaporation depends on the velocity of the fuel and oxidant stream and excessively high reactant flow contributes to fast evaporation. Evaporation is a common problem for all liquid electrolyte fuel cells and can only be lowered by using lower vapor pressure of the electrolyte and slower reactant flow in the anode and cathode compartment.

The corrosion of electrode causes a loss of electrolyte due to the participation and consumption of electrolyte in the corrosion process and also due to the increased pore volume of small pore sizes that draw in liquid electrolyte through capillary effect. Thus, electrode corrosion inevitably leads to electrode flooding as discussed in previous section. Clearly, performance decay may be relatively easily recovered by electrolyte replenishment if evaporation is the cause of electrolyte depletion; however, electrode corrosion almost always results in permanent damage.

6.7   FUTURE R&D

In the past (up to late 1970s and early 1980s), most of PAFC R&D activities were in the United States. Then the activities shifted to Japan with all major demonstration

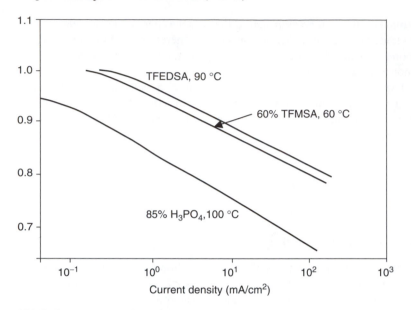

Figure 6.11 Performance comparison of oxygen-reduction reaction for oxygen at 1 atm and in 85% H_3PO_4 at 100 °C, in tetrafluoroethane disulfonic acid (TFEDSA) at 90 °C, and in 60% trifluoromethane sulfonic acid (TFMSA) at 60 °C, respectively. The electrode is commercial low-loading Pt electrode (0.3 mg/cm^2 Pt on Vulcan XC-72, Prototech, Inc.). Adapted from Reference 10.

projects there. The major issues preventing PAFC commercialization is the high cost, as well as the lack of convincing demonstration of the reliability and sufficient lifetime of PAFC power systems.

Future R&D should focus on

- Cost reduction through component and system optimization and integration as well as improved performance, including the fuel processor and the balance of plant. This is because for PAFC power systems, PAFC stacks may only account for no more than one-third (in fact, as low as a quarter) of the total system cost and weight (or volume), while the fuel processor and the balance of plant can account for the remaining two thirds (approximately one-third for each of them).
- Reliability and long lifetime, including the fuel processor and the balance of plant. Experience indicates that the disruption of PAFC power systems often occurs as a result of the component failures in the balance of plant, such as pumps, blowers, and so on.
- Better performance. Higher cell potential (hence higher energy conversion efficiency), but the cell potential is limited to no more than 0.8 V. Therefore, the main focus should be on the enhancement of current density, which requires the improvement of cathodic oxygen reduction reaction.
- Better catalysts. Cheaper and more effective catalysts.
- Further improvement of electrodes, which requires a better understanding of transport processes and electrochemical reactions. For many years, it is an art for

electrode design and fabrication. It is beginning now that mathematical modelling coupled with experimental measurements will contribute to the understanding and better electrode performance.

- Better acids as electrolyte. Much work has been conducted, as discussed earlier in Section 6.3.1. The best perhaps is fluorinated sulfonic acids such as "liquid Nafion," TFMSA (CF_3SO_3H), or TFEDSA (($CF_2SO_3H)_2$), as shown in Figure 6.11. Clearly, significant improvement in performance can be obtained for the fluorinated sulfonic acids, even at lower operating temperatures. However, this is no longer a PAFC! One has to start all over again from the very beginning and compare with other types of fuel cells, such as proton exchange membrane fuel cells, which is the subject of the next chapter.

It might be mentioned that as of today, there is indication of little commerical viability for PAFCs. A second generation with twice the present performance could open up for small stationary power plants such as onsite co-generation applications.

6.8   SUMMARY

This chapter focused on the phosphoric acid fuel cells (PAFC) and described the various advantages and disadvantages of the PAFC as well as the typical applications intended. Basic principles and operation of the PAFC are outlined. Overall half-cell and whole-cell reactions have been described as well as the various issues associated with the steam reforming of natural gas and the water–gas shift reaction. The calculation of the equilibrium composition in the reformed fuel stream for a given temperature and pressure has been outlined. The various components of the cell, stack and system, as well as their geometrical configurations are given along with the typical materials used. The effect of the various operating parameters on the PAFC performance, including the temperature, pressure, reactant concentration and utilization, the effect of various impurities, cell current density, and cell aging were described. The significance of carbon monoxide poisoning as well as the negative impact by the other impurities such as sulphur containing compounds has been provided along with the tolerance limits for various impurities. Long-term performance decay and the various factors influencing the PAFC lifetime were discussed. Finally various technical issues that are to be resolved for this type of fuel cell were stated briefly.

BIBLIOGRAPHY

1. Hiramoto, J. and R. Anahara. 1982. *Fuji Electric. J.*, 9: 555.
2. Mellor's *Comprehensive Treatise on Inorganic and Theoretical Chemistry*, 8(3): 669.
3. Kivisari, J. 1995. Fuel Cell Lecture Notes.
4. Anahara, R. 1993. Research, development and demonstration of phosphoric acid fuel cell systems. In Fuel Cell Systems, eds. L. J. M. J. Blomen and M. N. Mugerwa. New York: Plenum Press.
5. Hirota, T., Y. Yamazaki and Y. Yamakawa, 1981. *Fuji Electric. J.*, 61(2): 133–187.
6. Song, C. C. and X. Ma, 2004. Desulfurization processes for petroleum refining, *Inter. J. Gr. En.* 1(2).

7.   Jalan, V., J. Poirier, M. Desai and B. Morrisean, 1990. Development of CO and H_2S tolerant PAFC anode catalysts *Proc. Second Annual Fuel Cell Contractors Review Meeting.*

8.   Benjamin, T. G., E. H. Camara, and L. G. Marianowski, 1980. *Handbook of Fuel Cell Performance.* Institute of Gas Technology for the United States Department of Energy under Contract No. EC-77-C-03-1545, 40.

9.   Chin, D. T. and P. D. Howard, 1986. *J. Electrochem. Soc.*, 133: 2447.

10. Appleby, A. J. and F. R. Foulkes, 1988. *Fuel Cell Handbook.* New York: Van Nostrand Reinhold.

PROBLEMS

6.1   Describe briefly the advantages and disadvantages as well as the areas of applications for phosphoric-acid fuel cells.

6.2   Describe the operation principle such as half-cell and whole-cell reaction, the primary fuels expected to be used for phosphoric-acid fuel cells.

6.3   Discuss briefly the typical (or target) operating conditions such as cell voltage, cell current density, temperature, pressure, fuel and oxidant utilization, and chemical to electrical energy conversion efficiency, and so on.

6.4   Describe the effect of operating conditions on the cell performance.

6.5   Describe the geometrical configuration of cell and stack, typical materials used for, and the thickness as well as other dimensions of cell components.

6.6   Describe the components of the phosphoric-acid fuel cell system such as cooling (thermal management), product removal, electrolyte management, fuel processing, control, and so on.

6.7   Describe the factors affecting the short- and long-term performance of phosphoric-acid fuel cells.

6.8   Describe the critical technical barriers to be overcome for the commercialization of phosphoric-acid fuel cells, the possible solutions, and their pros and cons.

6.9   Consider the following reaction for the steam reforming of methanol

$$CH_3OH + H_2O \longrightarrow 3H_2 + CO_2$$

In order to determine the amount of heating required for the reforming reaction calculate the enthalpy of reaction for the above reaction at the standard temperature and pressure (25 °C and 1 atm).

6.10 For the steam reforming reaction of methanol shown in the previous problem, and assuming the methanol-to-steam ratio of 1 initially in the reactant mixture, determine the equilibrium composition for such a reaction by considering the water–gas shift reaction only so that carbon monoxide is the only other species in the product mixture, in addition to hydrogen and carbon dioxide. Generate the results, similar to Figure 6.1, for the temperature range from 100 °C to 500 °C and the pressure of 1 atm.

PROTON EXCHANGE MEMBRANE FUEL CELLS (PEMFCs)

7.1 INTRODUCTION

Proton exchange membrane fuel cells (PEMFCs) are also known as ion exchange membrane fuel cells (IEMFCs), solid polymer (electrolyte) fuel cells (SP(E)FCs), polymer electrolyte (membrane) fuel cells (PE(M)FCs), etc. They have increasingly become the most promising candidates as the zero-emission power source for transportation, stationary co-generation and portable applications in the last decade, and research, development and demonstration (RD&D) activities have also expanded into a variety of practical applications, ranging from small units of a few watts to power home electronic devices such as cellular phones, personal computers, to medium sizes of a few kilowatts for residential cogeneration (electricity and heat or hot water), and large systems of around 50 kW for electric passenger vehicles and a couple of hundred kilowatts for urban transit buses. In recent years, proton exchange membrane fuel cells have been in the spotlight due to significant technical advances and successful demonstration projects of PEMFC power systems for urban transit buses and passenger cars. At present, PEMFC is being intensively developed worldwide and its commercialization is dawning.

The electrolyte used in PEMFCs is a proton-conducting membrane cast in solid polymer form. Solid electrolyte offers a number of advantages as compared to systems with liquid electrolyte. It allows a simple and compact cell structure and operation, leading to relatively simple design and easy for manufacture. No free corrosive liquid electrolyte in the cell exists, giving rise to minimal corrosion of cell components and hence longer cell lifetime. The solid electrolyte can be made in a very thin sheet, as thin as 200 μm or thinner (50 μm or even thinner), to produce low internal resistance cells because the resistance to ion migration in the electrolyte typically accounts for the absolute majority of the entire cell's electrical resistance, usually over 95% or even more. Therefore, high energy efficiency and high output power density are obtained since practical fuel cells are almost always operated in the ohmic polarization dominated region of the cell polarization curve. Since the solid membrane can act as an electrode support and major cell structural component, the resulting PEMFC is able to withstand large pressure differentials between the anode and cathode compartment and large pressure fluctuations in the reactant gas supply lines. As large as over 5-MPa crossstream pressure differential has been reported for PEMFCs.[1] As a result, expensive precision sensors and control units can be avoided for the PEMFC operation. This unique characteristic, coupled with its insensitivity to orientation due to solid electrolyte, makes a PEMFC system ideal for mobile applications. The stability and lifetime of the membrane limits the cell operating temperature, typically to less than 100 °C. Such a low temperature operation offers almost instantaneous power output, resulting in easy and quick startup, making PEMFC system ideally suited for transportation applications with frequent on and off operations. In contrast, phosphoric-acid fuel cell system with a hydrocarbon reforming unit can take as much as two hours to warm up to the required operating temperature of 200 °C, a significant handicap for vehicular applications. Since the membrane electrolyte has sulfonic acid groups attached to the polymer backbone, it is essentially acid. As shown in previous chapters, sulfonic acid electrolyte has reasonably fast oxygen reduction kinetics, offering one of the best fuel cell performances among all the acid electrolytes suitable for fuel cell applications. As acid electrolyte fuel cell, PEMFC can tolerate the presence of carbon dioxide in both fuel and oxidant streams, thus capable of operation with hydrocarbon reformed fuels and atmospheric air. Other benefits are associated with the use of a solid-membrane electrolyte. For example, the acid concentration of the electrolyte is fixed at the time of membrane fabrication, it cannot be diluted by process water or product water, nor concentrated or depleted by the evaporation of the liquid electrolyte, thus there is no need of maintenance to refurbish or regenerate the electrolyte during the operation and potable liquid product water is obtained without any purification or cleanup required. The latter may be particularly important for space applications, and NASA is exploring the possibility of replacing existing alkaline fuel cells by the PEMFCs in their space shuttle orbiter.[1]

On the other hand, low temperature operation results in low-quality waste heat coming off the fuel cell stacks and it is difficult to integrate thermally with a fuel reforming processor for additional system efficiency enhancement. The acid nature of the electrolyte together with the low operating temperature requires the use of Noble metals such as platinum as the electrocatalyst, which is susceptible to the poisoning of carbon monoxide. However, carbon monoxide is always present in the reformed fuel stream in the order of a few percentages. The degree of carbon monoxide poisoning is much more severe for PEMFCs than for PAFCs which have a much higher operating temperature of around 200 °C. Therefore, carbon monoxide must be removed from the fuel stream to below a few ppm levels to avoid the considerable performance degradation. Section 7.6 focuses on the mechanism of carbon monoxide poisoning as well as mitigation methods available at our disposal. The use of platinum as catalyst has been traditionally regarded as a concern for the high capital cost of the PEMFC system, even though the catalyst loading has been reduced substantially to the acceptable levels of estimated $5 U.S./kW, at least confirmed in single cell laboratory testing.[2] But long-term high performance with the low catalyst loadings in the electrodes still needs to be demonstrated, especially for large stacks under practical conditions.

Important issues for PEMFC design and operation include thermal and water management, bipolar plate designs consisting of material selection, fabrication techniques, and flow distribution field layout, and the selection of operating conditions like cell temperature, pressure, flow rates, the choice of fuels, and so on. Even though air is used as oxidant for all terrestrial applications, some argument exists as to whether liquid methanol (with onboard steam reforming) or hydrogen fuel (in the form of pressurized gas, liquefied hydrogen, or metal hydride) should be used as the primary fuel. But it is most likely that natural gas is used as the primary fuel for residential co-generation and industrial combined heat and power applications, while liquid methanol is the choice of fuel for small portable systems.

It is important to emphasize that the polymer membrane used as electrolyte has acceptable conductivity to proton migration when it is fully humidified. The electrical resistance increases when the membrane is drying out, increasing ohmic polarization and Joule heating at the same time. Local heating leads to further water evaporation and accelerates the local drying, resulting in vicious self-accelerated destruction of cell performance. The membrane is unstable at high temperatures, and local heating also limits the lifetime of the membrane electrolyte, hence the cell lifetime as well. However excessive presence of water floods the electrode pore regions, giving rise to the so-called **water flooding phenomena**, which severely reduce the rate of reactant mass supply to the reaction sites and degrade the cell performance considerably. In fact in PEMFCs, mass transfer limitations are mostly likely caused by the water flooding of cathode electrode. Therefore, proper membrane humidification without causing electrode flooding by water, commonly referred to as water management, remains one of the major technical challenges of this type of fuel cell. Obviously, water management is closely related to thermal management due to water evaporation or condensation. These two technical challenges are the two critical issues in the PEMFC operation and design. Later in this chapter, the reason of membrane humidification, and proper techniques for water and thermal management are described in detailed.

Fuel cells with ion exchange membranes as electrolyte may be dated as far back as Haber and his associates in their investigation of the reversible cell potential of hydrogen–oxygen fuel cells.[3] However, General Electric Company (GE) is credited for the initial development of modern PEMFCs in 1950s[4,5,6] that resulted in the first practical application of fuel cells for the U.S. Gemini space missions from 1962 to 1966.[7] Major technical difficulties and problems persisted then, including water management under microgravity condition in space; limited power densities obtainable ($<50\,mW/cm^2$); high platinum loading (as much as $28\,mg\,Pt/cm^2$), and hence high capital cost; and limited lifetime due to the polystyrene sulfonate membrane used as electrolyte which was not sufficiently stable under the fuel cell operating condition. DuPont's introduction of the more stable sulfonated polytetrafluoroethylene (with Nafion as the trade name) in 1966, and the subsequent identification and deployment

Figure 7.1 Number of Ballard PEM fuel cell stacks required to provide 50 kW with increasing power density.[8] Clockwise from the upper left corner: Mk300 (1991), Mk500 (1993), Mk700 (1995), and Mk800 (1997).

of Nafion membrane as the solid polymer electrolyte was hailed as a major break-through in the PEMFC development. Nafion has since become the industry standard for PEMFCs until the present time. Although NASA's Space Shuttle Fuel Cell Technology Project (1972–1974) was based on Nafion membrane, other persistent technical difficulties described here eventually made the alkaline fuel cell the choice for NASA's subsequent space programs: the Apollo missions and space shuttle orbiters. By early 1980s, GE stopped further commercial development of PEMFC systems.

From the mid to late 1980s, a resurgence of interest in and work on the PEMFC technology occurred, and the significant progress and achievement in this technology must be recognized of Ballard Power Systems' technology advancement and Los Alamos National Laboratory's fundamental research and studies. In 1990s there was an explosive increase in the R&D activities of PEMFCs for terrestrial applications, that led to significant and accelerated technical achievement and technology advance in the last decade or so. For example, the PEMFC-stack power density, a key parameter for mobile applications, has been increased from 0.1 kW/L in 1989 to 1.31 kW/L in January 2000 (Ballard Power Systems' Mark 900). Ballard's previous automotive fuel cell, the Mark 700 stack design announced in 1995, achieved mass and volume power densities of 1 kW/kg and about 0.7 kW/L, respectively. The Mark 900 module was configured for automotive use with the designed power output of 75 kW, and occupied approximately half the space and weighed about 30% less than the Mark 700 stack. Figure 7.1 illustrates that advancement for Ballard's PEMFC stack from 1991 to 1997.[8] The most advanced fuel cell stack to date is Ballard's Mark 902 design — Ballard's fourth-generation fuel cell stack for transportation announced in October

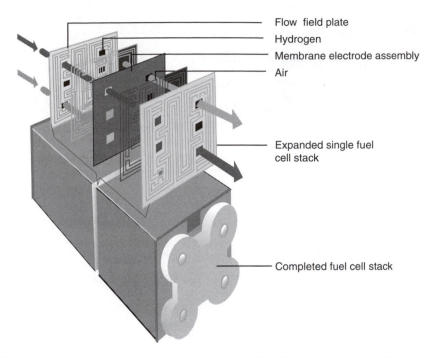

Flow field plate
Hydrogen
Membrane electrode assembly
Air

Expanded single fuel cell stack

Completed fuel cell stack

Figure 7.2 Schematic of typical proton exchange membrane fuel cell stack with unit cell architecture (one membrane electrode assembly between two flow-field plates) shown on the top [8].

2001. It allowed configurations for stationary power-generation applications and was scalable from 10 kW to 300 kW depending on specific requirements and applications. Typical power output for transportation is 85 kW for passenger vehicles and 300 kW for transit bus. The Mark 900 Series unit cell architecture, as shown in Figure 7.2, achieved power densities well over 2.2 kW/L.

Another significant advancement made is the platinum catalyst loading, which has been reduced from 8 mg/cm^2 per cell (including both anode and cathode) to well below a fraction of 1 mg/cm^2 (some single cell testing has shown ultra low loading of even 0.04 mg/cm^2). As shown in Section 7.3.3, various improved designs of bipolar plates with advanced flow field channels have been developed along with cheaper materials and manufacturing techniques of mass production capability. Associated with this technology advancement is the accompanied considerable reduction of the PEMFC cost and the commercialization of PEMFC technology. A detailed account of PEM fuel cell development within the last several decades is available elsewhere.[9]

7.2 BASIC PRINCIPLES AND OPERATIONS

A single proton exchange membrane fuel cell unit is shown schematically in Figure 7.3, and many such units connecting in series form a PEMFC stack. Each cell is composed of a solid polymer membrane acting as the electrolyte, which is sandwiched in between

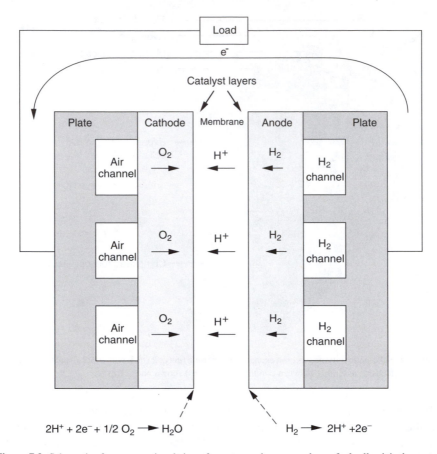

Figure 7.3 Schematic of a cross-sectional view of a proton exchange membrane fuel cell unit in the context of a fuel cell stack.

the two platinum-catalyzed carbon electrodes. The two electrodes and the membrane electrolyte are often mechanically compressed by screws or pneumatic pressure to form a single piece, commonly referred to as the membrane-electrode assembly (MEA). As will be explained in Section 7.4.3, the polymer membrane requires complete hydration for proper function as electrolyte in fuel cell operation and to prevent membrane dehydration due to water evaporation, both fuel and oxidant streams are usually fully humidified with 100% relative humidity.

Humidified anode- and cathode-feed gas are supplied to each electrode through the flow distribution channels produced on the bipolar plates positioned between each MEA in the stack. Therefore, the bipolar plate is also often referred to as the flow-field plate (or flow distribution plate). Convection mass transfer occurs between the flowing gas stream in the flow channels and the electrode backing-layer surface. The transport of a gas mixture through the porous backing layer is primarily by molecular diffusion in the direction of the cell thickness (i.e., transverse direction) for typical PEMFC operating conditions, even though convection in the flow channels penetrate into the porous backing layer for flow in the direction parallel with the flow channels (the longitudinal direction).

Because the electrolyte is essentially sulfonic acid fixed in the solid polymer matrix structure, the electrochemical reactions occurring in the anode- and cathode-catalyst layers are identical to those in the phosphoric-acid fuel cell. At the anode catalyst layer, electrooxidation of hydrogen occurs with the production of protons and electrons

$$\text{At the anode:} \quad H_2 \longrightarrow 2H^+ + 2e^- \tag{7.1}$$

The protons are transported from the anode to the cathode side through the hydrated membrane electrolyte under the effect of electric double layers near the two electrodes; therefore, the proton migration is influenced by the electric field effect, proton concentration gradients which would exist if the membrane is not fully hydrated or local drying of the membrane occurs, and convective motion if pressure differential exists between the anode- and the cathode-feed gas streams. The proton migration from the anode to the cathode side through the membrane electrolyte is in the form of hydronium ions, $(H_2O)_\xi H^+$, thus taking or dragging ξ number of water molecules per proton along with it — the phenomenon is often called **electroosmotic drag effect**. This often results in less molecules of water on the anode side of the membrane, especially at high current-density operation and increases water concentration on the cathode side of the membrane. This mechanism is responsible for the drying out of the membrane on the anode side and the accumulation of excess amount of water on the cathode side, leading to the water flooding phenomenon there, coupled with the fact that product water is also formed at the cathode catalyst layer as shown in Equation (7.2), although some water diffuses back to the anode side due to the water concentration gradient. The lower water concentration on the anode side causes the formation of local dry spots on the membrane, producing higher activation polarization, higher internal electrical resistance (normally over 95% of which is due to the resistance in the membrane), and crack formation on the membrane. Thus both PEMFC performance and lifetime are reduced accordingly.

On the other hand, the electrons are forced to be transported through the external electrical circuit and perform work on the electrical load at the same time because the membrane electrolyte has extremely large resistance to electron motion, although it has a very low resistance to proton motion through it. At the cathode catalyst layer, the oxygen molecules, supplied from the oxidant flow streams in the flow channels, combine with the protons and electrons originated from the anode catalyst layer to form product water:

$$\text{At the cathode:} \quad \frac{1}{2}O_2 + 2H^+ + 2e^- \longrightarrow H_2O \tag{7.2}$$

As a result, the entire cell reaction is obtained by summing up the two half-cell reactions shown in Equations (7.1) and (7.2):

$$\text{Overall cell reaction:} \quad H_2 + \frac{1}{2}O_2 \longrightarrow H_2O + \text{Waste Heat} + \text{Electrical Energy} \tag{7.3}$$

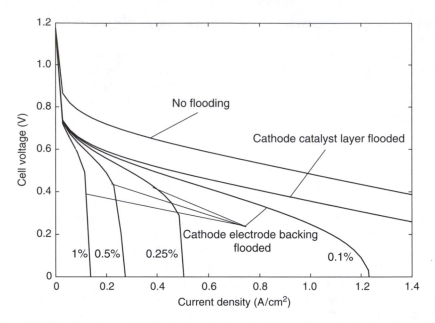

Figure 7.4 The effect of water flooding in the cathode catalyst layer and the cathode electrode-backing layer on the cell performance (based on a model prediction[10]). The numbers by the curves represent the percentage of the void space in the electrode backing region that is being flooded by liquid water.

The reaction product water is formed at the membrane–catalyst interface in the cathode catalyst layer, typically in the liquid form at the PEMFC's operating condition, and seeps into the porous structure of the electrode backing layer, water removal and control become one of the major issues in PEMFCs. Because the chemical energy stored in the reactant hydrogen and oxygen cannot be completely converted into the useful electrical energy, waste heat is produced in the conversion process due to both reversible and irreversible mechanisms as explained in Chapter 2. Therefore, water and waste heat are the two reaction byproducts accompanying with the production of electric power and they need to be properly managed for the optimal performance of the PEMFC stacks and systems. Since water may vaporize or condense with a large amount of heat absorption or release depending on the local thermodynamic condition, and ineffective cooling of the cell and stack can lead to the formation of local hot spots (approximately, over 100 °C), and consequent formation of local dry spots on the membrane, water and heat management strategies are closely related and they are the two critical issues in the proper design and operation of PEMFC systems with sufficient reliability and lifetime. Figure 7.4 shows the effect of water flooding in the catalyst layer and the cathode electrode backing layer on the cell performance, based on a model prediction.[10] It is clear that the cell performance is significantly reduced due to the water flooding entire cathode catalyst layer. The performance degradation is even more striking when the void region of the cathode electrode-backing layer is flooded even by a small amount, ranging from 0.1% to 1% shown

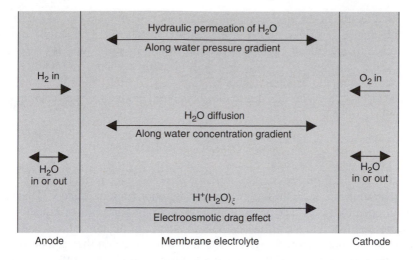

Figure 7.5 Various modes of water transport in a PEM fuel cell.

in Figure 7.4. The limiting current density is considerably reduced, showing clearly the concentration polarization arising from the mass transfer limitation of the oxygen due to the water flooding, as illustrated in Section 4.6.4. The prevention of the water flooding while maintaining membrane full hydration requires a careful consideration of the mechanisms involved in the water production, removal, and transfer during the dynamic operation of the cell. This is an extremely challenging task and majority of the design complexities in PEMFCs are related to it.

Figure 7.5 illustrates the various mechanisms of water transport in the membrane electrolyte. As pointed out earlier, water is produced at the cathode catalyst layer as a result of the electrochemical reaction (the product water) and can be brought into the cell by the humidified reactant gases (the process water). Water transport in the membrane can occur as a result of electroosmotic drag effect, diffusional mechanism due to concentration gradient and hydraulic permeation due to the presence of pressure gradient. The water flux due to the electroosmotic drag effect is associated with the protonic current (i.e., the transport of the protons from the anode to the cathode side) and can be expressed as

$$J_{w,\text{drag}} = \xi(\lambda) J_{\text{H}^+} = \frac{J\xi(\lambda)}{n_{\text{H}^+} F}, \quad \text{mol}/(\text{m}^2 \cdot \text{s}) \tag{7.4}$$

where J is the cell current density (A/m^2), $n_{\text{H}^+} = 1$ is the charge number of the proton, F is the Faraday constant, and $\xi(\lambda)$ is a dimensionless parameter called electroosmotic drag coefficient, which depends on the degree of the membrane hydration λ defined as the ratio of the number of water molecules per ion exchange site in the membrane or

$$\lambda = \frac{\text{Total number of water molecules in the membrane}}{\text{Total number of ion exchange sites in the membrane}} = \frac{N(\text{H}_2\text{O})}{N(\text{SO}_3\text{H})} \tag{7.5}$$

The direction of the electroosmotic water flux is from the anode to the cathode side, thus under normal operation of the cell the anode side tends to dry out, and water tends to accumulate in the cathode side where product water is produced as well. Excessive amount of water is then built up in the cathode, potentially causing the water flooding of the porous cathode.

The diffusional water flux due to the concentration gradient is a vector quantity and can be written as

$$J_{w,\text{diff}} = -D(\lambda)\nabla c_w, \quad \text{mol}/(\text{m}^2 \cdot \text{s}) \tag{7.6}$$

where $D(\lambda)$ is the diffusion coefficient (m^2/s) of water in the membrane at the membrane hydration of λ, c_w is the local water concentration (mol/m^3), and ∇ is the vector gradient operator (1/m). The negative sign on the right hand side of Equation (7.6) represents the fact that the diffusional flux is always in the direction of decreasing concentration. Since normally water concentration is high on the cathode side and low on the anode side, this diffusional flux tends to provide some offset, though not a complete relief, for the water accumulation at the cathode side.

The water flux associated with the hydraulic permeation of water due to pressure gradient is

$$J_{w,\text{hyd}} = -c_w \frac{\kappa_{\text{hyd}}(\lambda)}{\mu}\nabla P, \quad \text{mol}/(\text{m}^2 \cdot \text{s}) \tag{7.7}$$

where $\kappa_{\text{hyd}}(\lambda)$ is the hydraulic permeability of the membrane (m^2), an approximate measure of the average of the diameter squared for the pores in the membrane and again it is a function of the membrane hydration, and μ is the dynamic viscosity of liquid water, $\text{kg}/(\text{m} \cdot \text{s})$. The water flux $J_{w,\text{hyd}}$ is in the direction of decreasing pressure P, represented by the negative sign in Equation (7.7). In order to provide additional means for the reduction of water in the cathode, cells may be differentially pressurized such that the oxidant gas is supplied at a higher pressure than the fuel gas in the anode. Thus, the predominant direction of the diffusional and hydraulic water fluxes can be arranged opposite to that of the electroosmotic water flux. The water concentration distribution in the membrane, hence the membrane hydration level, is the balancing result of these three water-transport mechanisms in the membrane under a given cell operating condition. In practice, water transport in the membrane must be taken into account when considering and implementing strategies for reaction product water removal and process water addition for the maintenance of membrane hydration. The product water is usually removed on the cathode side by the oxidant stream flowing in the flow channels machined on the bipolar plates. Therefore, proper design of the flow distribution channels (or the bipolar plates) is critical to water removal and hence the high current density operations. This topic is specifically discussed in Sections 7.3.2 and 7.4.3. The fuel stream is normally required to be humidified before entering the cell because of the electroosmotic effect and to avoid the local dry spot formation due to convective evaporation on the cathode side of the membrane near the cell inlet, humidification of the oxidant stream is also necessary for large cells.

Typically a proton exchange membrane is operated at the temperature of 80 °C and from atmospheric to about 8 atm pressures. However, the typical operating pressure is around 3 atm for optimal performance for large PEMFC stacks, although the optimal operating pressure is under continual evaluation for specific design and operating conditions. PEM fuel cells can operate on pristine hydrogen or hydrocarbon reformed fuel gas mixtures and pure oxygen or air. The flow rate of the reactant supply is usually expressed in terms of stoichiometry, representing the ratio of the actual reactant supply to the ideal stoichiometric amount required for the electric power output, as defined in Chapter 2. When pure hydrogen and air are used for cell operation, optimal performance is obtained when the stoichiometry is usually around $1.1 \sim 1.2$ for hydrogen, and about 2 for oxygen. Since only 21% of the air is oxygen, the amount of airflow becomes very high, nearly 10 times that of the stoichiometric amount of oxygen flow required for the cell reaction. This high volume airflow requires significant amount of power for compression and water for humidification and air compression represents the majority of the parasitic power consumption for the stack output power. Therefore, small PEMFC systems are often operated at the atmospheric condition and only large systems are pressurized for higher power density and energy efficiency. To guarantee the sufficiently high energy conversion efficiency, the operating cell voltage is often set above 0.6 V, which corresponding to over 40% efficiency based on the higher heating value. On the other hand, the operating cell voltage is often limited to below 0.7 V in order to have sufficiently high power density, which is primarily determined by the operating current density. Therefore, the operating cell voltage is normally between 0.6 and 0.7 V, the corresponding current density is as high as possible for high output power density, and all the technological improvement is aimed to increase this operating current density. High power density is especially important for transportation applications, and it also represents fewer amounts of materials and labors needed for construction and manufacture, hence lower cost of the power generation system. At the current status of technology, the operating current density ranges approximately from 300 to 500 mA/cm^2 running on pure hydrogen and air, and it is as high as over 1 A/cm^2 when operating on pure hydrogen and oxygen.

The overall electrochemical reactions in PEMFCs, as described in Equations (7.1) and (7.2), involve the oxidation of molecular hydrogen to protons at the anode catalyst layer, and the reduction of molecular oxygen to water at the cathode catalyst layer. Both reactions require active catalyst sites to break the molecular bond in the diatomic gaseous reactant molecules because of the low temperature reaction environment. Therefore, they both are heterogeneous electrocatalytic reactions occurring at the surface of the catalyst. As pointed out in Chapter 3, the hydrogen oxidation reaction (HOR) is orders of magnitude higher than the corresponding oxygen reduction reaction (ORR) at the low temperature of PEMFCs, which translates to a much higher cathodic polarization for PEMFCs.

The energy loss associated with the resistance to the cathodic oxygen reduction reaction (ORR) represents the largest single source of energy losses in the PEMFCs, as in all other low-temperature fuel cells, because of the sluggish ORR kinetics. Due to the acid nature of the membrane electrolyte and low temperature operation, platinum or platinum alloys are the "best" known catalysts to facilitate and accelerate the reaction process. From Equation (7.2), it is clear that in order for the ORR to proceed steadily, three essential elements are required at the reaction sites in the cathode catalyst layer, that is, the presence and availability of

1. Oxygen gas molecules,
2. Protons, and
3. Electrons.

From Equation (7.1), it is realized that these same three elements are essential for the steady hydrogen oxidation reaction because the protons and electrons, as the reaction products, need to be transferred away to avoid the product accumulation effect.

Therefore, the membrane electrolyte must penetrate into the catalyst layer to surround the catalyst surface in order to provide a passage for proton to be transported in; the electronically conductive catalyst particles must be connected all the way to the electrode backing layer in order to provide a means for electron migration; and sufficient porosity must be provided for the catalyst layer in order for the oxygen gas to be transferred to the active sites. Good proton conduction requires the high loadings of the membrane in the catalyst layer, which would increase the thickness of the membrane layer covering the catalyst surface, through which the oxygen molecules need to cross. Hence a high rate of oxygen transfer through the surface membrane layer minimizes the mass transfer resistance for the oxygen gas to reach the active sites, however, the membrane electrolyte is desired to have low reactant gas permeability in order to separate the reactant gases from intermixing (crossover), yet excessively low reactant gas (i.e., oxygen) permeability yields excessively high resistance for the oxygen gas to cross through the membrane layer covering the catalyst surface. Low resistance to electron supply dictates the high connectivity of the catalyst particles, which would decrease some of the catalyst surface area that would otherwise be available for surface reaction; and high porosity for the catalyst layer provides ample space for oxygen gas transfer, but that reduces the space available for the solid catalyst particles. Clearly, the structure and composition of the catalyst layer can affect the PEMFC performance significantly; in fact, in laboratory testing no output electric power can occur if the catalyst layers are not properly prepared. Therefore the catalyst layer structure and composition should be optimized in order to achieve the best possible cell performance.

For the oxygen reduction reaction at the Pt/Nafion interface, the rate-determining step is the electron-charge transfer reaction. However, for the hydrogen oxidation reaction at the Pt/Nafion interface, three rate-determining steps are possible:

1. Electrooxidation of adsorbed hydrogen atoms (charge transfer reaction),
2. The dissociative chemisorption of H_2 molecules on platinum–catalyst sites to form adsorbed hydrogen atoms,
3. Mass transport-limited supply of H_2 gas to the platinum/membrane interface.

Fortunately, since hydrogen has a much higher diffusion coefficient for mass transport and the very high rate of electrochemical reaction, the associated polarization is very small, often negligible, under the PEMFC operating conditions. The exception is when the anode side of the membrane becomes dehydrated and/or the fuel stream

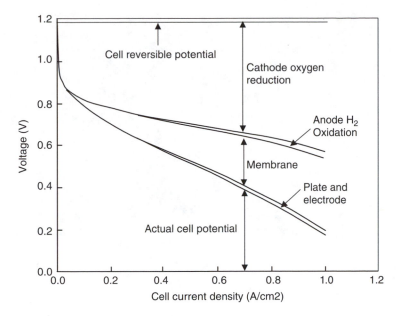

Figure 7.6 A typical performance of the PEM fuel cell in terms of the cell polarization curve with a breakdown of the various voltage-loss mechanisms identified.

contains contaminants such as carbon monoxide. These two issues are discussed in Sections 7.4 and 7.6, respectively. A typical PEMFC performance in terms of cell polarization is shown in Figure 7.6 with a breakdown of the various voltage loss mechanisms identified.

7.3 COMPONENTS AND CONFIGURATIONS

This section, describes the typical components and the geometrical configurations (or designs) involved for a single PEM fuel cell, a PEM fuel cell stack, and a PEM fuel cell system. While the materials and associated fabrication processes are described in the next section, those that are necessary for the description of the component designs are briefly mentioned in this section as well.

7.3.1 A Single PEM Fuel Cell

Figure 7.7a shows a single proton exchange membrane fuel cell typically used for laboratory testing, and Figure 7.7b provides the corresponding illustration of the cell components. Such a single cell often consists of a membrane electrolyte, two catalyzed electrodes, two Teflon masks and two endplates. The catalyzed electrode has a thin porous catalyst layer (about 5–50 μm thick) applied onto the gas-diffusion backing layer (typically 100–300 μm thick), and the membrane electrolyte is normally about 50–175 μm thick. The membrane electrode assembly (MEA) is fabricated by hot pressing two electrodes onto the membrane with the catalyst layer bonded to

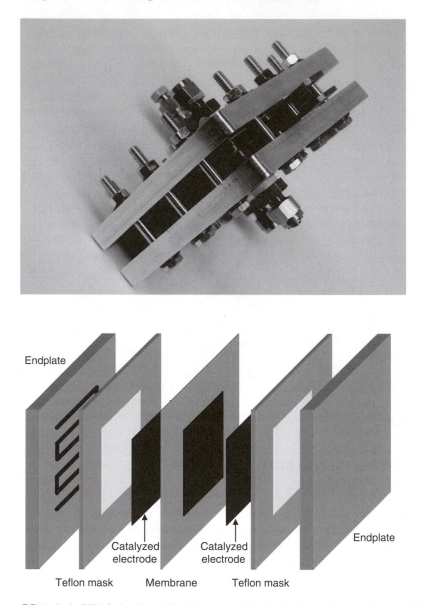

Figure 7.7 A single PEM fuel cell used for laboratory testing. (a) A photograph of such an assembled cell. (b) Schematic of typical components excluding the current collector plates.

the membrane. The catalyst layer may be considered macrohomogeneous, consisting of dispersed catalyst particles and membrane electrolyte surrounding the catalyst particles, as shown in Figure 7.8. Figure 7.9 illustrates a corresponding magnified view of the idealized structure of platinum catalyst particles supported on a larger carbon particle surrounded by membrane electrolyte. Such a catalyst is often called carbon-supported platinum or platinum-supported on carbon with the abbreviation of Pt/C. Many such structures form the catalyst layer.

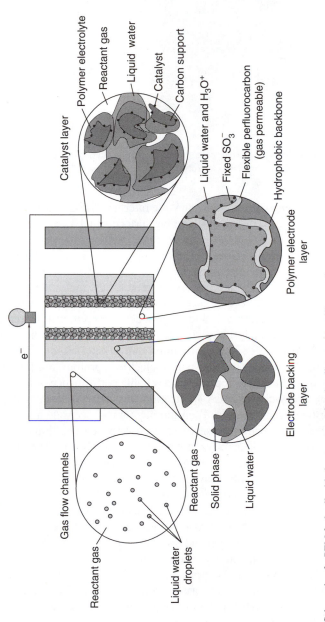

Figure 7.8 Schematic of a PEM fuel cell with the structure of each cell component illustrated.

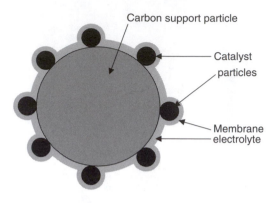

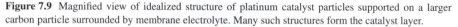

Figure 7.9 Magnified view of idealized structure of platinum catalyst particles supported on a larger carbon particle surrounded by membrane electrolyte. Many such structures form the catalyst layer.

The electrochemical reactions occur at the interface between the catalyst particle and the membrane (i.e., at the catalyst particle surface which is covered by the membrane). The solid catalyst particles must be connected over its surfaces to allow for the transport of electrons; therefore, not all the surfaces of the catalyst particles can be reaction sites. The membrane phase surrounding the catalyst particle has to be connected with the membrane in the electrolyte region for the migration of protons. The void region in the catalyst layer is connected to the void space in the electrode-backing layer and is used for reactant supply and the product water removal. The structure and the amount of the solid catalyst particles, the membrane, and the void region (often in terms of the porosity) are significantly important in determining cell performance and their optimization among the transport of electrons, protons, reactant supply, and product removal is essential for optimal and reliable performance of PEM fuel cells.

In another approach for MEA fabrication, the catalyst layer is applied onto the electrolyte membrane directly. Then the membrane electrode assembly (MEA) denotes the electrolyte membrane sandwiched in between two thin catalyst layers, or sometimes also called the **catalyzed electrolyte membrane**. Then the electrode only contains a gas-diffusion (or backing) layer. Irrespective of which method of fabrication is used, both the catalyst layer and the porous electrode-backing layer are wetproofed by treatment with polytetrafluoroethylene (PTFE), which has hydrophobic property and rejects water. Hence, PTFE is used for easy water removal and to avoid the water flooding of the electrode porous structure. The amount and distribution of the PTFE in the catalyst layer and backing layer affects PEM cell performance significantly via the water flooding phenomenon.

Whether the catalyst layer is applied to the electrode-backing layer first or to the membrane first, the complete and assembled single PEM fuel cell is identical, as shown in Figure 7.3. In laboratory testing, such a single cell is usually clamped by two endplates on either side of the cell. The endplate serves the function of current collector and reactant gas distribution along the electrode surface. The latter is accomplished by various flow fields (or flow distribution channels) machined on the plate. The assembled single cell is shown in Figure 7.3 for a cross-sectional view and in Figure 7.7

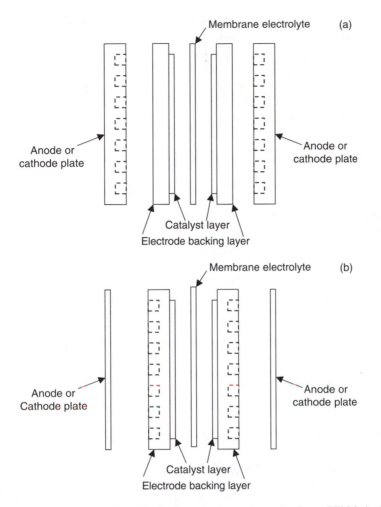

Figure 7.10 Two basic designs of the cell repeating units that can be used to form a PEM fuel cell stack. (a) Flow channels are built on the anode and cathode plates. (b) Flow channels are built on the electrode backing layers for easy access of reactant to the catalyst layer resulting in more uniform distribution of the reactant over the cell active area.

for a three-dimensional view. The Teflon masks shown in Figure 7.7 serve as an effective seal with the periphery of the membrane to prevent the reactant gas from leaking across the cell and out of the cell, so that the reactant gas is confined to the active area of the cell, which is highlighted as the black region in the center of the membrane in Figure 7.7. The endplates will be replaced by bipolar plates located between cells when such an individual cell is assembled into a fuel cell stack, and in that case the bipolar plate will have flow channels machined on its both major surfaces.

In the content of a stack, a basic repeating unit, equivalent to a single cell unit that is being repeatedly continued to form a stack, can have a number of different configurations, and two of typical design arrangements are shown in Figure 7.10 for comparison. The design shown in Figure 7.10a is the same as those described early in this section, that is, the flow distribution channels are located in the anode and cathode plates, similar to Figure 7.7. Whereas for the design shown in Figure 7.10b, the porous electrode-backing layer have channels for reactant fluid flow, although reactants can also flow within the interstices of the porous layers. Therefore, this design has the advantage of easy and more uniform distribution of the reactant over the entire active areas of the electrode surface yielding better cell performance. For this design, the electrode-backing layers become much thicker due to the need for the fabrication of the channels, but the two plates separating one cell from another in the stack can now be very thin. The structures shown in Figure 7.10 constitute the fundamental repeating unit that exists in a typical PEM fuel cell stack, the only component which is missing is the implementation of stack cooling arrangement, to be discussed in the next section.

7.3.2   A PEM Fuel Cell Stack

Figure 7.3 illustrates schematically a single PEM fuel cell unit in the context of a fuel cell stack with a cross-sectional view. Many such cells connected together in series, but separated by the bipolar plates, form an integral stack for desired power ratings, as shown in Figure 7.2. Majority of the PEM fuel cells and stacks are in planar form, although other configurations such as tubular structures have been proposed.[11] In this section, our description exclusively focused on planar structures, as they are the dominant practical designs.

In a typical PEM fuel cell stack, the membrane electrode assembly is pressed in good electric contact on its both major surfaces with two electrically conductive plates called bipolar plates. These plates have at least one flow passage (or channel) engraved or milled on the surface facing the MEA, therefore, these plates are also referred to as fluid flow-field plates. The flow channels direct and distribute the fuel to the anode and oxidant to the cathode electrode. The function of the bipolar plate is to serve as the collector for the electric current generated in the MEA, to provide mechanical support for the MEAs, to provide flow channels for the distribution of fuel and oxidant to the respective anode and cathode electrodes, for the effective removal of product water formed in MEAs. An effective design for the flow channels is to distribute the reactant gas over the electrode surface as uniformly as possible to utilize effectively all the active area of the electrode and to allow for the reactant gas stream to take the product water away avoiding the water flooding of the electrode during the operation of the cell. Appropriate materials with high electric conductivity and sufficient mechanical strength are normally selected for the bipolar plate in order to fulfill its function as the current collector and mechanical support for the thin cell. Once the material is chosen, the design of the bipolar plate is reduced to the design of the complex flow channels for effective reactant supply and water removal. The flow channel configurations are extremely important for the proper operation and hence the performance of a PEMFC stack, this is because the MEA is less than 1-mm thick, while the bipolar plate can be as thick as 6–8 mm. It might be claimed that the significant improvement achieved in the last decade or so for the practical stacks could be primarily attributed to the improvement made in the bipolar plate design with complex flow fields — this topic are discussed further in the next subsection (or Section 7.3.3).

To meet the overall output power requirement from a stack, many cells are, in general, connected electrically in series, although sometimes in parallel typically for small stack sizes. In the series connection, one side of the bipolar plate serves as an anode plate for the one cell and the other side of the same plate serves as the cathode plate for the adjacent cell. Therefore, the bipolar plate also acts as a separator preventing the mixing of the reactants, it has to be impervious for hydrogen and oxygen, and be stable and corrosion resistant in the oxidizing and reducing environment. The commonly used material for the bipolar plate, graphite, is not very impervious to the reactants; as a result, fairly thick graphite plates are generally used for the bipolar plates.

Because the electrochemical and physical processes occurring in the MEAs and in the bipolar plates degenerate the energy into heat, which, if not properly removed, increases the average cell temperature as well as produce temperature gradients. Both of these have undesirable effects on the cell performance and the lifetime. Normally the energy conversion efficiency is about 50% for PEMFC stacks, implying that for every watt of electrical power generated there is the same amount of waste heat produced in almost 1:1 correspondence. Therefore, cooling techniques are required for the thermal control. In practice, the cooling technique to be employed is considerably dependent on the size (output power) of the stack. Normally for stack sizes of a few kW to 10 kW or above, which are the case for stationary co-generation or transportation applications, liquid water cooling is almost a must for the effective cooling of the stack to avoid temperature gradients in and across the cell. The cooling water flows in a separate cell with its own flow channels, and cooling cell may be placed every or every other active cell; some of the typical cooling cell arrangements are shown in Figure 7.11.

Figure 7.11 shows the side cross-sectional views of some of the examples for basic fuel cell repeating units used in stack construction. It consists of three plates with fluid flow channels for each membrane electrode assembly for the configuration shown in Figure 7.11a, the three plates are the anode and cathode plates with fuel and oxidant flow field channels, and one cooling plate with coolant flow field channels. In this design, cooling plates are located for every active cell composed of an MEA sandwiched between the anode and cathode plates. Obviously, the fluid flow field channels are built only on one side of the plates which are fluid impermeable, and in this sense, there is no plate in the stack that really acts as the conventional bipolar plate which has reactant flow field channels on its two major surfaces. For the configuration shown in Figure 7.11b, two plates are used for each MEA, one of the plates has fluid flow-field channels made only on one of its two surfaces, while the other plates have flow channels made on both of its two sides. However, for the latter plate coolant flows in the channels facing away from the MEA, so that the plate has geometry similar to, but its function is really different from, the conventional bipolar plate. This second structure has only two plates, compared to the first structure shown in Figure 7.11a, and has one less plate to make and one less contacting surface. Thus, it is easy for fabrication and assembly of the stack, and also has less contact resistance

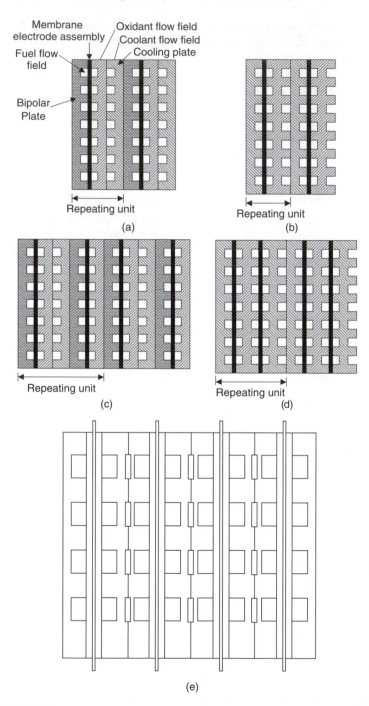

Figure 7.11 Typical cross-sectional view through the active part of a stack with the cooling cell placed at: (a) and (b) every cell apart or (c) and (d) every other active cell apart. The cooling flow fields are either formed asymmetrically (a)–(d) or symmetrically (e) between the anode and the cathode flow-field plates.[12]

for electrical and heat transport, resulting in better stack performance. The structures shown in Figure 7.11c and d are essentially the same as those shown in Figure 7.11a and b, respectively; the only difference is that the cooling cell is located at every other active cells, rather than for every active cell. The increase in stack performance, especially the stack power density, is obvious due to the reduction in the number of cooling cells, which are fairly sizeable in the stack structure. On the other hand, Figure 7.11e shows the cooling flow fields formed by the channels symmetrically located on the backside of both the cathode and anode flow-field plates,[12] as opposed to the case located only on one of the flow field plates, as shown in Figure 7.11b.

As shown in the next section, proton-conducting membranes require full hydration to maintain high conductivity and this is normally accomplished by humidifying the fuel and oxidant streams before they enter the fuel cells in the stack. Humidification of the gas streams can be achieved by introducing the gas stream at the bottom of a water container through a perforated plate — the so-called gas bubbling technique; injecting steam into the gas stream; spraying liquid water into the gas stream, and so on. Each of these techniques has its own problems and limitations; for example, the gas bubbling approach is easy to make and operate and is often used in laboratory testing of single or a small stack of a few cells, but for high gas-flow rates typical for large stacks and high current-density operations it tends to drag small water droplets into the electrode porous structure and cause electrode flooding by liquid water while the gas stream itself may be under humidified. This is caused by the short residence time and slow water-evaporation rate for high velocity gas flow, and requires special attention on the reactant supply manifolding sizes and configurations. Liquid water spray requires extra thermal energy available for the vaporization of water, otherwise, the inlet gas temperature may be below the cell operating temperature, resulting in temperature gradient in the cell and stack structure. Further, the same electrode flooding problem may arise at high gas-flow rates. Steam injection requires external steam generator and the heat supply for the water evaporation. Both water spray and steam injection belong to the class of active control, the exact degree of humidification can be controlled. This is useful during the operation, for example, full gas stream humidification is needed at the low current density (or partial load) operations when the product water produced in the cell is limited. On the other hand, at high current-density (or high load) operations significant amount of liquid water is formed inside the cell and the cathode stream can be under humidified to facilitate the water removal and avoid water flooding of the cathode. However, both techniques require precise and dynamic control during the operation, adding the complexity and difficulty to the developmental work.

Further, the water saturation pressure as shown in Chapter 1 depends exponentially on the gas temperature, suggesting that the ability of gases such as air to absorb water varies significantly with changes in temperature, especially at low operating pressures. In some laboratory testing, the anode stream is humidified at a temperature slightly above the cell operating temperature to provide extra water available for the anode side to avoid the membrane dehydration at the anode side due to the electroosmotic drag effect that depletes the water there. However, it is undesirable for practical applications since it requires special heating source to provide the extra high-temperature humidification unit. On the other hand, gas stream humidification at

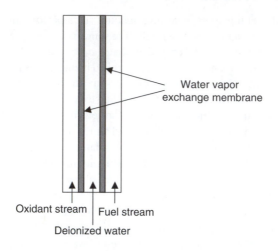

Figure 7.12   Schematic of humidification cell in the in-stack reactant humidification design.

a temperature significantly below fuel cell operating temperature could produce insufficient humidification that dehydrates the membrane when the gas stream enters the operating cell. Therefore, it is preferable to humidify the fuel and oxidant gas streams at, or as close as possible to, the operating temperature and pressure of the fuel cell. One passive technique that is commonly used to meet the above requirement is the in-stack humidification of the reactant gas streams as described in the following text.

Figure 7.12 illustrates the principle of the in-stack humidification cell design. The fuel and oxidant gas streams are humidified by flowing on one side of a water-vapor exchange membrane, while on the other side of the same membrane deionized water flows. The water permeates through the membrane and provides the source of water for gas humidification. Any proton-conducting membrane, which absorbs and allows for the transport of water such as perfluorinated sulfonic acid membrane like Nafion, can be used as the humidification membrane material and other commercially available water exchange membranes are suitable materials as well. Deionized water is used to prevent the contamination by undesired ions of both the water vapor exchange membrane and the proton conducting membrane in the active cells. Since deionized water is used for the cooling of the active cells for the same reason, cooling water is convenient to be utilized for the reactant humidification inside the same stack. This is because the cooling water, after cooling the active cells in the stack, reaches the high end of the cell operating temperature, then when it is circulated through the humidification cell before exiting the stack, it has sufficient thermal energy content to provide the vaporization of water for the reactant humidification as well as heat up the reactant gas stream to the cell operating temperature. Therefore, such an in-stack humidification arrangement serves two purposes. This passive design also has the advantage of accommodating variations in the operating conditions without the need of external sensing and control units. For example, when the stack output power changes with the load requirement, the reactant flow rates change accordingly. Then the convective heat and mass transfer coefficients change with the flow rate, adjusting

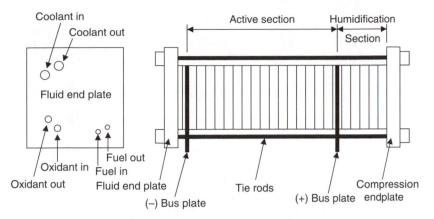

Figure 7.13 A PEM fuel cell stack arrangement incorporating in-stack humidification.

the rate of water vaporization and heat transfer to the reactant streams. A PEM fuel cell stack incorporating in-stack humidification is shown in Figure 7.13. Clearly, such a stack contains an active section for power generation and a humidification section containing many humidification cells for saturating the reactant streams. Figure 7.14 illustrates the in-stack fluid flow paths for the fuel and oxidant streams as well as the cooling water. Other in-stack arrangements have also been explored for the reactant humidification.[13]

For PEM fuel cell stacks, the reactant supply and exhaust manifolds are typically built inside the stack in what is called the internal manifolding design. Therefore, the stack contains the manifolds and inlet ports for directing the fuel and oxidant streams to the anode and cathode flow-field channels and exhaust manifolds and outlet ports for the expulsion of the unreacted fuel and oxidant exhaust streams. The cooling and humidification water flow manifolds are also built inside the stack. Therefore, the stack includes another set of manifolds and inlet port for distributing the coolant fluid (mostly deionized water, others are discussed later), to interior cooling channels built within the stack to absorb heat generated within the stack and an outlet port for the coolant water exiting the stack. The cooling channel design is substantially the same as the reactant distribution flow-field channels, however, the direction of the coolant flow and the layout of the cooling channels are important in the cooling of the stack, in the removal of product water, and in the establishment of the incell or across the cell temperature variations, which can influence the cell lifetime related to the membrane saturation, and proper sealing methods. These issues are explained later. The inlet and outlet ports of the intake and exhaust manifolds in the stack can be located at the same end of the stack, as shown in Figure 7.13, the opposite ends, and on the side around the middle of the stack or any combination of the arrangements as shown in Figure 7.15, and a complete three-dimensional view of the stack is shown in Figure 7.16.

Note that the previous description of the PEM fuel cell stack is just one typical configuration, and many other stack configurations are also possible, that differ from this description of the PEM fuel cell stack. For example, Vanderborgh and Hedstrom[14] disclosed a stack design with in-stack humidification, temperature

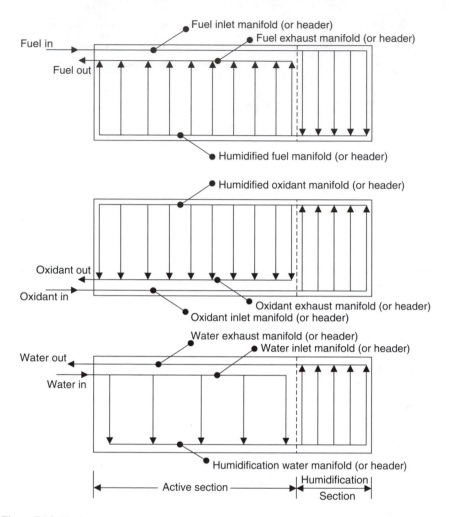

Figure 7.14 The in-stack fluid flow paths for the fuel and oxidant streams as well as the cooling and humidification water corresponding to the stack design shown in Figure 7.13.

control, and product water removal. It is accomplished by locating the humidification, cooling, and water removal cell adjacent to and around the periphery of the active cell with all these units built on the same plane as the active cell itself. Off-stack humidification of the reactant gas has also been proposed, that requires a separate device external to the stack for injecting water into the reactant gas streams (e.g., Fleck and Hornburg[15]).

The manifolding structure and the location of the inlet and outlet ports for a stack can have significant influence on the stack performance, especially for the performance variation from one cell to another. Figure 7.17 shows the effect of stack manifolding configuration on the stack performance for a 5-kW stack made up of 50 cells operating on pure hydrogen and air at 0.6 A/cm^2, the results shown in the

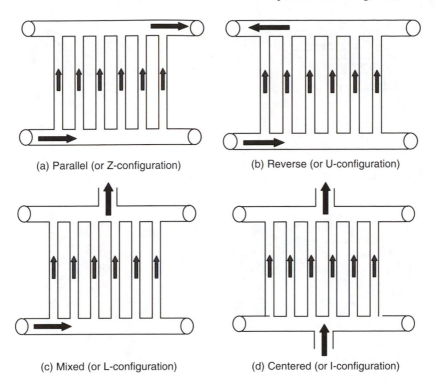

(a) Parallel (or Z-configuration)   (b) Reverse (or U-configuration)

(c) Mixed (or L-configuration)   (d) Centered (or I-configuration)

Figure 7.15 The manifold arrangement along with the various possible locations of the inlet and outlet ports in the stack.

figure are based on a stack modelling study.[16] The cell voltage for each of the 50 cells in the stack is shown in Figure 7.17a. Clearly the cell voltage decreases towards the middle of the stack and then increases again until it recovers the performance of the first few cells for the Z configuration, which is schematically shown in Figure 7.15, whereas the cell voltage decreases monotonically for the U configuration. Clearly, the cell-to-cell variation is the largest for the U configuration. Since the best stack performance is obtained when every cell in a stack has the same voltage (uniform cell voltage distribution), we can quantify stack performance by defining a cell-to-cell variation of the cell performance within a stack, in terms of the cell voltage E, as

$$s = \frac{|E_{max} - E_{min}|}{E_{max}} \qquad (7.8)$$

For the results shown in Figure 7.17a, $s \approx 9\%$ for the Z configuration and 18% for the U configuration — almost double the amount of variation. The reason for this significant cell-to-cell variation arises from the considerable uneven oxidant (air) flow distribution among the cells, as shown in Figure 7.17b, caused by the fluid mechanics.

Further analysis indicates that the effect of the manifold configuration is related to the sizes of the manifold cross section compared to the cross-sectional area of the fluid flow channels on the anode and cathode plates facing the MEA. The cell-to-cell

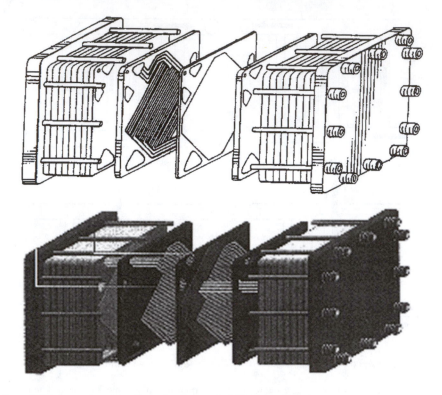

Figure 7.16 A complete three-dimensional view of a PEM fuel cell stack.

variation s decreases as the manifold cross-sectional area A_m increases, and s reduces effectively to zero, i.e., uniform performance among the cells when A_m becomes larger than a critical value, $A_{m,c}$. Also this critical value becomes very small (smaller than typical dimensions encountered in practice) when pure hydrogen and pure oxygen are used as reactants. For this case the cell-to-cell variation disappears altogether for any reasonable size of manifolds selected. However, this critical value is within the practical size range when air and/or reformed hydrogen-rich gas mixture is used as the reactant, therefore, a stack designed for pure hydrogen and oxygen operation may not and usually cannot, function properly when either air or reformed fuel is employed. Furthermore, when the size of the flow channels in the anode and cathode is increased, the critical manifold size, $A_{m,c}$, also increases, leading to more delicate consideration in the stack design.

Another point should be emphasized is that these results do not necessarily suggest that the Z-manifold configuration always produces better performance than the other configurations shown in Figure 7.15. The appropriate answer to this question is that it depends. It depends on the flow condition and relative sizes of the manifold and flow channels on the plates. In fact, each of the configurations shown could produce a more uniform and better performance for the active cells in the stack under a specific operating and design condition. This may explain the variety of the design arrangements in practice.

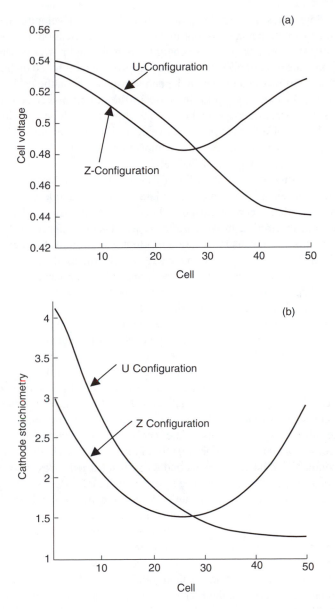

Figure 7.17 Effect of stack manifolding configuration on the stack performance for a 5-kW stack made up of 50 cells operating on pure hydrogen and air at 0.6 A/cm^2. (a) Cell voltage for each cell. (b) Air flow rate to each cell in terms of cathode stream stoichiometry.[16]

7.3.3   Design of Fluid Flow-Field Channels on Plates

Despite the rapid progress being made, substantial cost reduction and cell performance improvement are required before PEMFC can reach widespread commercial use.[2] It has been recognized[17] that one of the main obstacles to large-scale commercialization are (i) gas flow fields and bipolar plates, including the development of low-cost lightweight construction materials, optimal design and fabrication methods and their impact on PEMFC performance (i.e., energy efficiency and power density); and (ii) the most promising direction for performance improvement is based on the minimization of all transport resistances, which depend substantially on the design of reactant gas flow fields. As much as a 50% increase in the output power density has been reported just by appropriate distribution of gas flow fields alone.[18,19] In spite of all the industrial R&D efforts, the time-effective design and optimization of the gas flow fields and bipolar plates remain one of the important issues for the cost reduction and performance improvement of PEM fuel cells.

For an operating fuel cell, the MEA (<1-mm thick) is interposed between two fluid-impermeable electrically conductive plates, called the anode and cathode plates, respectively. The plates serve as current collectors, provide structural support for the porous and thin electrodes, provide a means for reactant supply to the respective electrode, and provide means for reaction product water removal. When the reactant gas-flow channels are formed in the anode and cathode plates, the plates are normally called gas flow-field plates as well. In a stack, one side of a given gas flow-field plate is the anode plate for one cell, and the other side is the cathode plate for the adjacent cell. In such an arrangement, the gas flow-field plate is also called a bipolar plate. The cross section of the gas flow channels is typically rectangular, with the channel width and depth in the range of 1~2 mm, and each bipolar plate is almost an order of magnitude thicker than the MEA itself. A fuel cell stack usually has a coolant fluid, typically water, flowing in interior channels (or cooling layers) within the stack to absorb the heat generated. The cooling layers are located at periodic intervals along the stack, usually every or every few fuel cells. Therefore, the heat produced in each MEA within a stack is transferred by convection to reactant streams and conduction through the solid stack components before reaching the coolant in the cooling layers.

The proton conducting polymer membrane must be fully hydrated to have adequate ion conductivity. It becomes nonconductive when dried excessively and then not useful for ion transport in fuel cells. The membrane in fuel cells is subject to moisture removal by evaporation due to heat generated in the electrochemical reaction and current transport (i.e., joule heating), and proton migration through the membrane, which drags water molecules along with it from the anode to the cathode (the electroosmotic effect). Excess water is then accumulated on the cathode side, also due to the formation of liquid water there as the reaction product. Some excess water diffuses back to the anode due to concentration gradient, but it is not always sufficient to prevent excessive membrane drying under high current-operating conditions. As a result, the fuel cell must be operated under conditions where water removal must be balanced by water supply and the reactant gases, both hydrogen and oxygen, need to be humidified before entering each cell to maintain the saturation of the membrane within the MEA. Thus, water and thermal management become critical for efficient cell performance, are fairly complex, and require dynamic control to match the varying operating conditions of the fuel cell. Because of these limitations, the operating temperature of PEMFCs is usually less than 120 °C, typically at 80 °C. Current polymer electrolyte is made of perfluorinated sulfonic acid membrane, such as Nafion from DuPont.

As described here, water accumulates in the cathode, and is expelled into the oxidant stream as a liquid in the form of small droplets. The accumulation of excessive liquid water in the porous cathode interferes with the access of oxygen to the cathode active sites. Inadequate drainage of liquid water in the cathode flow channels results in poor oxidant gas flow distribution on the cathode side. These phenomena reduce the cell performance significantly. On the other hand, excessive water removal will cause membrane dehydration, leading to increased resistance to proton migration, and the formation of local hot spots due to higher Joule heating resulting from the higher ohmic resistance of the dehydrated membrane. This self-accelerated phenomenon decreases the cell performance, potentially leading to the cracking of the dry membrane, causing the reactant crossover as well as reducing cell lifetime.

Therefore, the design of the reactant gas flow fields must allow for gaseous and liquid water flow while maintaining adequate electronic conductivity. But the design of large-area flow fields ($>500 \text{ cm}^2$) demonstrates scaling problems in thermal and water management not experienced in small single cells ($\sim 5 \text{ cm}^2$) of laboratory testing. A typical problem is liquid water accumulation inhibiting gas transport. Thus one of the important tasks that need to be addressed is the optimal flow geometry of the reactant gas flow fields. In order to obtain high fuel cell performance, it is essential to maintain a sufficiently high reactant concentration while preventing either excess water hydration or dehydration. Also of importance is the distribution of heat, which has a strong impact on fuel cell performance by affecting, for example, water and species transport as well as the rate of electrochemical reactions in the electrodes. Hence, proper thermal control and water management are necessary to maintain stable high performance.

Various flow-field channel layouts have been proposed to address the often-conflicting requirements and comprehensive review is also available for various designs that have been proposed.[20] The early work on the bipolar plates focused on the reduction in electrical resistance and increase in the mechanical strength, as described by Pollegri and Spaziance[21] and Balko and Lawrance.[22] As to the geometrical configurations of the gas flow fields, a variety of different designs are known, and conventional designs typically comprise either pin, straight, or serpentine designs of flow field channels. Examples of the pin-type flow fields are illustrated by Reiser and Sawyer[23] and Reiser,[24] and an example is shown in Figure 7.18. The flow field network is formed by many pins arranged in a regular pattern, and these pins can be in any shape, although cubical and circular pins are most often used in practice. Normally both cathode and anode flow-field plates have an array of regularly spaced cubical or circular pins protruding from the plates and the reactant gases flow across the plates through the intervening grooves formed by the pins. The actual fluid flow thus goes through a network of series and parallel flow paths. As a result, pin-design flow fields result in low reactant-pressure drop. However, reactants flowing through such flow fields tend to follow the path of least resistance across the flow field, which may lead to channeling and the formation of stagnant areas, thus uneven reactant distribution, inadequate product water removal, and poor fuel cell performance. Furthermore, relatively stable recirculation zones may arise behind each pin since the reactant flow is very slow in such a small flow channel, and the Reynolds number for

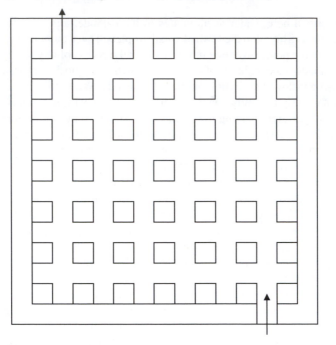

Figure 7.18   Flow field plate with pin-type flow fields, as illustrated by Reiser and Sawyer[23] and Reiser.[24] Normally cubical and circular pins are used.

the reactant flow remains small, particularly for the fuel stream Reynolds numbers may range from a few tens to low hundreds. Reactant concentration may be depleted in the stable recirculation zones as well, decreasing the cell and stack performance. These issues may become particularly problematic with flow fields having certain geometric shapes.

Pollegri and Spaziance[21] showed a straight flow field design, which is further exemplified by General Electric and Hamilton Standard LANL No. 9-X53-D6272-1 (1984). In this design, the gas flow-field plate includes a number of separate parallel flow channels connected to the gas inlet and exhaust headers, which are parallel to the edges of the plate. An example is shown in Figure 7.19a, the flow channel cross-sectional shape is shown in Figure 7.19b. Figures 7.19c, d, and e show a few variations of the basic flow-field channel configurations illustrated in Figure 7.19a. When air is used as the oxidant, it is found that low and unstable cell voltages occur after extended periods of operation because of cathode gas flow distribution and cell water management. As the fuel cell is operating continuously, the water formed at the cathode accumulates in the flow channels adjacent to the cathode, the channels become wet, and the water thus tends to cling to the bottom and sides of the channels. The water droplets also tend to coalesce and form larger droplets. A force, which increases with the size and number of the droplets is required to move the droplets through the channel and out of the cell. Since the number and size of the water droplets in the parallel channels are likely different, the reactant gas then flows preferentially through the least obstructed channels. Water thus tends to collect in the channels in which little or no gas is passing. Accordingly, stagnant areas tend to form at various areas throughout the plate. Hence, poor cell performance arises from the inadequate water drainage and poor gas flow distribution on the cathode side. This problem is similar to that occurs in the pin-type flow field, as discussed earlier.

Another problem associated with this design is that the straight and parallel channels in the bipolar plates tend to be relatively short and have no directional changes. As a consequence, reactant gas has a very small pressure drop along these channels, and the pressure drop in the stack distribution manifold and piping system, which is normal to the bipolar plates, tends to be large in comparison. This inadequate pressure loss distribution results in nonuniform flow distribution of reactant gases among various active cells in the stack, usually first few cells near the manifold inlet have more flow than those towards the end portion of the inlet manifold, as illustrated

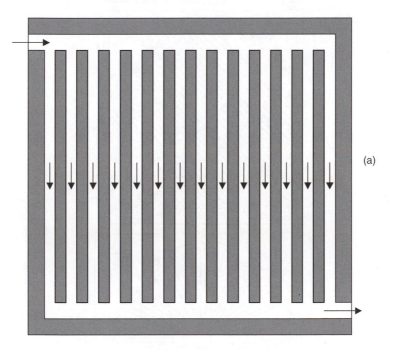

(a)

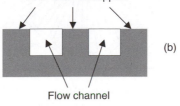

(b)

Figure 7.19 (*Continued*)

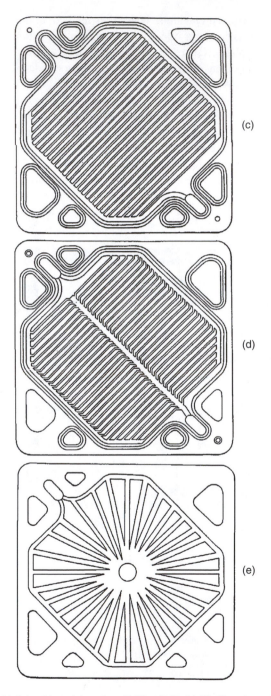

Figure 7.19 Flow field plate with straight and parallel flow fields (a). The flow channel cross section (b) can be square, rectangular, trapezoidal, triangular, semicircular, or any other shape. Trapezoidal cross section is more common. A few variations of the basic flow-field configurations are shown in (c), (d), and (e).

in Figure 7.17. One possible solution is to artificially place some restrictions at the inlet and the exit of these parallel flow channels to increase the pressure drop in the channels and hence improve the flow distribution among the active cells. However, this complicates the design and fabrication, thus, the cost. A further problem with this design is the possible and often arisen non-uniform distribution of a compressive load carried across the fuel cells within the stack when the flow channels on the anode and cathode plates are aligned in parallel to allow for the concurrent and countercurrent flow arrangements for the fuel and oxidant stream. The contact area, as defined by the overlap of the ribs on the anode and the cathode plates, depends on the manufacturing tolerances affecting the width of the ribs, the smoothness of the rib surface, the exact location of the ribs, rib edge machining and assembly alignments (plate to plate), etc. The variation in the contact areas of the ribs results in the variation in the local stress and the associated cell strain. A minimum local stress is necessary to maintain minimum electrical (as well as thermal) contact resistance. Whereas a significantly high local stress may lead to the damage and premature failure of cell components. To ensure uniform compression load across the cell, it is necessary to have even distribution of both parallel and perpendicular contact areas (i.e., crossflow and co-flow arrangements).

In an attempt to tackle the aforementioned problems, Spurrier, et al.[25] and Granata, et al.[26] described a modified serpentine gas flow field across the plate surface, as shown schematically in Figure 7.20. The channels are generally linear and arranged parallel to one another, but skewed to the edge of the plate, while the spaced

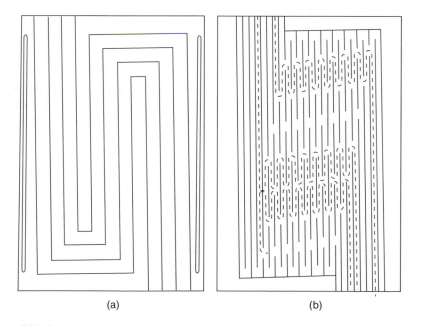

(a)   (b)

Figure 7.20 Two variations of flow-field channel configuration that may improve reactant gas distribution and stack compression load across the cell surface.[25]

slots allow cross-channel flow of the reactant gas in a staggered manner, which creates a multiple of mini-serpentine flow paths transverse to the longitudinal gas flow along the channels. Thus, adjacent pairs of the channels are interconnected by the spaced slots. The flow channels on the anode and cathode plates are skewed in opposite directions in such a manner that exact co-flow arrangement is avoided and some cross flow and some nearly co-flow configuration is achieved. So that it is claimed that this design can improve reactant flow distribution across the electrode surface of the fuel cells and produce a uniform distribution of stack compression loading on each fuel cell within the stack. In reality, this design may incur high reactant-pressure loss with potential formation of stagnant areas due to the cross-channel flow by the spaced slots, shown in Figure 7.20b.

To resolve the problem of water flooding resulting from the inadequate water removal from the cells, Watkins, et al.[18] proposed using a continuous fluid-flow channel that had an inlet at one end and a outlet at the other and typically followed a serpentine path. A schematic diagram is shown in Figure 7.21. Such a single serpentine flow field forces the reactant flow to traverse the entire active area of the corresponding electrode, thereby eliminating areas of stagnant flow. However, this channel layout results in a relatively long reactant flow path, hence a substantial pressure drop and significant concentration gradients from the flow inlet to outlet. In addition, the use of a single channel to collect all of the liquid water produced from the electrode reaction may promote flooding of the single serpentine, especially at high current densities. Hence, for higher current density operations, especially when air is used as the oxidant or with very large gas flow-field plates, Watkins, et al.[19] pointed out that several continuous separate flow channels might be used in order to limit the pressure drop and thus minimize the parasitic power required to pressurize the air, which can be as much as over 30% of the stack power output. This design, shown schematically in Figure 7.22, ensures adequate water removal by the gas flow through the channel, and no stagnant area formation at the cathode surface due to water accumulation. Watkins, et al.[19] reported that under the same experimental conditions, the output power from the cell could be increased by almost 50% with this new type of flow field plates. Although multiple serpentine flow-field designs of this type reduce the reactant pressure drop relative to single serpentine designs, the reactant pressure drop through each of the serpentines remains relatively high due to the relatively long flow path of each serpentine channel, thus the reactant concentration changes significantly from the flow inlet region to the exit region for each active cell.

Although reactant pressure losses through the flow distribution fields increase the parasitic load and the degree of difficulty for hydrogen recirculation, they are actually helpful for the removal of product water in vapor form. Assuming ideal gas behavior, the total reactant gas pressure $P_T = P_{vap} + P_{gas}$, where P_{vap} and P_{gas} are the partial pressure of the water vapor and reactant gas in the reactant gas stream, respectively. Then the molar flow rate of the water vapor and the reactant (either hydrogen or oxygen) is related as

$$\frac{\dot{N}_{vap}}{\dot{N}_{gas}} = \frac{P_{vap}}{P_{gas}} = \frac{P_{vap}}{P_T - P_{vap}} \tag{7.9}$$

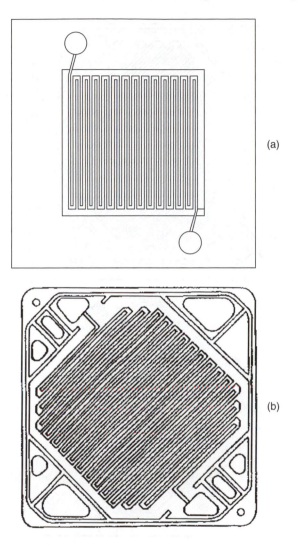

(a)

(b)

Figure 7.21 Flow field configuration with a single serpentine flow channel traversing the entire active area.[18,27]

Hence, the total pressure loss along a flow channel will increase the amount of water vapor that can be carried and taken away by a given amount of the reactant gas flow if the relative humidity is maintained. This approach can be used to enhance water removal by both oxidant and fuel streams. In fact, a sufficient pressure loss in the anode flow channels can even draw water through the membrane from the cathode side, and remove the excess water by the anode stream, so that the fuel cell performance at high current operations can be improved significantly, as demonstrated by Voss, et al.[28,29]

    Equation (7.9) also indicates that an increase in the water-vapor partial pressure can enhance the ability of the reactant gas stream to remove water and water vapor

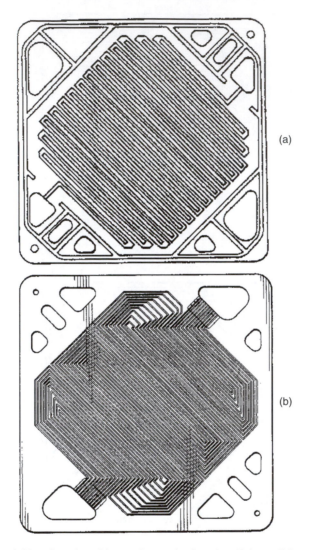

Figure 7.22 Flow field configuration with several serpentine flow channels in parallel traversing the entire active cell.[19]

pressure is limited by the saturation pressure determined by the gas stream temperature. Hence, liquid water can flood the serpentine channels and the electrodes after the cathode gas stream has been saturated. However, if the reactant gas temperature is increased along the flow direction from the inlet to the outlet of the fuel cell, the capacity of the gas stream to absorb water also increases. Fletcher, et al.[30] described a stack arrangement where the coolant flow is substantially parallel to the reactant flow, such that the coolest region of each cooling layer coincides with the inlet region of the adjacent reactant layer where the gas stream has the lowest temperature and water content, and the warmest region of each cooling layer coincides with the outlet

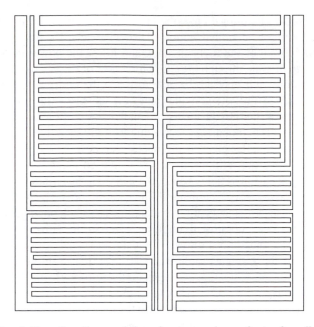

Figure 7.23 Flow field configuration consisting of sectors made up of sets of parallel flow channels connected in series.[31] It is a combination of straight and parallel channel design with the serpentine flow configuration.

region of the adjacent reactant layer where the reactant gas stream has the highest temperature and water content. Thus, the temperature increase along the cooling path is also used to increase the cathode stream temperature, enhancing the cathode stream's capability of absorbing and removing the reaction product water in the vapor form. However, this creates an undesirable nonuniform temperature distribution throughout the cell.

Because of the large pressure drop across the cell, hence the large parasitic power required for air compression, as well as significant uneven distribution of the reactant over the active cell area that are associated with the serpentine flow channel configuration shown earlier, Cavalca, et al.[31] presented a gas flow-field plate design that has ability to distribute the reactants more uniformly over the entire flow field with higher average reactant concentrations, while maintains a low pressure drop for the reactant streams and prevents the formation of stagnant flow areas. The flow field consists of a multiple of symmetric flow sectors having separate inlets and outlets connected to the supply and exhaust manifolds, respectively, as shown in Figure 7.23. The flow field is divided into several sectors and each flow sector includes a number of parallel flow channels, which are further subdivided into a few sets of channels connected in series and the flow channels within each set are arranged in parallel flow relationship. This design is an attempt to combine the advantageous characteristics of the previously mentioned pin, straight, and serpentine designs.

In all the previously mentioned and conventional designs of bipolar plates, typically made of rigid, resin-impregnated graphite, with gas-flow distribution channels have a number of shortcomings. First, the plates must be made sufficiently thick to accommodate the engraved, milled or molded flow channels. The thick plates increase the weight and volume of the fuel cell stack and lower the power density. Second, the reactant has to migrate through the entire thickness of the electrode before reaching the catalyst layer and may not be able to reach the entire catalyst layer if the electrode is too thin. Third, the graphite plates are expensive in raw material costs and the fabrication of the flow channels of small cross sections in close proximity on graphite plates is difficult and expensive due to significant tool wear and process time, resulting in the bipolar plates being one of the most expensive components of a PEMFC stack. Therefore, Wilkinson, et al.[32] disclosed a lightweight MEA design that incorporates the gas flow fields within the electrode material. The bipolar plate, free of any flow passages, can be made thinner and lightweight with less material used and less expense associated with manufacture. The stack power density is hence improved. Because reactants flow through passages in the porous electrode itself, the distance reactants must travel to reach the catalyst layer is shortened. This facilitates the access of the reactants to the catalyst layer, hence improves the fuel cell performance. Furthermore, the flow passages on the porous electrode can be formed relatively easily, thus the cost of a stack can be reduced.

Recognizing the difficulty associated with the conventional methods of gas flow field fabrication on rigid graphite plates, Washington, et al.[33] and Wilkinson, et al.[27] described, respectively, a laminated and an embossed fluid flow field plate for use in PEMFCs. However, other difficulty arises in the fabrication process of these plates and the plates remain significantly thick. The laminated plate contains additional contact resistances that decrease the cell performance. Also the stack assembly process is time-consuming as well to properly align the various layers of the cells and plate components.

Chow, et al.[34] released a bipolar plate design which possesses both reactant gas flow field and cooling flow field on the same plate surfaces, as shown in Figure 7.24. The gas flow field faces directly the electrochemically active area of the adjacent MEA, while the cooling flow field surrounds the gas flow field. This integrated reactant and coolant flow field plate design eliminates the need for a separate cooling layer in a stack, thus significantly improves the stack power density. In the same spirit, Ernst and Mittleman[35] described a fluid flow field plate assembly, which is divided into a multiple of fluid flow subplates, illustrated in Figure 7.25. Each subplate is electrically insulated from all other subplates of the same plate assembly, and has its own reactant flow field adjacent to the electrochemically active area of the nearby MEA. A cooling flow field may be positioned in between and around each of the gas flow subplates. However, these designs cannot maintain a uniform temperature distribution over the entire fuel cell surface.

Conventional bipolar plates are typically made from graphite, which is lightweight, corrosion resistant and electrically conductive in the PEMFC environment. However, graphite is quite brittle making it difficult to handle mechanically and has a relatively low electrical and thermal conductivity compared with metals. It is also quite porous making it virtually impossible to make thin gas impervious plates desirable for low weight, low volume, and low internal-resistance fuel cell stacks. Therefore,

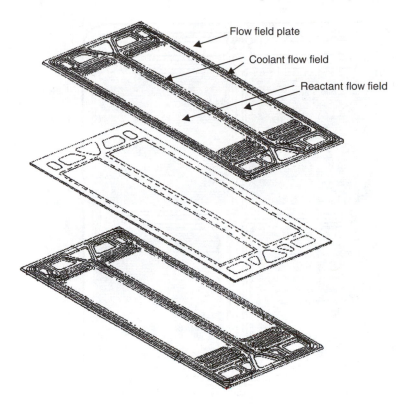

Flow field plate

Coolant flow field

Reactant flow field

Figure 7.24 Both reactant gas flow field and cooling flow field designed and built on the same bipolar plate surface.[34]

proposals have been made to fabricate bipolar plates from metals such as titanium, chromium, stainless steel, niobium, and so on.[36,37] Figure 7.26 illustrates a few possible configurations for the reactant and cooling flow fields. The plates comprise corrosion-resistant thin metal sheets brazed together to provide a cooling flow field between the sheets and reactant gas flow fields on the two outside surfaces of the sheets. Such a bipolar plate design eliminates the need for a separate cooling plate, decreases material usage for stack construction, and reduces the weight and volume of the stack.

For all these designs of flow fields, the flow channels are fabricated on the flow distribution plates (or bipolar plates), or to the lesser degree, on the porous electrode-backing layers and they provide continuous flow passages, from the stack inlet manifold to the exit manifold, while traversing through the electrode surface of the active areas of the cell. In this configuration, as schematically shown in Figure 7.27, the dominant reactant flow is in the direction parallel to the electrode surface, and the reactant flow to the catalyst layer, required for electrochemical reaction and electric power generation, is predominantly by molecular diffusion through the electrode-backing layer. Not only molecular diffusion is a slow process, easily leading to the

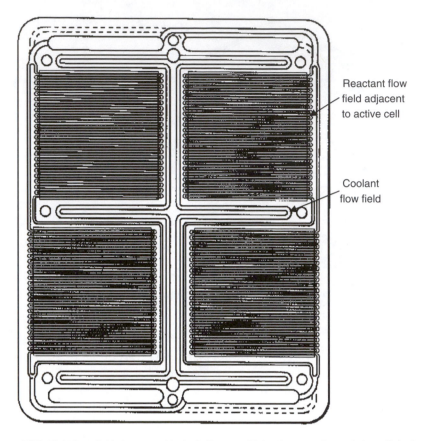

Reactant flow field adjacent to active cell

Coolant flow field

Figure 7.25 Fluid flow-field plate assembly, including a multiple of reactant flow subplates (four shown above) and cooling flow field shown as the plus sign.[35]

occurrence of large concentration gradients across the backing layer and mass transfer limitation phenomenon for the cell operation, but also it is difficult to remove liquid water which exists in the porous region of the backing layer. This difficulty is compounded by the fact that typical flow in the flow channels is laminar due to the small gas velocity and the small flow channel dimensions. Therefore, interdigitated flow fields have been explored to provide convection velocity normal to the electrode surface for better mass transfer and convection flow in the porous backing layer for enhanced water removal capability. An interdigitated flow field, as shown in Figure 7.28, consists of deadended flow channels built on the flow distribution plates. The flow channels are not continuous from the stack inlet manifold to the exit manifold, so that the reactant flow is forced under pressure to go through the porous electrode-backing layer to reach the flow channels connected to the stack exit manifold, thus developing the convection velocity towards the catalyst layer and convection flow in the backing layer itself. Such flow field design can remove water effectively from the electrode structure, preventing water flooding phenomenon and providing enhanced

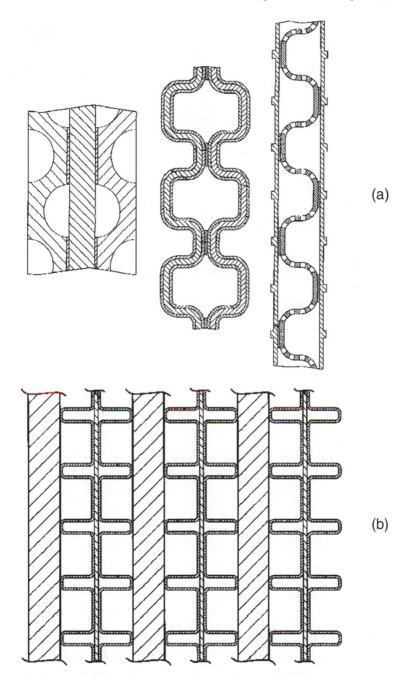

(a)

(b)

Figure 7.26 Cross-sectional view of a few possible configurations for the reactant and cooling flow fields made by metal sheets; (a) from Neutzler[36] and (b) from Vitale.[37]

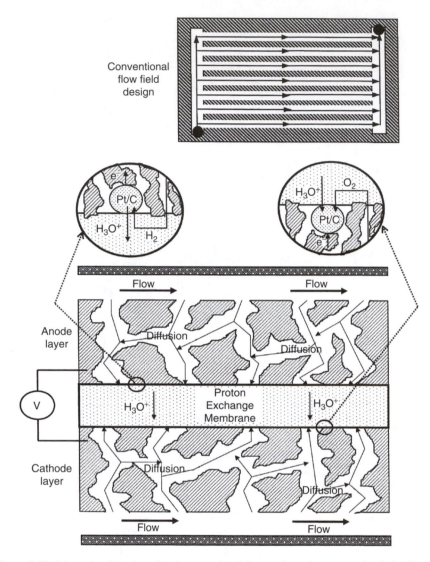

Figure 7.27 Schematic of dominant flow in conventional flow configurations: convection in the direction parallel to the electrode surface and molecular diffusion across the electrode-backing layer (from Wood, et al.[38])

performance at high current density operation. However, a large pressure loss occurs for the reactant gas flow, especially the oxidant air stream. The parasitic power required for air compression may limit the application of this flow field design to smaller stack sizes.

From the previous discussion and description, it becomes clear that the industry effort to develop a commercially viable PEMFC system has resulted in many patents. At this moment, fuel cell designers are continually searching for improved reactant

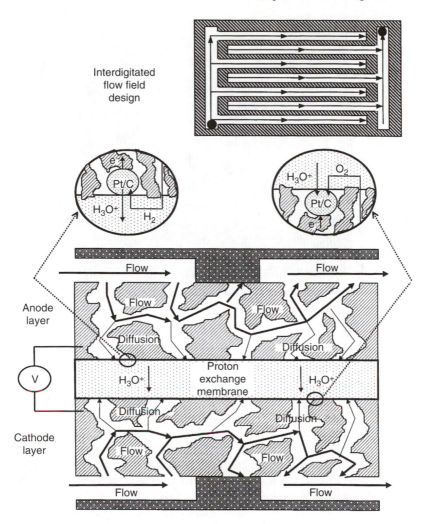

Figure 7.28 Schematic of interdigitated flow field configuration and convection flow through the porous electrode backing layer (from Wood, et al.[38])

gas flow fields and bipolar plate designs in order to lower the cost and optimize the performance of PEMFCs (i.e., higher efficiency and power density on both volume and weight basis). The design and optimization of gas flow fields and bipolar plates remain a significant focus of technology development for practical PEM fuel cell stacks, one of the latest efforts is illustrated by Oko and Kralick[39] and Nelson,[40] as given in Figure 7.29. It seems that most of flow field designs remain in the form of a few parallel serpentine channels of various possible layouts.

It should be pointed out that the aforementioned complex reactant flow-field configurations are elaborated to address the uniform reactant distribution over the active cell surface and product water removal as an important part of water management,

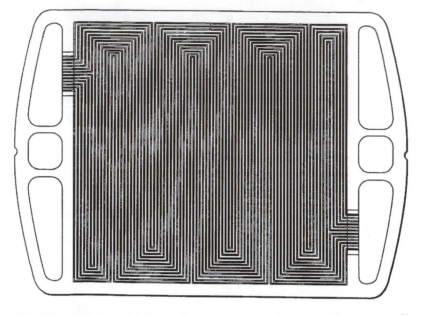

Figure 7.29 Schematic of one of the latest efforts on the design of flow fields and bipolar plates.[39]

which is one of the two critical issues in the PEM fuel cell stack design and operation. The other critical issue, thermal management for the stack, is accomplished mainly by cooling flow field plates strategically located periodically throughout the stack. The design and configuration of the cooling flow fields are substantially the same as the reactant flow field described in this section, and one latest example of the cooling flow-field design is illustrated in Figure 7.30. Also the cooling flow can be arranged in many ways relative to the reactant flows, for example, in concurrent flow with the oxidant flow so that the oxidant stream is gradually heated by the coolant which increases its temperature as heat is absorbed along the flow direction. Then more water can be removed in vapor form by the oxidant stream, as the increase in the oxidant temperature enhances the rate of reaction, more than compensating the effect of oxidant depletion along the flow direction. Thus the cell performance is improved. On the other hand, it is preferable to keep the coolant and the fuel stream in a countercurrent arrangement, so that the fuel stream is gradually cooled along its flow direction. This will gradually lower the water saturation pressure, and hence maintain high relative humidity of the fuel stream, preventing the anode side of the membrane from dehydration, despite water depletion caused by the water migration from the anode side to the cathode side due to the electroosmotic drag effect. Another approach is to introduce the relative low temperature coolant into the cooling cell near the cell periphery where the cell sealant is located, in order to prevent the sealant from overheating. Overheating of the sealant can significantly reduce its lifetime and sealant breakdown basically spells the end of the cell or stack lifetime. Hence, this arrangement improves the reliability and the lifetime of the entire cell or stack.

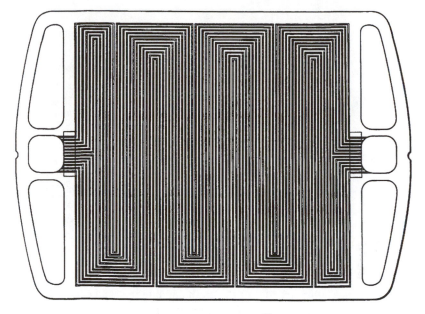

Figure 7.30 Schematic of exemplary cooling flow field design.[39]

7.3.4   A PEM Fuel Cell System

A PEM fuel cell system usually has at least one or multiple PEM fuel cell stacks for the generation of DC electric power. If a number of stacks are employed, they may be connected electrically either in series or in parallel, or a combination of them. Series connection provides higher voltage output while the electric current is relatively low, leading to lower ohmic losses both inside and outside the stacks and among all the electrical connections. However, a series connection can result in a system breakdown if any of the cell or stack components break down, thereby increasing maintenance requirements and potentially limiting the reliability and lifetime of the entire system. On the other hand, if all the stacks are connected electrically in parallel, the system can still provide partial power if any one of the stacks breaks down, thus providing time for response and avoiding potential accidents or even catastrophic events. But a parallel connection among the stacks yields a relatively high electric current for the system, thus increasing the ohmic losses. A specific design decision usually has to be made based on the particular application for the system that is being designed for, the total electrical power, voltage and current requirement for the application, and the whole integrated system including the fuel cell power system and the power consuming devices as well as their interconnection and any control and monitoring devices.

In addition to the stack(s), a PEM fuel cell system includes a fuel and oxidant supply and conditioning subsystem, thermal and water management, control and monitoring subsystem, and possibly a DC-to-AC inverter, which may be avoided depending on the application for which whether DC or AC power is needed. Figure 7.31 shows the schematic of a PEM fuel cell system. It is noticed that if multiple fuel cell

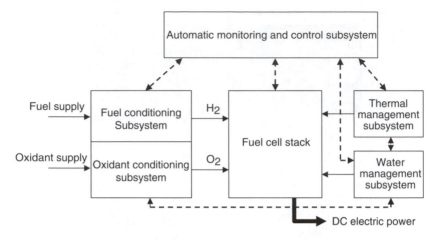

Figure 7.31 Schematic of a PEM fuel cell system with its major components.

stacks are used, then the reactant and coolant flow through the stacks can also be connected by pipelines either in series or in parallel among the stacks, or a combination of them. The flow pipelines connected in series result in each stack being operated in successively lower pressure and lower reactant concentrations due to the pressure loss incurred in flowing through, and the reactant consumption in, the successive stack structures. Thus each of the stacks may not produce the same amount of the power as designed for, and the reactant supply pressure needs to be higher, leading to higher parasitic losses associated with the reactant compression. But the reactant flow rate is low for series flow connection. This especially causes problems for the air (oxidant) supply because typical industrial compressors or blowers can provide high flow rate with low pressure increases and conventionally requires specially designed, rather than off-the-shelf, compressors for fuel cell applications, which certainly adds to the cost of the fuel cell power system. On the other hand, a parallel flow arrangement results in low pressure losses through all the stacks, but the total flow rate required from the compressors is much higher, benefits result clearly from using the standard off-the-shelf compression products. However, such a parallel arrangement can lead to significant flow nonuniformity among the stacks, similar to the phenomena shown in Figure 7.17 for flow distribution among cells within a stack. The humidification of such a high flow stream may require further attention, especially for off-stack humidification design. Therefore, some combinations of the series and parallel flow connection among all the stacks may be beneficial, although judicial selection is required for the specific fuel cell system and geometrical configurations of the flow pipe network.

Fuel Conditioning Subsystem   It is simple if pure hydrogen is used as the fuel and the only consideration is water addition for the humidification and temperature control (heating up the fuel stream to the stack operating temperature) since pure hydrogen in storage is normally at a temperature lower than the stack environment. Because of the sensitive dependence of water saturation pressure (or relative humidity) on temperature, heating and humidification process are implemented in a single

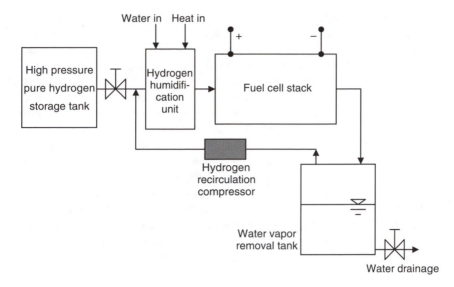

Figure 7.32 Flow loop for pure hydrogen used in a PEM fuel cell system.

synchronized process. Figure 7.32 illustrates the hydrogen flow loop that might be implemented in a practical fuel cell system design. It might be noticed that if the in-stack humidification is employed, the external humidification unit can be removed along with the heat and water addition requirement, greatly simplifying the system design and control. This also reduces the size of the external radiator needed to reject the stack waste heat carried out by the cooling stream, as it is shown later for the thermal management. Otherwise, it is not trivial to provide the provision and control for the addition of heat and water as well as the humidity and temperature of the fuel stream entering the fuel cell stack. Hydrogen exhaust is normally recirculated back to the fuel cell stack.

On the other hand, if natural gas or methanol is used as the primary fuel, as proposed for the stationary residential or mobile transportation applications, it is necessary to convert them into the hydrogen-rich gas mixture first before being fed into the fuel cell stacks for power generation. The conversion process is typically accomplished in a fuel-reforming device through endothermic steam reforming and the fuel stream exhaust is normally not recirculated back to the fuel cell stack to avoid the diluent carbon dioxide and the impurities, especially carbon monoxide, from accumulating in the fuel stream. Instead, the fuel stream exhaust is burned in a catalytic combustor to provide the heat needed for the reforming process, and the details of the various reforming processes are presented in Chapters 6 and 8. For the time being, it is sufficient to note that a sulphur removal stage is necessary before the reformer if natural gas is used as the primary fuel. After the reformer, the gas mixture contains mostly hydrogen gas (up to 70%–75%), a good portion is carbon dioxide (up to 25%), and some carbon monoxide (up to a few percentages). Due to the acid nature of the electrolyte membrane, carbon monoxide decreases greatly the electrochemical activity of the noble metal catalyst used. It therefore becomes imperative to remove the carbon monoxide impurity from the hydrogen-rich gas mixture to a few ppm levels, which imposes significant technical challenges for today's technology. The issue of carbon monoxide poisoning and the mitigation methods is described in further detail later in this chapter, but it suffices to state that usually two stages of carbon monoxide removal are implemented after the reformer, one is at higher temperature and another is at lower temperature. This also requires heating and cooling of the hydrogen-containing gas mixtures and a system integration approach is needed for optimal energy efficiency. Roughly speaking, the fuel conditioning subsystem including the reformer and CO removal devices can amount to a third of the weight, volume, and the cost of an entire PEM fuel cell system; another third is accounted by the PEM fuel cell stacks, and a last third by the rest of accessories. Thus, the nonfuel cell stack components constitute approximately two-thirds majority for the size, weight, and cost of a fuel cell power system; contribution from the fuel cell stacks has been reported as low as 24% by some designs. Understandably, significant technical progress can be obtained by improving the nonfuel cell stack components.

Because of the technical challenges associated with the reforming process and cleaning up of the reforming gas mixture, pure hydrogen is generally intended for the majority of the PEM fuel cell development programs, even though significant efforts are being spent on the processing of hydrocarbon fuels and conversion to the hydrogen-rich gas mixtures. Chapters 6 and 8 provide more information on the hydrocarbon fuel reforming and cleanup processes.

Oxidant Conditioning Subsystem   It is most simple if pure oxygen is used, as for space applications, where oxygen is typically dead-ended instead of circulated during operation. Then, proper pressure regulation to direct the oxygen in the high-pressure storage tank to the operating stacks is needed along with temperature control and proper humidification. For terrestrial applications, it is imperative that air be used instead of pure oxygen. Then atmospheric air needs to be compressed to the stack operating pressure and air compression can consume as much as 30% of the power generated in the stack — a significant, and in fact, the largest single parasitic power loss in the system. Also air temperature from the compressor is usually much higher than the stack operating temperature and heat exchangers are needed to cool the air stream and capture the heat for other purposes, such as heating up the fuel stream to improve the energy efficiency of the system. Clearly thermodynamic analysis is valuable to integrate various components and to optimize the use of the quantity and quality of the energy in the system. Figure 7.33 shows the schematic flow loop for the air stream. Note that air stream exhaust is still at a pressure higher than the atmospheric pressure and a turbine may be used to expand the air stream to the atmospheric pressure and the power thus derived can be used to help, but not enough by itself, to drive the air compressor. Suggestion has been made that some fuel be burned in the stack effluent air stream to increase its energy content, so that the gas turbine has enough power output to drive the compressor without the need of additional electric power from the stack. This would simplify the system monitoring and control subsystem.

Since approximately only 21% of the air is oxygen, plus the sluggish oxygen reduction kinetics, both concentration and activation polarization on the cathode are

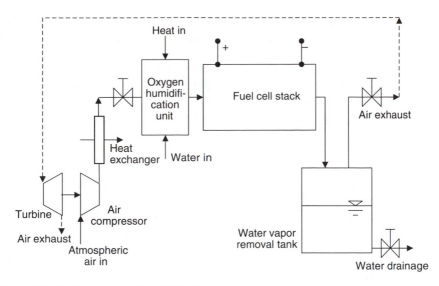

Figure 7.33 Air flow loop used in a PEM fuel cell system.

considerable. An attempt has been made to increase the concentration of oxygen in the air stream in order to enhance the cell and stack performance. Pressure swing absorption technique may be used, but the size, weight, and cost added tend to overshadow the performance gain. Membrane purification has also been tried to enrich the air stream and high purity oxygen gas can be obtained. However, it tends to induce excessive pressure loss across the membrane and the rate of purification is low compared to the rate of oxygen supply needed for the stack operation. Further, investigation suggests that good performance improvement can be obtained for oxygen concentration in the oxidant air stream up to about 40%. Further increase of oxygen concentration in the oxidant stream in excess of 40% does not bring about sufficient performance improvement to justify the effort. Replacing air with pure oxygen can increase the power output by about 30% in a well-designed system and it is noted that a stack designed for air can operate on pure oxygen very well, but the reverse is usually not true.

Thermal Management Subsystem   Typically PEM fuel cells are designed to operate at the average cell voltage E of $0.6 \sim 0.7$ V. For pure hydrogen as fuel reacting with oxygen to form product water, the thermal neutral cell voltage $E_{tn} = 1.48$ V if the product water is in the liquid form (HHV), or 1.25 V if the product is in the vapor state (LHV). Therefore, the chemical to electrical energy conversion efficiency is

$$\eta = \frac{E}{E_{tn}} = 48 \sim 56\% \qquad (7.10)$$

based on the LHV. The rest of the chemical energy is converted into sensible thermal energy, or commonly referred to as the waste heat produced in the fuel cells. Therefore, the amount of the waste heat generation is approximately the same as the stack's electrical power output. It should be removed from the cells continuously for steady operation, efficiently for optimal system performance, and effectively for near isothermal temperature distribution across the cells; this is because best fuel cell performance is achieved for isothermal cells, as discussed in Chapter 2.

Depending on the stack size (power output) and the specific applications, waste heat removal from the PEM fuel cell stack can be accomplished in a variety of ways by using a cooling flow, attaching the entire stack to a cooler or fixing extended heat transfer surfaces (fins) to the stack's outside surfaces.[24] However, cooling circulation is most practical for commercial applications and any gas or liquid can be used as a coolant, although air and water are the two common cooling agents. In principle, evaporative cooling is the best to achieve isothermal cell operating condition.[41,42] Since PEM fuel cell stacks operate at temperatures below the normal boiling temperature of water, alternative liquids have to be used, for example, dielectric liquids commonly used for the cooling of electronic devices have boiling temperatures ranging from 30 °C to well over 100 °C, depending on the specific liquids. Therefore they can be used for the cooling of PEM fuel cell stacks. However, these dielectric liquids have smaller thermal capacity than water and they are either fluoro- or chlorocarbons chemically. In case leakage occurs inside the stack, their compatibility with the membrane electrolyte needs to be fully investigated. Dielectric liquids are also either ozone depleting agents or have global warming potentials. But most importantly, the use of dielectric liquids as coolant requires two liquid circulation loops around the stack because water is needed for reactant humidification and is removed from the stack as reaction byproduct. Consequently, common practice is that small stacks are cooled by air, through either natural or forced convection, and large stacks are cooled by circulating water, which is integrated with the stack water management. Also water cooling is usually employed for systems designed for combined heat and power (the so-called co-generation) applications such as those for residential and industrial use. The following describes first cooling load for a stack (i.e., the rate of heat generated within a stack) and then the mode of cooling for different stack sizes.

Cooling Load The rate of heat generation in a single cell has been derived in Chapter 2 as a function of the cell operating condition, the cell current density, and the cell voltage losses. This is convenient for the determination of waste heat generation during the actual cell operation. But for a stack design, it is more useful to express the rate of heat generation in a stack in terms of the stack power rating, P_s, and the average cell voltage, E, which represents the efficiency of the energy conversion in the stack — these are the two most important parameters in the stack design. Recognizing that the cell voltage losses due to both reversible and irreversible mechanisms can be written as $(E_{tn} - E)$, and the stack is composed of m active cells, the stack heat-generation rate becomes:

$$Q_s = m\,JA(E_{tn} - E) \tag{7.11}$$

where Q_s represents the rate of heat generation (W), J is the current density (A/cm²), and A is the active electrode area (cm²). Since the stack power output is $P_s = m\,JAE$, the rate of heat generation becomes

$$Q_s = P_s \left(\frac{E_{tn}}{E} - 1 \right) = P_s \frac{(1 - \eta)}{\eta} \tag{7.12}$$

where the stack energy conversion efficiency η is given in Equation (7.10).

Natural Air Cooling   Let us first consider the case where stack cooling is achieved by combined natural convection of air and thermal radiation on the outer surface of the stack. Then the rate of cooling Q_c is

$$Q_c = h A_0 (T_0 - T_\infty) \tag{7.13}$$

where h is the equivalent heat transfer coefficient combining both the effect of air natural convection and radiation, A_0 is the outer surface of the stack, and T_0 and T_∞ are the temperature of the stack's outer surface and the ambient air, respectively. Denoting P_s''' as the stack volumetric power density (W/m^3), the volume of the stack becomes

$$V_s = \frac{P_s}{P_s'''} \tag{7.14}$$

Typically the stack has a square cross section, with the side length denoted by a, and the stack length is $2a$. Then the stack volume is $V_s = 2a^3$, and the outer surface area of the stack, including the two end and four side surfaces, becomes $A_0 = 2a^2 + 4(a \times 2a) = 10a^2$. Considering Equation (7.14), we have

$$A_0 = 10 \left[\frac{1}{2} \left(\frac{P_s}{P_s'''} \right) \right]^{2/3} \tag{7.15}$$

Now combining the Equations (7.14) and (7.15), we have the steady-state temperature at the outer surface of the stack as

$$T_0 = T_\infty + \frac{\left(2 P_s''' \right)^{2/3}}{10h} \left(\frac{E_{tn}}{E} - 1 \right) P_s^{1/3} \tag{7.16}$$

Equation (7.16) indicates that the stack surface temperature increases with the stack power rating. For small stacks of low power ratings, average cell voltage E between 0.6 and 0.7 V is a reasonable design target. The product water is normally all vaporized in vapor form, hence $E_{tn} = 1.25$ V and power density $P_s''' = 100$ W/ℓ or 100×10^3 W/m^3 is a reasonable assumption. The natural convection heat transfer coefficient for air is approximately 5~10 W/(m$^2 \cdot$°C), and the thermal radiation heat-transfer coefficient is roughly the same as for the air natural convection. Therefore, $h = 10$ W/(m$^2 \cdot$°C) would be reasonable for the combined natural air-convection and radiation heat-transfer coefficient. Figure 7.34 shows the stack surface temperature as

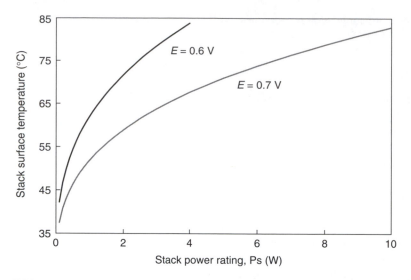

Figure 7.34 Stack outer surface temperature as a function of stack power rating at the steady state for heat loss from the stack outer surfaces according to Equation (7.16).

a function of stack power rating as calculated from Equation (7.16) with the ambient air temperature of 25 °C. Clearly, the stack outer surface temperature increases quickly for low stack-power ratings, and then the increase becomes more gradual. It reaches about 80 °C at the stack power rating of about 4 W and 10 W for the average cell voltage of 0.6 V and 0.7 V, respectively. Due to heat generation inside the stack, the stack interior temperature is slightly higher because the stack volume is quite small for the power ratings up to a few watts. Therefore, heat transfer from the outside surface due to natural air convection and radiation is sufficient for the cooling of stacks up to a few watts without any specially designed cooling mechanism.

For PEM fuel cell stacks intended to operate on the ambient air, the active cells in the stack may be arranged in the form of fins, imitating the compact heat exchange design. An example of such is disclosed by Fletcher, et al.[43] Active cells are not connected by bipolar plates, rather they are connected in series with edge current collection and the cells do not physically contact each other on its two major surfaces. Instead, active cells are separated from each other with spacing between them. This spacing is for ambient air flow by natural convection, which is aided by the water vapor in the air, formed from the evaporation of reaction product water. This is because water molecular weight is smaller than the dry air. The optimal spacing depends on the cell sizes, especially the vertical dimension. In principle, better heat and mass transfer occur when the boundary layer formed on the adjacent cell surfaces does not meet at the top edge of the cell and the optimal spacing ranges approximately from 2 mm to 10 mm in practice. Under this condition, each cell is essentially isolated thermally, and is cooled separately by the natural air currents, and to a lesser degree, by thermal radiation since now each cell is faced with another cell at the same temperature except for the two edge cells. However, such a design yields very low power density, and has very limited practical applications.

Forced Air Cooling by Cathode Air Stream   For PEM fuel cell stacks of more than a few watts, the air stream in the cathode may be used for cooling as well, and sometimes even excessive amount of the air flow may be implemented to achieve the cooling purpose as long as the product water evaporation can maintain relative humidity of 100% at the air-stream exit from the stack. Otherwise, membrane electrolyte dehydration occurs and the stack does not operate properly. For this case, the cooling and water management need to be considered at the same time.

Let us consider the amount of air flow needed for a stack of given power rating P_s. According to Faraday's law of electrochemistry, the rate of oxygen consumption in terms of moles per second can be expressed as a function of time for each cell:

$$\dot{N}_{O_2} = \frac{I}{n_{O_2} F} = \frac{JA}{n_{O_2} F}; \quad \text{(mol/s)} \tag{7.17}$$

where n_{O_2} is the number of moles of electrons transferred per mole of oxygen consumed and for the cathode oxygen-reduction reaction shown in Equation (7.2), $n_{O_2} = 4$; $F = 96,487$ C/mol. electrons transferred in the reaction. For a stack composed of m active cells, the stack oxygen consumption rate is given by

$$\dot{N}_{O_2} = \frac{mI}{n_{O_2} F} = \frac{m JA}{n_{O_2} F} = \frac{P_s}{n_{O_2} FE}; \quad \text{(mol/s)} \tag{7.18}$$

Then the mass consumption rate for the oxygen within the stack becomes:

$$\dot{m}_{O_2} = W_{O_2} \left(\frac{P_s}{n_{O_2} FE} \right); \quad \text{(kg/s)} \tag{7.19}$$

where the molecular weight of oxygen is $W_{O_2} = 31.999$ or 32 g/mol $= 32 \times 10^{-3}$ kg/mol. Since for commercial terrestrial applications, air is normally used instead of pure oxygen, and there are 4.773 mol of air for every mole of oxygen in the air, as shown in Table 1.2, the rate of air mass consumption corresponding to Equation (7.18) can be written as follows

$$\dot{m}_{air} = 4.773 W_{air} \left(\frac{P_s}{n_{O_2} FE} \right); \quad \text{(kg/s)} \tag{7.20}$$

where $W_{air} = 28.964$ g/mol or 28.964×10^{-3} kg/mol, given in Table 1.2. As described in Chapter 2, more oxygen (or air) is provided to the active cells or stacks than what is necessary for stoichiometric reaction in order to minimize the Nernst losses. The excess amount is expressed in terms of a parameter defined as stoichiometry, S_t. Therefore, the actual rate of air flow provided to the stack becomes

$$\dot{m}_{air,in} = 4.773 S_{t,O_2} W_{air} \left(\frac{P_s}{n_{O_2} FE} \right); \quad \text{(kg/s)} \tag{7.21}$$

Similar derivation process yields the rate of product water formation as well as the rate of hydrogen supply to the stack as follows

$$\dot{m}_w = W_w \left(\frac{P_s}{n_w FE} \right); \quad \text{(kg/s)} \tag{7.22}$$

$$\dot{m}_{H_2,\text{in}} = S_{t,H_2} W_{H_2} \left(\frac{P_s}{n_{H_2} FE} \right); \quad \text{(kg/s)} \tag{7.23}$$

where n_w and n_{H_2} are the number of moles of electrons transferred per mole of water formed and per mole of hydrogen consumed, respectively; and for the cathode oxygen-reduction reaction shown in Equation (7.2), $n_w = 2$; for the hydrogen oxidation reaction shown in Equation (7.1), $n_{H_2} = 2$. Water and hydrogen molecular weight are $W_w = 18.02 \times 10^{-3}$ kg/mol and $W_{H_2} = 2.016 \times 10^{-3}$ kg/mol, respectively. The stoichiometry for hydrogen S_{t,H_2} is typically 1.2.

At the steady state, the energy balance for the heat transfer becomes, assuming the stack's outer surface temperature T_o, as well as the exit temperature of the air and hydrogen streams, is the same as the stack operating temperature T

$$Q_c + \dot{m}_{\text{air,in}} C_{p,\text{air}} (T - T_\infty) + \dot{m}_{H_2,\text{in}} C_{p,H_2} (T - T_\infty) = Q_s \tag{7.24}$$

where the specific heat at constant pressure, $C_{p,\text{air}} = 1{,}004$ J/(kg · °C) for air, and $C_{p,H_2} = 14{,}320$ J/(kg · °C) for hydrogen.

Figure 7.35 shows the stack temperature as a function of stack power rating and various oxygen stoichiometry for the air stream, as calculated from Equation (7.24). The average stack voltage E is taken as 0.65 V and all the water produced as shown

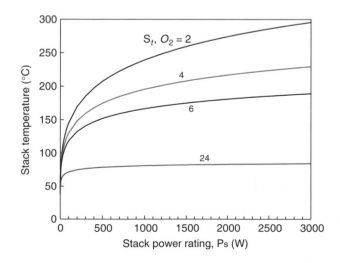

Figure 7.35 Stack temperature as a function of stack power rating at the steady state for various oxygen stoichiometry for the air stream according to Equation (7.24).

in Equation (7.22) is evaporated into vapor form so that the thermal neutral voltage $E_{tn} = 1.25$ V. Clearly, the results shown in the figure suggest that at the practical oxygen stoichiometry of 2–6, the stack temperature would become much higher than what is practically feasible unless the stack size is limited to below 100 W and the outside surfaces of the stack are also exposed to forced air convection. Only at the very high air-flow rate (oxygen stoichiometry of 24) does the stack temperature remain practically reasonable, around 84 °C even at the stack power of 3000 W. For such high air-flow rate, the air stream at the stack exit is severely unsaturated, even all the product water is in the vapor form, indicating severe membrane dry out occurs unless appropriate measure is taken to prevent it. The easiest measure is to separate the cathode-reaction air stream from the cooling air stream and have them flow in their own respective flow channels. That is, the reaction air is supplied to the cathode flow channels and the cooling air is provided to the cooling channels built on the other side of the cathode plate, as described earlier in this chapter. The alternative measure of humidifying the entire reaction and cooling air for air flow in the cathode channels is no longer feasible because of the excessive size of the humidifier and excessive amount of water needed.

Forced Air Cooling by Cooling Air Stream Separate From the Cathode   The need to separate the air stream into the reaction air and cooling air streams can be understood from the following consideration of the relative humidity in the air stream at the stack exit. Since the mass flow rate of the air at the stack exit is the difference between the air flow into the stack given in Equation (7.21) and the amount of oxygen mass consumed in the stack reaction given in Equation (7.19), we have

$$\dot{m}_{air,exit} = \dot{m}_{air,in} - \dot{m}_{O_2} \tag{7.25}$$

Then, based on the definition of the humidity ratio introduced in Equation (1.5) and considering the amount of water vapor present in the incoming air stream, $\dot{m}_{w,in}$, and the amount of water produced in the stack reaction as given in Equation (7.22), the humidity ratio at the stack exit for the air stream becomes

$$\gamma = \frac{\dot{m}_w + \dot{m}_{w,in}}{\dot{m}_{air,exit}} = \frac{\dot{m}_w + \dot{m}_{w,in}}{\dot{m}_{air,in} - \dot{m}_{O_2}} = \gamma_r + \frac{\dot{m}_{w,in}}{\dot{m}_{air,in} - \dot{m}_{O_2}} = \gamma_r + \frac{\gamma_{in}}{1 - \frac{1}{4.773 S_{t,O_2}} \frac{W_{O_2}}{W_{air}}} \tag{7.26}$$

where γ_r and γ_{in} are the humidity ratio due to the product water and the water vapor in the incoming air stream, respectively. If the corresponding water vapor partial pressure is denoted by $P_{w,r}$ and $P_{w,in}$, respectively, and taking into account of the relation between the humidity ratio and water partial pressure, as shown in Equations (1.7) and (1.9), or

$$\gamma = \frac{W_w P_w}{W_{air} P_{air}} = \frac{W_w}{W_{air}} \frac{P_w}{P - P_w}$$

Equation (7.26) can be simplified, after some algebraic manipulation, to the following expression

$$P_w = \frac{\frac{n_{O_2}}{n_w} + \frac{4.773 S_{t,O_2} P_{w,in}}{(P-P_{w,in})}}{\frac{4.773 S_{t,O_2} P}{(P-P_{w,in})} - \frac{W_{O_2}}{W_{air}} + \frac{n_{O_2}}{n_w}} P \qquad (7.27)$$

where P represents the total pressure of the air stream and the total pressure loss from the stack inlet to the outlet has been neglected. Equation (7.27) shows the dependence of the water vapor partial pressure as a function of the air stream flow rate and the total pressure. Since the effect of the water vapor in the inlet air stream is minimal, unless the stack operating temperature is sufficiently close to the inlet air temperature, neglecting $P_{w,in}$, Equation (7.27) becomes

$$P_w = \frac{n_{O_2}/n_w}{4.773 S_{t,O_2} - W_{O_2}/W_{air} + n_{O_2}/n_w} P \qquad (7.28)$$

Substituting the known values of the relevant parameters, Equation (7.28) reduces to

$$P_w = \frac{0.4190}{S_{t,O_2} + 0.1876} P \qquad (7.29)$$

Now corresponding to the stack temperature shown in Figure 7.35 due to the cooling by air flow through the stack's cathode flow channels, the relative humidity at the stack exit can be determined as follows

$$RH_{exit} = \frac{P_w}{P_{w,sat}(T)} \qquad (7.30)$$

where the water partial pressure P_w is calculated from Equation (7.27), and the inlet water partial pressure is obtained from

$$P_{w,in} = RH_{in} P_{w,sat}(T_\infty) \qquad (7.31)$$

The air inlet temperature has been taken as the same as the ambient air temperature T_∞, at which the inlet water saturation pressure is calculated. The water saturation pressure is obtained from Appendix A. Assuming the inlet air relative humidity of $RH_{in} = 70\%$, the air stream relative humidity at the stack exit has been obtained and shown in Figure 7.36 for various air flow rate expressed as the oxygen stoichiometry which is labeled in the figure for the various curves. The solid curves are for the total air stream pressure of 1 atm, and the dashed curves for 3 atm. The calculated relative humidity of over 100% is a direct result of assuming all the product water is

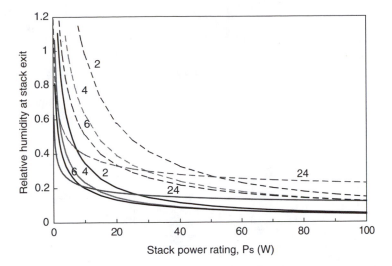

Figure 7.36 Relative humidity at cathode exit as a function of stack power rating at the steady state for various oxygen stoichiometry for the air stream, as labeled in the figure, under the condition of the air stream cooling as shown in Figure 7.35. The results are calculated according to Equations (7.28) and (7.30) assuming all the product water is in the vapor form. The solid curves are for the total pressure of 1 atm and the dashed curves are for the total pressure of 3 atm.

in the vapor state in the calculation, which is physically impossible. In reality, liquid water condenses to form small droplets, which can be carried out of the stack by the oxidant stream if the "theoretical" relative humidity is only slightly above 100%. However, liquid water flooding of the cathode electrode occurs if the theoretical relative humidity is much higher over 100% as for the small stack power sizes shown in Figure 7.36. Clearly, it is also shown that the relative humidity decreases quickly as the stack power rating is increased. Even though the relative humidity at the stack exit increases with the total air stream pressure, the water vapor available is still not sufficient to provide adequate humidification, at the practically reasonable oxygen stoichiometry of 2–6 due to the high stack temperature as shown in Figure 7.35, and for the case of $S_{t,O_2} = 24$ due to excessively high air flow rate, even though stack temperature is reasonable for the PEM stack operation as noticed earlier in Figure 7.35. Therefore, the practical approach is to provide only sufficient amount of air, which can maintain sufficient humidification levels, to the cathode flow channels for the stack reaction and supply the cooling air stream to separate cooling channels for the stack cooling purpose.

As shown in Figure 7.35, air cooling is sufficient for stacks of up to at least a few kW sizes. The high air flow can be provided by fans or blowers and the air pumping required by fans or blowers may consume up to 1% of the stack power output. Such low parasitic-power consumption is quite acceptable, even though the air flow rate may seem to be very high.

Forced Water Cooling   As the stack sizes are further increased, the amount of cooling air flow becomes excessively high and it is known that the pumping power required

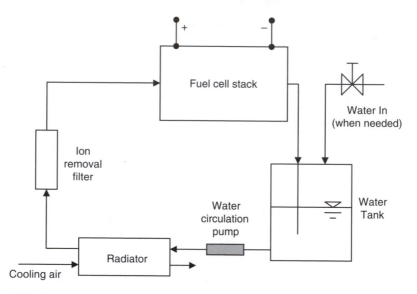

Figure 7.37 Water cooling flow loop used in a PEM fuel cell system.

for air is orders of magnitude higher than for liquid, and the flow channel size for air flow must be much larger than for liquid as well, both due to the orders of magnitude smaller density of air. Therefore, for stacks sizes of over 5 kW or so, liquid water cooling becomes more energy efficient with more compact stack design, while for stack sizes of below 2 kW or so, air cooling is more convenient. Both air and water cooling can be used for stack sizes between 2 and 5 kW and other considerations may decide the eventual choice of the cooling method for specific design and application, for example water cooling is more popular for combined heat and power applications such as for residential or industrial co-generation applications.

Figure 7.37 illustrates the flow loop for cooling water in a large PEM fuel cell system. Because any impurity in the cooling water may contaminate the membrane electrolyte with more details described in the next section, high quality water must be used for the cooling purpose. The required purity should be no less than the water used in typical industrial boilers. In practice, deionized or distilled water should be used as cooling water. Also cooling water may be contaminated by impurities existing in the cooling loop, especially the positively charged metal ions due to the corrosion of the flow piping components; thus an ion-removal filter is necessary in the cooling loop before the coolant entering the stack.

Water Management Subsystem   PEM fuel cells are conceptually very simple and compact, but in practice, consisted of complex structures due to the need to remove reaction product water from the cathode side on one hand, and while at the same time providing adequate humidification to prevent the membrane electrolyte from dehydration, especially at the anode side of the membrane, yet avoiding excessive water from flooding the porous electrode structure. Therefore the management of proper water at the right place with the right amount is critical to the optimal performance of PEM fuel cells.

Water management is clearly composed of two interrelated issues: product water removal and complete hydration of the membrane electrolyte. Product water removal is accomplished often by employing wicks, gravity fed sumps, wet-proofed electrode structures, and oxidant gas-flow stream. In reality, a number of these methods can be combined to their best effect. For example, a wet-proofed electrode is almost always used in the PEM fuel cells for all designs. The porous electrode is made of carbon powders with appropriate amount of hydrophobic agents like polytetrafluoroethylene (PTFE) to avoid water from wetting the electrode surface due to capillary action. Thus water at the cathode catalyst layer tends to flow through such a hydrophobic electrode structure and beads up on the outside surface of the electrode adjacent to the oxidant flow channels. This beaded water can be either wicked off of the electrode surface by a fibrous wick which extends around the edges of the electrode and directs water to a porous ceramic blocks or drained by gravity to a water collection reservoir. These two methods are common when pure oxygen is used as the oxidant, that is, when the oxidant flow rate is low. Therefore, PEM fuel cell stacks of the former design have a fibrous wick for every active cell and the porous ceramic block common to all the active cells for water collection. However for this design, the edge of the active cells is very difficult to seal properly, consequently there was frequent reactant crossover at the cell edges with resultant fires and other hazards. Gravity drainage of water is relatively easy to implement and system vibrations inevitable in some applications, such as motor vehicles, can even enhance the water removal capability. However, it is much too sensitive to the configuration and orientation of the fuel cell system and not precisely controlled and reliable at all times under all circumstances.

Therefore, it is practically more feasible to allow the oxidant stream to carry away the water accumulated at the electrode surface adjacent to the oxidant flow channels. The water can be carried out mostly in the vapor form, and some may be even in the form of small droplets suspended in the oxidant stream at high current-density operations. This technique is often employed for air as the oxidant, that is, at the relatively high oxidant-stream flow rate, and requires a significantly well-structured flow fields, thus leading to many patents on the flow field designs as described in the previous section.

One important issue is the possibility of the product water evaporation, providing the needed air stream humidification without the requirement of the external humidi-fication arrangement. This approach simplifies the water management and hence the system makeup. However, if the air flow rate is low, more water may be present in the cathode, leading to the cathode flooding; if the air stream flow rate is too high, not sufficient amount of water in the cathode results and membrane dry out ensues. As shown earlier in this section, a balance between the air flow and water formation may be accomplished, primarily depending on the air flow rate and the stack operating temperature. Assume some stack control mechanism (i.e., cooling) is implemented to maintain the stack temperature at a desired value, then the theoretical humidity levels at the stack exit can be calculated based on Equation (7.30) by considering all the product water is in the vapor state. Figure 7.38 shows two sets of such calculated

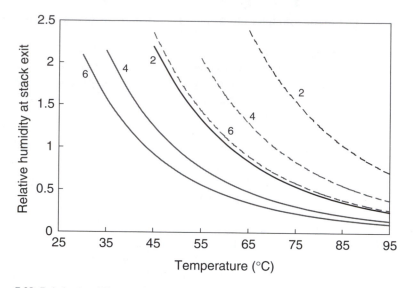

Figure 7.38 Relative humidity at cathode exit as a function of stack operating temperature for various oxygen stoichiometry for the air stream, as labeled in the figure. The results are calculated according to Equations (7.28) and (7.30) assuming all the product water is in the vapor form. The solid and dashed curves are for the total pressure of 1 and 3 atm, respectively.

results for the total pressure of 1 and 3 atm, respectively. Although the relative humidity at the stack inlet is assumed as 70% in the calculation, it has negligible effect on the results except when the stack temperature is very close to the inlet air temperature.

Figure 7.38 indicates that for a given oxidant stream pressure, the relative humidity decreases with the stack operating temperature as well as with the oxidant flow rate. As explained earlier, water flooding occurs in practice for relative humidity much in excess of 100% and membrane dehydrates if the humidity level is too low. Only for a small range of humidity levels around 100% does a PEM fuel cell stack operate satisfactorily. At the total pressure of 1 atm, this self-water sufficiency occurs at the stack temperature of approximately 60, 50, and 43 °C, respectively, corresponding to the oxygen stoichiometry of 2, 4 and 6. For the total pressure of 3 atm, the optimal stack temperature becomes about 85, 70, and 63 °C, respectively, for $S_{t,O_2} = 2$, 4 and 6. Clearly an increase in the operating pressure has increased the optimal stack temperature in order to balance the water production and water evaporation in the oxidant stream. Alternatively, if the operating temperature is fixed, then the water content in the oxidant stream increases almost linearly with the total pressure and the tendency to liquid water flooding is enhanced. This effect is consistent with general operating experience of PEM fuel cell stacks.

These results also suggest that at the total pressure of 3 atm, an operating temperature of 85 °C yield sufficient water to maintain the proper humidification levels for the oxidant stream near the stack exit. Since most of the PEM fuel cells are operated around this temperature in practice, it might suggest that no external humidification is needed. However for large stack sizes of large active cell areas, the oxidant stream has low relative humidity for the majority of the flow paths within the stack, therefore, a significant portion of the membrane may still dry out from the stack inlet onwards, even though full hydration is possible near the stack exit. This usually leads to the formation of local drying and then local hot spots due to the increasing membrane ohmic losses as a result of dehydration. The problem is exacerbated significantly if air is used as the oxidant because of the high air-flow rate needed due to the low oxygen concentration in atmospheric air. Consequently, for large stack sizes both the oxidant and fuel streams are fully humidified with process water before their entry into the active portion of the stack.

The humidification of the fuel and oxidant stream with process water is usually achieved either in an in-stack or off-stack arrangement. The former has been described in the previous section. External humidification can be achieved by employing water columns, reactant gas recirculation, and direct water injection into either the anode or/and cathode reaction compartment or into the reactant streams outside the stack. For water column technique, process water is stored in a container and reactant gas is introduced at the bottom of the container. Then the gas stream is broken down into small bubbles by a series of perforated or porous plates before entering into the water column. The saturated reactant gas stream leaves the container at or near the top of the container. The amount of water that can be picked up by the gas stream and brought into the stack therefore depends on the humidification temperature. Low temperatures result in low water partial pressures, while high temperatures result in low reactant partial pressures. This technique works well for low reactant gas-flow rate, such as pure hydrogen stream. For air at high current density operations, high air flow rate tends to carry with it small water droplets and bring them into the anode electrode backing layer, causing electrode flooding; while air stream itself may be unsaturated due to the short residence time, resulting in membrane drying out locally.

The reactant gas-recirculation design is often implemented when pure hydrogen is used as the fuel, as shown in Figure 7.32. Satisfactory stack operation can be achieved and no external process water is required during the operation when the product water from the oxidant exhaust stream is utilized for the humidification of both the fuel and oxidant streams. However, this method requires an external piece of equipment, often a compressor, to recirculate the hydrogen gas. Since the recirculation compressor is usually powered by the electric energy from the stack, compressor power consumption represents parasitic power loss for the stack. Such a power loss depends on the compression ratio as well as the amount of hydrogen gas that is needed for circulation, which is determined by the hydrogen stoichiometry for the stack operation. Therefore, compression power will change with the stack power output, or hydrogen flow rate. It is most convenient to express the compression power as an equivalent voltage loss, on average, for each cell, ΔE_{H_2}, as

$$\mathcal{P}_{H_2} = m I \Delta E_{H_2} \tag{7.32}$$

where, as defined early, m is the number of the active cells in the stack, and I is the total electrical current output from the stack. On the other hand, assuming hydrogen is an ideal gas and the hydrogen compression process is steady and adiabatic with negligible kinetic and gravitational potential energy changes, energy balance provides the following expression for the hydrogen compression power requirement if all the hydrogen flow into the stack is compressed by the recirculation compressor

$$
\mathcal{P}_{H_2} = \dot{m}_{H_2,in} \frac{C_{P,H_2} T_{out}}{\eta_{comp}} \left[\left(\frac{P_{in}}{P_{out}} \right)^{(k-1)/k} - 1 \right] \tag{7.33}
$$

where C_{P,H_2} is the specific heat at constant pressure for hydrogen (J/(kg · K)), k is the specific heat ratio, T_{out} and P_{out} are the temperature and pressure, respectively, at the fuel cell stack outlet, which are considered as the inlet condition for the hydrogen recirculation compressor. Similarly, P_{in} is the stack inlet pressure, which has been taken as the same as the outlet pressure of the recirculation compressor, which has an isentropic efficiency of η_{comp}. Now substituting Equation (7.23) into Equation (7.33) and combining Equation (7.32) with Equation (7.33) result in the final expression for the equivalent cell-voltage loss due to the hydrogen recirculation compression

$$
\Delta E_{H_2} = \frac{S_{t,H_2} W_{H_2} C_{P,H_2} T_{out}}{n_{H_2} F \eta_{comp}} \left[\left(\frac{P_{in}}{P_{out}} \right)^{(k-1)/k} - 1 \right] \tag{7.34}
$$

As an example, for typical operation of PEM fuel cell stacks, $T_{out} = 80\,°C$, $P_{in} = 3$ atm, $P_{out} = 2.5$ atm, $\eta_{comp} = 0.55$, $C_{P,H_2} = 14{,}320$ J/(kg · K), $k = 1.4$, and $S_{t,H_2} = 1.2$, then Equation (7.34) yields the loss associated with hydrogen recirculation compressor of only about 6 mV, a small voltage loss indeed. However, for the technique of anode water removal to avoid water flooding of the cathode,[28,29] high hydrogen flow rate is necessary, in fact, the optimal hydrogen stoichiometry could range as much as from 5 to over 20. Assume the same condition as what is shown above, the voltage loss now ranges from 25 mV for $S_{t,H_2} = 5$ to 102 mV for $S_{t,H_2} = 20$. Clearly, the voltage loss for hydrogen circulation may not be neglected at high hydrogen flow rates.

Now consider the more realistic hydrogen recirculation loop shown in Figure 7.32, the amount of hydrogen, or the rate of hydrogen flow, that needs to be recirculated is in reality the amount of hydrogen in the fuel exhaust stream, or

$$
\dot{m}_{H_2,exhaust} = \frac{S_{t,H_2} - 1}{S_{t,H_2}} \dot{m}_{H_2,in} \tag{7.35}
$$

Substituting Equation (7.23) into the above equation yields the mass flow rate for the hydrogen in the exhaust stream for the recirculation, and recall $P_s = m J A E = m I E$, we obtain

$$
\dot{m}_{H_2,exhaust} = \left(S_{t,H_2} - 1 \right) W_{H_2} \left(\frac{P_s}{n_{H_2} F E} \right) = \left(S_{t,H_2} - 1 \right) W_{H_2} \left(\frac{m I}{n_{H_2} F} \right) \tag{7.36}
$$

Replacing the rate of hydrogen flow in Equation (7.33) by the above equation, the voltage loss for the case of realistic hydrogen recirculation becomes

$$\Delta E_{H_2} = \frac{(S_{t,H_2} - 1)\, W_{H_2} C_{P,H_2} T_{out}}{n_{H_2} F \eta_{comp}} \left[\left(\frac{P_{in}}{P_{out}} \right)^{(k-1)/k} - 1 \right] \qquad (7.37)$$

For the previous example of $T_{out} = 80\,^{\circ}C$, $P_{in} = 3$ atm, $P_{out} = 2.5$ atm, $\eta_{comp} = 0.55$, $C_{P,H_2} = 14{,}320\,J/(kg{\cdot}K)$, $k = 1.4$, and $S_{t,H_2} = 1.2$, Equation (7.37) indicates the voltage loss associated with hydrogen recirculation amounts to only about 1 mV. This voltage loss is sufficiently small, and in fact, it may be avoided altogether since hydrogen is most likely stored at high pressures, alternative methods such as jet pump mechanism can recirculate this small amount of the hydrogen back to the stack. However, for the technique of anode water removal,[28,29] realized by excessively high hydrogen flow rate of $S_{t,H_2} = 5{\sim}20$ or over, a compressor seems to be the only method for the recirculation. Assume the same condition as what is shown here, the voltage loss now ranges from 20 mV for $S_{t,H_2} = 5$ to 97 mV for $S_{t,H_2} = 20$. Again the voltage loss for hydrogen recirculation may not be neglected at such high hydrogen flow rates.

In contrast, the parasitic power required for the compression of air stream entering the stack, as given in Equation (7.21), yields much larger voltage losses. Similar derivation results in the equivalent average cell voltage loss associated with the air stream pressurization as

$$\Delta E_{O_2} = \frac{4.773\, S_{t,O_2} W_{air} C_{P,air} T_{\infty}}{n_{O_2} F \eta_{comp}} \left[\left(\frac{P_{in}}{P_{\infty}} \right)^{(k-1)/k} - 1 \right] \qquad (7.38)$$

As an example, the ambient air temperature and pressure can be taken as $T_{\infty} = 25\,^{\circ}C$ and $P_{\infty} = 1$ atm, and the stack inlet pressure $P_{in} = 3$ atm, the compressor's isentropic efficiency $\eta_{comp} = 0.55$, $C_{P,air} = 1{,}004\,J/(kg{\cdot}K)$, the specific heat ratio $k = 1.4$, and $S_{t,O_2} = 2$, the voltage loss is about 144 mV according to Equation (7.38), much higher than the losses associated with hydrogen recirculation. The air compression power consumption increases rapidly with the oxygen flow stoichiometry, hence a deterrent to pressurized operation for air as the oxidant.

In the direct water-injection design, water can be injected in vapor (steam) or liquid form and injected directly into the anode and cathode compartment, or to the fuel and oxidant stream before their entry to the stack. Steam injection outside the stack is easy to accomplish, while into the reaction compartment directly requires complex electrode designs, and components needed to achieve this tends to be expensive to construct and difficult to incorporate into a practical stack. Liquid water injection external to the stack can be easily implemented by spray technique, which is simple and efficient to operate and control, and the amount of the water injection can be controlled accurately. Liquid water can also be introduced into the reaction compartment directly, for example, by separate smaller water-distribution channels interspersed between large reactant gas-flow channels built on the same flow distribution plates.[40] For direct liquid water injection, more water can be introduced into the reaction compartment to maintain the membrane hydration than what is possible by just saturating the incoming gas streams. As water in the form of vapor is lost from the anode gas by the net transport of water from the anode to the cathode (electroosmosis minus back diffusion), more water vapor can be generated by the evaporation of liquid water, which simultaneously absorbs waste heat in the cell and is a very effective thermal management system. However direct liquid water injection in PEM fuel cells is difficult to implement in reality because of the extremely narrow range of water content allowed. Too little water can produce local drying of the membrane, while a little too much water results in electrode flooding. The direct liquid water-injection technique may work well with the interdigitated flow-field design described in the previous section.

Other techniques that combine the product water removal at the cathode and process water addition to the anode have been proposed. For example, the impervious cathode and anode flow distribution plates can be replaced by porous hydrophilic plates[23,24] or some variations of them as described by Koncar and Marianowski.[12] With appropriate matching of the pore sizes among the anode electrode, anode flow-distribution plate, cathode electrode and cathode flow-distribution plate, a judicially controlled set of pressure differences between the oxidant stream and the adjacent cooling water stream and between the cooling water and the fuel stream transfers the product water in the cathode to the cooling water stream, then cooling water is transported to the anode for the humidification of the fuel stream. Thus the direction of water transport is from the cathode to the cooling stream, then to the anode by the pressure differentials between the various streams involved, and from the anode to the cathode via electroosmotic drag effect. The liquid water vaporization in the anode also contributes to the cooling of the cells. This approach for the management of water and cell cooling is idealistic and difficult to implement in practice because of the tight quality control requirements of the fine and finer pore sizes for the various stack components, and is also expensive for the manufacture of these stack components of different pore sizes. Precision monitoring and control of pressure differentials adds the cost of the system and operation and maintenance.

On the other hand, one possible advantage may be achieved by introducing liquid water directly into the anode and allowing it to contact the anode membrane interface, which can result in possibly higher membrane hydration and therefore higher conductivity. This is because membranes immersed in liquid water have higher water content than those in contact with water vapor, and more discussion is provided on this topic in the next section.

Hydration of the membrane electrolyte can also be maintained if sufficient catalyst particles are embedded in the membrane layer. Since a small amount of hydrogen and oxygen does dissolve in and diffuse into the membrane interior and the availability of the catalyst particles in the membrane catalyzes the hydrogen–oxygen reaction to yield water, which is right in the membrane and is used to keep the membrane humidified. Such a design may avoid the need for external humidification of the reactant gas stream altogether. However, care should be taken to avoid the catalyst particles from connecting with each other and electrically short circuit the cell. This implies that the catalyst particle concentration should be sufficiently low and the catalyst should be embedded in a thin region adjacent to the anode electrode surface. Then, the hydrogen and oxygen reaction does not produce useful electric energy accompanying the water production; this represents a parasitic power loss and reduces the fuel utilization and cell energy conversion efficiency. Therefore, this technique has not been widely employed for practical PEM fuel cells.

PEM Fuel Cell Systems   From the flow diagrams for the fuel, oxidant, and cooling water presented in Figures 7.32, 7.33, and 7.37, one may easily construct a flow diagram for a PEM fuel cell system using pure hydrogen as the fuel and air as the oxidant, which is pressurized or at the atmospheric pressure supplied by a blower or fan. Figure 7.39 shows a schematic diagram of a PEM fuel cell power system developed by Siemens. The fuel cell module shown includes a PEM fuel cell stack of about 70 active cells and the ancillary devices, such as humidifiers, product water separator from the reactant gas stream as well as other support hardware. The power output from the module is rated 34 kW at 53 V and the individual active cells have a dimension of 400 mm × 400 mm.

Figure 7.40 shows a conceptual design of a 10-kW PEM fuel cell system running on methanol reformed fuel and pressurized air and targeted for transportation applications, developed between 1991 and 1993 for the U.S. Department of Energy.[46] PEM fuel cell stacks are only about 25% of the overall system weight and volume for the chosen system running on steam-reformed methanol on board the vehicle. Also shown are the devices for the reduction of the carbon monoxide concentration in the

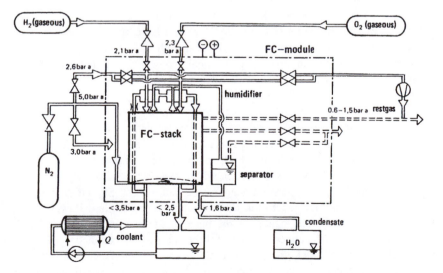

Figure 7.39 Schematic of Siemens fuel cell system running on pure hydrogen and oxygen stored in pressurized containers.[45]

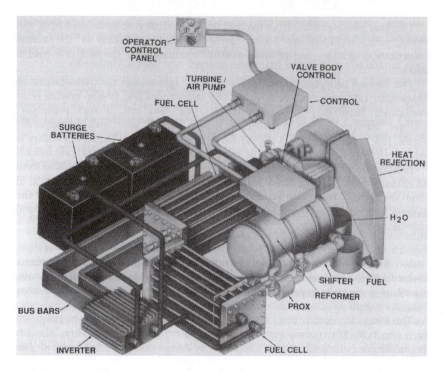

Figure 7.40 Conceptual design of a 10-kW PEM fuel cell system running on methanol-reformed fuel and pressurized air and targeted for transportation applications.[46]

reformed gas mixture, referred to as the shifter and PROX (preferential oxidation) in Figure 7.40. Notice that the shown system is actually a hybrid fuel cell/battery power system and the battery is intended for cold start, for peak power, and for energy storage by regenerative braking. Although the PEM fuel cell stack has, in principle, the capability of providing the peak power, it in general cannot store the energy from the regenerative braking. Hence the need exists for the battery for this purpose.

However, most of the PEM fuel cell based power systems for transportation application that are being developed are based on the use of pure hydrogen as fuel, stored on board the vehicle. The hydrogen is mostly stored in the form of high-pressure gas in fiberglass-wound aluminum cylindrical containers. Figure 7.41 shows a small racecar powered by a PEM fuel cell stack where clearly the stack only accounts for a very small portion of the power system's size and weight. Shown in Figure 7.42 is a schematic of the first-generation urban transit bus by Ballard Power Systems. The 120-kW power system located at the back of the bus was based on 24 PEM fuel cell stacks of 5 kW each, consisting of three stack groups connected electrically in parallel; while each stack group had eight stacks connected electrically in series. Compressed hydrogen storage cylinders were placed under the bus floor. The state-of-the-art Ballard transit bus is now powered by a single 300-kW PEM fuel cell stack based on Ballard's Mark 900 Series technology, as shown in Figure 7.2.

PEM fuel
cell stack

Figure 7.41 A small race car powered by a PEM fuel cell stack running on pure hydrogen and air as demonstrated during the 7th Grove Fuel Cell Symposium held in London, in September 2001.

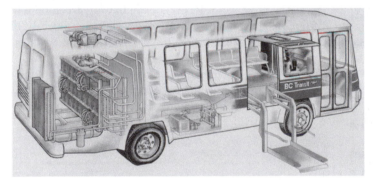

Figure 7.42 Schematic of the first-generation urban transit bus by Ballard Power Systems. The 120-kW power system seen at the back was based on 24 PEM fuel cell stacks of 5-kW each. Compressed hydrogen storage cylinders were placed under the bus floor.

7.4   MATERIALS AND MANUFACTURING TECHNIQUES

As described early in this chapter, just like any other types of fuel cells discussed so far, the heart of a PEM fuel cell power system is the PEM fuel cell stacks, which is composed of membrane electrode assembly (MEA), the bipolar plates for reactant distribution over the cell active surface and the cooling plates for the thermal management of the stacks. The MEA is in turn made of two catalyzed electrodes and membrane electrolyte. As discussed in Section 7.3, the cooling plate design is essentially the same as the bipolar plate, in general, with the same materials and construction methods used, with very similar, if not identical, geometrical configurations. Therefore, this section, is devoted to the components which make up the MEA and the generalized bipolar plates with the understanding that the latter really refers to the conventional bipolar plates and cooling plates, as well.

7.4.1   Electrodes

The electrodes for PEM fuel cells are porous in structure to allow for reactant gas transport through to reach the catalyst layer, the water transfer to the anode side of the membrane and the product water removal from the cathode catalyst layer. Electrodes are also used to collect electrical current generated in the cell. By electrode, we mean electrode-backing layer within this subsection. It is usually made of carbon powder (for good electron conduction and corrosion resistance), mixed with hydrophobic agents such as PTFE, in the form of porous carbon cloth or carbon paper. As described earlier, hydrophobic treating of the electrode structure is necessary to prevent the liquid water flooding of the electrode especially important for large stacks where some of water removed from the cathode is in the liquid state. PTFE in the electrodes (as well as in the catalyst layers discussed later) serves two functions: binding the high surface area carbon particles into a cohesive layer and imparting some hydrophobic property to the layer. The presence of PTFE is essential for structural integrity and product removal from the electrode pore region. However, too much PTFE content lowers the carbon proportion, hence lowers the electrical conductivity of the electrodes. The amount of PTFE content in the electrode structure may vary from 15% to over 40% by weight,[47,48] and the optimal value is somewhere in the middle of this range although the exact optimal amount also depends on many other factors, such as fabrication techniques and operating conditions. Typical electrode such as Vulcan X-72 is available commercially.

7.4.2   Catalysts

Due to the acid nature of the membrane electrolyte and the low temperature operation of the PEM fuel cells, platinum remains the most effective, or the "best" possible, catalyst for the facilitation of both the hydrogen oxidation and oxygen reduction reaction in the PEM fuel cells, although other materials such as ruthenium, palladium as well as nonprecious metals may be used as the catalyst with reduced performance, to be described later in association with the tolerance with impurities in the reactant streams, especially carbon monoxide. Because the slow kinetics of the oxygen reduction reaction (ORR) plays the key role in the performance limitations of the PEM fuel cells when pristine hydrogen is used as the fuel, various measures have been taken to increase the ORR kinetics. The most effective measure is to increase significantly the effective surface area of the platinum catalyst, which has been achieved by using dispersed and small platinum particles and maximizing the platinum particle utilization in the reaction.

In the early PEM fuel cell technology, including recent demonstration projects for the technical feasibility of the PEM fuel cells, pure platinum (or platinum black) has been used as the catalyst. Platinum black has a spectrum of particle sizes with a minimum of approximately 10 nm in diameter, resulting in the limited active surface area per unit mass of the platinum, A_s. Designating the particle size distribution as $f(D)$, which is the function of the particle diameter D, the platinum particle surface area on a per-unit mass basis can be calculated by assuming the platinum particles are spherical

$$A_s = \frac{\text{Surface area of platinum particle}}{\text{Mass of platinum within particle}}$$

$$= \frac{\int f(D)\pi D^2 dD}{\int f(D)\rho_{Pt}\left(\frac{\pi D^3}{6}\right) dD} = \frac{6}{\rho_{Pt}D_{32}} \tag{7.39}$$

where ρ_{Pt} is the density of the platinum black, taken as a constant, and D_{32} represents a volume-to-surface-area mean diameter of all the particles. The active surface area per-unit mass can be determined from a knowledge of the mean diameter D_{32} and the platinum density, and clearly the smaller the particle sizes, the larger the active surface area for a given amount of platinum catalyst. A typical value is 28 m^2/g Pt. This limited degree of platinum dispersion, along with low catalyst utilization in the early PEM fuel cell technology, dictates that high platinum loading be employed for the performance, reliability, and durability of the PEM fuel cells, typically 4 mg Pt/cm^2 in order to achieve satisfactory cathode performance, especially when air is used as the oxidant. For the ease of fabrication, the anode catalyst loading is typically kept the same as the cathode side, even though the fast hydrogen-oxidation reaction kinetics allows the use of lower catalyst loadings. This is acceptable because of the specialized space and military applications as well as demonstration of technical feasibility, where cost is relatively insignificant when compared to other system measures of success such as the performance and the reliability.

With such high loadings of platinum, the catalyst cost is prohibitively high for commercial applications. For example, the cost for platinum as the catalyst can be estimated as follows for vehicular applications: Assume the very optimistic design performance, that is, the PEM fuel cells are operated at $E = 0.6$ V and $J = 500$ mA/cm^2, or the power density of

$$P_s''' = EJ = 0.6 \text{ V} \times 0.5 \text{ A/cm}^2 = 0.3 \text{ W/cm}^2$$

on a per-electrode surface area basis. Since each cell requires a total of $m_{Pt} = 8$ mg Pt/cm^2 (anode and cathode combined) of the catalyst loadings, an automobile may require a minimum of $P_s = 50$ kW for the stack power output (recall the net system power is lower due to parasitic power losses) and the platinum price is taken as

$600/oz.t., the cost of the platinum can then be calculated as follows

$$
\begin{aligned}
\text{Platinum Cost} &= \frac{P_s}{P_s'''} \times m_{Pt} \times \text{Price} \\
&= \frac{50,000 \, W}{0.3 \, W/cm^2} \times 8mgPt/cm^2 \times \$600/oz.t. \times \frac{1 \, oz.t.}{31,103 \, mg} \\
&= \$25,721
\end{aligned}
$$

which is too expensive for platinum catalyst alone; and at least an order of magnitude reduction in the platinum cost should be achieved. This is because the conventional gasoline and diesel engines, with which the fuel cell power system competes in the transportation applications, has a very low capital cost with an often-quoted cost of $30/kW. Consequently, considering other cell components as well as fabrication costs, the platinum catalyst cost of $5/kW has been mentioned as the target. Therefore, the platinum mass loading should be reduced considerably, but in order to maintain the same or similar performance, the platinum surface area over which the electrochemical reaction occurs needs to be maintained. The successful approach so far has been using much smaller platinum particles supported on a much larger carbon particle.

For carbon-supported platinum catalyst, the support-carbon particles are in the size range of a few microns with the maximum diameter of approximately $20 \, \mu m$, the platinum particles as small as 2 nm in diameters are made routinely, in contrast to the minimum particle size of about 10 nm obtainable in platinum blacks. According to Equation (7.39), at least five times more platinum surface areas could be obtained in principle for the carbon-supported platinum than the platinum black. In practice, different amount of platinum can be deposited on the larger carbon particle surface, leading to various percentage of platinum on the carbon particles by weight (or mass). For example, 10% Pt/C catalyst implies that in the platinum-carbon particle mixture, 10% of it is the platinum and 90% is the carbon by weight (or mass). Figure 7.43 shows transmission electron microscopy (TEM) images of a typical carbon support (Vulcan X-72) and of a carbon (Vulcan 72)-supported platinum catalyst, and Table 7.1 shows the practically achieved platinum surface area per-unit mass of platinum in various supported forms.[49] As a result, platinum mass loading can be reduced significantly.

Raistrick[50] was the first to lower platinum loading to as low as 0.4 mg/cm^2 by using the carbon-supported platinum, while obtaining good PEM fuel cell performance. Since then, various improvements have been made, and the platinum loadings as low as 0.05 mg/cm^2 have been reported.[51] However, considering practical fabrication process of MEAs, platinum loading of 0.1–0.2 mg/cm^2 seems to be desirable to maintain a good repeatability and reproducibility of the MEAs.[46]

Normally, the catalyst layer in MEAs consists of a mixture of carbon-supported platinum and PTFE for water rejection purpose and often PTFE is also used as binder as well. Investigations to optimize the PTFE content in the catalyst layer have also been conducted.[52] In general, the optimal PTFE content in the gas diffusion layer can be different from that in the catalyst layer. The use of supported platinum as catalyst and the PTFE for hydrophobic property of the catalyst layer is very similar to those used in alkaline and phosphoric acid fuel cells.

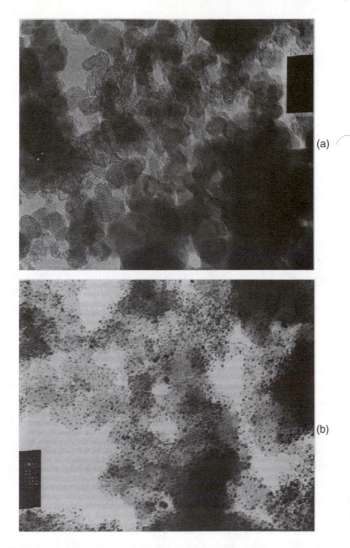

Figure 7.43 Transmission electron microscopy images of (a) a typical carbon support (Vulcan XC-72) and (b) a carbon-supported platinum catalyst (10% Pt/Vulcan XC-72). The platinum particle size is typically in the range of 1.5–2.5 nm.[46]

It might be mentioned that carbon blacks are highly dispersed materials by themselves, composed of practically pure carbon. The structure of the individual particles is characterized by graphitized zones and amorphous interregions. Since the 1920s, large-scale industrial fabrication of carbon black has been by the thermal decomposition of hydrocarbons (natural gas) or aromatic hydrocarbons, although carbon blacks have been used for preparing Indian ink since ancient times. About 95% of the carbon black fabricated in the world, which amounts to 6.1 million tonnes per year,[53] is by the furnace carbon-black process,[54] and the produced carbon black has specific

Table 7.1   Platinum Surface Areas for Various Types of Platinum Catalysts[49]

Type of Catalyst	Surface Area/Pt Mass, A_s (m^2/g)
10% Pt on carbon black	140
20% Pt on carbon black	112
30% Pt on carbon black	88
40% Pt on carbon black	72
60% Pt on carbon black	32
80% Pt on carbon black	11
Platinum black	28

surface area ranging from a few tens up to more than 1500 m^2/g. At present, 90% of the carbon black produced worldwide is used in the rubber industry, mostly for the reinforcement of tires. Carbon black is also used as a filler for conducting polymers as well. It has negligible corrosion in acids and is relatively cheap (on the orders of a few dollars to $10 per kg, of course, the specific price depends on the specific surface area of the carbon-black particles).

Since oxygen reduction kinetics is orders of magnitude slower than the hydrogen oxidation kinetics, the cathode catalyst loading is typically much higher than the anode counterpart. To reduce the cost associated with the overall catalyst loadings for PEM-FCs, alternative cathode catalysts have been sought for, such as some nonprecious metals such as the macrocyclics of the type FeTMPP (iron tetramethylphenylporphyrin) or CoTMPP,[55] which have been quite successfully developed in the past for the cathodic oxygen-reduction reaction in alkaline fuel cells. However, their much lower catalytic activity and limited long-term chemical stability in the acid environment of PEMFCs make them unsuitable for PEMFC applications, especially considering the ultra-low platinum loadings needed for PEMFCs, as described here. On the other hand, some supported platinum-alloys, such as Pt-Ni, Pt-Co and Pt-Cr supported on carbon, exhibit higher cathodic activity per-unit mass of platinum.[56,57] The highest cathodic activity (per mg Pt) is displayed by the PtCr/C alloys, as much as five times higher than that of Pt/C when operating on pure oxygen at 90 °C. This is equivalent to a voltage gain of about 40 mV at constant current density.

7.4.3   Solid Polymer Membrane: The Electrolyte

Introduction   As described in the classification of fuel cells, the characteristics of the PEM fuel cells, including its construction, material selection, and operation, are defined by the electrolyte employed, the so-called proton-conducting (exchange) membrane. It is essentially the sulfonic acid type, solidified in the polymer matrix backbone, hence another name of solid polymer membrane. Several such polymer membranes are available for use as electrolyte in the fuel cells because they meet the general requirements of excellent electronic insulator, excellent ion conductor (typically hydrogen ions or protons), and adequate barrier for the separation of reactants from mixing. This is achieved by their high protonic conductivity and low gas permeabilities. In addition, they are chemically and mechanically stable in the fuel cell environment, which is necessary for long lifetime of PEM fuel cells considering the acidic nature of the membranes themselves and the oxidizing environment to which it is exposed on the cathode side of PEMFCs.

The proton exchange membrane belongs to a class of materials called ionomers or polyelectrolytes, which contain, in their polymeric structure, many ionizable groups or functional groups (e.g., sulphonic acid, SO_3H or in alkali cationic salt forms like SO_3Li, SO_3Na, SO_3K, SO_3Rb, etc.) that dissociate in the presence of water. This dissociation produces two ionic components of these groups: One ionic component is fixed to or retained by the polymeric structure (the so-called fixed ion, such as SO_3^-) and the other is a mobile, replaceable and simple ion (the so-called counter ion, such as H^+, Li^+, Na^+, K^+, Rb^+). The counter ion is electrostatically associated with the fixed ion and can freely exchange with ions of the same sign from the solution (in the presence of water), hence the name of ion–exchange membranes. These membranes are permeable to the simple counter ion, but do not allow the transport of direct flow of liquids, gases and ions of opposite charge. This permselectivity with respect to charge sign is an utmost important characteristic of the membranes for ion transport in an electric field, and hence for the practical use of the membranes.

Depending on whether the fixed ions are negatively or positively charged (or the counter ions are positively or negatively charged), the membrane is referred to as cation or anion–exchange membrane, respectively. Although anion–exchange membranes can result in a number of advantages for fuel cell applications as described in Chapter 1, they have not yet been fabricated until very recently[58]; almost all the past and present membranes are cation–exchange types, that is, the current available polyelectrolytes have cation as the counter ion. In the case of fuel cell application with hydrogen as fuel, the cation is proton (H^+). Therefore, the membrane must be fully hydrated in order to have adequate ion conductivity. As a result, the fuel cell must be operated under conditions where the product water does not evaporate faster than it is produced and the reactant gases, both hydrogen and oxygen, need to be humidified. Therefore, water and thermal management in the membrane become critical for efficient cell performance and are fairly complex and require dynamic control to match the varying operating conditions of the fuel cell. Because of the limitation imposed by its membrane and problems with water balance, the operating temperature of PEMFCs is usually limited to about 100 °C, and typically at 80 °C. This rather low operating temperature requires the use of Noble metals as catalysts in both the anode and cathode side with generally higher catalyst loadings than those used in PAFCs. Because the acid molecules (SO_3^-) are fixed to the polymer and cannot be leached out and the protons on these acid groups are free to migrate through the membrane when fully hydrated, foreign cations in any form would block these active sites and therefore metals are not allowed to come into contact with the membrane, and high purity water, free of metal ions, should be used for humidification purpose in order to avoid the degradation of the membrane's protonic conductivity.

Requirements of High-Performance Membranes   As pointed out earlier, the performance, construction and fabrication of PEM fuel cells depend critically on the membrane used as the electrolyte. Although the general criteria and procedures leading to the selection and fabrication of the membranes with desirable properties are hard to define, the essential requirements defining good performance membranes are known. They include high ionic conductivity, no electronic conductivity, no reactant gas crossover, adequate chemical and thermal stability, and sufficient mechanical strength under operating conditions.

These requirements are mutually conflicting, and the practical membranes of today are a result of compromise satisfying these requirements. For example, high protonic conductivity for low ohmic polarization (or low internal resistance) suggests that the membrane structure be porous and thin; while the ability to separate reactant gases from intermixing requires that the membrane be of low gas permeability (i.e., low porosity and small pore sizes) and high fixed-charge concentration; and sufficient mechanical strength manifests that the membrane be reasonably thick. PEM fuel cell performance in general improves at high temperatures due to higher conductivity and faster reaction kinetics, yet these membranes are generally not stable in hot corrosive and oxidative environments of PEMFC operating condition. Consequently, membranes are usually tailor-made for specific applications, optimized for cell performance and lifetime[59] and the specific fabrication methods and procedures remain proprietary.

Historical Development and State-of-the-Art Membranes   The idea of using an ion-exchange membrane as electrolyte was first introduced by Grubb,[4-6] and significant progress was made in early 1960s by the General Electric Company and others as well.[60-62] The membranes used in early PEMFCs such as those for U.S. Gemini space program employed hydrocarbon-type polymers, including crosslinked polystyrene-divinylbenzene sulfonic acid (PSSA) and sulfonated phenol-formaldehyde. Figure 7.44 illustrates their chemical structures, both linear and crosslinked. The useful lifetime of these membranes were quite limited because hydrocarbon-type membranes are unstable due to C–H bond cleavage, especially at the sites of α-hydrogen atom where the functional groups are attached.[63-65]

However, the lifetime can be significantly extended by replacing all the hydrogen atoms with fluorine atoms — the so-called perfluorination process — because of the much stronger C–F bond. Thus came the critical technical breakthrough in the membrane technology, the development, and synthesis of Nafion[TM 66], which is the registered trademark of E.I. DuPont de Nemours. It exhibits excellent stability over crosslinked PSSA membranes. Since then, improvements have been made for its chemical stability against peroxide degradation and modifications have been made to a basic Nafion homogeneous polymer film for special material properties. For example, open weave Teflon fabric can be laminated into the polymer film for increased strength; multiple layers of different equivalent weights (which is defined in Equation 7.40 later) of polymer film can be laminated together to make composite membranes; even surface treatments can be made to modify the surface layer properties.

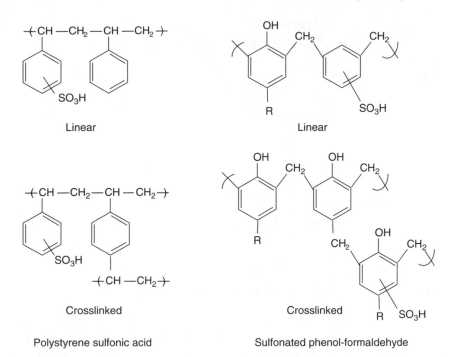

Figure 7.44 The chemical structure of polystyrene sulfonic acid (PSSA) and sulfonated phenol-formaldehyde (linear and crosslinked structure).

In general, Nafion membrane consists of a fluorocarbon-polymer backbone with sidechains terminating in sulfonic acid groups, which are bonded chemically. The fluorocarbon polymer backbone is essentially PTFE, which is strongly hydrophobic and used in fuel cell electrodes for water removal and prevention of electrode water flooding. The strong C–F bonds give rise to the thermal and chemical stability of the material. The sulfonic acid groups at the end of the sidechains are highly hydrophilic and they attract and retain water in the membrane structure — the so-called membrane humidification or hydration. With the presence of water, the proton from the sulfonic acid group becomes mobile and can be transported through the membrane structure in the form of hydronium ions, $H^+(H_2O)_\xi$, where ξ represents the number of water molecules per proton that is being transported through the membrane and it depends on the history of the thermal treatment of the membrane, the degree of membrane hydration, and whether liquid or vapor water is used for the humidification. This is explained with more details later in this section.

Nafion and other perfluorosulfonic acid (PFSA) membranes are used commercially in the chlor-alkali industry, and they represent the state-of-the-art membranes for PEMFCs at present, with Nafion as the industry standard against which other membranes are being compared routinely. All other PFSA membranes have remained at the developmental stage, such as Dow experimental membrane from Dow Chemical U.S.A., (now owned by DuPont) Flemion^{TM} from Asahi Glass Co., Ltd. (Japan),

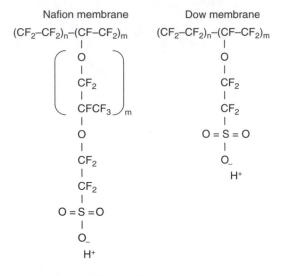

Figure 7.45 The chemical structure of Nafion and Dow membranes (m ≥ 1, n = 6 ∼ 10).

Aciplex™ from Asahi Chemical Industry Co., Ltd. (Japan), and Gore-Select™ from W.L. Gore and Associates (U.S.A.), etc. All these PFSA membranes for PEMFCs have similar PTFE-like polymer backbone, with sidechains ending with the sulfonic acid group. The difference lies in the length of the sidechains, as shown in Figure 7.45. The long sidechain, such as Nafion™, Flemion™ and Aciplex™, has two ether oxygen atoms between the sulfonic acid functional group and the polymer backbone, whereas the short sidechain has only one oxygen atom, such as the Dow membrane. The Gore-Select™ membrane is a microcomposite material, providing higher mechanical strength; thus it can be made much thinner in order to reduce the ohmic losses at the high PEMFC current operations.

These PFSA membranes possess a number of desirable characteristics to enhance the cathodic oxygen-reduction kinetics, which represents the single largest source of energy losses in PEMFCs. They include high solubility of oxygen in the membrane electrolytes (as much as nearly 10-times higher than the solubility of oxygen in water); their mixed hydrophilic/hydrophobic property; and the immobilized (or fixed) anions. Because the rate of oxygen reduction reaction is typically proportional to the oxygen concentration (first-order reaction), higher oxygen solubility reduces the mass transfer limitation and increases the oxygen concentration at the reaction sites, thereby reducing the cathodic activation polarization. The interspersion of the molecular-scale hydrophilic and hydrophobic regions could be beneficial for the different steps of the oxygen-reduction reaction and the immobilization of the anions eliminates (or at least minimizes) the site-blocking effects of absorbed anions at the platinum catalyst surface.

As mentioned early, these polymer membranes are generally highly conductive to protons when fully hydrated (water content is typically about 20%–30% by weight for PEMFC operating environment, but could be as much as 50%). Because the acid molecules are fixed to the polymer backbone and cannot be leached out, the acid concentration of these PFSA membranes is fixed as well and cannot be diluted by product or process water — this is a significant advantage as electrolyte for fuel cell applications as it simplifies greatly the management of electrolyte during the operation and the lifetime of the fuel cell units. Therefore, the membrane protonic conductivity, as a direct result of the acid presence, remains invariable over the fuel cell lifetime.

Ion-Exchange Capacity   The ability for the exchange of ions by these membranes under the fuel cell operating conditions depends on the membrane's acid concentration. The acid concentration of the membrane is closely related to the amount of ionic groups in the membrane, and is conventionally characterized by two important properties: the equivalent weight, EW (or ion–exchange capacity) and the level of hydration of the functional sulfonic acid groups. Equivalent weight is defined as

$$EW = \frac{\text{Dry polymer mass in g}}{\text{Mole of ion−exchange sites (i.e., the fixed } -SO_3^-)}$$
$$= \frac{1000}{\text{Ion−exchange capacity}} \tag{7.40}$$

Equivalent weight has a significant impact on the protonic conductivity of the membrane, the amount of water uptake by the membrane and the thermal stability of the membrane. In general, the higher the equivalent weight, the more stable the membrane becomes the lower the equivalent weight, the higher the protonic conductivity and water uptake. In fact, for sufficiently low equivalent weights, the membrane is no longer in the solid form, rather it becomes aqueous. Membranes with lower equivalent weight and smaller thicknesses yield better cell performance. But thinner membranes increase parasitic losses associated with higher rates of reactant crossover. Since membrane expands in volume when hydrated, membrane hydration lowers the acid concentration of the membrane electrolyte. But more importantly the membrane protonic conductivity is highly dependent on the degree of the membrane hydration, which in turn depends on the thermal history of pretreatment as well as other factors to be explored later in this section.

Nafion and other PFSA membranes were selected as the PEMFC electrolyte because of their desirable properties such as high oxygen solubility, high protonic conductivity (when hydrated), high chemical stability, high mechanical strength, and low density. Commercial Nafion products are denoted by numbers, such as Nafion 117, where the first two digits are the equivalent weight divided by 100 and the last digit is the thickness in mils (or thousandth of inches) of the dry membrane. For years, the standard membrane was Nafion 120 (1200 EW and 10 mils thick), and then followed by Nafion 117 (1100 EW and 7 mils or about 175 μm-thick), which remains the industry standard for the electrolyte in practical PEMFCs, although Nafion 115 (125-μm-thick) and 112 (50-μm-thick) yield higher cell performance due to reduced membrane thickness. The Dow membrane has an equivalent weight of 800, hence it gives much better cell performance, but the lifetime is considerably reduced since it is less stable. The other membranes have an equivalent weight in general between that of Nafion and Dow membranes. In general, for the sulfonate membranes, the EW of greatest interest for practical electrochemical applications ranges from 1100 to 1350, corresponding to the ion–exchange capacity of 0.741–0.909 meq/g.

Structure of Perfluorinated Sulfonic Acid Membranes   Structural properties of these membranes affect directly the transport properties, hence the performance of the membranes. Extensive studies have been devoted to the three-dimensional structure of these membranes.[67,68] A simple structural representation of an ionomeric membrane has been given by Mauritz and Hopfinger,[67] as shown in Figure 7.46. The performance of the membrane depends on the (ionizable) functional groups pendant at the end of the sidechains, which are attached to the organic polymer backbone. The common functional group for fuel cell application is sulfonic acid (SO$_3$H) with the fixed ion of SO$_3^-$, which is hydrophilic and responsible for the swelling of the hydrophobic polymer matrix on exposure to water. The matrix for the conventional ion–exchange membranes is usually made insoluble by chemical crosslinking between the macromolecular chains, however, for the perfluorinated membranes the matrix is not crosslinked; rather it is thermoplastic polymer with pendant acid groups with concentrations less than about 15% by mole, exceeding which the perfluorinated membranes would become soluble.

Ion aggregates form in the perfluorinated membranes and large aggregates, termed clusters, are composed of a nonionic backbone material and many ion pairs. Thus the structure of the membrane may be viewed as consisting of a microphase-separated system having a matrix of low ion content interspersed with ion-rich clusters, as shown in Figure 7.47. As pointed out earlier, the backbone matrix is hydrophobic and the hydrophilic ion clusters uphold water. The ionic aggregates are very stable and act

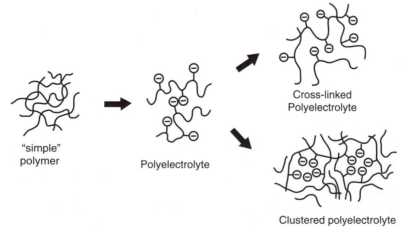

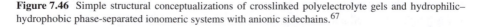

Figure 7.46 Simple structural conceptualizations of crosslinked polyelectrolyte gels and hydrophilic–hydrophobic phase-separated ionomeric systems with anionic sidechains.[67]

as crosslinkage sites; and highly polar solvents, which interact with the ionic groups, do not solvate the polymer. Hence, the ion clusters remain intact once hydrated. The crystalline domains originating from tetrafluoro ethylene (TFE) backbone behave like crosslinked points. Consequently, both the ionic and the nonionic regions act as crosslinkages, making insoluble for the membranes with equivalent weights of 1100 and higher. Nafion membrane has the aggregation of the ionogenic side chains in hydrophilic clusters embedded in the hydrophobic organic phase, hence is insoluble

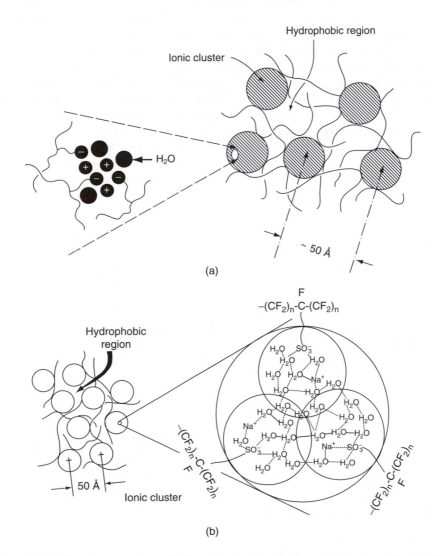

Figure 7.47 Microphase separation of Nafion into a hydrophobic fluorocarbon matrix and hydrophilic ionic cluster regions: (a) The negative ion is the fixed ion (SO_3^-) and the positive ion is the counter ion which should be proton for Nafion for PEMFCs[67]; (b) The counter ion Na^+ should be proton for PEMFCs.[69]

even without the chemical crosslinks. On the other hand, membranes with EW of 970 or lower have a low TFE content, become weak in strength once hydrated and is even soluble.

Various theoretical models have been proposed for the microstructure of the per-fluorinated sulfonic membranes.[67] Figure 7.48 illustrates the molecular organization of the dry and hydrated structures for Nafion, leading to the ion cluster formation, as postulated in the theory. For the PFSA membranes intended for PEM fuel cell applications, the concentration of sulfonic acid moieties is high and two or more microphase-separated regions may exist. In fact, three regions have been suggested as important in defining the polymer transport properties and such a three-region structural model, consistent with various ionic diffusional results and spectroscopic evidence of different water environments in these polymers, is depicted in Figure 7.49.[70] Region A consists of fluorocarbon backbone material, some of which is in a microcrystalline form, and region C is the ion clusters (inverted micelle-like) where the majority of the polar groups (sulfonate exchange sites, counterions, and sorbed water) exist. The interfacial region B is an amorphous hydrophobic region of lower ionic content. The hydrophilic ionic clusters (C) and the interfacial region (B) are responsible for ionic conduction. A network of these ion clusters has been considered interconnected by short and narrow channels in the fluorocarbon backbone network, as shown in

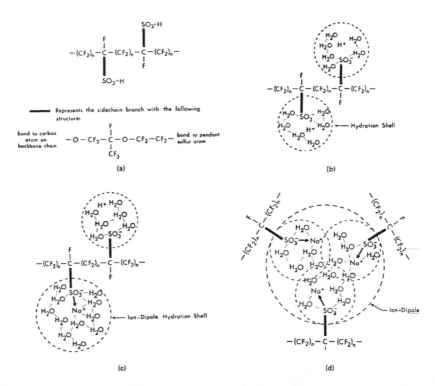

Figure 7.48 Local schematic representations of individual Nafion chain structures. (a) Dry, sulfonic acid form; (b) Wet, sulfonic acid form; (c) Wet, ion-dipole formation; (d) Wet, ion-dipole cluster formation.[67]

Figure 7.50.[73] This structure is illustrated as an inverse micelle in which the polar sulfonic exchange sites surround a spherical water domain from the polymer-rich perfluorophase, and the two clusters is connected by a channel or pore, whose diameter is about 1 nm estimated from hydraulic permeability and water diffusion data. Ion transport through the membrane in an electrical field is considered to occur from one cluster to another through these interconnected intercluster channels.

More sophisticated models have been developed, supported by experimental evidences. More recent work[46] suggested that the ionic clusters in Nafion membranes are most probably spherical with a size distribution and spacing that does not vary much with equivalent weight. The spatially distributed and small ion aggregates could be sufficiently close to coalesce upon swelling once hydrated and thus could provide percolation pathways for ionic transport, instead of the nanoscale intercluster channels described here. It should be noted that it is generally believed that ionic clusters exist in the PFSA membranes, but the details of the microstructures are still under active research.

Two issues need further attention. First, the microstructure of the membrane depends greatly on the thermal history of the material and the hydration levels and two-, three- phase regions or deviations from them may be realized, giving rise to a range of metastable structures with various water content, ionic cluster size, and

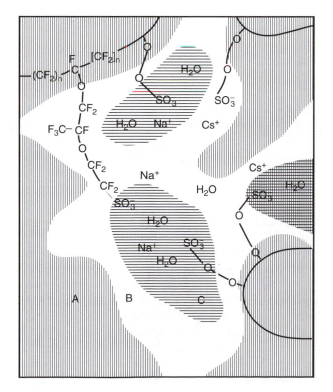

Figure 7.49 (*Continued*).

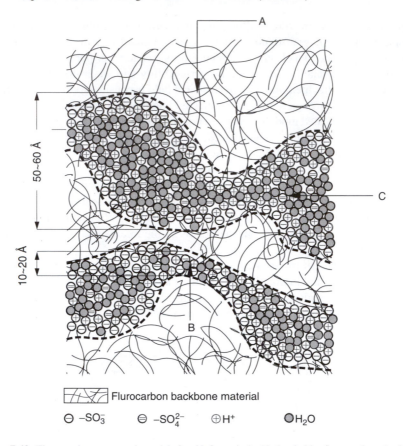

Figure 7.49 Three-region structural model for Nafion. **A** is Hydrophobic fluorocarbon backbone material[70]; **B** is Interfacial zone, an amorphous hydrophoic region of lower ionic content[71,72]; and **C** are Ionic clusters.

cluster connectivity. Consequently, transport properties of the membrane vary as well, such as protonic conductivity of the membrane. Therefore, the normal practice is to have some heat pretreatments for the membrane as received from manufacturer, such as boiling in water and/or dilute acid described in literature. It not only sets up the desired microstructure for the membrane functionality, but also removes impurities that may have contaminated the membrane. Hence, the immersion/boiling treatments are necessary before the membrane is made into the membrane-electrode assembly. To describe properly the performance of the membrane or the PEM fuel cells, it is important to specify the exact pretreatments made for the membrane.

The second issue is the various conditions under which the PFSA membranes can be made soluble, and hence should be avoided during the fabrication and operation of the PEMFCs. For example, the partially microcrystalline fluorocarbon backbone can be melted at high temperatures. Nafions with equivalent weights of 1100 and 1200 can be dissolved in 50:50 propanol-water and 50:50 ethanol-water at 250 °C

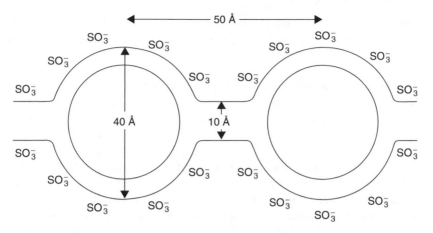

Figure 7.50 Illustration of the cluster-network model applied to Nafion.[73]

and elevated pressures due to the elimination of the crystallites. The ionic membrane can also be chemically converted into the nonionic precursor (sulfonyl fluoride) form, which dissolves under relatively mild conditions.

Water Sorption and Membrane Hydration   Significant variations arise for the amount of water sorption in the PFSA membranes depending on the state of water and the thermal pretreatment history of the membranes, and the amount of water presence determines the transport properties of given membranes. The amount of water uptake by the membranes is conventionally expressed in terms of the ratio of the number of water molecules per ion exchange site in the membrane, λ, as defined in Equation (7.5), which is convenient for fundamental characterization of the membranes. For experimental measurements and practical applications, it is more useful to express the water sorption of the membranes by the ratio of the weight (or mass) of the water absorbed to the weight (or mass) of the membrane in dry state, i.e.,

$$Y_w = \frac{\text{Weight (or mass) of water absorbed by membrane}}{\text{Weight (or mass) of membrane in dry state}} = \frac{m_w}{m_m} \qquad (7.41)$$

It is well known that water uptake (λ or Y_w) can vary greatly when the uptake occurs from the liquid water or water vapor, which has important implications since both situations of liquid and vapor water can be encountered during PEMFC operation. Other important parameters affecting the water uptake include membrane equivalent weights, which represent the number of hydrophilic ion–exchange sites available for water sorption; the type of counter ions present in the membrane, important for contaminated membranes; and the mechanical resistance to the membrane volumetric expansion accompanying the water sorption. In the following, the water uptake from the liquid phase will be presented first, followed by that from the water vapor phase for various PFSA membranes under a variety of conditions. The impact of membrane hydration on the membrane transport properties such as protonic conductivity, water diffusion coefficient and hydraulic permeability are discussed in the next subsection.

Water Sorption From Liquid Phase The amount of liquid water a membrane initially, as received from the manufacturer, can absorb increases considerably with temperature, as shown in Figure 7.51 for Nafion of equivalent weight of 1200 in the sulfonic acid form. Water uptake of as much as 100% of the dry membrane weight can be achieved for Nafion 120 when it is in contact with liquid water at 180 °C. The as-received form membrane is often referred to as the normal (N-form) or standard membrane. At room temperature, N-form Nafion 120 can absorb 16.5 water molecules per sulfonate group when the counter ion is proton. However, the water uptake is reduced substantially when the counter ion is alkali or alkaline-earth metal ion, in accordance with hydrative capacity of the counter ion. For alkali counter ion forms, the water absorption capacity is reduced with increasing size of the counter ion according to $Li^+ > Na^+ > K^+ > Rb^+ > Cs^+$. For example, Li^+-form Nafion at room temperature has a value of $\lambda = 14.3$ water molecules per sulfonic acid molecule and the Cs^+-form membrane only has $\lambda = 6.6$.

Water absorption is usually accompanied with the volume expansion of the membrane. Membranes with high mechanical strength have higher resistance to mechanical deformation, hence resistance to water absorption. Increased valence for the counter ion (i.e., Ca^{2+} in place of H^+) increases the polymer strength and reduces the amount of water uptake and the expansion of a given membrane. This is due to a number of reasons: First, the number of counter ions in the polymer is reduced because of the electroneutrality requirement; second, triplet associations

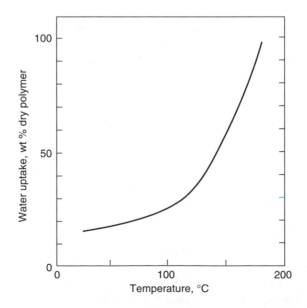

Figure 7.51 Water uptake for N-form Nafion of 1200 equivalent weight as a function of temperature.[68]

such as $-SO_3^- \ldots Ca^{2+} \ldots SO_3^- -$ can form in the membrane, which lead to ionic crosslinks and decrease the ionic hydrative capacities. As a result, the water absorption capacity becomes a weaker function of the counter ion types as the valence is increased. It has been determined that $\lambda = 11.6$ for the Ba^{2+} form and 14.1 for the Zn^{2+} form of the Nafion 120.

The water absorption capacity typically decreases as the equivalent weight is increased (i.e., the concentration of the hydrophilic functional groups is reduced), and a result for the sulfonic acid form of Nafion is shown in Figure 7.52. The dashed curve represents the theoretical predictions based on the ion cluster formation of the microstructure described in the previous subsection. Although the theoretical prediction deviates from the experimental data as the EW is increased the difference is within about $\pm 6\%$ for the range of practical importance (i.e., EW = 1000–1200).

The PFSA membranes also absorb large amounts of other solvents, including acid and basic solutions as well as alcohols and other protic solvents. Figure 7.53 shows the water absorption as a function of NaOH concentration for the sulfonate PFSA membranes of different equivalent weights and Figure 7.54 shows the dependence of the acid electrolyte uptake of the sulfonic acid membrane on the acid concentration for three different acid electrolytes. In general, the solvent absorption decreases as the acid or basic concentration is increased and as the equivalent weight is increased. The latter is expected as explained earlier that higher equivalent weight yields lower concentration of ionic groups per-unit polymer weight.

Having shown that the PFSA membranes can absorb a large number of solvents with different absorption capacity in different solvents with membranes in sulfonic acid or alkaline salt form, hence the importance of keeping the membranes clean and

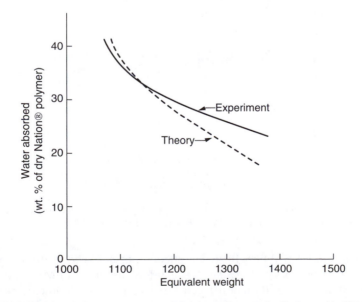

Figure 7.52 Effect of equivalent weight on the water absorption capacity Y_w for standard Nafion in the sulfonic acid form.[67]

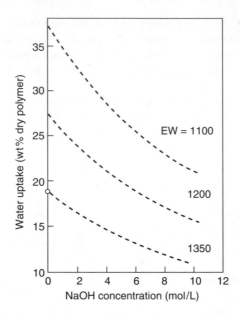

Figure 7.53 Water uptake for perfluorinated sulfonate membranes of different equivalent weights as a function of NaOH concentration (adapted from [68]).

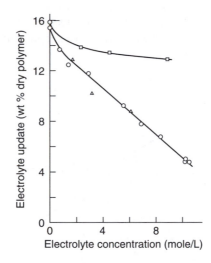

Figure 7.54 Electrolyte uptake for Nafion of equivalent weight of 1200 as a function of electrolyte concentration for three acid electrolytes: HCl (o), HBr ($\triangle$), and H_3PO_4 ($\square$).[68]

uncontaminated, we will now focus our attention exclusively on the sulfonic acid form of the membranes which are the particular form useful for PEMFC applications.

As mentioned earlier, the thermal pretreatment can change the microstructure, and hence the water sorption capacity of the PFSA membranes. As shown in Figure 7.51, the water uptake increases substantially at high temperatures, approximately over

Table 7.2   Effect of thermal Treatment History on Water Uptake by Three Different Membranes from Liquid Water[46]

Membrane	Equivalent Weight	No Thermal Treatment		Dried at 105 °C[a]					
		Hydration Temperature		Redydration Temperature					
		27 °C < T < 94 °C		27 °C		65 °C		80 °C	
		λ	Y_w	λ	Y_w	λ	Y_w	λ	Y_w
Nafion 117	1100	21	0.34	12	0.20	14	0.23	16	0.26
Membrane C[b]	900	21	0.42	11	0.22	15	0.30	15	0.30
Dow Membrane	800	25	0.56	16	0.36	23	0.52	25	0.56

[a] Membrane dehydrated completely at this temperature; incompletely dried membranes behave as in the "No Thermal Treatment" column.
[b] From Chlorine Engineers.

100 °C, the membrane retains the same amount of water at lower temperatures, and even after drying out at room temperature, it will maintain that same amount of water content when in contact with liquid water at temperatures at or below the pretreatment temperature. This state of the membrane is often called the expanded (or E-form) and the memory effect of the pretreatment can be destroyed by drying at elevated temperatures. On the other hand, much-reduced water absorption is obtained if the membrane is heated and dried completely at elevated temperatures; such a membrane is in a shrunken state, often referred to as the S-form. The standard or N-form membrane has less water uptake than the E-form, but has more water absorption capacity than the S-form, that is $\lambda = 22.3$ for the E-form and 16.5 for the N-form has been reported. Table 7.2 shows the effect of the thermal treatment history on the amount of water uptake for three different membranes. It is important to realize that these membranes are particularly sensitive to these thermal pretreatments if they are in the acid form, which is the one used in the PEMFCs. As a result, special attention must be paid to the thermal history of the membrane used in practical PEMFCs.

The physical mechanism leading to the previous different water uptake capacities under different thermal treatments can be attributed to the combined effect of temperature and hydration on the microstructure of the membrane, which changes successively to a rubbery, then a viscous state as temperature is raised and its mechanical strength is weakened dramatically. The membrane is then in the expanded (or swollen) state due to the high water uptake from the liquid into the ionic clusters, if water, or another polar swelling solvent, is absorbed into the membrane from the liquid at this high temperature. Even the membrane is subsequently cooled or partially dried at a lower temperature, the swollen state remains and high water uptake results if the membrane is re-immersed in the liquid water (or other solvent). On the other hand, when the membrane is dried completely at elevated temperatures, the ionic clusters collapse or shrink to smaller sizes, and the shrunk state is maintained when cooled.

Thus, less amount of water is absorbed by the membrane when the shrunk membrane is re-immersed into the liquid water again.

Table 7.2 also shows a substantial increase in water uptake in terms of the weight percentage, Y_w, when the equivalent weight is lowered. Since higher water uptake is usually accompanied with a higher protonic conductivity, hence less ohmic polarization, membranes of low equivalent weights are preferred for PEMFC applications, and the limitation imposed is due to the deteriorating structural stability. It has been reported that Dow membrane with equivalent weight of 597 can even absorb as much as about 550% of water compared to the membrane dry weight, but the membrane is on the verge of dissolution!

Water Sorption From Water Vapor   Water vapor is the primary form of water for the external hydration of the membrane in a practical PEMFC, particularly on the anode side. Membrane hydration or water uptake by the membrane from water vapor depends on the activity of water in the gas mixture and the temperature. Water vapor activity is defined as the ratio of the water-vapor partial pressure to the saturation pressure of water at the gas mixture temperature

$$a_w = \frac{P_w}{P_{\text{sat}}(T)} \tag{7.42}$$

Figure 7.55 shows typical water sorption isotherms for water uptake from the vapor phase at the temperature of 30 °C. At the low vapor activity of $a_w = 0.15$–0.75, the amount of water absorbed by the membranes increases slowly with the water vapor activity and the enthalpy of absorption is large, about 52 kJ/mol, higher than the enthalpy of water condensation at the same temperature, which is about 44 kJ/mol. This high absorption enthalpy is attributed to the fact that water absorption is by the ions in the membrane. On the other hand, at the high water-vapor activity of 0.75–1.0, water uptake is increased substantially with the activity a_w, along with a much smaller enthalpy of sorption, about 21 kJ/mol at $a_w = 1$, which is lower than the enthalpy of water condensation. This behavior results from absorbed water filling the submicropores and swelling the membrane, with the accompanied lower enthalpy of the sorption caused by a number of factors such as a weaker water–ion interaction, endothermic deformation of the polymer matrix on swelling, and a decrease in the degree of hydrogen bonding in the polymer matrix relative to the state of pure liquid water.[46] Figure 7.55b indicates that the water sorption in terms of λ is almost identical for the three different membranes shown, the difference in Y_w shown in Figure 7.55a for the three membranes can hence be entirely attributed to the difference in the equivalent weight of the membranes. One may have noticed that the water vapor activity defined in Equation (7.42) is the same as the relative humidity defined previously, that is the reason of the importance of relative humidity in the reactant gas streams discussed extensively in the previous sections on the reactant stream humidification and the reactant gas distribution flow-field channel designs.

It is also noticed that $\lambda = 14$ shown in Figure 7.55b (water uptake from the vapor phase), whereas λ is over 20 given in Table 7.2 for water uptake from the liquid

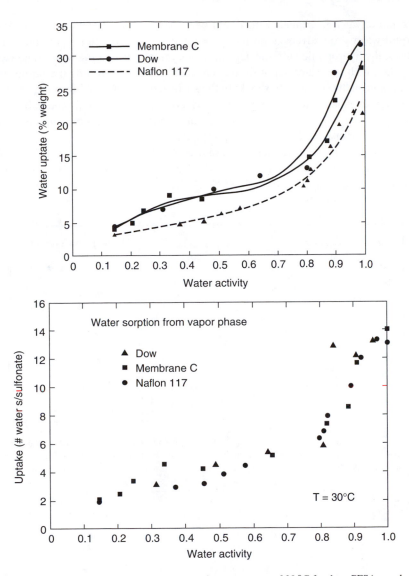

Figure 7.55 Water uptake from the vapor phase at the temperature of 30 °C for three PFSA membranes as a function of water vapor activity: (a) Weight percentage, Y_w, (b) Number of water molecules/sulfonate group, λ.[46]

phase (for no thermal treatment). Clearly, the amount of water absorption from the liquid ($a_w = 1$) and from the saturated water-vapor state ($a_w = 1$) is not the same for a similarly prepared membrane, even though water activity is the same. Similar behavior has been observed for several polymer/solvent systems, and was initially reported by Schroeder in 1903 — hence the phenomenon has been called Schroeder's paradox. Evidence suggests that the membrane surface is strongly hydrophobic Teflon-like when it is in contact with water vapor (whether saturated or not), and it becomes hydrophilic when in contact with liquid water, as illustrated in Figure 7.56. When a liquid water drop advances on the membrane surface, wetting more areas of the surface, the hydrophilic sulfonic acid moieties initially inside the membrane (when in contact with the water vapor on the surface) spring out towards the liquid water now spread on the surface, thus making the surface more hydrophilic and more water uptake occurs when liquid water wets the surface. On the other hand, water absorption from the vapor phase involves water condensation on the hydrophobic surface, leading to less water uptake when from the vapor phase.

The water uptake at higher temperatures has been measured and Figure 7.57 shows the results for Nafion 117 at both 30 °C and 80 °C, the latter corresponds to the typical operating temperature of PEMFCs. Although the amount of water absorption

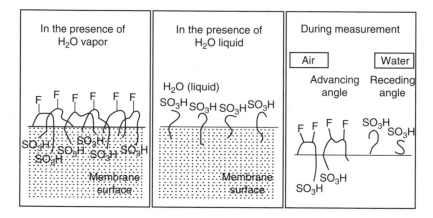

Figure 7.56 Illustration of PFSA-membrane surface morphology when it is in contact with water vapor and liquid water in explaining the difference in the water uptake by the membranes from the liquid and the vapor state of water.[46]

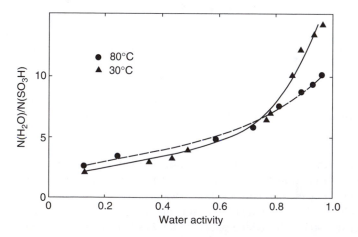

Figure 7.57 Water sorption as a function of water vapor activity for Nafion 117 at the temperature of 30 °C and 80 °C, respectively.[46]

increases slightly with temperature at low water vapor activity, it decreases with increasing temperature above approximately 70 °C at high water activity, especially for the hydration from saturated water vapor. This behavior has also been observed for other PFSA membranes.

In the modelling and analysis of PEMFCs, an empirical correlation based on the adsorption isotherm of Nafion 117 has been often used for the water absorption by the membrane. It provides an allowance for the water activity to exceed unity, recognizing that the maximum hydration of $\lambda = 14$ is obtained for water uptake from the water vapor phase, as shown in Figure 7.55b, but a maximum hydration of $\lambda = 16.8$ is possible in the presence of liquid water[74] for Nafion 117 immersed in water at 80 °C. The hydration λ is assumed to vary linearly from 14 to 16.8 when the mole fraction of water vapor x_w exceeds saturation up to $3x_{w,sat}$, as follows

$$\lambda = \begin{cases} 0.043 + 17.81a - 39.85a^2 + 36.0a^3, & 0 < a = \frac{x_w P}{P_{w,sat}} \leq 1 \\ 14 + 1.4(a-1), & 1 < a = \frac{x_w P}{P_{w,sat}} \leq 3 \end{cases} \quad (7.43)$$

Although the above correlation has been widely used in the modelling of PEM-FCs, it must be cautioned that the temperature dependence of the water uptake, as shown in Figure 7.57, is absent in the empirical correlation. In particular, the maximum hydration of $\lambda = 14$ for water uptake from the saturated water vapor is lowered at the PEMFC operating temperature of about 80 °C, as pointed out earlier.

It should be mentioned that the humidification of the membranes with water vapor is a very slow process; care should be taken in the measurement to ensure that complete hydration is indeed obtained. Also the water uptake is influenced by the thermal pretreatment of the membranes. It should be therefore cautioned when comparing the experimental results in literature.

The mobility of water in the membrane structure depends on the level of membrane hydration. It is not significantly different from that of bulk water at high levels of water uptake, suggesting bulk water like morphology in the membrane. But water motion is increasingly hampered as the water content is decreased. At low levels of membrane hydration, the water in the membrane is absorbed by the ions, hence ion–water and ion–ion interactions constrain the water motion considerably and the liquid in the membrane behaves like a concentrated acid solution. The dynamic orientation, or flipping, of the hydrophilic sidechains on the wall surface of the submicropores in the membrane, similar to those illustrated in Figure 7.56, hinders the water motion as well. Consequently, the transport properties such as protonic conductivity, water diffusion coefficient and electroosmotic drag coefficient are expected to vary with the water content and the submicro- or even nano-structure of the membranes and they are the subject of the next subsection.

Transport Properties of the Membrane   Protonic conductivity is the most important property of the membranes for fuel cell applications as it directly relates to the ohmic polarization and the efficiency of the energy conversion. However, the protonic conductivity of the PFSA membranes is sensitively dependent on the level of membrane hydration, which in turn depends on the various mechanisms of water transport in the membrane under the dynamic operating conditions of PEMFCs, as described in Section 7.2. The instantaneous water distribution in the membrane is determined by the water diffusion, hydraulic permeation and electroosmotic drag effect, hence the transport properties of primary importance in determining the membrane hydration are the water diffusion coefficient, hydraulic permeability, and electroosmotic drag coefficient. In addition, the ability of hydrogen and oxygen gas to permeate through the membrane is important for the electrochemical reaction on the catalyst surface and in separating the reactant gases from intermixing. Hence, either the permeability, or diffusion coefficient, of the hydrogen and oxygen gas in the membrane is another determinant of the membrane's suitability for fuel cell applications. All these properties are described in this subsection.

Protonic Conductivity   The PFSA membranes behave like an insulator in the dry state, and become conductive when hydrated. The Nafion polymer is found to be conductive when exposed to the humid atmosphere and absorbing approximately $\lambda = 6$ H_2O/SO_3^- [68]. Similar to other conducting media, protonic conductivity of the membranes is determined by the product of the density and mobility of the charge carrier (proton). The proton density in the membrane with an equivalent weight of 1100 is equivalent to that in 1-M aqueous sulfuric acid solution, and the proton mobility in a fully hydrated membrane is about one order of magnitude lower than that in the aqueous solution. Therefore, the conductivity of a fully hydrated membrane is at least three-to-four orders of magnitude higher than what is achieved for solvent-free ionically conducting polymers at similar temperatures.[46]

Figure 7.58 shows the protonic conductivity of the PFSA membranes as a function of hydration λ at 30 °C. As the water content is decreased, the protonic conductivity reduces as well for all three membranes shown in Figure 7.58a and the reduction is almost linear until $\lambda \approx 6$, then a steeper drop in conductivity is observed. This may be attributed to the fact that at such low water content, some of the protons may be bound to the sulfonic acid groups due to the electrostatic effect (i.e., some of the sulfonic acid groups may be undissociated), so that proton mobility is reduced drastically. For λ between approximately 6 and 22, the proton mobility is strongly tied to the mobility of water in the membrane, which behaves similarly to the bulk water. Whereas the polymer matrix merely serves as the porous structure hindering the motion of protons and water in terms of volumetric blockage. This suggests that the variation of protonic conductivity in the PFSA membranes with the membrane hydration may follow Bruggeman-type relation:

$$\kappa_m = 0.54\kappa_e(1 - V_P)^{1.5} \tag{7.44}$$

where κ_m is the membrane conductivity, κ_e is the conductivity of an equivalent sulfuric acid solution having the same acid concentration as the sulfonic acid in the membrane, and V_P is the volume fraction of polymer in the hydrated membrane. As the membrane

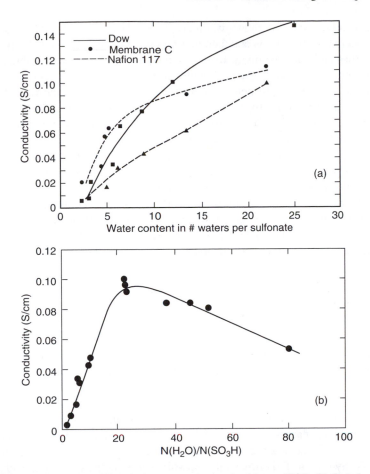

Figure 7.58 Effect of membrane hydration on proton conductivity at 30 °C. (b) Includes data for membrane pretreated in hot glycerol ($\lambda > 22$).[46]

is hydrated, its volume expands, and κ_e decreases slightly. But at the same time, V_P is also decreased, and the strong dependence on V_P makes the membrane conductivity κ_m increase with the hydration level λ. This relation has been verified and can explain the protonic conductivity variations with the hydration very well for the range of λ from about 6 to 22.

In reality, the proton motion in the membrane is different from that in aqueous acid solutions because the proton in the membrane needs to be stabilized by the fixed sulfonic acid group before and after each fundamental jump or hop from one fixed anionic site to another. Up to $\lambda = 22$, a combination of sufficiently high concentration of the sulfonic acid groups and the mobility of sidechains carrying them seems to meet the proton stabilization requirement along the path of the proton motion. Further beyond $\lambda = 22$, the membrane volume expansion lowers the acid concentration and increases the separation distance between the sidechains, the requirement of proton stabilization is now increasingly not satisfied, leading to a gradual reduction in the membrane conductivity as shown in Figure 7.58b. The practical implication is that the maximum conductivity is obtained by immersing the membrane in boiling water (the E-form), and other exotic means to imbibe more water in the membranes turns out to be unnecessary for the fuel cell applications.

Recalling that the equivalent weight of Nafion 117, Membrane C, and Dow is 1100, 900, and 800, respectively, Figure 7.58a also reveals that at high water contents, the membrane conductivity increases as the equivalent weight is reduced. This is expected since the proton (acid group) concentration increases with decreasing equivalent weights. However, for the partial hydration of λ below about 10, the conductivity for the Dow membrane becomes less than the corresponding Membrane C. It indicates that the effect of protonic mobility has overweighed the concentration change of the proton, implying more complexities in the proton-hopping process. The exact mechanism of proton transport in the membrane is still under active research.

An empirical correlation has been developed for the conductivity of Nafion 117 as a function of temperature T (in K) and membrane hydration[74]

$$ \kappa_m = \exp\left[1268\left(\frac{1}{303} - \frac{1}{T}\right)\right] \times (0.005130\lambda - 0.00326), \quad \text{(S/cm)} \qquad (7.45) $$

where the membrane hydration can be determined by Equation (7.43). This equation can be used for Nafion 117 humidified by water vapor phase, as normally encountered in PEMFCs.

Water Diffusion Coefficient   The diffusion coefficient of water in the membrane is required to calculate the water diffusion from the cathode side to the anode side of the membrane, the so-called back-diffusion, and is sensitively dependent on the membrane hydration. Experimentally, self-diffusion coefficient is relatively easily measured by tracking the tracer in a homogeneously hydrated membrane, due to the random molecular motion. A representative of such measured results for the self-diffusion coefficient of water is shown in Figure 7.59 at the temperature of 30 °C and 80 °C. In general, the water self-diffusion coefficient is similar in the PFSA membranes for the same hydration levels, in terms of the magnitude, the trend of variations with the water content and temperature. For Nafion 117 pretreated in glycerol at elevated temperatures, excessively high amount of water can be absorbed, for example, as much as $\lambda = 80$ H_2O/SO_3^- can be achieved if pretreatment temperature is raised to 225 °C, and the corresponding water diffusion coefficient has been measured to be as high as 1.7×10^{-5} cm^2/s for such Nafion subsequently immersed in liquid water at 30 °C, which is only slightly less than the counterpart in liquid water, which is approximately 2.2×10^{-5} cm^2/s at 30 °C. On the other hand, water self-diffusion coefficient is merely about four times lower than the value in bulk liquid water when Nafion membrane is hydrated by saturated water vapor.

Caution has been raised that the self-diffusion coefficient data shown in Figure 7.59 may be too high due to instrumental error, especially at the lowest water contents shown for which the value shown may be too high by as much as a factor

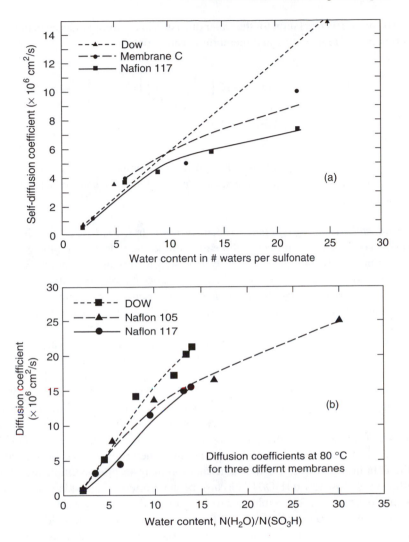

Figure 7.59 Self-diffusion coefficient of water in several PFSA membranes as a function of membrane hydration at (a) 30 °C and (b) 80 °C.[46]

of two.[46] The self-diffusion coefficient is measured when the membrane hydration is uniform and homogeneous for the entire membrane often referred to as the intradiffusion coefficient, which is applicable for the completely hydrated membrane in PEMFC operations. In practice, the membrane during the dynamic operation of PEMFCs may be partially dried out on the anode side and yet still maintain full hydration on the cathode side. In the presence of such water activity gradient, the appropriate coefficient describing the water diffusion through such a membrane is the interdiffusion (or Fickian diffusion) coefficient, which is related to the self-diffusion coefficient given in Figure 7.59.[46]

An empirical correlation for the water diffusion in Nafion 117 has been developed as a function of temperature and membrane water content[74]

$$D_{w-m} = 10^{-6} \exp\left[2416\left(\frac{1}{303} - \frac{1}{T}\right)\right]$$

$$\times \left(2.563 - 0.33\lambda + 0.0264\lambda^2 - 0.000671\lambda^3\right) \quad (7.46)$$

where the diffusion coefficient of water in the membrane, D_{w-m}, is in cm²/s, and the temperature T is in K and the membrane hydration λ can be determined by Equation (7.43).

Electroosmotic Drag Coefficient   Electroosmotic drag coefficient $\xi(\lambda)$, like water diffusion coefficient, is a property of the membrane and it depends on the water content of a uniformly and homogeneously hydrated membrane. In another word, the electroosmotic drag coefficient is defined as the ratio of moles-of-water per mole-of-protons transported through the membrane in the absence of concentration and pressure gradient. For Nafion membranes immersed in liquid water, early measurements indicate that $\xi = (2 - 3)$ H₂O/H⁺ for $15 \le \lambda \le 25$ and then decreases linearly with lower water content and that ξ decreases slightly with decreasing equivalent weight.[75] However, more recent measurements give $\xi = 2.5$ H₂O/H⁺ for fully hydrated Nafion membranes with equivalent weight of 1100 ($\lambda = 22$); and 0.9 for $\lambda = 11$.[76] For the same study, ξ as high as 4 H₂O/H⁺ was measured for fully hydrated membrane C.

For membranes humidified by water vapor, Fuller and Newman[77] reported $\xi = 1.4$ H₂O/H⁺ for $5 \le \lambda \le 14$ and a gradual decrease to zero from $\lambda = 5$ to 0 for Nafion 117. However, a similar measurement by Zawodzinski, et al.[78] yields $\xi = 1.0$ H₂O/H⁺ for $1.4 \le \lambda \le 14$ for the same Nafion 117 membrane. The difference has been attributed to the data-fitting process for the same measurement technique employed. A drag coefficient of close to 1.0 H₂O/H⁺ has also been reported over a wide range of water contents for several PFSA membranes with water uptake from the water vapor.[79] This invariance of the drag coefficient for different membranes at widely different water contents may be interpreted as that a similar proton transport mechanism occurs for all the membranes, such as proton solvation and local water structure.[46]

Hydraulic Permeability   The hydraulic permeability, κ_{hyd}, as defined in Equation (7.7), is associated with the superficial velocity of the water through the membrane, and is often expressed as the product of the hydraulic permeability, κ_P, associated with the water velocity in the pore regions and the volume fraction of water in the membrane, $\epsilon_{w,m}$, that is,

$$\kappa_{hyd} = \kappa_P \epsilon_{w,m} \quad (7.47)$$

where κ_P is often taken as a constant value of 1.8×10^{-14} cm² [80], while the water volume fraction depends on the membrane hydration λ. The following linear relation is often assumed

$$\epsilon_{w,m} = \epsilon_{w,m,\max}\left(\frac{\lambda}{\lambda_{\max}}\right) \qquad (7.48)$$

For Nafion 117, $\lambda_{\max}$ can be as large as 16.8, as discussed earlier, for humidification from the water vapor phase with the presence of liquid water. The maximum water-volume fraction, $\epsilon_{w,m,\max}$, occurs at the full hydration of the membranes and can vary considerably for different membranes.[81] For Nafion 117, $\epsilon_{w,m,\max}$ may be taken as 0.35 based on the measurements of Wakizoe, et al.[81]

Diffusion Coefficients of Hydrogen and Oxygen Gases   The requirement for the transport of reactant gases through the membrane is self-conflicting: on one hand, low diffusion coefficients, hence low rates of the reactant transfer, through the membrane are mandatory to separate the fuel and oxidant gas from mixing in order to avoid the degradation of cell performance and the occurrence of potential hazards; on the other hand, the electrochemical reaction for electric energy generation in the catalyst layers is heterogeneous, occurring at the surface of the catalyst which is surrounded by the membrane electrolyte, as depicted in Figure 7.9 for the idealized structure. A membrane electrolyte covering the catalyst surface is essential for the steady reactions to proceed orderly because it provides a medium for the protons to be transported away avoiding reaction product accumulation in the anode catalyst layer and to transfer the protons in as reactants for the cathode reactions. This creates a significant challenge for the reactant gases (H_2 and O_2) to reach the catalyst surface and high values of hydrogen and oxygen diffusion coefficients are preferred in order to minimize cell performance losses related to mass transfer resistance (or the depletion of the reactants at the reaction sites).

Therefore, an optimization is needed for the practical PEMFC design. For the catalyst layers, too little membrane presence would increase resistance to proton transport, while too much membrane loading would increase resistance to reactant transfer, thus an optimal membrane loading for the catalyst layers exists. There are experimental and modelling efforts in literature to determine this optimal membrane loading under a variety of operating and design conditions, such as [82]. For a given membrane with the fixed diffusion coefficients for the reactants, the thickness of the membrane electrolyte region can be optimized for cell performance between the reactant crossover and the ohmic polarization arising from the resistance to proton transport in the membrane. However, any reactant crossover represents the loss of fuel efficiency; furthermore, the crossover would continue during the offload period unless the reactants are purged by some inert gas like nitrogen. Inert gas purging is easily implemented for stationary generation applications, but is not feasible for transportation applications where vehicles would have to make frequent stops (offload) in urban driving conditions.

Recognizing the importance of the diffusion coefficients for oxygen and hydrogen, experimental measurements have been conducted and semi-empirical correlations have been developed. For fully hydrated Nafion 117, Srinivasan, et al. gave for the oxygen diffusion coefficient[83]

$$D_{O_2-m} = 2.88 \times 10^{-6} \exp\left[2933\left(\frac{1}{313} - \frac{1}{T}\right)\right], \quad cm^2/s \qquad (7.49)$$

Similarly, the diffusion coefficient for hydrogen in the fully hydrated Nafion 117 membrane is available[80]

$$D_{H_2-m} = 4.1 \times 10^{-3} \exp\left(-\frac{2602}{T}\right), \quad cm^2/s \qquad (7.50)$$

In Equations (7.49) and (7.50), the temperature is in Kelvin.

It should be noted that the diffusion of oxygen and hydrogen through the membrane occurs after the oxygen and hydrogen gas have dissolved in the hydrated membrane, the dissolution process represents the resistance to mass transfer at the interface between the membrane and reactant gas. Since the membrane is hydrated with water, the dissolution is similar to gas dissolution in the water, following the Henry's law, since the reactant gas is only weakly soluble

$$C_i = \frac{P_i}{H_i}, \quad mol/cm^3 \qquad (7.51)$$

where C_i is the concentration of the gas species i at the gas–membrane (or gas–liquid) interface on the membrane (or liquid) side, P_i (atm) is the partial pressure of the gas species i at the same interface on the gas side, and the Henry's constant H_i has a unit of atm $\cdot$ cm^3/mol.

For oxygen dissolving in the hydrated Nafion 117 membrane, the Henry's constant is[80]

$$H_{O_2-m} = \exp\left(-\frac{666.0}{T} + 14.1\right) \qquad (7.52)$$

which yields $H_{O_2-m} = 2.015 \times 10^5$ atm $\cdot$ cm^3/mol at the typical PEMFC operating temperature of 80 °C, which is about 50% smaller than the corresponding Henry's constant for hydrogen gas dissolution in the same membrane: $H_{H_2-m} = 4.5 \times 10^5$ atm $\cdot$ cm^3/mol at the same temperature of 80 °C. Note that H_{H_2-m} is a much weaker function of temperature, compared to H_{O_2-m}, hence may be approximated as a constant value within a small temperature range.

Henry's constant for oxygen and hydrogen gas dissolving in the fully hydrated membrane is about one-order-of-magnitude smaller than the corresponding values for oxygen and hydrogen gas dissolving in liquid water, which are[84,85]

$$H_{H_2-w} = 1.43 \times 10^6 \text{ atm} \cdot cm^3/mol, \quad H_{O_2-w} = 1.24 \times 10^6 \text{ atm} \cdot cm^3/mol \quad (7.53)$$

Therefore, oxygen and hydrogen have higher solubility in the Nafion membrane than in liquid water. As mentioned earlier, high solubility in the membrane electrolyte lowers the resistance to mass transfer, enhancing the rate of electrochemical reaction.

Volumetric Expansion and Concentration in a Hydrated Membrane   It is known that the PFSA membranes expand in volume when water is absorbed (or hydrated). As a result of the volume change, the concentration of the fixed charge sites, hence, the acid concentration also changes, exhibiting a dependence on the degree of the membrane hydration, λ. Zawodzinski, et al.[74] measured the membrane swelling for Nafion 117 from a dry state to a fully hydrated condition. They assumed a linear variation of the volumetric expansion with the membrane hydration, then the concentration of the fixed charge sites and water in the hydrated swollen membrane can be expressed as follows

$$C_{fc} = \frac{\rho_m^{\text{dry}}}{\text{EW}} \frac{1}{(1 + \delta\lambda)}, \quad C_w = \lambda C_{fc} = \lambda \frac{\rho_m^{\text{dry}}}{\text{EW}} \frac{1}{(1 + \delta\lambda)} \tag{7.54}$$

where ρ_m^{dry} is the density of the dry membrane, EW is the equivalent weight of the membrane, and the volumetric expansion coefficient δ is often determined experimentally. For Nafion 117, δ has been determined as 0.0126.[74]

In summary, we have focused on the characteristics and transport properties of Nafion, since it is the best or standard membrane available so far, even though it is not the ideal membrane as electrolyte for PEMFCs. The quest for better membranes is continuing, and the properties of other membranes such as Flemion, Gore-select, and so on are available in the literature.[86]

7.4.4   Fabrication of the Membrane Electrode Assembly

The membrane electrode assembly (MEA) forms the basic structure for a single PEM fuel cell and it is the central component for a PEMFC stack because it is the sole place that electric energy is produced in a PEMFC stack, and all other stack components are just put in place to facilitate and harness the continual and steady production of electric energy there. The combined effect of electrochemical reactions, ohmic losses, and the mass transport phenomena there primarily determines the stack performance. In order to tackle the slow oxygen reduction kinetics and reduce the cost of the PEMFC system, highly dispersed platinum-based catalysts are often used with large surface-to-volume (mass) ratio, implying the use of extremely fine nanoscale catalyst particles. During the fabrication process, it is critically important to make sure these expensive catalysts are utilized effectively during the electrochemical reactions, and a number of design and fabrication techniques have been developed in order to achieve the highest catalyst utilization possible. Table 7.3 summarizes the two general techniques developed for the MEA fabrication with some variations, and the techniques are classified based on how catalyst layer is prepared and applied to the electrode backing layer (made of carbon cloth/paper) or the membrane electrolyte first during the process of MEA fabrication. It is important to emphasize that good bonding between electrode and membrane is essential since extremely high contact resistances can arise from any water film and gas layer between the electrode and membrane surface. The following description of the MEA fabrication primarily follows.[46]

Table 7.3   Techniques (or Modes) of Catalyst Layer Preparation and Application[46]

(A) Bonding to the Membrane First		(B) Bonding to the Carbon Cloth/Paper First	
Mode	Application	Mode	Application
A1	Hot-pressed Pt black/PTFE layers	B1	Ionomer-impregnated Pt/C//PTFE
A2	Electroless deposition of Pt on membrane	B2	B1 + sputtered Pt layer
A3	Hot-pressed Pt/C//ionomer layers	B3	Pt catalyst electrodeposited at carbon/ionomer interface

Platinum Black Catalysts Bonded to the Membrane   Bonding platinum black catalysts to the ion–exchange membrane directly, corresponding to the mode A1 shown in Table 7.3, is the oldest practice for MEA fabrication, as described in the original U.S. patent,[87] and is still in practice due to the high reliability of MEAs produced by this technique. Finely comminuted platinum black powders are mixed with a binder first, usually polytetrafluoroethylene (PTFE). The resulting Pt/PTFE mixture forms a film from an emulsion, which is directly applied to the membrane surface to form the active layer. The desired bonding between the active layer and the membrane is achieved under a suitable condition of temperature and pressure, normally referred to as hot pressing. The resulting MEA suffers from two major transport-related problems: A large fraction of catalyst particles, located away from the membrane electrolyte, is not surrounded by the membrane electrolyte due to the limited penetration of the thermoplastic nature of the membrane (less than 4-μm penetration of the total catalyst layer thickness of 20 to 25 μm), thus effective means of providing protons to these catalyst particles does not exist and these catalyst particles are not utilized during the cell operation due to the lack of protons. On the other hand, for those catalyst particles deeply embedded into the membrane, reactant gas is difficult to reach there due to the limited gas solubility and low diffusion coefficient. Hence these catalyst particles are again not utilized during the cell operation due to the lack of reactant gas; the overall platinum catalyst utilization is lower than 5%. Further, the platinum black particles are larger (or low surface area per-unit mass), as described earlier in Section 7.4.2. The combined effect of the low platinum dispersion and utilization results in the high platinum-catalyst loading, typically 4 mg Pt/cm^2. Further development and improvement of this technique can be found in literature.[88]

Electroless deposition techniques have also been developed for the direct bonding of platinum black catalysts to the membrane surface, corresponding to the mode A2 in Table 7.3. One approach is to expose the membrane to an anionic salt of the metal ($PtCl_6^{2-}$) on one side and a reducing agent (e.g., N_2H_4) on the other. The reducing agent diffuses through the membrane, and reacts with the metal salt solution to form a platinum film on the membrane surface.[89] This technique has been modified into a two-step process: First the membrane is charged with Pt cation ($Pt(NH_3)_6^{4+}$) and then one surface of the membrane is exposed to an anionic reductant, $NaBH_4$. The Pt cations in the membrane interior diffuse to the membrane surface and reacts with the reductant to form a platinum film on the surface.[90,91] The platinum particles formed are relatively large, in the range of 10–50 nm in diameter, and are embedded within the first few microns of the membrane surface. Hence, reactant gas, electrons and protons are difficult to reach a large portion of the platinum catalyst particles so embedded during the cell operation.

Carbon-Supported Platinum Catalysts Bonded to the Electrode Backing Layer
For this technique, carbon-supported platinum catalyst particles are mixed with PTFE, Nafion solution, and some organic solvents, the resulting mixture is applied onto the surface of porous carbon paper/cloth. The organic solvents are burned off (or evaporated) in the subsequent heat treatment of the electrode to create hydrophobic pores. The Nafion impregnated electrode, after heat treatment, is then hot pressed onto the membrane to make the MEA. Nafion impregnation of the catalyst layer by recast ionomer guarantees an effective passage for proton transport to or from the active sites. The reactant gas supply to the active sites is provided by the PTFE-treated hydrophobic pores in the catalyst layer and the electron transport to or from the active sites is ensured by the good connectivity within the highly electron-conductive carbon support. Nafion impregnation tackles successfully the problem of the lack of effective proton access to the active sites. Raistrick[50] first recognized the importance of effective proton access to the catalyst surface and pioneered this technique for the MEA fabrication.

The application of the catalyst layer on the carbon paper/cloth can be achieved by paint brushing, spraying, screen printing, and so on. Each method of application requires a different formulation of the catalyst particles, PTFE, and ionomer mixture. For example, the mixture is made paste-like before paint brushing, dilute solution with catalyst particle suspension before spraying, and ink-like solution before screen printing. The exact formulation of the mixture is often held confidential for obvious reason. In some projects of demonstrating PEMFC technology feasibility, platinum black is used as catalyst in order to guarantee the reproductibility, the durability, and the high performance of MEAs. In general, carbon-supported platinum such as 10% wt. Pt/C and 20% wt. Pt/C is used as the catalyst. Supported platinum particles can be made much smaller than the platinum black counterparts; particle diameter of 2 nm is easily obtained in carbon-supported form, as compared with 10–50 nm typical for platinum blacks. Such increased platinum dispersion and the effective access of the protons, electrons, and reactant gas to the active sites provide a PEMFC performance with much reduced platinum loadings, as low as 0.4 mg Pt/cm^2, an order of magnitude lower than the conventional platinum loadings. The catalyst layer thickness is typically 100 μm for 10% Pt/C and 50 μm for 20% Pt/C. These thicknesses for the catalyst layer are substantial as oxygen may only penetrate a fraction of the layer thickness due to the limited rate of mass diffusion, hence the catalyst utilization is still fairly low.[82]

Clearly, optimizing the composition and structure of the catalyst layers is required to obtain good cell performance with low catalyst loadings. Such attempts have been made, for example, for the amount of Nafion impregnant,[92,93] catalyst types (percentages of platinum supported on carbon),[94] sputtering a thin film of platinum on the carbon electrode surface before ionomer impregnation,[56] and PTFE content in the catalyst layer,[52] etc. A more recent study by a team of researchers from Johnson Matthey and Ballard Power Systems[2] indicates that 40% Pt/C is the optimal catalyst type, with the catalyst layer thickness of approximately 20 to 25 μm and the platinum utilization of over 60%. The much thinner catalyst layer, as compared to the case for 10% or 20% Pt/C catalyst, allows for more effective access of the protons and reactant gases to the active sites. It is shown that a low platinum catalyst loading of 0.11 mg Pt/cm^2 is achievable for over 0.4 W/cm^2 of power density running on H$_2$ and air, which is compatible for the target set for light duty vehicle application. Ultra-low platinum catalyst loading can be achieved by electrodeposition of platinum directly onto a porous carbon electrode,[51] with catalyst loadings as low as 0.05 mg Pt/cm^2.

A related issue is the condition of hot-pressing process in bonding the catalyzed electrodes to the membrane, such as the bonding temperature, pressure, and duration should be optimized. Usually a good bondage can be achieved between the electrodes and the membrane by using hot-pressing technique. However, as the resulting MEAs, after being assembled into the cell/stack structure, are humidified during the cell/stack operation, the membrane expands significantly with the hydration, while the carbon support and the platinum particles do not change their dimensions on exposure to water. Therefore, local cracks and partial delamination may occur, enhancing cell internal resistance and affecting the reproductibility and reliability of MEAs so produced.

Carbon-Supported Platinum Catalysts Bonded to the Membrane   For this technique,[95] the catalyst layer structure is bound together by using recast ionomer, instead of the conventional PTFE, in order to guarantee the effective access of protons to the active sites, and very low catalyst loadings (0.12–0.16 mg Pt/cm^2) can be achieved with a very thin catalyst layer (<10-μm thick). For such PTFE-free thin-film catalyst layers, gas penetration of about 5–7 μm thick is achievable even in the absence of hydrophobic components to yield good catalyst utilization and good cell performance.

The fabrication process typically includes three steps. First, a catalyst (i.e., 20% Pt/C) and an appropriate amount of 5%-solubilized Nafion is mixed well (mixing for a few hours) into ink. The relative proportion of the supported catalyst to Nafion ranges from 5:2 to 3:1 by weight for the cathode catalyst layer, and somewhat lower for the anode (as low as 1:1 by weight). Glycerol may be added to the mixture in the amount equivalent to Nafion solution in order to improve the stability of the catalyst particle suspension and the paintability of the resulting ink. The Nafion in the ink is in the proton form initially, and is subsequently converted to the TBA$^+$ form (tetra-butylammonium) by adding 1-M TBAOH in methanol to the ink. The large TBA$^+$ counterions in the membrane structure effectively cloak the anionic sites and minimize the ionic interactions between the polymer sidechains — a protective measure. Then, a thin-film catalyst layer can now be made by either a complex decal process or simply casting the ink directly onto the membrane. Finally, the catalyzed membranes are rehydrated and the TBA^+ form of the membrane in the catalyst layer is converted back to the protonic form. This is achieved by immersing the catalyzed membranes in lightly boiling 0.5 M sulfuric acid for several hours, followed by rinsing in deionized water. The MEA is eventually made by hot-pressing two carbon electrodes onto the cleaned catalyzed membrane.

Figure 7.60 illustrates schematically the structure of the catalyst layer made by different fabrication processes, and the corresponding catalyst utilization is compared in Figure 7.61 for cathode catalyst layer running on air. Clearly the technique of

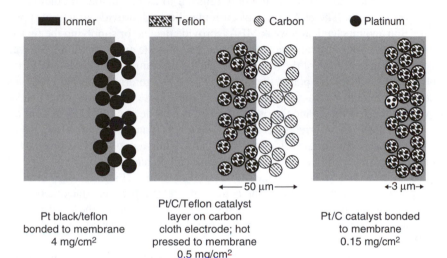

Figure 7.60 Schematic of the structures of the catalyst layer fabricated by different fabrication techniques.[46]

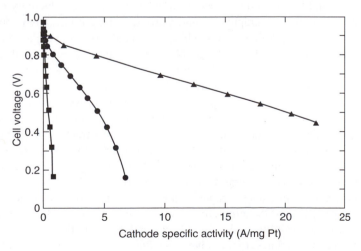

Figure 7.61 Air cathode catalyst utilization for different types of catalyst layers. Platinum black/PTFE (4 mg Pt/cm^2) ■; Ionomer-impregnated gas-diffusion electrodes (0.45 mg $Pt/cm2$) ●; Thin film of Pt/C//ionomer composite (0.13 mg Pt/cm^2) ▲.[46]

thin-film catalyst layer bonded to the membrane yields much higher catalyst utilization, especially at higher cell power density (i.e., higher cell current density or lower cell voltage), due to better access of protons and oxygen to the active sites (or less resistance to mass transport processes).

7.4.5   Bipolar Plates

In PEM fuel cells, each of the MEAs is interposed between two fluid-impermeable, electrically conductive plates, commonly referred to as the anode and the cathode plates, respectively. The plates serve as current collectors, provide structural support for the thin and mechanically weak MEAs, provide means for supplying the fuel and oxidant to the anode and cathode, respectively, and provide means for removing water formed during operation of the fuel cell. When the reactant flow channels are formed on the anode and cathode plates, the plates are normally referred to as fluid flow-field plates. When the flow channels are formed on both sides of the same plate, one side serves as the anode plate and the other side as the cathode plate for the adjacent cell, the plate is normally referred to as bipolar (separator) plate. It is more often that one of the reactants flows on one side of such a plate, while a cooling fluid flows on the other side of the same plate, and these plates collectively have to keep the fuel, oxidant and cooling fluid apart. Because the MEAs are very thin (<1 mm for each of them), the bipolar plates make up almost the entire volume of a fuel cell stack, about 80% of the stack mass and as much as 60% of the stack cost. Therefore, the bipolar plate is the major engineering challenge for power density enhancement and cost reduction of a PEMFC stack.

The most common material used for bipolar plates in PEMFC stacks is graphite, which has good electronic conductivity, corrosion resistance, and also low density (hence, light weight). However, graphite is quite brittle, making it difficult to handle mechanically and has a relatively low electrical and thermal conductivity compared with metals. Furthermore, graphite is quite porous, making it virtually impossible to make very thin gas-impervious plates, which is desirable for low weight, low volume, low internal-resistance fuel cell stacks. Normally graphite is resin-impregnated to improve mechanical strength and fluid impermeability. The exact composition and fabrication of the graphite plates are available in literature.[12]

The fluid flow channels are typically rectangular in cross section, even though other configurations such as trapezoidal, triangular, semicircular, and so on, have been explored. The flow channel dimensions range from a fraction of 1 mm to about 2 mm in width and depth as a low limit for a reasonable fluid pressure loss due to frictional losses. The most common methods of fabricating fluid-flow channels on the bipolar plates require the engraving or milling of flow channels into the surface of the rigid, resin-impregnated graphite plates. These methods of fabrication place significant restrictions on the minimum achievable cell thickness due to the machining process, plate permeability, and required mechanical properties. For example, the minimum practical thickness for a double-sided flow field plate is approximately 2 mm (about 6 mm in reality). Further, the resin-impregnated graphite plates are expensive, both in raw material costs and in machining costs. The machining of the flow channels and the like into the graphite plate surface causes significant tool wear and requires significant processing times. Therefore, one may conclude that graphite is not the ideal material for the bipolar plates in PEMFC stacks.

Alternative materials and accompanying fabrication methods have been explored for bipolar plates, such as injection moulding of carbon-carbon composites followed by graphitization achieved through heat treatment at 2500 °C; injection moulding and compression moulding of graphite-filled polymer resin; and screen printing technique of an ink-like solution of graphite powders. However, significant problems remain, such as the high-temperature heat treatment is not only expensive, but also difficult to precisely control the plate dimensions due to thermal deformation and distortion; and low electronic conductivity for conducting polymers, and low corrosion resistance for screen-printed plates. Consequently, metals have been proposed as the possible lightweight low-cost materials for the bipolar plates.[36,96]

Metals have much higher mechanical strength and are fluid impervious, hence bipolar plates can be made thin and easy for handling and stack assemblage; metals have higher thermal and electronic conductivity, potentially improving the stack performance. Bipolar plates can be made by three-piece metallic components assembled together or unitized metallic sheet stamped into shape. A metallic sheet can be made from corrosion-resistant metals or from corrosion-susceptible metals which have been coated with a corrosion-resistant outer layer. Particularly effective corrosion-resistant metals for PEMFC operating environment include titanium, chromium, stainless steel, and so on. These metals develop a dense, passive, oxide barrier layer over the surface to resist corrosion and prevent dissolution into the coolant flow. However, these oxide layers have low electrical conductivity, thus increases the internal resistance of the fuel cell. Furthermore, the metal bipolar plates are not easy to make with the flow channels and with sufficient flatness, which is required in order to avoid uneven compression of the MEAs in the stack. Perforated or foamed metal has been explored as the flow fields and current collectors — another low-weight and low-cost alternative.[97] However, it may be applicable for small cells, for large cells the fluid pressure losses are excessive. In fact, a niobium metal screen was used as the current collector, with electrical contact made at the edge of the electrode, by General Electric in 1970s for their 12-W power unit (under the trademark "PORTA-POWER"), and the flow field plate is not electrically conductive, made of aluminum plate coated with a plastic. It seems much more progress needs to be made before metals become a viable option in practice for commercial applications.

7.5   PERFORMANCE

The performance of a PEMFC depends on the design parameters, operating conditions and the fabrication processes. The electrode-backing layer and the catalyst layer may be made by a number of different techniques such as screen-printing, rolling, brushing, filtering, spraying, and so on, as discussed earlier, with various procedures for heat treatment. The fabrication of the catalyst layer includes further whether the platinum catalyst is deposited by sputtering, electrochemical deposition, and so on.

The performance of the membrane electrolyte may depend on its thickness, its equivalent weight, the history of thermal pretreatment, in addition to water contents and any foreign contaminants. The major design parameters include:

For the electrode:

1. The type of carbon (carbon paper/carbon cloth),
2. The amount of PTFE,
3. The porosity and the tortuosity of the layer, and
4. The thickness of the layer.

For the catalyst layer:

1. The type of catalyst support (carbon paper/carbon cloth);
2. The type of catalyst (platinum black, supported platinum, platinum alloys, macrocycles, platinum amount or percentage on support, platinum particle sizes, or distribution, etc.);
3. The amount and distribution of catalyst (catalyst loading and sputtered platinum film);
4. The amount of PTFE;
5. The type of ionomer for impregnation (Nafion, Dow, recast, etc.);
6. The amount of ionomer (ionomer loading);
7. The thermal pretreatment of the ionomer;
8. Thickness of the layer;
9. The porosity and the tortuosity of the layer.

For the membrane electrolyte:

1. The type of membrane (e.g., the type of backbone matrix, equivalent weights, etc.);
2. The thermal pretreatment of the membrane;
3. The hydration level;
4. The thickness of the layer.

In addition, the cell size (the active area), the flow field channel size and configuration, the flow plate materials, and surface condition (smoothness), can all influence cell performance.

The major operating conditions include the operating temperature and pressure (including whether the fuel and oxidant stream are operated at the same pressure or with a pressure differential; are humidified at the same temperature or with a temperature differential, etc.), since it is known that if an anode is humidified at a higher temperature than the cell operating temperature, more water is available on the anode side to prevent the dehydration of the membrane, hence a better cell performance results. Other operating parameters are reactant concentration and flow rate (stoichiometry or utilization), the level of humidification, the presence of any contaminant, and so on. Since almost all the fuel cell performance would decay over the time of operation (to be discussed later), the time at which the performance is measured needs to be specified as well. Normally, unless specifically stated, the cell performance is referred to the performance after about 100 hours of the continuous operation of a newly fabricated MEA.

It is clear from this impartial list of the relevant parameters that have significant influence on the cell performance, the determination of the cell performance is very lengthy and time-consuming indeed, if all the parameters are properly characterized and controlled. Unfortunately, not all the relevant parameters are provided for the performance data available in the literature, therefore, the performance data shown here are only intended to be indicative, and the appropriate original reference should be consulted for further information. In addition, the good performance of the PEMFCs was demonstrated, for the feasibility for practical applications, no later than earlier 1990s using Nafion and Dow membrane as the electrolyte, after that the emphasis was shifted to the cost reduction (such as lowering platinum loadings and seeking alternative membranes) and practical design issues (i.e., flow field design and scaling up).

Extensive measurements of the cell performance as a function of operating temperature and pressure have been carried out, for example, by Kim, et al.[98] for cells running on both pure oxygen and air. Additional data are shown in.[99] It is generally accepted that the optimal operating temperature for PEMFCs employing Nafion membrane is about 80 °C–85 °C. The operating pressure has varied considerably and no consensus has been reached for an optimal pressure. Generally, small PEMFC stacks tend to operate at the atmospheric pressure, while large stacks are typically pressurized and 3 atm seems to be employed often. The selection of the operating pressure is mainly determined between the performance gain (relatively smaller humidifier and better cell voltage) and the parasitic power for reactant pressurization, such an analysis is illustrated in.[100]

Figure 7.62 shows a single cell performance when operating on pure oxygen and air. Clearly lower oxygen concentration in the air reduces the cell voltage and the presence of nitrogen barrier to the oxygen mass transfer results in the mass transfer limitation to occur at the smaller current density as indicated by steeper drop in the cell voltage beyond the current density of about 1400 mA/cm^2. Therefore, the effect of lowering the reactant concentration is not simply the dilution of reactant gas.

Figure 7.63 shows the effect of membrane electrolyte on the cell performance. The active cell area is 1,180 cm^2, which is very large compared to most of the PEMFCs being developed. The reactants are hydrogen and oxygen at approximately 2 bar. Clearly, the thinner Nafion 115 gives much better performance than the thicker Nafion 117 and the lower equivalent weight Dow membrane yields even better performance. At the reference voltage of 0.684 V, Dow membrane gives the power density, which is almost three times that of Nafion 117.

Long-term performance of the PEMFCs has also been demonstrated for several thousands of hours, and the main possible factors causing the performance degradation may be the membrane degradation, catalyst agglomeration leading to a loss of active surface areas, which has bigger impact on the cathodic oxygen-reduction reaction, and long-term anode poisoning by impurities such as carbon monoxide. It has been demonstrated that the most significant performance loss over the long term may arise from the poisoning of anode catalyst by impurities, even when running on "99.99% pure" bottled hydrogen because the platinum catalyst can only tolerate the carbon

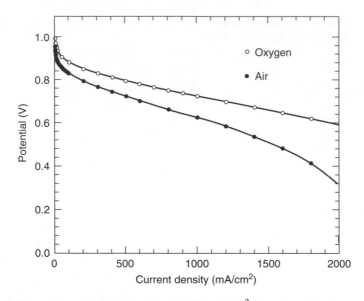

Figure 7.62 The performance of a single cell with 0.40 mg Pt/cm^2 electrodes. Operating temperature is 95 °C and pressure is 5 atm.[92]

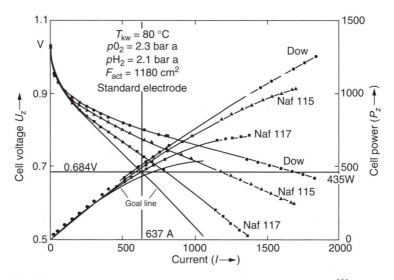

Figure 7.63 Effect of the thickness and type of membranes on the cell performance.[101]

monoxide concentration in the order of a few ppm levels. Since PEMFCs for mobile applications may be required to operate on hydrocarbon reformed fuels which may contain carbon monoxide in a few percentage levels, the significant performance loss and possible mitigation methods for such poisoning effect become one of the technical challenges facing the commercialization of the PEMFCs and we focus on this topic in the next section.

7.6   CARBON MONOXIDE (CO) POISONING AND METHODS OF MITIGATION

As mentioned earlier, PEMFC performance degrades significantly when carbon monoxide is present in the fuel gas; this phenomenon is referred to as CO poisoning, first documented by Gottesfeld and Pafford[102] for a range of low CO concentrations, as shown in Figure 7.64, and has been reviewed by a number of authors, including.[103] Even though the polarization associated with the cathodic oxygen reduction is the largest single source of energy losses in well-behaved PEMFCs operating on pristine hydrogen, significant performance degradation could occur due to two anode-related problems: the local drying of membrane electrolyte arising from the electroosmotic drag of water discussed in Section 7.4.3 and even more importantly, the poisoning of anode catalyst by impurities, most notably carbon monoxide. The maintenance of the membrane hydration has been successfully addressed in practical PEMFC systems by proper humidification of the reactant gas streams, and possibly by using thinner membranes of higher water permeability (the latter is met with the challenge of higher reactant gas crossover). Since carbon monoxide concentration that could have significant adverse impact on the cell performance is very low, on the order of a few ppm levels, and even the nominal 99.99%-pure bottled hydrogen could have long-term performance degradation due to anode platinum-catalyst poisoning effect, especially when the fuel stream is recirculated.

Unfortunately, a fuel gas stream with CO content may not be avoided for practical applications, whether mobile or stationary. This is because the use of pure hydrogen as a fuel source currently has a number of formidable limitations.[104] The major limitation is the availability of high-purity hydrogen gas and onboard hydrogen storage. The

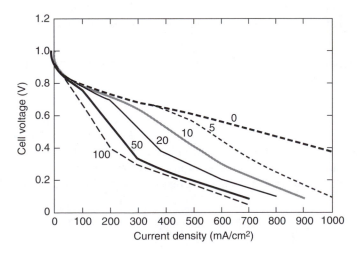

Figure 7.64 Effect of carbon monoxide concentrations (in ppm) in the fuel-feed stream on cell performance at 80 °C with both electrodes based on an ionomer-impregnated Pt/C catalyst and thin sputtered-platinum film, of total loading 0.45 mg Pt/cm^2.[102]

refueling of a vehicle using hydrogen would be slow and the major storage schemes of compressed hydrogen, cryogenic hydrogen, compressed and cryogenic hydrogen, and metal hydride adsorption each have significant disadvantages.[105] These onboard storage issues are exacerbated by the lack of infrastructure for hydrogen distribution and the fact that hydrogen is not abundantly available on earth in natural form. Thus, an attractive and practical option is to produce hydrogen onboard by reforming liquid hydrocarbon or alcohol fuels, with the most likely candidate being methanol for mobile applications and natural gas for stationary co-generation. The reformation of methanol results in a gas mixture of about 74% hydrogen, 25% carbon dioxide and 1% to 2% CO.[106] Using a selective oxidation process, the CO concentration can be reduced further to about 2~100 ppm. But even at this level, CO poisoning would plague PEMFC operation and performance, leading to a reduction in the energy conversion efficiency.[107] In addition to the carbon monoxide originally present in the fuel feed stream, a small amount of CO can be produced in situ at the platinum catalyst surface by the reduction of carbon dioxide, as analyzed later. Currently, CO poisoning is a significant hurdle to the commercialization of PEMFC technology for practical applications.

The mechanism of CO poisoning of the anode reaction is the much higher affinity of the CO molecule to the platinum surface at the relatively low temperature of 80 °C, as compared with the hydrogen molecule, thus CO molecules preferentially adsorb onto the platinum catalyst surface, blocking active sites from available for hydrogen oxidation reaction. Since CO electrooxidation occurs at much higher anode potential than the corresponding hydrogen electrooxidation, the absorbed CO molecules remain on the platinum surface as long as hydrogen oxidation occurs, which is so fast that it maintains a low anode potential, thus CO coverage of the platinum surface accumulates. Even CO partial pressure as low as 10^{-6} atm can result in a high platinum-surface coverage by CO. This section is devoted to the kinetics of anode reactions for a mixture of CO and H_2, the poisoned PEMFC performance characteristics and the available methods for the tolerance of CO presence in the fuel-feed stream.

7.6.1   Kinetics of Carbon Monoxide and Hydrogen Reaction at Anode Catalyst Surface

In order to understand the behavior of a PEMFC performance in the presence of CO, the electrochemistry of CO and hydrogen on the surface of platinum must be understood. The electrochemistry of CO in an acidic environment has been studied extensively.[108] CO oxidation and adsorption occurs on the (100) and (110) sites of platinum in an acid electrolyte, the adsorption of CO involves linearly bonded CO to platinum and the adsorption isotherm is that of a Tempkin isotherm, which can be written as[109]

$$\theta_{CO} = \frac{-\Delta G_{CO}^0}{r} - \frac{RT}{r} \ln H_{CO} + \frac{RT}{r} \ln\left(\frac{[CO]}{[H_2]}\right) \tag{7.55}$$

where θ_{CO} is the surface coverage of CO on platinum catalyst, a dimensionless parameter; ΔG_{CO}^0 is the standard free energy of adsorption, kJ/mol; R is the universal gas constant, T is the temperature, K; H_{CO} is the Henry's law constant for CO solubility, (atm $\cdot$ L/mol); and the square bracket represents the concentration of the species shown inside the bracket. The interaction parameter, r (kJ/mol) and the free energy of adsorption are a function of temperature, and in addition, the interaction parameter is highly dependent on catalyst structure. Equation (7.55) is only valid for relatively high temperatures of greater than 130 °C and for low temperatures, such as those encountered in a PEMFC, the coverage θ_{CO} appears to be a function of the anode potential as well as concentration, in addition to the temperature and catalyst structure. The enthalpy of CO adsorption–desorption is 134 kJ/mol[110] and the Gibbs free energy of adsorption on platinum in H_3PO_4 was found to be -50.7 kJ/mol at 190 °C and -60.3 kJ/mole at 130 °C[109].

The oxidation of CO occurs from a voltage range of 0.6 to 0.9 V, depending on the voltage sweep rate used in the voltammetry experiment.[108] The rate of oxidation of CO at low coverage is rapid while the rate is poisoned by a high coverage of CO. This dependence on coverage could be explained if the electrochemical oxidation of CO involved adjacent surface sites, or the so-called reactant pair mechanism for CO oxidation[111]:

$$
CO_{dissolved} + M + H_2O \longrightarrow \overbrace{
\begin{array}{cc}
CO & H_2O \\
| & | \\
M & M \\
\left\{\begin{array}{c} \text{linear} \\ \text{adsorbed} \\ CO \end{array}\right\} & \left\{\begin{array}{c} \text{adsorbed} \\ H_2O \end{array}\right\}
\end{array}}^{\text{Adjacent Sites}} \qquad (7.56)
$$

$$
\overbrace{\begin{array}{cc} CO & H_2O \\ | & | \\ M & M \end{array}}^{\text{Adjacent Sites}} \longrightarrow \overbrace{\begin{array}{cc} CO\ -\ -\ H_2O\ -\ -\ e^- \\ | \qquad\quad | \\ M \qquad\quad M \end{array}}^{\text{Activated Complex}} \longrightarrow \begin{array}{cc} CO\ -\ -\ OH\ +e^-\ +H^+ \\ | \qquad\quad | \\ M \qquad\quad M \end{array} \quad (7.57)
$$

$$
\begin{array}{cc} CO\ -\ -\ OH \\ | \qquad\quad | \\ M \qquad\quad M \end{array} \longrightarrow CO_2 + e^- + H^+ + 2M \qquad (7.58)
$$

where M is a metal adsorption site. Equation (7.56) is an adsorption step, which is very rapid and, in most experimental situations, the rate of CO adsorption is controlled by the diffusion of CO to the metal catalyst surface. This step determines the initial surface concentration of the reactant pairs. Equation (7.57) is the electron-transfer step which is rate determining for any fixed concentration of reactant pairs. Equation (7.58) is the final electron transfer reaction and is assumed to be rapid compared to Equation (7.57). This mechanism of CO oxidation can be thought of as CO lowering the energy of activation for the dissociation of water and is similar to the catalyzed gas-phase oxidation of CO by oxygen.

The oxidation of hydrogen and the effect of CO in an acid environment have also been extensively studied. The mechanism for electrochemical hydrogen oxidation over smooth platinum surface in an acid electrolyte is the slow dissociation of adsorbed hydrogen molecules to hydrogen atoms, known as the Tafel reaction (or the so-called **dissociative chemisorption**), followed by the fast charge-transfer reaction of the adsorbed hydrogen atoms, known as the Volmer reaction,[112] is illustrated

$$H_2 + 2M \rightleftharpoons 2MH \qquad \text{Tafel Reaction} \qquad (7.59)$$

$$2MH \rightleftharpoons 2M + 2H^+ + 2e^- \quad \text{Volmer Reaction} \qquad (7.60)$$

The heat of adsorption of hydrogen on platinum, with zero hydrogen coverage, is -87.9 kJ/mol[113] and the Gibbs free energy of adsorption for hydrogen on platinum at 22 °C in H_2SO_4 is -54.4 kcal/mol.[114] Oxidation of hydrogen on platinum surface occurs in the potential range of 0 to 0.2 V,[108] which is much lower than the potential of 0.6–0.9 V required for the corresponding electrooxidation of CO. The dissociative chemisorption of hydrogen on platinum is independent of both surface geometry and crystallite size, so that the exchange current density for the hydrogen reaction is independent of catalyst structure.[115]

The process of CO poisoning of the hydrogen oxidation reaction has been found to follow the following sequence[115]: CO chemisorbs on the platinum sites to the exclusion of hydrogen. This is possible because CO is more strongly bonded to platinum than hydrogen, as indicated by a greater potential required for the oxidation of CO than hydrogen and a sticking probability of CO on platinum of 15 times higher than that of hydrogen on platinum. Also, if the Gibbs free energy of adsorption of hydrogen and CO are compared and it is assumed that the Gibbs free energy of adsorption for CO continues to become more negative as temperature decreases, it can be seen that CO preferentially or much more strongly adsorbs to platinum due to the more negative Gibbs free energy of adsorption. The result is that even a relatively small concentration of CO can lead to the complete coverage of the platinum surface, to the exclusion of the hydrogen. In spite of CO preferentially adsorbing on the platinum surface, the rate of hydrogen oxidation on even the few remaining platinum sites is so rapid that it controls the surface potential or free energy of the catalyst. Unfortunately, because this potential is less than the potential needed to oxidize CO, the coverage of CO remains at that dictated by the CO adsorption isotherm, despite the continuous oxidation of hydrogen on the same platinum surface. Thus, the mechanism of CO poisoning of hydrogen oxidation is that linearly bonded CO blocks sites for the dissociative chemisorption of hydrogen and the current density, or reaction rate, of hydrogen in the presence of CO is reduced and can be written as

$$J_{H_2/CO} = J_{H_2}(1 - \theta_{CO})^2 \qquad (7.61)$$

where $J_{H_2/CO}$ is the current density for hydrogen oxidation reaction in the presence of carbon monoxide, and J_{H_2} is the current density for hydrogen reaction without the presence of carbon monoxide. Clearly, a high coverage of CO on the platinum surface can degrade the anode reaction rate significantly. In the experimental studies cited early, it was concluded that the coverage of CO is a function of potential as well as CO partial pressure. Also of interest, CO and hydrogen do not interact during their co-adsorption at temperatures of less than 200 °C. However, it has been noted that the apparent activation energy for hydrogen oxidation increases with increasing CO concentration, indicating an increased difficulty for the hydrogen to oxidize in the presence of CO.[109] This was speculated to be because either CO preferentially adsorbs to the *better* active sites or hydrogen is involved in the CO oxidation process.

7.6.2   Cell Performance With Carbon Monoxide Poisoning

With the above discussion of the CO poisoning mechanism, we can now explain the behavior of the cell performance poisoned by carbon monoxide. Figure 7.64 shows that the cell performance loss is very small below some threshold current density, which is about 100–300 mA/cm^2 depending on the amount of CO in the fuel stream. However, as the current density is increased beyond the threshold value, cell voltage decreases significantly to a level unacceptable for practical applications. This characteristic of the cell performance with carbon monoxide poisoning effect may be interpreted by the following kinetic based considerations.[46]

The kinetics of the hydrogen- and carbon monoxide-oxidation reaction shown in Equations (7.56)–(7.60) may be simplified by assuming the following set of reactions being the dominant process at the anode platinum-catalyst surface

$$M + CO \longrightarrow CO - M \tag{7.62}$$

$$H_2 + 2Pt \longrightarrow 2(H - M) \tag{7.63}$$

$$H - M \longrightarrow M + H^+ + e^- \tag{7.64}$$

$$CO - M + OH_{ads} \longrightarrow M + CO_2 + H^+ + e^- \tag{7.65}$$

Here the adsorption of carbon monoxide, Equation (7.62) and the dissociative chemisorption of hydrogen in Equation (7.63) are assumed to occur in parallel. At the small anodic overpotentials, reaction in Equation (7.64) is a fast and potential-driven process; hydrogen chemisorption in Equation (7.63) may become increasingly rate determining when the CO coverage on the platinum catalyst surface increases, even though the electrode current is primarily generated by hydrogen electrooxidation in Equation (7.64). Finally, the electrooxidation of the adsorbed carbon monoxide with the water-derived adsorbed OH, Equation (7.65), is extremely slow at low anodic overpotentials and may be neglected, even though the CO surface coverage may be high. Then the rate of variations for the surface coverage by the adsorbed hydrogen,

θ_H, and the anodic current density J generated become

$$\frac{d\theta_H}{dt} = 2k_{ads}P_{H_2,cat}(1 - \theta_H - \theta_{CO})^2 - k_{e,H}\theta_H\left[\exp\left(\frac{\eta}{b}\right) - \exp\left(-\frac{\eta}{b}\right)\right] \quad (7.66)$$

$$J = Fk_{e,H}\theta_H\left[\exp\left(\frac{\eta}{b}\right) - \exp\left(-\frac{\eta}{b}\right)\right] \quad (7.67)$$

where k_{ads} is the reaction rate constant for dissociative chemisorption of hydrogen Equation (7.63), $P_{H_2,cat}$ is the partial pressure of hydrogen at the catalyst site, $k_{e,H}$ is the rate constant for electrooxidation of adsorbed hydrogen, b is the Tafel slope, and η is the anodic overpotential from the equilibrium potential of the H_{ads}/H^+ system.

At low anodic overpotentials, $\theta_H << 1$ and $\theta_H << \theta_{CO}$, and the steady-state assumption for the surface coverage requires $d\theta_H/dt = 0$. Now solving Equation (7.66) for θ_H and substituting into Equation (7.67) gives the anodic current density arising from the hydrogen electrooxidation

$$J = 2Fk_{ads}P_{H_2,cat}(1 - \theta_{CO})^2 \quad (7.68)$$

where the surface coverage by CO is determined by the CO adsorption, Equation (7.62), and electrooxidation, Equation (7.65), and is an invariant at low anodic overpotentials and dictated by the adsorption isotherm, modified by larger values of η. For a CO-free fuel stream, $\theta_{CO} = 0$, the current density produced for the oxidation of hydrogen only yields $J = 2Fk_{ads}P_{H_2,cat}$. Substituting back into Equation (7.68) results in Equation (7.61). Equation (7.68) also indicates that the current density is independent of the anodic overpotential at low η, so that the cell voltage does not deviate from the CO-free values until the current density exceeds the value given by Equation (7.68). Then, a significant anodic overpotential is required to increase the rate of electrooxidation of CO, Equation (7.65), in order to free up sufficient active sites for hydrogen oxidation and to produce the higher current density, hence, excessive cell voltage losses occur for current densities beyond the limiting value set by Equation (7.68). The prediction of the cell voltage at larger current densities requires a more thorough analysis of the kinetics at the anode electrode, and can be found elsewhere.[116]

7.6.3   Mechanism of Carbon Dioxide Poisoning of Anode Platinum Catalyst

The PEMFCs are essentially acid-electrolyte fuel cells, and the membrane electrolyte can tolerate the presence of carbon dioxide. However, CO_2 concentrations may be in excess of 25% for hydrocarbon reformed-fuel gas mixtures and experimental measurements have been reported that the voltage losses in the anode are greater than what can be accounted for through the reactant dilution, or the so-called Nernst losses, as shown in Figure 7.65.[95] The additional performance loss beyond the reactant dilution effect of carbon dioxide has been referred to as the carbon dioxide poisoning of the anode

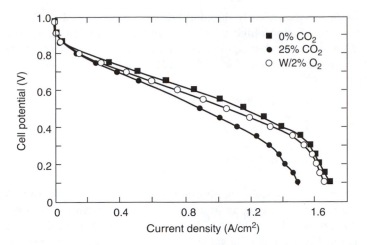

Figure 7.65 Effect of carbon dioxide and of oxygen addition to the fuel gas on the performance of a H_2/air PEMFC.[95]

platinum catalyst. Evidence suggests that this effect of carbon dioxide is due to in situ production of CO from CO_2 on the platinum surface and the subsequent adsorption of CO on the active platinum sites through either the reverse water–gas shift reaction

$$CO_2 + H_2 \rightleftharpoons CO + H_2O \tag{7.69}$$

or by the electroreduction of CO_2

$$CO_2 + 2H^+ + 2e^- \rightleftharpoons CO + H_2O \tag{7.70}$$

Wilson, et al.[95] discovered that the severity of CO_2 poisoning decreased if all of the anode platinum catalyst was tied up within an electroactive structure.[95] Since Equation (7.69) requires only a catalyst site in contact with the gas phase, this indicates that Equation (7.69) is more prevalent than Equation (7.70), which requires a site with good gas, ionic, and electronic access. The equilibrium concentration of CO due to water–gas shift reaction can be calculated through the following relationships

$$n_{CO_2,in}CO_2 + n_{H_2,in}H_2 + n_{H_2O,in}H_2O$$
$$\longrightarrow n_{CO_2,out}CO_2 + n_{H_2,out}H_2 + n_{H_2O,out}H_2O + n_{CO,out}CO \tag{7.71}$$

$$K_P = \frac{\left(P_{CO,out}\right)\left(P_{H_2O,out}\right)}{\left(P_{H_2,out}\right)\left(P_{CO_2,out}\right)} \tag{7.72}$$

where $n_{i,in}$ is the initial number of moles of species i in the fuel feed stream, $n_{i,out}$ represents the number of moles of species i at the final equilibrium for the water–gas shift reaction, K_p is the equilibrium constant for partial pressure, and P_i is the partial pressure of the species i. Clearly, the equilibrium CO concentration will be reduced if water vapor concentration is high, which is fortunately the case in the fuel feed stream.

The formulation for the equilibrium composition, $n_{i,out}$, is completed by the conservation of atomic species for C, O, and H. That is,

$$C: \quad n_{CO_2,out} + n_{CO,out} = n_{CO_2,in} \tag{7.73}$$

$$O: \quad 2n_{CO_2,out} + n_{H_2O,out} + n_{CO,out} = 2n_{CO_2,in} + n_{H_2O,in} \tag{7.74}$$

$$H: \quad 2n_{H_2,out} + 2n_{H_2O,out} = 2n_{H_2,in} + 2n_{H_2O,in} \tag{7.75}$$

The concentration of CO at equilibrium has been calculated, by using K_P values from,[117] for temperatures ranging from 25 °C to 95 °C, pressures varying from 1 to 3 atm and the relative humidity from 0% to 100%. The result of these calculations is presented in Table 7.4 and the values obtained for the equilibrium concentration of CO is consistent with.[104] As is illustrated in the table, an increase in temperature increases the equilibrium concentration of CO, if the initial relative humidity is zero. However, with the initial presence of water vapor in the fuel stream, an increase in temperature decreases the equilibrium concentration of CO. This is because the partial pressure of water increases at higher temperatures, which drives Equation (7.69) to the left and results in a decreased CO concentration. An increase in total pressure results in an increased equilibrium concentration of CO due to the larger initial partial pressures of CO_2 and H_2, which drives Equation (7.69) to the right. The majority of equilibrium concentrations of CO calculated in Table 7.4 are well in excess of the 2 to 10 ppm CO so that a PEMFC cannot tolerate without suffering from an appreciable performance loss, and thus operation with 25% CO_2 would result in significantly inferior performance if the water–gas shift reaction reaches equilibrium. The fact that the CO_2 poisoning effect is far less than what would otherwise be expected from

Table 7.4 Equilibrium Concentration, in ppm, of CO with Varying Temperature, Relative Humidity and Total Pressure

Temperature (°C)	Initial Relative Humidity of H_2O			
	0	**50**	**80**	**100**
25	1310 (1310)	106 (308)	65.2 (198)	51.5 (159)
50	2590 (2590)	98.0 (315)	56.6 (194)	42.9 (153)
80	4920 (4920)	62.2 (269)	26.0 (152)	15.0 (113)
95	6500 (6500)	35.7 (233)	7.25 (119)	1.44 (82.0)

CO concentration for a total pressure of 3 atm is given in parenthesis while the values outside the parenthesis are the result for the total pressure of 1 atm. The initial concentration of dry gas is 25% CO_2 and 75% H_2.

the equilibrium CO concentrations suggest that the water–gas shift reaction does not proceed rapidly at temperatures experienced by a PEMFC. As a result, the actual concentration of CO may be much less in an operating PEMFC than the equilibrium value shown in Table 7.4. This may also explain why the effect of this CO_2 poisoning is minimal if CO is already present as the effect of CO is much greater than the CO_2 poisoning.[104]

7.6.4   Characteristics of PEMFC CO Poisoning

The polarization curves of a PEMFC in the presence of various CO concentrations are shown in Figures 7.64 and 7.66. Both results are similar in that the CO poisoning effect is small for the small current densities below about $50 \, mA/cm^2$; and for the larger current densities a considerably large anodic overpotential required to keep up the high rate of reaction lowers the cell potential significantly. Oetjen, et al.[118] examined the effect of CO on a PEMFC performance shown in Figure 7.66. The cell operating temperature was 80 °C and CO concentrations of 25, 50, 100 and 250 ppm were used in the fuel gas. It is seen that for hydrogen and 25-ppm CO, the cell polarization curve looked more similar to the curve without CO only with a more negative slope. However, for CO concentrations greater than 100 ppm, the polarization curve had two distinct slopes. The lower slope was explained by the adsorption and oxidation kinetics of hydrogen and CO at the anode. At higher current densities, the potential of the anode increased to values at which adsorbed CO could be oxidized to CO_2 and thus leading to higher reaction rates for hydrogen adsorption and oxidation. It was also found that the poisoning effect takes a significant amount of time to reach steady state, as illustrated in Figure 7.67. CO poisoning could be reversed by operation at open circuit voltage with pure hydrogen for 2 to 3 hours. Therefore, the CO poisoning effect must be taken as a transient phenomenon in the operation of PEMFCs used in

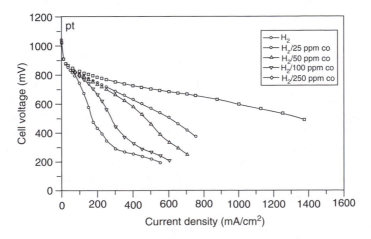

Figure 7.66 The effect of CO in the fuel stream on the performance of a PEM fuel cell operating at 80 °C[118].

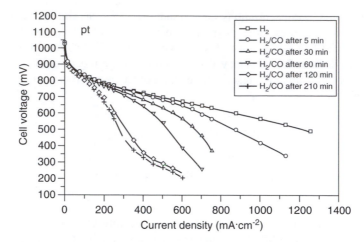

Figure 7.67 Illustration of the transient nature of CO poisoning of a PEM fuel cell.[118] CO concentration is 100 ppm.

transportation applications. The presence of CO in the anode gas does not affect the open-circuit cell potential. Similar PEMFC performance data with CO in the fuel gas can be found in.[95,119,120]

Zawodzinski, et al. compared their PEMFC poisoning data with others and found discrepancies in the amount of CO that could be tolerated.[121] It was surmised that these discrepancies might have been due to different flow rates. Lower flow rates allow better CO tolerance, if oxygen crossover from the cathode to the anode is present, through the oxidation of CO; thus freeing up sites for the hydrogen oxidation reaction. This conjecture may be confirmed by an estimate of the amount of oxygen crossover from the cathode and the details of the estimate are presented in the next section in connection with the introduction of a small amount of oxygen into the fuel stream to improve CO tolerance.

7.6.5   Methods Available for the Mitigation of CO Poisoning

From the above description, it is required that the mitigation of CO-poisoning effect be achieved at low anodic overpotentials without significant CO electrooxidation rates at the catalyst surface and for the full range of operating cell current densities in the presence of significant site coverage by CO. There are three methods available to mitigate the effect of CO poisoning in PEMFCs. These are the use of Pt alloy catalysts, an increase in fuel cell operating temperature and the introduction of oxygen into the fuel gas stream for the chemical, instead of electrochemical, oxidation of adsorbed CO. These methods are described in the following sections.

CO-Tolerant Catalysts: Platinum Alloy Catalysts   The first mitigation method for the alleviation of CO poisoning that is discussed is the use of platinum alloys as a CO-tolerant anode catalyst. The Pt-Ru alloys have been shown as the most

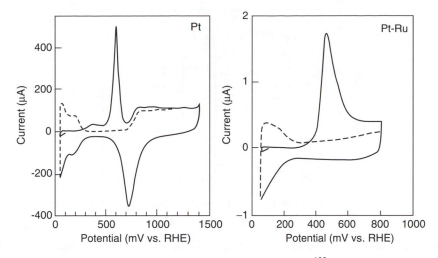

Figure 7.68 Comparison of the oxidation of CO on Pt and Pt-Ru catalyst.[122]

promising candidate.[120] Experiments illustrate that the oxidation of CO on a Pt-Ru alloy electrode occurs at potentials of 170 to 200-mV lower than that of a pure Pt electrode,[122] as shown in Figure 7.68. The increased CO-oxidation activity observed on the Pt-Ru alloys can be understood in terms of Ru having a lower oxidation potential than Pt. This leads to the preferential adsorption of water on the Ru atoms, creating Ru–OH and allowing the CO adsorbed on the Pt atoms to oxidize with the adjacent Ru–OH through the reactant pair mechanism

$$Ru-OH + Pt-CO \longrightarrow Ru + Pt + CO_2 + H^+ + e^- \qquad (7.76)$$

Thus, it would be expected that the best performance would occur for an electrode where Ru and Pt atoms are always adjacent to one another. This expectation is confirmed by experimental data, which find that $Pt_{0.5}Ru_{0.5}$ alloys exhibit the best performance.[120] Although the Pt-Ru alloys give better performance than pure Pt alloys, the performance improvement is achieved mainly at lower cell current densities (typically, $<300\ mA/cm^2$) and considerable losses still persist at higher current densities. Further, existing experimental data indicate a considerable spread in the degrees of CO tolerance for different Pt-Ru anode catalysts, implying that the exact Pt-Ru catalyst formulation, preparation, and structure, as well as the test conditions may all be important in determining the level of CO-tolerance and performance improvement, and hence more investigation is warranted. The long-term stability of the Pt-Ru alloys is an open question[106,118] although some long-term tests indicate negligible degradation over a time period of 3000 hours.[2]

In addition to Pt-Ru, other Pt alloys have been examined for CO tolerance. The Pt-Mo alloy electrocatalyst has been shown to have higher CO tolerance than pure Pt in a PEMFC environment.[123] The mechanism for CO tolerance is similar to that of Pt-Ru in that the adsorbed CO on Pt sites is oxidized by oxygenated species activated at the neighboring Mo atoms. This is believed to occur at low anodic overpotentials and

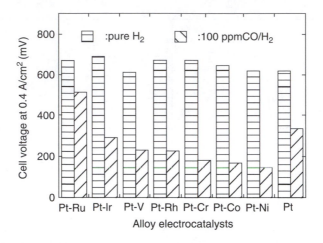

Figure 7.69 Comparison of the CO tolerance of various Pt-alloy catalysts for use in a PEM fuel cell.[119]

the oxidation of CO by Pt–OH occurs at higher overpotentials. Another Pt alloy found to mitigate CO poisoning is the Pt-Sn electrode.[124,125] This electrode system is of interest because CO does not adsorb onto Sn and thus CO and OH do not compete for adsorption sites on Sn as with Ru. However, the surface composition of this electrode is unstable under extreme anodic polarization. The Pt-W alloy electrocatalyst has also been studied for CO tolerance by Tseung, et al.[126] Pt-W alloy catalysts exhibit a co-catalytic activity due to tungsten being active as a redox catalyst, present in an oxidized state such as tungsten bronze, at the potential of a PEMFC anode. The cocatalytic activity is said to be due to a rapid change of the oxidation state of W, which renders the tungsten sites to be active for the dissociation adsorption of water.[127]

An exhaustive study on the effect of different Pt alloys on the CO tolerance of PEMFCs has been carried out by Iwase and Kawatsu.[119] PEMFCs operating at pressures of 0.15 MPa and temperatures of 80 °C were employed in their experiments. The Pt-Ru, Pt-Ir, Pt-V, Pt-Rh, Pt-Cr, Pt-Co, Pt-Ni, Pt-Fe, Pt-Mn, and Pt-Pd alloys supported on carbon were investigated as the anode electrocatalyst in order to determine which alloys are more CO tolerant. It was found that the only alloy that exhibited greater CO tolerance than pure Pt was the Pt-Ru alloy. In fact, it was found that the Pt-Ru alloy operating in 100 ppm CO gave the same performance as a pure Pt electrode operating in a pure H_2 environment, as shown in Figure 7.69. Gotz and Wendt[127,128] also examined Pt alloy catalysts in a PEMFC operated at 75 °C. Figure 7.70 shows the results for Pt-Ru, Pt-W, Pt-Sn, and Pt-Mo binary alloys as well as Pt-Ru-W, Pt-Ru-Mo, and Pt-Ru-Sn ternary alloys with 150-ppm CO present in the fuel gas. It is seen that the Pt-Ru-W anodes gave superior performance to Pt-Ru electrodes for current densities below about 300 mA/cm^2 while Pt-Ru-Sn electrodes gave superior performance to Pt-Ru electrodes only at current densities in excess of 300 mA/cm^2. However, at the practically important current densities (larger than approximately 25 mA/cm^2 in Figure 7.70b), all the performance with the presence of CO in the fuel gas is significantly lower than the performance for the pure hydrogen gas as the fuel, no matter whether the binary or ternary Pt alloys are used as the anode catalyst.

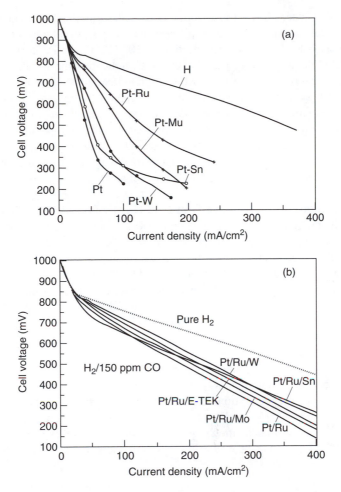

Figure 7.70 Performance of cells operated on H_2/150-ppm CO with (a) binary anode catalysts[127] and (b) ternary anode catalysts.[128]

Therefore, the use of platinum alloys as the CO-tolerant anode catalyst alone is not sufficiently effective for hydrocarbon-reformed fuel mixture as the fuel-feed stream for practical PEMFCs.

Higher Operating Temperatures   Another easy method to mitigate CO poisoning is to increase the cell temperature. This was examined for a PEMFC by Zawodzinski et al.[121] In these experiments, a PEMFC was used and its temperature was varied from 80 °C to 120 °C. The Pt-Ru system was examined and it was found that higher Pt-Ru catalyst loading and higher cell temperature resulted in better CO tolerance. At a CO concentration of 100 ppm, a cell temperature of 100 °C was required to maintain CO tolerance. The dependence of CO tolerance on cell temperature was explained by the strong temperature dependence of the adsorption equilibrium constant of CO on Pt. The Pt surfaces being freed from CO by either marginal thermal desorption or electrochemical oxidation rates explained the increase in tolerance with respect to the increased loading. It was found that at 100 °C, a pure Pt electrode showed the same CO tolerance as a Pt-Ru alloy at 80 °C, indicating the significant effect of temperature. Unfortunately, with a PEMFC, high temperatures are not feasible due to membrane dehydration and using elevated temperatures to mitigate CO poisoning is currently not practical and remains so unless the membrane electrolyte is significantly improved for long-term stability at the expected elevated temperatures. Even the supposed high-temperature membranes are available, higher temperature operation brings other undesirable effects, such as the startup of the power system as encountered in the higher temperature phosphoric-acid fuel cells.

Chemical Oxidative Removal of CO in Fuel Gas by Oxygen   The best state-of-the-art method to mitigate the CO-poisoning effect is to introduce a small amount of oxygen gas into the anode gas stream to clean up CO presence through the chemical oxidation of CO into CO_2. Currently there are two techniques to accomplish this. One of them is referred to as O_2 bleeding,[102] and both pure oxygen or air can be used for this purpose. One typical result is shown in Figure 7.71. The injection of 2% to 5% oxygen into the anode gas stream can result in CO tolerance at concentrations of up to 500 ppm. Wilson, et al.[95] found that O_2 bleeding was less effective when using thin-film electrodes, indicating that excess catalyst was required in order to oxidize the CO with the O_2 in the fuel gas. It was also found that O_2 bleeding and the use of the thin-film catalyst were beneficial in mitigating the CO_2 poisoning of the electrode. Zawodzinski, et al.[121] found that O_2 bleeding was the only effective way to achieve CO tolerance at CO concentrations of greater than 100 ppm. However, one must bear in mind that the lower explosion limit for oxygen–hydrogen mixture is about 5% O_2 in hydrogen, and safety consideration requires that the amount of oxygen introduced in the anode gas mixture must be well below this limit. Consequently, 100-ppm CO may

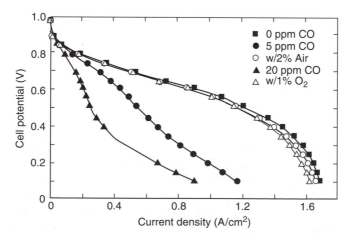

Figure 7.71 Effect of CO poisoning and oxygen bleeding on the performance of a PEMFC with platinum anode catalyst.[95]

be the maximum concentration of CO in the fuel gas that can be effectively cleansed by the oxygen bleeding technique at the PEMFC operating temperature of 80 °C.

Another technique for the introduction of oxygen into the anode gas stream is through the use of hydrogen peroxide (H_2O_2) in the anode humidifier.[129] The H_2O_2 decomposes to hydrogen and oxygen in the anode and the oxygen acts to oxidize the CO and improves the CO tolerance of the anode catalyst. The main advantage for the use of H_2O_2 as opposed to O_2 bleeding is that H_2O_2 has none of the safety problems associated with mixing hydrogen and oxygen gas.[106]

As mentioned earlier, Zawodzinski, et al.[121] surmised that the oxygen crossover from the cathode could be the cause of different CO tolerances being reported in the literature. The amount of oxygen crossover from the cathode of a fuel cell to the anode can be approximated at low current densities in the following manner. A fully humidified H_2/O_2 fuel cell operating at a pressure of 3 atm and a temperature of 80 °C is assumed for these calculations. As illustrated in Figure 7.72, the flux of oxygen through the membrane can be approximated as

$$J_{O_2} \cong \frac{D_{O_2-m}}{\delta_m} \Delta C_{O_2} \tag{7.77}$$

The maximum flux of oxygen would occur if the concentration of oxygen was zero at the anode side of the membrane and if the concentration of oxygen at the cathode side of the membrane was at its maximum, corresponding to the near open-circuit operation. Then the maximum concentration at the cathode side can be determined with the partial pressure of oxygen in the flow channel and Henry's law of oxygen

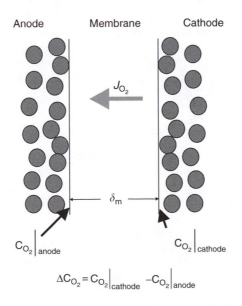

Figure 7.72 Schematic of oxygen crossover from the cathode to the anode through the membrane electrolyte region of a PEM cell.

solubility in the hydrated membrane so that the oxygen concentration difference between the anode and cathode side of the membrane is

$$\Delta C_{O_2} = \frac{P_{O_2}}{H_{O_2}} \tag{7.78}$$

Thus, the flux of oxygen through the membrane, with the previous assumptions and using D_{O_2-m}, H_{O_2} and the membrane thickness δ_m values for Nafion 117 as given in previous sections, is 6.72×10^{-10} mol/(s · cm^2). Assuming a current density of 100 A/m^2, the flux of H$_2$ entering the anode catalyst layer can be calculated using Faraday's law to be 5.18×10^{-8} mol/(s · cm^2). Thus, the ratio of oxygen flux over hydrogen flux in the anode is approximately 1 percent which is close to the amount of oxygen injection needed to mitigate CO poisoning by O$_2$ bleeding.[102] Thus, it is reasonable to expect that the CO tolerance of a fuel cell can be influenced, particularly at low current densities, by oxygen crossover from the cathode to the anode side, particularly for thinner membranes.

From this description, it becomes clear that the practical method for CO tolerance is to employ a combination of the platinum alloy catalysts along with the introduction of oxygen (air) into the fuel gas for the chemical cleansing of the carbon monoxide present. Figure 7.73 shows the effect of such a combination on the performance of a PEMFC poisoned by carbon dioxide. The same cell is employed under the same operating condition as those results shown in Figure 7.65, clearly the use of Pt-Ru anode catalyst reduces the degree of CO$_2$-poisoning effect, and also enhances the effect of oxygen bleeding. However, such combination does not completely recover the PEMFC performance with the typical amount of CO contents in a reformed fuel gas mixture, and the search for a practical CO-tolerant technique is continuing.

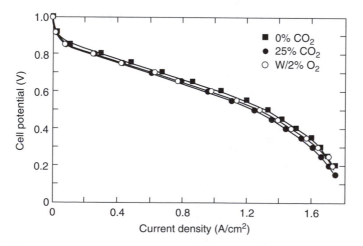

Figure 7.73 Effect of CO and O$_2$ bleeding on the performance of a PEMFC with a Pt-Ru anode catalyst.[95]

7.7 FUTURE R&D

Significant advances have been made since the initial practical use of PEM fuel cells in space program back in 1960s.[130,131] The main obstacle for the commercial application of PEM fuel cells is its high capital cost. Further, the reliability of performance and the lifetime of PEM fuel cells are still to be demonstrated in a convincing manner for potential consumers. These can be achieved through better understanding of the various transport phenomena involved in the PEM fuel cell, that leads to improvements in the design, materials, and manufacturing of PEM fuel cell components and systems as a whole, improvements in the performance of PEM fuel cells, especially stack power density which reduces the amount of materials usage and manufacturing effort for a given power requirement.

From a fundamental point of view, slow oxygen-reduction kinetics result in the largest single overpotential in PEM fuel cells operating on pure hydrogen and air (or oxygen) and hence alternative catalysts are of practical importance that are cheaper than the platinum-based Noble catalysts and that are more active electrocatalytically. The second largest single overpotential is due to the ohmic loss in the membrane electrolyte. It is necessary to develop alternative membranes that have high ionic conductivity and have sufficient mechanical strength and chemical stability to be made as thin as possible. Since absolute majority of the stack volume is occupied by the bipolar plates, which also comprise a significant portion of the stack cost (both materials and manufacturing), alternative light-weight materials for bipolar plates highly desirable to allow the reduction of the stack volume and weight are for easy and cheap manufacturing with volume production potential.

From engineering point of view, stack design is of uttermost importance since the two critical issues in PEM fuel cells, the water, and thermal management, must be integrally dealt with at the stack level. Water and thermal management in PEM fuel cell stacks are currently tackled by proper design and layout of reactant and coolant flow channels on the bipolar and cooling plates. The design and fabrication of membrane electrolyte assembly are critical since they are closely related to the cell performance and the reduction of catalyst loadings. Membranes that require less attention on water balance and that can operate without on-purpose humidification will certainly lead to a revolution for PEM fuel cells. The balance of plants is also a significant concern for the overall cost and performance of the PEM fuel cell systems.

For practical applications, especially transportation, it is important for PEM fuel cells to be able to operate on hydrocarbon reformed fuel gases in order to deal with the difficulty associated with hydrogen onboard storage and to fit in with the existing infrastructure for fuel supply. Then the tolerance of CO in the fuel gases becomes another critical issue. It is urgently needed to develop methods and techniques that can effectively mitigate the effect of CO poisoning on PEM fuel cell performance. The three existing methods of CO-tolerant Pt-alloy catalysts, high-temperature membranes, and O_2 bleeding in the fuel stream need a technological breakthrough for practical applications.

System optimization and integration will be required, and that requires a better understanding of the various physical and electrochemical processes occurring in the

PEM fuel cells. Some phenomena that might be secondary for small or single cells turn out to be significant for large or a stack of many cells. Understanding of the scale effect will be necessary for both design and control of PEM fuel cells.

7.8 SUMMARY

In this chapter, we first outlined the advantages and disadvantages of the PEM fuel cell that inevitably determined the kind of practical applications for PEM fuel cell. Then we proceeded to describe the basic operation and principle of PEM fuel cells. In particular, various modes of water transport through the electrolyte membrane are outlined along with the need for the proper membrane humidification in order to maintain adequate ionic conductivity, hence good cell performance. As a result, the significance and the complexities of water management in PEM fuel cells are highlighted.

Next we present the components and geometrical configurations of an individual PEM fuel cell, stack, and system. A single cell includes an electrolyte membrane sandwiched in between two catalyzed electrodes (anode and cathode), two (anode and cathode) flow distribution plates, two current collector plates and two endplates. In addition, an insulation layer is necessary between the endplate and the current collector and good sealing between the adjacent cell components. Ideal catalyst structure is illustrated for good cell performance, and two basic cell designs are shown for reactant distribution through flow channels made on either flow distribution plates or electrode-backing layers. A PEM fuel cell stack is composed of a basic cell repeating unit that may include a separate cooling cell interspersed throughout the stack for thermal management. A reactant humidification unit may be integrated as an integral part of a stack, or may stand alone, for water management. A variety of stack flow arrangement is shown possible that would provide different stack performance. Proper stack design with appropriate water and thermal management is the key to good stack performance, reliability, and lifetime and is of practical importance. Water and thermal management are accomplished by appropriate designs of the flow channels for the reactant streams and coolant flow paths, and various configurations of the flow channels are reviewed and summarized, citing many patents illustrating various innovative designs. A PEM fuel cell system is illustrated with various subsystems for the reactant conditioning and supply, cooling recirculation for thermal management, and the inherent interconnection between the thermal and water management. In addition, different stack cooling options are analyzed, including air-natural, air-forced, and water-forced cooling.

Section 7.4 describes the materials and manufacturing techniques for various cell/stack components illustrated earlier, including electrodes and catalysts. A detailed coverage on the electrolyte membrane is provided, from the basic requirement, historical development, to the chemical and microstructure. Then we provide comprehensive coverage on the physical and transport properties of the membrane, emphasizing the different water sorption capability of the membrane by liquid and vapor water. These coverages highlight the origin and practical difficulty in the water and thermal management — the two critical issues of PEM fuel cells. Then, a number of techniques in the fabrication of membrane-electrode assembly (MEA) are considered and the reduction of catalyst loading is discussed for these different techniques.

The we present in Section 7.5 typical performance of a PEM fuel cell for a variety of operating and design conditions when pure hydrogen is used as fuel. For hydrocarbon-reformed fuel gases containing carbon monoxide, the cell performance can be severely reduced due to the phenomenon called carbon monoxide poisoning, which is covered in Section 7.6. The degradation of cell performance is shown with analysis on the mechanism of the CO-poisoning phenomenon. Three methods are described that can mitigate the effect of CO poisoning on the performance degradation. Then the phenomenon of carbon dioxide poisoning is described and its mechanism is shown through CO production via the water–gas shift reaction. The chapter concludes with a description of future directions for further R&D required for the commercialization of PEM fuel cells.

BIBLIOGRAPHY

1. Warshay, M., P. Prokopius, M. Le and G. Voecks. 1997. The NASA fuel cell upgrade program for the space shuttle orbiter. *Proc. of the Intersociety Energy Conversion Engineering Conf.*, 1: 228–231.
2. Ralph, T.R., G. A. Hards, J. E. Keating, S. A. Campbell, D. P. Wilkinson, M. Davis, J. St-Pierre and M. C. Johnson. 1997. *J. Electrochem. Soc.*, 144: 3845–3857.
3. Haber, F. 1906. *Z. Anorg. Allg. Chem.*, 51: 245–288, 289–314, 356–368.
4. Grubb, W. T. 1957. *Proc. of the 11th Annual Battery Research and Development Conf.*, Red Bank, NJ: PSC Publications Committee, 5.
5. Grubb, W. T. 1959a. U.S. Patent No. 2,913,511.
6. Grubb, W. T. 1959b. *J. Electrochem. Soc.*, 106: 275.
7. Cohen, R. 1966. Gemini Fuel Cell System *Proc. of 20th Power Sources Conf.*, pp. 21–24.
8. St-Pierre, J. and D. P. Wilkinson. 2001. Fuel cells: a new, efficient and cleaner power source. *AIChE J.*, 47: 1482–1486.
9. Stone C. and A. E. Morrison. 2002. From Curiosity to Power to Change the World, *Sol. State Ionics* 152–153: 1–13.
10. Baschuk, J. J. and X. Li. 2000. Modelling of polymer electrolyte membrane fuel cells with variable degrees of water flooding. *J. Pow. Sour.*, 86: 181–196.
11. Dodge, C. E. 1995. Tubular fuel cells with structural current collectors. U.S. Patent 5,458,989.
12. Koncar, G. J. and L. G. Marianowski. 1999. Proton exchange membrane fuel cell separator plate. U.S. Patent 5,942,347.
13. Chow, C. Y. and B. M. Wozniczka. 1995. U.S. Patent 5,382,478.
14. Vanderborgh, N. E. and J. C. Hedstrom. 1990. Fuel cell water transport. U.S. Patent 4,973,530.
15. Fleck, W. and G. Hornburg. 1995. Process and apparatus for humidifying process gas for operating fuel cell systems. U.S. Patent 5,432,020.
16. Baschuk, J.J. and X. Li. 2004. Modeling of polymer electrolyte membrane fuel cell stacks based on a hydraulic network approach. *Int. J. Energy Res.*, 28: 697–724.
17. Gamburzev, S., C. Boyer and A. J. vAppleby. 1998. *Proc. of 1998 Fuel Cell Seminar*, pp. 556–559.
18. Watkins, D. S., K. W. Dircks and D.G. Epp. 1991. U.S. Patent 4,988,583.
19. Watkins, D. S., K. W. Dircks and D.G. Epp. 1992. U.S. Patent 5,108,849.
20. Li, X. and I. Sabir 2004. Review of bipolar plates in PEM fuel cells: flow field designs. *Int. J. Hydro. Energy*, forthcoming.
21. Pollegri, A. and P. M. Spaziante. 1980. U.S. Patent 4,197,178.

22.   Balko, E. N. and R. J. Lawrance. 1982. U.S. Patent 4,339,322.
23.   Reiser, C. A. and R. D. Sawyer. 1988. Solid polymer electrolyte fuel cell stack water management system. U.S. Patent 4,769,297.
24.   Reiser, C. A. 1989. Water and heat management in solid polymer fuel cell stack. U.S. Patent 4,826,742.
25.   Spurrier, F. R., B.E. Pierce and M. K. Wright. 1986. U.S. Patent 4,631,239.
26.   Granata, Jr., J. Samuel and B. M. Woodle. 1987. U.S. Patent 4,684,582.
27.   Wilkinson, D. P., G. J. Lamont, H. H. Voss and C. Schwab. 1996. Embossed fluid flow field plate for electrochemical fuel cells. U.S. Patent 5,521,018.
28.   Voss, H. H., D. P. Wilkinson and D. S. Watkins. 1993. Method and apparatus for removing water from electrochemical fuel cells. U.S. Patent 5,260,143.
29.   Voss, H., D. P. Wilkinson, P. G. Pickup, M. C. Johnson and V. Basura. 1995. Anode water removal: A water management and diagnostic technique for solid polymer fuel cells. *Electrochimica Acta* 40: 321–328.
30.   Fletcher, N. J., C. Y. Chow, E. G. Pow, B. M. Wozniczka, H. H. Voss, G. Hornburg and D. P. Wilkinson. 1996. Canadian Patent 2,192,170.
31.   Cavalca, C., S. T. Homeye and E. Walsworth. 1997. U.S. Patent 5,686,199.
32.   Wilkinson, D. P., H. H. Voss and K. B. Prater. 1993. U.S. Patent 5,252,410.
33.   Washington, K. B., D. P. Wilkinson and H. H. Voss. 1994. U.S. Patent 5,300,370.
34.   Chow, C. Y., B. Wozniczka and J. K. K. Chan. 1999. Integrated reactant and coolant fluid flow field layer for a fuel cell with membrane electrode assembly. Canadian Patent 2,274,974.
35.   Ernst, W. D. and G. Mittleman. 1999. U.S. Patent 5,945,232.
36.   Neutzler, J. K. 1998. U.S. Patent 5,776,624.
37.   Vitale, N. G. 1999. U.S. Patent 5,981,098.
38.   Wood, D. L., J. S. Yi and T. V. Nguyen. 1998. Effect of direct liquid water injection and interdigitated flow field on the performance of proton exchange membrane fuel cells, *Electrochimica Acta*, 43: 3795–3809.
39.   Oko, U. M. and J. H. Kralick. 2001. Fuel cell system having humidification membranes. U.S. Patent 6,284,399.
40.   Nelson, M. H. 2001. Fuel cell channeled distribution of hydration water. U.S. Patent 6,303,245 B1.
41.   McElroy, J. F. 1989. High power density evaporatively cooled ion exchange membrane fuel cell. U.S. Patent 4,795,683.
42.   Sonai, A. and K. Murata. 1994. Solid polymer electrolyte fuel cell apparatus. U.S. Patent 5,344,721.
43.   Fletcher, N. J., G. J. Lamont, V. Basura, H. H. Voss and D. P. Wilkinson. 1995. Electrochemical fuel cell employing ambient air as the oxidant and coolant. U.S. Patent 5,470,671.
44.   Meyer, A. P., G. W. Schelffer and P. R. Margiott. 1996. Water management system for solid polymer electrolyte fuel cell power plants. U.S. Patent 5,503,944.
45.   Vellone, R. 1988. *Program and Abstracts 1988 Fuel Cell Seminar*, Long Beach, CA, p. 168.
46.   Gottesfeld, S. and T. A. Zawodzinski. 1997. Polymer electrolyte fuel cells. *Advances in Electrochemical Science and Engineering*, eds., R. C. Alkire, H. Gerischer, D. M. Kolb and C. W. Tobias, 5, Wiley-VCH, New York: pp. 195–301.
47.   Paganin, V. A., E. A. Ticianelli and E. R. Gonzalez. 1996. *J. Appl. Electrochem.* 26: 297.
48.   Giorgi, L., E. Antolini, A. Pozio and E. Passalacqua. 1998. Influence of the PTFE content in the diffusion layer of low-Pt loading electrodes for polymer electrolyte fuel cells. *Electrochimica Acta* 43: 3675–3680.
49.   E-TEK. 1995. Gas Diffusion Electrodes and Catalyst Materials, 1995 Catalogue.
50.   Raistrick, I. D. 1986. Diaphragms, separators, and ion exchange membranes, eds., J. W. Van Zee, R. E. White, K. Kinoshita, and H. S. Burney *The Electrochemical Society Softbound Proc. Series*, PV 86–13, Pennington, NJ: p. 172.
51.   Taylor, E. J., E. B. Anderson, N. R. K. Vilambi. 1992. *J. Electrochem. Soc.* 139: 145.
52.   Rocco de Senna, D., E.A. Ticianelli and E.R. Gonzalez. 1990. *Abstracts of the Fuel Cell*, Seminar in Phoenix, AZ, Washington, DC: Courtesy Associates Inc., 391.
53.   Khner, G. and M. Voll. 1993. Manufacture of carbon black, *Carbon Black*, 2nd eds., J.-B. Donnet, R. C. Bansal, and M. -J. Wang New York: Marcel Dekker, pp. 1–66.
54.   Collin, G. and M. Zander. 1991. *Chem.-Ing.-Tech.* 63: p. 539.

55. Shukla, A. K., P. Stevens, A. Hamnett and J. B. Goodenough. 1989. *J. Appl. Electrochem.* 19: 383.
56. Mukerjee, S. and S. Srinivasan. 1993. *J. Electroanal. Chem.* 357:201.
57. Mukerjee, S., S. Srinivasan, M. P. Soriaga and J. McBreen. 1995. *J. Phys. Chem.*, 99: 4577.
58. Agel, E., J. Bouet and J. F. Fauvarque. 2001. Characterization and use of anionic membranes for alkaline fuel cells. *J. Pow. Sour.* 101: 267–274.
59. Kesting, R. E. 1971. *Synthetic Polymeric Membranes*, New York: McGraw-Hill.
60. Grubb, W. T. and L. W. Niedrach. 1960. *J. Electrochem. Soc.* 107: 131.
61. Cairns, E. J., D. L. Douglas and L. W. Niedrach. 1961. *AIChE J.* 7: 551.
62. Niedrach, L. W. and W. T. Grubb. 1963. *Fuel Cells*, ed. W. Mitchell, Jr., New York: Academic Press, p. 253.
63. Hodgdon, R. B., J. R. Boyack and A. B. LaConti. 1966. The degradation of polystyrene sulfonic acid. TIS Report No. 65DE5.
64. D'Agostino, V., J. Lee and E. Cook. 1978a. U.S. Patent 4,107,005.
65. D'Agostino, V., J. Lee and E. Cook. 1978b. U.S. Patent 4,012,303.
66. Connolly, D. J. and W. F. Gresham. 1966. U.S. Patent 3,282,875.
67. Mauritz, K. A. and A. J. Hopfinger. 1982. Structural properties of membrane ionomers. *Modern Aspects of Electrochemistry* No. 14, eds. J. O'M Bockris, B. E. Conway, and R. E. White, New York: Plenum Press.
68. Yeo, R. S. and H. L. Yeager. 1985. Structural and transport properties of perfluorinated ion-exchange membranes. *Modern Aspects of Electrochemistry*, No. 16, eds. B.E. Conway, White R.E. and J.O'M Bockris, New York: Plenum Press.
69. Gierke, T. D., G. E. Munn and F. C. Wilson. 1981. *J. Polym Sci. Polym. Phys.* 19: 1687.
70. Yeager, H. L. and A. Steck. 1981. *J. Electrochem. Soc.* 128: 1880.
71. Kreuer, K. D., 2001. *J. Membrane Sci.* 185:29–39.
72. Fang, C., B. Wu, and X. Zhou, 2004. *Electrophoresis*, 25: 375–380.
73. Gierke, T. D. 1977. Ionic clustering in Nafion perfluorosulfonic acid membranes and its relationship to hydroxyl rejection and chlor-alkali current efficiency. *152nd National Meeting*, The Electrochemical Society, Atlanta, October 1977.
74. Zawodzinski, T. A., T. E. Springer and S. Gottesfeld. 1991. Polymer electrolyte fuel cell model. *J. Electrochem. Soc.* 138(8):2334–2342.
75. LaConti, A. B., A. R. Fragala, J. R. Boyack. 1977. *Electrode Materials and Processes for Energy Conversion and Storage*, eds., J. D. E. McIntyre, S. Srinivasan, and F. G. Will. The Electrochemical Society Softbound Proc., Pennington, NJ, PV 77–6, p. 354.
76. Zawodzinski, T. A., Jr., T. A. Springer, J. Davey, R. Jestel, C. Lopez, J. Valerio and S. Gottesfeld, 1993. A comparative study of water uptake and transport through ionomeric fuel cell membranes. *J. Electrochem. Soc.*, 140: 1981–1985.
77. Fuller, T. and J. Newman. 1992. *J. Electrochem. Soc.*, 139: 1332.
78. Zawodzinski, T. A., Jr., T. Springer, F. Uribe and S. Gottesfeld. 1993. *Solid State Ionics*, 60:199.
79. Zawodzinski, T. A., J. Davey, J. Valerio and S. Gottesfeld. 1995. *Acta Electrochimica*, 40:297.
80. Bernardi, D. M. and M. W. Verbrugge. 1992. A mathematical model of the solid-polymer-electrolyte fuel cell. *J. Electrochem. Soc.* 139: 2477–2491.
81. Wakizoe, M., F. N. Buchi and S. Srinivasan. 1996. *J. Electrochem. Soc.*, 143(3): 927–932.
82. Marr, C. and X. Li. 1999. *J. Pow. Sour.*, 77: 17–27.
83. Srinivasan, S., A. Parthasarathy and A. J. Appleby. 1992. Temperature dependence of the electrode kinetics of oxygen reduction at the platinum/Nafion interface - a microelectrode investigation. *J. Electrochem. Soc.* 139:2530–2537.
84. Denny, V. E., D. K. Edwards and A. P. Mills. 1973. *Transfer Process*. New york: Holt, Rinehart & Winston.
85. Prausnitz, J. M., R. C. Reid and T. K. Sherwood. 1977. *The Properties of Gases and Liquids*. 3rd ed. New York: McGraw-Hill.
86. Yoshida, N., T. Ishisaki, A. Watakabe and M. Yoshitake. 1998. Characterization of Flemion membranes for PEFC. *Electrochimica Acta* 43: 3749–3754.
87. Niedrach, L. W. 1967. U.S. Patent 3,297,484.
88. Lawrence, R. J. 1981. U.S. Patent 4,272,353.

89. Takenaka, H. and E. Torikai. 1980. Japanese Patent 55, 38943.

90. Fedkiw, P. S. and W.-H. Her. 1989. *J. Electrochem. Soc.* 136: 899.

91. Aldebert, P. F. Novel-Cattin, M. Pineri and R. Durand. 1989. *Solid State Ionics* 35: 3.

92. Ticianelli, E. A., C. R. Derouin, A. Redondo and S. Srinivasan. 1988. *J. Electrochem. Soc.* 135: 2209.

93. Lee, S. J., S. Mukerjee, J. McBreen, Y. W. Rho, Y. T. Kho and T. H. Lee. 1998. Effects of Nafion impregnation on performances of PEMFC electrodes. *Electrochimica Acta*, 43: 3693–3701.

94. Ticianelli, E. A., C. R. Derouin and S. Srinivasan. 1988. *J. Electroanal. Chem.* 251: 275.

95. Wilson, M. S. 1993. U.S. Patents 5,211,984 and 5,234,777.

96. Davies, D. P., P. L. Adcock, M. Turpin and S. J. Rowen. 2000. Bipolar plate materials for solid polymer fuel cells. *J. Appl. Electrochem.* 30: 101–105.

97. Murphy, O. J., A. Cisar and E. Clarke. 1998. Low cost light weight high power density PEM fuel cell stack. *Electrochimica Acta* 43: 3829–3840.

98. Kim, J., S. Lee and S. Srinivasan. 1995. Modeling of proton exchange membrane fuel cell performance with an empirical equation. *J. Electrochem. Soc.* 142: 2670–2674.

99. Kordesch, K. and G. Simader. 1996. *Fuel Cells and Their Applications*. New York: VCH.

100. Larminie, J. and A. Dicks. 2003. *Fuel Cell Systems Explained*. 2nd ed., New York: Wile.

101. Strasser, K. 1991. PEM fuel cells for energy storage systems, *IECEC'91*, 26th Intersociety Energy Conversion Engineering Conference, Boston, August 4–9, 1991, p. 636.

102. Gottesfeld S. and J. Pafford. 1988. A new approach to the problem of carbon monoxide poisoning in fuel cells operating at low temperatures. *J. Electrochem. Soc.* 135(10): 2651–2652.

103. Baschuk, J. J. and X. Li. 2001. Carbon monoxide poisoning of proton exchange membrane fuel cells. *Int. J. Energy Res.*, 25: 695–713.

104. Bellows, R. J., E. P. Marucchi-Soos and D. Terence Buckley. 1996. Analysis of reaction kinetics for carbon monoxide and carbon dioxide on polycrystalline platinum relative to fuel cell operation. *Ind. Eng. Chem. Res.* 35(4): 1235–1242.

105. Aceves, S. M. and G. D. Berry. 1998. Thermodynamics of insulated pressure vessels for vehicular hydrogen storage. *J. Energy Res. Tech.* 120: 137–142.

106. Divisek, J., H. -F. Oetjen, V. Peinecke, V. M. Schmidt and U. Stimming. 1998. Components for PEM fuel cell systems using hydrogen and CO containing fuels. *Electrochimica Acta*, 43(24): 3811–3815.

107. Watkins, D. S. 1993. Research, development and demonstration of solid polymer fuel cell systems. In *Fuel Cell Systems*, eds. L. Blomen and M. Mugerwa New York: Plenum Press pp. 493–530.

108. de Becdelievre, A. M., J. de Becdelievre and J. Clavilier. 1990. Electrochemical oxidation of adsorbed carbon monoxide on platinum spherical single crystals: Effect of anion adsorption. *J. Electroanal. Chem.* 294(1–2): 97–110.

109. Dhar, H. P., L. G. Christner and A. K. Kush. 1987. Nature of CO adsorption during H_2 oxidation in relation to modeling for CO poisoning of a fuel cell anode. *J. Electrochem. Soc.* 134(12): 3021–3026.

110. Kohlmayr, G. and P. Stonehart. 1973. Adsorption kinetics for carbon monoxide on platinum in hot phosphoric acid. *Electrochimica Acta.* 18(2): 211–223.

111. Gilman, S. 1964. The mechanism of electrochemical oxidation of carbon monoxide and methanol on platinum II: the "reactant pair" mechanism for electrochemical oxidation of carbon monoxide and methanol. *J. Phys. Chem.* 68(1):70–80.

112. Stonehart, P. and P. Ross. 1975. The commonality of surface processes in electrocatalysis and gas-phase heterogeneous catalysis. *Catalysis Reviews-Science and Engineering.* 12(1): 1–35.

113. Sakellaropoulos, G., 1981. Surface reactions and selectivity in electrocatalysis. *Advances in Catalysis*, eds. D. Eley, H. Pines, P. Weisz, Volume 30. New York: Academic Press pp. 217–333.

114. Dhar, H. P., L. G. Christner, A. K. Kush and H. C. Maru. 1986. Performance study of a fuel cell Pt-on-C anode in presence of CO and CO2, and calculations of adsorption parameters for CO poisoning. *J. Electrochem. Soc.* 133(8): 1574–1582.

115. Vogel, W., J. Lundquist, P. Ross, and P. Stonehart. 1975. Reaction pathways and poisons-II. The rate controlling step for electrochemical oxidation of hydrogen on Pt in acid and poisoning of the reaction by CO. *Electrochimica Acta.* 20(1): 79–93.

116. Baschuk, J.J. and X. Li, 2003. Modeling CO poisoning and O_2 bleeding in a PEM fuel cell anode. *Int. J. Energy Res.*, 27: 1095–1116.

117. Kuo, K.K. 1986. *Principles of Combustion*, New York: Wiley.

118.   Oetjen, H. -F., V. M. Schmidt, U. Stimming and F. Trila. 1996. Performance data of a proton exchange membrane fuel cell using H_2/CO as fuel gas. *J. Electrochem. Soc.* 143(12): 3838–3842.

119.   Iwase, M. and S. Kawatsu. 1995. Optimized CO tolerant electrocatalysts for polymer electrolyte fuel cells. In Proton Conducting Membrane Fuel Cells I. eds. S. Gottesfeld, G. Halpert, and A. Landgrebe, *Electrochemical Society Proc. Volume 95–23*. Pennington, NJ: pp. 12–23 The Electrochemical Society.

120.   Schmidt, V. M., R. Ianneillo, H. -F. Oetjen, H. Reger, U. Stimming and F. Trila. 1995. Oxidation of H_2/CO in a proton exchange membrane fuel cell. Proton Conducting Membrane Fuel Cells I, eds. S. Gottesfeld, G. Halpert, A. Landgrebe, *Electrochemical Society Proc. Volume 95–23*. Pennington, NJ: pp. 1–11. The Electrochemical Society.

121.   Zawodzinski, T. A., C. Karuppaiah, F. Uribe and S. Gottesfeld. 1997. Aspects of CO tolerance in polymer electrolyte fuel cells: Some experimental findings. Electrode Materials and Processes for Energy Conversion and Storage IV, eds. S. Srinivasan, J. McBreen, A.C. Khandkar V.C. Tilak *Proceedings of the Electrochemical Society Volume 97–13*. Pennington, NJ: pp. 139–146. The Electrochemical Society.

122.   Ianniello, R., V. M. Schmidt, U. Stimming, J. Stumper and A. Wallau. 1994. CO adsorption and oxidation on Pt and Pt-Ru alloys: dependence on substrate composition. *Electrochimica Acta.* 39(11/12): 1863–1869.

123.   Mukerjee, S., S. J. Lee, E. A. Ticianelli, J. McBreen, B. N. Grgur, N. M. Markovic, P. N. Ross, J. R. Giallombardo, and E. S. De Castro. 1999. Investigation of enhanced CO tolerance in proton exchange membrane fuel cells by carbon supported PtMo alloy catalyst. *Electrochem. Solid-State Let.*, 2(1): 12–15.

124.   Gasteiger, H. A., N. M. Markovic and P. N. Ross. 1995. Electrooxidation of CO and H_2/CO mixtures on a well characterized Pt_3Sn electrode surface. *J. Phys. Chem.* 99(22): 8945–8949.

125.   Wang, K., H. A. Gasteiger, N. M. Markovic, and P. N. Ross. 1996. On the reaction pathway for methanol and carbon monoxide electrooxidation on Pt-Sn alloy versus Pt-Ru alloy surfaces. *Electrochimica Acta.* 41(16): 2587–2593.

126.   Tseung, A. C. C., P. K. Shen, and K. Y. Chen. 1996. Precious metal/hydrogen bronze anode catalysts for the oxidation of small organic molecules and impure hydrogen. *J. Pow. Sour.*, 61(1-2): 223–225.

127.   Gotz, M. and H. Wendt. 1998a. Binary and ternary anode catalyst formulations including the elements W, Sn and Mo for PEMFCs operated on methanol or reformate gas. *Electrochimica Acta*, 43(24): 3637–3644.

128.   Gotz, M. and H. Wendt. 1998b. Preparation and evaluation of cocatalyst systems for anodic oxidation of methanol in PEM fuel cells. *1998 Fuel Cell Seminar Abstracts*: 616: 619.

129.   Schmidt, V. M., H. -F. Oetjen, and J. Divisek. 1997. Performance improvement of a PEMFC using fuels with CO by addition of oxygen-evolving compounds. *J. Electrochem. Soc.* 144(9): L237–L238.

130.   Paola, C. and S. Supramaniam. 2001. Quantum jumps in the PEMFC science and technology from 1960s to the year 2000. *J. Pow. Sour.*, 102: 253–269.

131.   Viral, M. and C. J. Smith. 2003. Review and analysis of PEM fuel cell design and manufacturing. *J. Pow. Sour.*, 114: 32–53.

PROBLEMS

7.1   Describe briefly the advantage and disadvantage as well as the areas of applications for PEM fuel cells.

7.2   Describe the operation principle such as half-cell and whole-cell reaction, the primary fuels expected to be used for PEM fuel cells.

7.3   Discuss briefly the typical (or target) operating conditions such as cell voltage, cell current density, temperature, pressure, fuel and oxidant utilization, and chemical-to-electrical energy conversion efficiency, etc. for PEM fuel cells.

7.4   Describe the effect of operating conditions on the cell performance.

7.5   Describe the geometrical configuration of cell and stack, typical materials used for, and the thickness as well as other dimensions of the cell components.

7.6   Describe the mechanism of water transport through the membrane electrolyte and how water management is implemented at the cell, stack, and system level.

7.7   Describe the various designs for the flow channels on the bipolar plates and summarize the pros and cons for the various types of design.

7.8   Describe the stack cooling considerations, provide an analysis for the cooling of stacks via air natural, air forced and water cooling.

7.9   Describe the structure and properties that characterize the Nafion membrane.

7.8   Describe the major factors affecting the lifetime of PEM fuel cells.

7.9   Describe the critical technical barriers to be overcome for the commercialization of PEM fuel cells, the possible solutions and their pros and cons.

MOLTEN CARBONATE FUEL CELLS (MCFCs)

8.1 INTRODUCTION

The molten carbonate fuel cell (MCFC) was initially developed with the intention of operating directly on coal, which has not materialized; the primary fuel for modern MCFCs is either coal-derived gases or more often natural gas. This contrasts with phosphoric acid fuel cells, as discussed in Chapter 6, which prefer natural gas as primary fuel. Despite the initial expectation in the late 1960s and early 1970s that MCFCs would be commercialized a few years, following PAFCs lead in commercialization (hence sometimes referred to as "the second-generation fuel cell," MCFCs are still under development and have not demonstrated technical maturity and market acceptance. It is a common belief that MCFCs have currently reached the precommercial demonstration stage.

The MCFC, as a concept, is almost a century old. It was probably first described in 1916 in a patent filed by W. D. Treadwell with Swiss patent number 78591 K1.109 and German patent numbers 325783 and 325784 K1.21b in 1917. Although the MCFC was investigated at the beginning of the last century, it is generally considered that it was conceived in Europe in 1940s as a result of R&D efforts for converting coal to electricity in carbonate media. The idea for the present MCFC stemmed from Davtyan and was further pioneered by Broers and Ketelaar, who perfected the idea so that current research is largely based on their work. The first MCFCs were demonstrated by Broers and Ketelaar in the 1950s. In 1960s, various R&D activities associated with MCFCs were initiated in Europe and the United States (as the backup technology for the phosphoric acid fuel cells for the TARGET program, which initiated PAFCs). The first pressurized MCFC stacks were operated in the early 1980s. Much of our knowledge about the MCFC performance was established during the period of 1975–1985, due to intensive, well-focused, and controlled research efforts.

The MCFC operates at higher temperature than all the fuel cells described in the previous three chapters. The operating temperature of the MCFC is generally around 600 °C~700 °C, typically 650 °C. Operation at such a high temperature allows MCFCs to not only tolerate carbon dioxide, but also carbon monoxide as well, eliminating the problems associated with the low-temperature fuel cells such as the carbon dioxide poisoning of alkaline fuel cells and the carbon monoxide poisoning of the acid electrolyte fuel cells. This paves the way for the carbonaceous fuels to be utilized directly in the MCFC. Furthermore, the high operation temperature of the MCFC results in high-quality waste heat which is suitable for co-generation or combined-cycle operation, for example; the MCFC can serve as the topping cycle with waste heat fed to a gas turbine system as bottoming cycle which leads to higher electric efficiency. Such high operation temperature yields the possibility of utilizing carbonaceous fuels directly, relying upon the so-called internal reforming to produce the fuel ultimately used by the fuel cell electrochemical reactions. This is because the MCFC operating temperature matches closely the temperature needed for the steam reformation of natural gas, as illustrated in Figure 6.1. This results in simpler MCFC systems (i.e., without the cumbersome external reforming or fuel processing subsystem), less parasitic load and less cooling power requirements, hence higher overall system efficiency as well. Another advantage of the MCFC is that at its operating temperature, the activation polarization has been reduced to the extent that it does not require expensive catalysts as low-temperature fuel cells do.

The operating temperature of the MCFC (at 650 °C) is lower than the other high-temperature fuel cell — the solid oxide fuel cell (SOFC), which operates at about 1000 °C. This relatively lower temperature was perceived to allow for simpler method of construction and easier sealing on the cells which was expected to give MCFC an edge over the SOFC in the prospects for commercialization, although this still remains to be confirmed. It has been estimated that the MCFC can achieve an energy conversion efficiency of 52%–57%, with potential for over 60% (from chemical energy to electrical energy) with internal reforming and natural gas as the primary fuel.

The MCFC is being developed for its potential as baseload utility applications, as well as dispersed or distributed electric-power generation with heat co-generation (combined heat and power). Due to its high operation temperature, it only has very limited potential for mobile applications. This is because of its relatively low power density and long startup times. However, it may be suitable as a powertrain for large surface ships and trains.

8.2   BASIC PRINCIPLES AND OPERATIONS

A single molten carbonate fuel cell is shown in Figure 8.1, exhibiting its operational principle. At the anode, hydrogen oxidation reaction combines with carbonate ions, producing water and carbon dioxide and releasing electrons to the external circuit. At the cathode, oxygen is reduced to carbonate ions by combining with carbon dioxide and electrons from the external circuit. Therefore, the overall electrochemical reaction

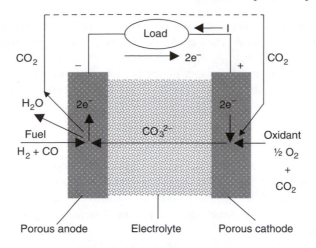

Figure 8.1 Schematic of a MCFC illustrating its operational principle.

occurring at the cathode is

$$\frac{1}{2}O_2 + CO_2 + 2e^- \longrightarrow CO_3^=$$ (8.1)

While at the anode the following electrochemical reaction occurs

$$H_2 + CO_3^= \longrightarrow H_2O + CO_2 + 2e^-$$ (8.2)

Therefore, the net cell reaction is

$$H_2 + \frac{1}{2}O_2 \longrightarrow H_2O + Heat + Electric\ Energy$$ (8.3)

Beside the hydrogen oxidation reaction at the anode, other fuel gases such as carbon monoxide, methane and higher hydrocarbons are also oxidized by conversion to hydrogen. Although direct electrochemical oxidation of carbon monoxide is possible, it occurs very slowly compared to that of hydrogen. Therefore, the oxidation of carbon monoxide is mainly via the water–gas shift reaction

$$CO + H_2O \rightleftharpoons CO_2 + H_2$$ (8.4)

which, at the operation temperature of the MCFC, equilibrates very rapidly at catalysts such as nickel. It is seen that accompanying the oxidation of carbon monoxide is the production of more hydrogen that is the fuel for anodic reaction. Therefore, it becomes clear that carbon monoxide is no longer a poison for the MCFC; instead it becomes a fuel indirectly through the water–gas shift reaction in the presence of water vapor. However, water is the product of the hydrogen oxidation reaction at the anode. As hydrogen is consumed, water is produced, which then drives the water–gas shift reaction forward to produce even more hydrogen from the carbon monoxide oxidation.

Direct electrochemical reaction of methane appears to be negligible. Hence methane and other hydrocarbons must be steam-reformed (often called **methanation equilibrium**):

$$CH_4 + H_2O \rightleftharpoons CO + 3H_2 \tag{8.5}$$

which can be accomplished either in a separate reformer (external reforming) or inside the anode chamber of the MCFC itself (the so-called **internal reforming**). Once again, water produced through the anodic electrochemical reaction (8.2) contributes to the steaming reforming of methane, that produces more hydrogen available for the reaction (8.2).

Therefore, it is evident that water and carbon dioxide are important components of the feed gases to the MCFCs. Water, produced by the main anodic reaction (8.2), not only helps to shift the equilibrium reactions (8.4) and (8.5) to produce more hydrogen for the anodic electrochemical reaction, but also it is required in the feed gas streams, especially in low-Btu (i.e., high CO) fuel gas mixtures, to avoid the formation of carbon particles that can deposit in the fuel gas-flow channels supplying the cell, or even inside the cell structure itself via the Boudouard equilibrium

$$2CO \rightleftharpoons CO_2 + C \tag{8.6}$$

From the half-cell reactions given in Equations (8.1) and (8.2), carbon dioxide is produced at the anode, while it is consumed at the cathode (i.e., carbon dioxide is one of the oxidants for the MCFC). To maintain carbon dioxide self-sufficient during the operation, carbon dioxide from the fuel exhaust gas is usually recirculated to the cathode, as shown schematically in Figure 8.1. The carbonate ions ($CO_3^=$) formed at the cathode move through the electrolyte toward the anode, carrying the electric current and completing the carbon dioxide circuit. The recycling of carbon dioxide from the anode to the cathode can often be achieved through the following two possible approaches: (i) the anode exhaust gas is burned with excess air, the heat released is used for the steam-reforming of the carbonaceous fuel and the combustion product is mixed with the cathode inlet gas after the water vapor is removed and (ii) the carbon dioxide is separated from the anode exhaust stream, such as through a carbon dioxide selective membrane or even with pressure swing absorption technique. This approach can provide a high concentration carbon-dioxide gas to the cathode, and thereby a better cell performance. However, the concentrations of the carbon dioxide are different in the anode and cathode compartments and the cell performance is affected by it accordingly.

The electrons formed at the anode move through the external electric load, thus yielding electric energy output, and arrive at the cathode, so completing the electronic circuit. The electric circuit is completed by the carbonate ions transport through the electrolyte. Therefore, the MCFC's carbon dioxide circuit is inherently related to the ionic–electronic current circuit.

It might be pointed out that the molten carbonate fuel cell is typically operated at the average cell temperature of approximately 650 °C, in reality, both reactant streams absorb the waste heat generated in the cell, thus the cell temperature and the reactant gas streams will increase along the direction of the reactant flows. In practice, to control the cell temperature around the average value, the reactant streams enter the cell at a lower temperature, as low as ~540 °C, and leave the cell at a much higher temperature, as high as nearly 700 °C. The electrolyte and the matrix may have a temperature anywhere between 600 °C–700 °C. As a result, significant temperature gradients may exist in incell and across-the-cell directions.

The factors involved in determining the operating condition for a MCFC are the same as those for other types of fuel cells: They include stack size, heat transfer rate, energy conversion efficiency, cell potential level, load requirement, and cost, and so on. The MCFC nominally operates at the current density of $100 \sim 200$ mA/cm^2 (typically 160 mA/cm^2) and cell potential of $0.75 \sim 0.95$ V (typically 0.75 V) at the atmospheric pressure and 75% fuel (hydrogen) utilization. The cell performance is improved when operated under pressurized conditions. However, high pressure operation favors the formation of solid carbon particles according to the Boudouard reaction (8.6), which has adverse effect on the cell performance and the cell lifetime. Hence, atmospheric operation is more common for MCFCs.

The design goals for the MCFCs are a cell operation lifetime of up to 40,000 hours (approximately 4½ years) with cell performance of 0.85 V at 160 mA/cm^2 for the low Btu gases as the primary fuel. Many of the current R&D activities are spent on scaling up of the cells to stacks with about 100 cells per stack and each cell size up to 1 m^2.

8.2.1   Internal Reforming

For low-temperature fuel cells described in previous chapters, hydrocarbon fuels must be first converted to a mixture of hydrogen, water vapor, carbon monoxide, and carbon dioxide in a fuel processor (or reformer) and the reformate gas must be further cleaned before the hydrogen-rich gas can be provided to the fuel cell for electrochemical reaction and electric energy production. The cleanup process depends on the type of fuel cell involved, for example, carbon dioxide must be removed for alkaline fuel cells, and carbon monoxide must be removed for acid electrolyte fuel cells. These fuel processing steps occur outside of the fuel cells, and hence called "external reforming." As shown in Figure 6.1, the operating temperature of the MCFCs is sufficiently high to allow the fuel reforming steps to be integrated within the fuel cell itself. As noticed in Section 8.2, carbon monoxide acts as an indirect fuel for the anode oxidation of hydrogen via the water–gas shift reaction of Equation (8.4), and carbon dioxide can be recycled to the cathode for the reduction reaction. Water produced from the anodic electrochemical reaction effectively maintains the reforming and water–gas shift reaction. In this approach, the reformation of hydrocarbon fuels occurs inside the fuel cell, hence called internal reforming, or IR-MCFC. Compared to the external reforming design, the IR-MCFC design is considered highly efficient and reliable in operation, simple, and compact in construction with low cost.

Natural gas is commonly considered as the primary fuel for the IR-MCFCs, and methane is the major constituent of natural gas. Because the methane steam-reforming reaction is favored by high temperature and low pressure, IR-MCFC is best suited to operation at low, near atmospheric, pressure.

External reformers typically operate at 800 °C~900 °C with a steam-to-carbon ratio of 2.5~3.0, and have about 95%~99% conversion of methane to hydrogen. For internal reforming at the MCFC operating temperature of 650 °C, a catalyst (i.e., Ni-supported on MgO or γ-LiAlO$_2$) is required to accelerate the reforming kinetics and achieve sufficiently fast conversion to avoid any additional cell performance degradation. Close to equilibrium conversion of about 85% at 650 °C is achievable at the open cell condition and the conversion quickly approaches 100% at fuel utilization of over 50% when the current is drawn from the cell with hydrogen consumption and water generation. Although the internal reforming design offers a potential of over 60% electrical energy efficiency, it also faces special technical challenges on the thermal management, which is further described in the next section.

8.3   CELL COMPONENTS AND CONFIGURATIONS

8.3.1   Single Cell

Similar to other types of fuel cell described in the previous chapters, a molten carbonate fuel cell is composed of molten carbonate electrolyte, immobilized in the lithium aluminate matrix (or tile) by capillary action, and two porous gas-diffusion electrodes, called the cathode and anode, that are held on each side of the electrolyte matrix. Figure 8.1 illustrates the basic cell structure of a molten carbonate fuel cell. Since the electrolyte is in liquid form and molten carbonate is very corrosive, the molten carbonate electrolyte is held in place by the porous electrolyte matrix as an integral part of the cell structure, instead of circulating the molten carbonate electrolyte by a circulating pump. Proper insulation (or even some sort of heating) might be required to prevent the solidification of the molten carbonate electrolyte in the circulation loop if electrolyte circulation is employed. Clearly, a much more elaborate electrolyte circulation loop would be needed compared to the lower temperature fuel cells discussed so far. Thus, mobile electrolyte through circulation is not commonly used for this type of fuel cell.

The molten carbonate fuel cell has sufficiently high operating temperature to allow the electrochemical reactions to proceed very fast on the metal electrode surfaces. Thus, no Noble metals are required separately for catalysis with small activation polarization. The actual electrochemical reactions occur in the three-phase zones near the interface between the electrode and the electrolyte matrix, similar to the case shown in Figure 6.3 for the phosphoric acid fuel cell.

The electrodes are porous for better gas transport for the reactant supply to and product removal from the reaction sites. Due to the corrosive nature of the molten carbonate electrolyte and the oxidizing/reducing environment of the cell, a wetproofing material similar to the PTFE used in low-temperature fuel cells does not exist. Therefore, the establishment of the three-phase zone for the electrochemical reactions, the prevention of the porous electrode structure from being flooded by the molten carbonate electrolyte and the maintenance of the liquid electrolyte in the electrolyte compartment (avoiding electrolyte being punctured) are achieved through a delicate control over the pore-size distributions of both the electrodes and electrolyte matrix. The pore-size distributions determine the distribution of electrolyte among these three components and therefore the degree of gas–liquid contact inside the porous electrodes, which influences the electrode performance.

Other important issues for molten carbonate fuel cell are:

Cell Sealing   In addition to cathode, anode and electrolyte, each cell structure also contains an electrolyte matrix used to hold the liquid electrolyte in place. The matrix structure is composed of a mixture of ceramic powder (usually lithium aluminate, $LiAlO_2$) and carbonate electrolyte. The mixture is semisolid (pastelike) and the molten carbonate electrolyte is immobilized by the capillary force. The resulting matrix structure is impermeable to the reactant gases and it is stiff, but also deformable. This plastic property of the matrix is utilized to provide a gastight seal around the edge (or periphery) of the cell, as gas sealing is a major challenge in high-temperature fuel cells. Such edge sealing technique is often called a **wet seal**. The concept of the wet seal is very similar to the sealing technique used in PEM fuel cell in that both use the electrolyte itself as the sealing material to provide the gas-tight sealing because the electrolyte itself is gas impermeable, and is compatible with the rest of the cell components. In the molten carbonate fuel cell, however, wet sealing of the cell is practically the only feasible sealing technique when the cell housing is made of metals. This is because the carbonate electrolyte is very corrosive and very few other materials can remain stable under MCFC operating conditions. Although high-density alumina and other dense ceramics are suitable as the sealing material, they cannot withstand thermal cycling.

Current Collectors   Current collectors are often used to enhance the electric current collection and reduce the ohmic losses. They are usually made of stainless steel or nickel metal screens and are located between the electrodes and the cell housing for good electrical contact between the two. The cell housing is often made of metal shells with flow distribution channels built on its inside surface for the proper distribution of the gas supply to the respective electrode.

Electrolyte Management   Another unique feature of the molten carbonate fuel cell structure arises from its unique method of electrolyte management. For electrolyte management in PAFCs and PEMFCs discussed in previous chapters, hydrophobic material such as PTFE is used. The dispersed PTFE in the porous electrodes serves both as a binder for the integrity of the electrode structure and as a wet-proofing agent for the establishment of a stable gas–liquid interface. However, such a method cannot be used for MCFCs because similar dewetting materials do not exist in molten carbonate under oxidizing conditions. Hence, capillary equilibrium is used as a means of controlling electrolyte distribution in the porous electrodes, and stable electrolyte/gas interfaces in MCFC porous electrodes (the so-called **three-phase zone**).

For a liquid–gas interface inside a circular tube of diameter d, the capillary pressure can be expressed as

$$P_\sigma = \frac{4\sigma \cos \theta}{d} \tag{8.7}$$

where σ is the surface tension, and θ the contact angle of the electrolyte. Clearly, the capillary pressure is inversely proportional to the tube diameter and commonly small electrode pores of microns or submicron sizes are used to have sufficient capillary pressure to keep the electrolyte in place, even during the possible transients of the gas-stream pressures.

At the steady-state operation, the capillary pressure in the two electrodes and the electrolyte matrix must be balanced to prevent the electrolyte from flowing from one cell component to another. The largest pore sizes in each of the porous cell components that are flooded or filled by the liquid electrolyte is related, according to Equation (8.7)

$$\left(\frac{\sigma \cos \theta}{d} \right)_{cathode} = \left(\frac{\sigma \cos \theta}{d} \right)_{electrolyte} = \left(\frac{\sigma \cos \theta}{d} \right)_{anode} \tag{8.8}$$

Therefore, proper pore sizes of the electrodes and matrix can be devised based on the previous relation and the values of the surface tension and contact angle in the different cell components. The electrolyte distribution in the cell components, and its control, are critical for high cell performance and endurance. As with any liquid-electrolyte fuel cell, electrolyte loss during the cell operation would cause gradual performance decay as well.

Bubble Pressure Layer   One function of the electrolyte is to maintain the separation of the fuel and oxidant stream; this ability of the electrolyte in MCFCs depends on the degree of the matrix filled by the liquid electrolyte. If electrolyte filling becomes too low locally, the large pores where the capillary pressure is low may become void of the electrolyte and reactant gas crossover may occur. Therefore, it is necessary to maintain a continuous liquid electrolyte layer that will keep the fuel and oxidant stream separated under all circumstances — this is important not only for good cell performance, but also for long cell life and safety consideration. Such a continuous liquid electrolyte layer is formed by filling a dense, fine-pore-sized metal (nickel) or ceramic ($LiAlO_2$) structure, and is usually located between the anode and the electrolyte matrix. This layer forms a barrier to prevent gas crossover from one electrode to the other, and is usually called **bubble pressure barrier**. The electrolyte filling of the bubble pressure barrier is guaranteed by the fine-pore sizes in the layer and the enhanced capillary action.

In practice, the bubble pressure layer is usually made an integral part of the tape-cast anode structure from the fine nickel powders, and its ability to maintain the continuous liquid electrolyte layer depends on its pore sizes relative to those of the anode and electrolyte matrix. An appropriate matching of the relative pore sizes can prevent electrolyte drainage from the layer, even under the circumstance of long-term electrolyte loss and cracks developed in the matrix. Typically, if the median pore size of the anode is 4~8 μm, and the median pore size of the electrolyte matrix is 0.5~0.8 μm, then the appropriate median pore size for the bubble pressure layer is 1.0~1.5 μm.

8.3.2   Stack

Similar to any stack of low-temperature fuel cells described earlier, a MCFC stack consists of many basic single cell units like the one shown in Figure 8.1. Cell-to-cell electronic contact is provided by the bipolar plate (also called a **separator plate**). The bipolar plate also separates the reactant gas supply between the adjacent cells, accomplished by the gas flow channels made on both sides of the plate. The metal separator plates provide the main structural support for the stack and they have high mechanical strengths for the purpose. Together with the plastic nature of the electrolyte matrix, the molten carbonate fuel cell can be made in large sizes still free of excessive mechanical stress that is the major limiting factor for the cell size scale-up. Therefore, a typical cell in the stack may have over 1-m^2 size with less than 0.5-cm thick; more than a hundred of such large cells can be found in a stack.

This unusually large cell size as compared to any other type of practical fuel cells results in a number of special technical challenges to MCFC stack design. This is because significant variations of the current density exist and temperature over the cell active surface (in-cell) as well as in the cross-the-cell direction. Therefore, a typical MCFC stack exhibits strong two-dimensional characteristics over the surface of each cell in the stack, and three-dimensional characteristics for the entire stack, leading to the importance and difficulty in the stack thermal management. Recall in Chapter 2, it is shown that a fuel cell operates most efficiently when isothermal. In addition to the performance degradation, the nonuniform distribution of the current density and temperature also results in large thermoelastic stresses, that might cause mechanical failure of the stack components.

The in-cell nonuniformity of current density and temperature distribution is largely affected by the reactant gas-flow distribution in cells such as crossflow or co-flow arrangement for the fuel and oxidant stream; it occurs even when the reactant gas-flow distribution is uniform over the cell active surface. The distribution of the reactant gas flows among cells in a stack largely affects the distribution of temperature and current density among the cells in that stack, hence the heat exchange between the cells. This is because in MCFC stacks, the reactant gas-flow distribution affects not only the local heat generation through the various polarizations, but also the rate of heat removal by convection. An optimal gas flow distribution can provide stable performance and avoids local hot spots and excessive water vapor concentration in the fuel gas, which cause rapid performance decay. Therefore, reactant gas-flow distribution and thermal management are the two most important issues in the design and operation of practical MCFC stacks.

Reactant gases are distributed among the cells in a stack by the stack manifolds and the in-cell flow distribution is provided by the flow channels on the bipolar plates. Therefore, the flow channelling and the manifolding design are the key to the control of reactant flow distribution.

Manifolding   As for the phosphoric acid fuel cells, stack manifolds can be internal or external to a stack, or a combination of them (i.e., internal for one reactant gas and external for another reactant gas) which is rare in practical stack design. Stack manifolding design not only affects the reactant gas distribution among the cells but also determines the overall structure of a stack.

For external manifolding, the reactant gas supply and exhaust removal are on the opposite side of the stack with crossflow configuration for the reactant gases. This offers a simple and symmetrical cell design, leading to low manufacturing cost and allows large manifold cross section with a minimum amount of material, resulting in low-pressure drop in the manifolds — beneficial for uniform gas distribution among the cells, desirable for good stack performance. Further, this design yields large active area per cell because all cell surface area can be active for this construction, hence better utilization of the material and space. As a result, external manifolding with crossflow stack design is very popular with many developers.

On the other hand, externally manifolded stacks are subject to reactant gas leakage and electrolyte displacement by migration from the cells near the positive end towards the cells near the negative end of the stack. These two phenomena are caused by the modified relatively inelastic manifold gas seal and thermal cycling. This is because between the external manifold and the side of the stack, a gas seal is used to prevent gas leakage around the edges of the stack. The gas seal has to be electronic insulator in order to avoid the cells from shorting through the metal manifold. Irrespective of the stack design, stack compression in the direction normal to the cell active surface is necessary for gas sealing around the edge of each cell in the stack and for the minimization of contact resistance between stack components, especially as the stack ages over the operation lifetime. Under this compressive load, the thickness of the porous cell components reduces, appreciably in the early stage of the stack operation, but at a slower rate afterwards for the rest of the stack lifetime. This continuous reduction in the stack thickness cannot be accommodated by the metal manifold and the relatively stiff gasket and gaps or imperfect contacts develop between the manifold and the side of the stack and between stack components. As a result, gas leakages occur along with increased contact resistance among the stack components.

The actual stack gas seal is made of zirconia felt filled with ceramics and a small amount of electrolyte (less than 1% by volume). The gas seal can be made more flexible with an increased amount of the electrolyte ceramic paste in order to reduce the amount of the gas leakage. It is noticed that the gas leakage may be reduced, but cannot be eliminated completely. However, the electrolyte in the gas seal forms an ionic path among all cells in the stack and under the electric field set up by the stack voltage, ion migration results in a net bulk motion of the electrolyte from one end of the stack to another; this not only reduces the stack performance, but also shortens the stack lifetime as well. The ion migration effect is enhanced if more electrolyte is contained in the stack gas seal and a compromise must be made between the need to reduce the gas leakage and the electrolyte displacement.

To avoid these two problems, internal manifolding design has been developed. Usually the manifolds for the fuel and oxidant stream are located around the periphery with the active area of the cell in the interior region. Gas sealing can be achieved by two different techniques: (i) the electrolyte matrix extends to the manifold area in an arrangement similar to the wet seal described earlier, such that the entire periphery of the cell is wet-sealed. This is often referred to as the penetrated electrolyte design and

(ii) the electrolyte matrix only covers the active portion of the cell area with separate gaskets (i.e., dry ring type) located in the manifold area that seal the periphery of the cell, in an arrangement often called nonpenetrated electrolyte design. For internally manifolded stacks, a gas-tight seal is achieved without electrolyte migration.

In addition, internal manifolding also allows flexible flow channelling on the bipolar plates, such as parallel flow (co- or counterflow), crossflow, or any other complex flow configuration as described in the previous chapter for PEM fuel cells, in order to achieve the desired in-cell flow distribution. Stack capacity can also be easily enlarged by stacking more cells, while it is not easy to do so without making a different external manifolds for externally manifolded stack design. Further, stack compression is only in the direction normal to the cell, instead of in all three directions for the externally manifolded stacks. On the other hand, internal manifolding complicates the cell structure, resulting in higher manufacturing and operation cost.

Flow Configuration   Flow configuration over the active cell surface is determined by the flow channel layout made on the bipolar plate. For externally manifolded stacks, crossflow is the only flow arrangement used so far with parallel and straight flow channels so that each flow passes through the cell once. On the other hand, for internal manifolding design, various types of flow channel layout are possible, as mentioned earlier. Because of the large cell sizes, the manifold cross section needs to be large as well if single manifold is used for each of the reactant gas flows. More often, multiple manifolds are employed with parallel flow arrangement with rectangular shape for the cell, in order to provide uniform flow over the cell surface with minimal pressure loss in the stack.

As pointed out earlier, flow configuration has significant impact on the distribution of the cell current density and temperature through the reactant gas distribution. For example, the crossflow arrangement tends to give rise to complicated temperature distributions, thus more difficult in the thermal control and management. The parallel flow leads to small current and temperature gradients in the transverse flow direction, thus the cell length transverse to the flow can be maximized, or hence the rectangular cell design with large aspect ratio. For this flow arrangement, multiple manifolds are necessary. The most predictable temperature distribution in the cell is achieved with co-flow arrangement, which provides uniform temperature distribution in the transverse flow direction with the maximum temperature located at the flow outlet. As a result, thermal control and management is relatively easy because the temperature distribution is known beforehand, and the temperature distribution primarily determines the distribution of the the the thermal stress in the cell structure.

Stack Scaleup   Stack scaleup is necessary in the development of fuel cell technology, typically based on the small-scale model stack developed for earlier test in laboratory. Stack scaleup can be achieved by either increasing the active area of the cell or simply stacking more cells. This is because the stack power output is directly proportional to the full stack voltage and the current drawn from the stack. Increasing the cell size increases the total current drawn from the stack, but the cell size is limited by the allowable maximum temperature and current density differences from the gas flow inlet to the outlet, as well as the maximum mechanical and thermal stress in the cell and stack components that may occur during the assembly and thermal cycling of the stack. Putting more cells into a stack increases the full stack voltage, and it is relatively easy to achieve for internal manifolding design as long as stack compression can be reasonably applied; the major limitation is how to provide uniform gas flow among all the cells in the stack and this normally requires large manifolds which may not be feasible for practical purpose. On the other hand, for external manifolding design changing the external dimensions of a stack requires the accompanying changes in the dimensions of the externally located manifolds and gas seals, implying the modifications of the design and fabrication of these relevant components.

Another practically feasible approach for stack scaleup is to make a stack composed of a number of smaller substacks, often four for symmetry reason. The substack consists of a smaller cell size and number for easy flow arrangement and smaller nonuniformity in the current and temperature distribution, and the overall stack can have a large power output. The substacks can be connected either in parallel or in series or in a combination both electrically and for gas flow. However, this design increases the fabrication and assembly effort, as well as the periphery needed for gas sealing. Therefore, this approach is often simply replaced by using a number of identical stacks for the desired application.

Separator Plate (or Bipolar Plate)   As for other type of fuel cell, the cost of separator plates, including the materials and fabrication, has significant impact on the total cost of the stack itself. For external manifolding design, the separator plate has a simpler structure with straight through flow channel layout. Therefore, the separator plate can be a corrugated metal sheet, or flat with ribbed electrodes similar to the case shown in the previous chapter for PEM fuel cells. A flat separator plate can also be used together with a ribbed current collector next to the electrode. A corrugated separator plate may also be used in conjunction with a permeable flat current collector (i.e., a perforated flat plate).

In contrast, separator plates in internally manifolded stacks have much more complex structures with provisions for the manifolding holes and gas-tight seals. Even provisions are made to accommodate the shrinkage of the porous cell components as the stack ages, hence, the so-called **hard-rail and soft-rail separators**, as shown in Figure 8.2. For hard-rail design, the flow channel on the separator plate usually has a rectangular or square cross section and the thickness of the plate, which determines

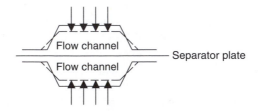

Figure 8.2 Cross-sectional view of a typical separator plate configuration for the soft-rail design.

the intercell separation distance, is fixed. Therefore, the stack may gradually lose compression over its lifetime, leading to poor contact between stack components (hence, higher contact resistance) and even in severe cases reactant gas leakage to the ambient and between the cells in the stack. The soft-rail separator plate has trapezoidal cross section for the flow channels with the channel height, hence the thickness of the plate, variable to accommodate changes in the thickness of the cell components. Thus the soft-rail design are able to maintain good contact between the stack components, and hence low contact resistance over the lifetime of the stack. Other more complex separator plates have also been considered; for example, a separator plate may consist of as many as seven layers of different structures and materials for different purposes.

For externally manifolded stacks, it is extremely important to have separator plates that maintain overall stack dimensions in order to have a good match with the external manifolds throughout the stack lifetime. As a result, thermal expansion of the stack components including the separator plates must be tightly controlled, and proper electrolyte management is essential to maintain the adequate amount of electrolyte in each cell.

8.3.3 System

The MCFC system consists of many ancillary components and subsystems, in addition to fuel cell stacks, to meet the power generation requirements similar to other type of fuel cell systems described so far. Typical performance targets for MCFC in power plants using liquid natural gas as primary fuel is the average cell voltage of 0.8 V at about $150 \, mA/cm^2$ for a fuel utilization of 80%, decay rate of 8 mV/kh or less in long-term operation, and net plant efficiency of 45% based on higher heating value in pressurized operation. To meet these target performances, MCFC efficiency needs to be 60%. That is, about 40% of the chemical energy of the fuel becomes waste heat, which needs to be removed efficiently for stable and continuous operation. Usually anode or cathode gas recycling is adopted as a cooling strategy. Anode exhaust-gas recycling to the cathode inlet is also necessary to close the carbon dioxide balance of the MCFC. These recycling operations must be carried out under pressurized conditions. Other system issues are common to all types of fuel cell and only two issues special to MCFC are described in the following subsections.

Distribution of Current Density and Temperature Because of the high operating temperature, the MCFC stack cooling is often achieved by either anode or cathode gas stream or both. Then the two considerations become important: (i) that the heat capacity of the gas streams is small as compared to liquid water cooling, and significant temperature gradients may exist along the gas flow direction as well as in the transverse direction, or in and across the cells; (ii) that cell potential losses due to activation, ohmic resistance, and mass transfer decrease with cell operation temperature. Local heat generation increases more than linearly with the current density since irreversible (ohmic) heat generation is proportional to the square of the current density J. If heat is not conducted or convected away from high current-density areas, the current density will be amplified, and local "hot spots" will be formed. This

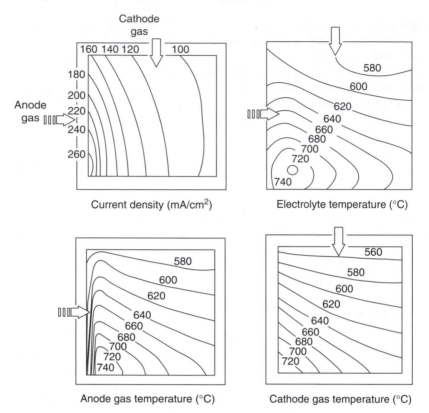

Figure 8.3 Model prediction of the current density and temperature distributions for a 3600-cm^2 cell. Reactant gas composition: $H_2/CO_2/H_2O = 77\%/18\%/10\%$ for the fuel stream and $O_2/CO_2/N_2 = 15\%/30\%/55\%$ for the oxidant stream; utilization of 60% for both fuel and oxidant; inlet temperature of 550 °C for both fuel and oxidant streams. The cell is operated at the atmospheric pressure and the cell voltage of 0.8 V at the average cell current density of 135 mA/cm^2 in the gas surrounding of 600 °C.[1]

self-accelerated mechanism would be detrimental to the cell operation and lifetime. As a result, the temperature distribution throughout the stack will be highly nonuniform, as compared to the low-temperature fuel cells discussed so far in the previous chapters, and thermal management is an important issue. Figure 8.3 shows the distributions of the current density and temperature for over a single cell operating on externally reformed fuel gas. It is seen that for a single cell, the cathode gas has a dominant influence on the temperature distribution in the cell components, including the anode gas and the electrolyte; while the hydrogen concentration in the anode plays the central role in determining the current density distribution. Whereas for a stack, significant deviations can occur from these single-cell patterns of temperature and current density distributions, as shown in Figure 8.4 for a 12-cell short stack. The temperature distribution in a stack is strongly influenced by the thermal conductivity of the separator plate and the thermal conduction between the separator plate and the electrode–electrolyte unit.

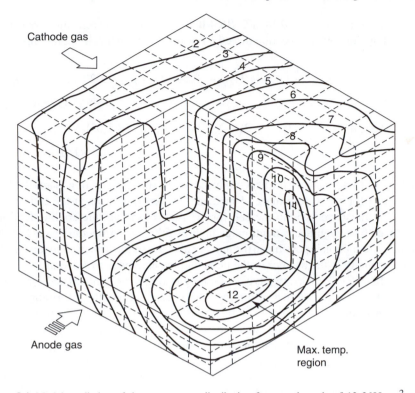

Figure 8.4 Model prediction of the temperature distribution for a stack made of 12 3600-cm^2 cells. Reactant gas composition: $H_2/CO_2/H_2O = 77\%/18\%/10\%$ for the fuel stream and $O_2/CO_2/N_2 = 15\%/30\%/55\%$ for the oxidant stream; utilization of 60% for the fuel and 20% for the oxygen; inlet temperature of 550 °C for both fuel and oxidant streams. The cell is operated at the atmospheric pressure and the cell voltage of 0.74 V at the average cell current density of 150 mA/cm^2 in the gas surrounding of 600 °C [1]. The temperature distribution is: 590.9 °C for Curve 1; 604.0 °C for Curve 2; 617.2 °C for Curve 3; 630.3 °C for Curve 4; 643.4 °C for Curve 5; 656.6 °C for Curve 6; 669.7 °C for Curve 7; 682.8 °C for Curve 8; 699.0 °C for Curve 9; 709.1 °C for Curve 10; 722.2 °C for Curve 11; and 735.4 °C for Curve 12.

Internal Reforming Stacks   A possible simplification in the stack thermal management is to integrate the reforming process inside the stack itself. In conventional MCFCs with external reforming, the heat required for the endothermic reforming reaction is usually provided by the combustion of the anode exhaust gas followed by heat exchange. Then the waste heat produced by the fuel cell reaction in the stack needs to be removed by a cooling medium.

On the other hand, the waste heat from the stack reaction can be used for the endothermic reforming reaction if the reforming occurs inside the stack itself, hence the concept of internal-reforming stacks. It might also be pointed out that the stack waste heat is difficult to be utilized directly for the reforming reaction outside the stack, due to the high temperature required for the reforming process and the inevitable temperature decrease during the heat exchange process. Of course, the indirect use is always possible, such as for the preheat of the reactant gas streams.

For internally reformed MCFC (IR-MCFC), the heat for reforming reaction is provided by the fuel cell reaction directly. For reversible fuel cell operation,

$$H_2 + \frac{1}{2}O_2 \longrightarrow H_2O + E_r + Q; \quad E_r = 1.020\,V;$$

$$Q = -50.53\,kJ/mol \quad (8.9)$$

$$CH_4 + 2O_2 \longrightarrow CO_2 + 2H_2O + E_r + Q; \quad E_r = 1.038\,V;$$

$$Q = 0.25\,kJ/mol \quad (8.10)$$

at 650 °C and 1 atm. Hence the heat produced by the fuel cell reaction in an IR-MCFC is sufficient for steam-reforming of methane. Since the heat from the cell reaction is directly transferred within the cell for the fuel reforming, the cooling load may be reduced by a significant fraction, as high as 50% for an operating stack, thus a significant reduction is possible in the number and size of the heat exchangers, leading to the simplicity in the balance of the plant and lower capital cost overall.

For IR-MCFC stacks, two types of configuration are often adopted for internal reforming. The direct internal-reforming (DIR) design allows the reforming reaction to occur in the anode compartment directly, thus provides direct reforming product for anode electrochemical reaction, direct heat exchange between the heat produced in the cell reaction and the heat needed for steam reforming, and fast reforming reaction and high methane conversion (approaching the equilibrium limit). However, the catalyst particles (and their supports) promoting the reforming reaction are easily attacked by the electrolyte vapor with degraded performance. Furthermore, the reforming reaction is strongly endothermic and causes a pronounced minimum in temperature distribution at the anode inlet, resulting in the need to optimize the temperature distribution or thermal management

The indirect internal-reforming (IIR) design provides a separate reforming chamber, next to but separated from the anode compartment by a partition. Therefore, the reforming catalyst is protected from the electrolyte attack, yielding significantly longer lifetime. However, significant temperature variations occur for IIR stacks because the IIR chamber is usually located every five or six cells, and the endothermic reforming reaction causes temperature gradients.

8.4   MATERIALS AND MANUFACTURING

8.4.1   Cathode

As mentioned earlier, molten carbonate is extremely corrosive, and under the oxidizing atmosphere of air– (or oxygen–) carbon dioxide mixtures in the cathode side, only a few Noble metals are stable as the cathode material. Practically, semiconducting oxides are the only feasible materials for the cathode from the cost point of view. The current choice for the cathode material is the lithiated nickel oxide. It is formed from porous nickel by oxidation and lithiation which occur spontaneously when nickel is in contact with a melt containing lithium carbonate under an oxidizing atmosphere during the initial cell operation. Nickel oxide is slightly soluble, which becomes the limiting factor for the cell lifetime. Hence, alternative cathode materials are being explored.

MCFC cathodes consisting of lithiated NiO start as nickel plaques or sinters which typically have a pre-oxidation porosity of 70%~80%. This is then reduced to 55%~65% by in situ oxidation. They have initially a mean pore size of $10\,\mu m$ but develop a bimodal distribution with median diameter of 5~7 μm. The smaller pores (or micropores) are filled (or flooded) with the liquid electrolyte and form the three-phase zone needed for electrochemical reactions and cross section for the ionic conduction path. The larger pores (or macropores) remain open and provide the path for the diffusion of gas into the interior of the electrode. Lithiation of the in situ formed NiO yields electronic conductivity of about $5/(\Omega \cdot cm)$. It is also possible to fabricate NiO ex situ from NiO powders.

The cathodic polarization is affected by the thickness of the cathode. Ohmic losses in the liquid conductor (electrolyte) and solid conductor (NiO) increase with thickness; gas diffusional losses in the gas phase also increase with the thickness. However, losses due to the liquid diffusion resistance and kinetic activation decrease as the small pore area increases. Therefore there exists an optimal cathode thickness, which yields minimum cathodic polarization. The optimal thickness also depends on gas composition and current density as well as other operating conditions. At present the optimum thickness ranges from 0.4 to 0.8 mm.

The cathodic potential losses are also influenced by the degree of electrolyte filling and macropore gas channelling of the cathode. The optimal filling, or the fraction of the cathode pore volume filled by the liquid electrolyte, is around 15%~30%. It should be noted that lithiated NiO is wetted completely by the molten carbonate electrolyte. For such a wetting surface, the cathode is easily flooded by the electrolyte if the pore size distribution is not properly controlled.

8.4.2   Anode

The anode usually operates under a reducing atmosphere at a low potential, typically about 0.7~1.0 V more negative than that of the cathode. Many metals are suitable as electrocatalyst for hydrogen oxidation under the above conditions. Nickel, cobalt, and copper can be used as anode materials, often in the form of powdered alloys and composites with oxides. As a porous metal structure, it is subject to sintering and creeping under the compressive force necessary for stack operation. Additives such as chromium or aluminum form dispersed oxides and thereby increase the long-term stability of the anode with respect to sintering and creeping.

MCFC anodes are at present made from porous sintered nickel, containing a few percent of chromium or aluminum which is oxidized in situ and forms submicron particles of $LiCrO_2$ or $LiAlO_2$ on the surface of the Ni particles forming the sinter. The function of the oxides is to prevent sintering of the Ni particles and to stabilize the sinter against creeping, which tends to occur in a stack under compression. Sintering leads to increased particle size and therefore reduced capacity of the anode to hold electrolyte. Creep is the microdimensional deformation under mechanical load, results in decreased porosity, increased contact resistance and increased risk of anode-to-cathode gas leaks. Dispersed oxide strengthening, such as by chromium or aluminum addition, has been shown to be effective against both sintering and creep.

Nickel is not wetted completely by molten carbonate electrolyte in a reducing atmosphere, though Ni wetting depends largely on the distribution of wettable oxide particles over the pore walls. The anode needs less internal surface area than the cathode because the electrode kinetics of hydrogen oxidation and mass transfer in the micropores of the Ni anode are more rapid than the corresponding processes in the cathode. Therefore, the thickness of the anode could be less than that of the cathode, the minimum thickness might be 0.4~0.5 mm for atmospheric operation. Because the anode is much less sensitive to overfilling than the cathode, the anode can serve as an electrolyte reservoir. Therefore, the anode is usually made thicker and filled to 50% ~ 60% of its pore volume with electrolyte for its storage. Typically the anode is about 0.8~1.0 mm in thickness.

8.4.3   Electrolyte

The electrolyte of choice is a mixture of lithium carbonate (Li_2CO_3) and potassium carbonate (K_2CO_3) and possibly smaller amounts of sodium carbonate (Na_2CO_3) and carbonates of earth-alkaline metals, with a melting point of approximately 500 °C. Optimization of the electrolyte composition and cell operating temperature is extremely important, as they affect ohmic resistance of the cell, cell polarization (gas solubility and oxygen reduction kinetics) and nickel oxide solubility (which limits cell lifetime).

Cell performance (i.e., the cell potential at a given current density) depends on the ohmic resistance of the cell and the activation polarization at the electrodes. Li-rich electrolytes have higher ionic conductivities and hence lower ohmic losses, that is, resistance for Li_2CO_3 is smaller than that for Na_2CO_3 and K_2CO_3. But the solubility and diffusivity of reactant gases, such as H_2, O_2, H_2O, and CO_2 are lower in Li_2CO_3-rich melts. The dissolution of the cathode material NiO is lower in Li–Na than in Li–K carbonate melt. Li–Na in eutectic or off-eutectic composition has better performance with respect to NiO cathode dissolution, electrolyte creepage, and volatilization as well as conductivity and cathode polarization. Therefore, significant interest remains in the optimization of the electrolyte composition, particularly Li–Na carbonate eutectic, despite the fact that most developers after 1975 have been using the electrolyte composition of 62% Li_2CO_3/38% K_2CO_3 (by mole) eutectic. For such eutectic mixture, the temperature at the melting point is 488 °C.

Usually a porous electrolyte matrix is used to hold molten carbonate in place through capillary effect, such that the molten carbonate-filled matrix occupies the electrolyte compartment, which prevents the reactant gases from crossover to the opposite electrodes and it is ionically conductive but electronically insulating. The porous matrix is typically made of ceramic power, such as lithium aluminate ($LiAlO_2$). It is known that $LiAlO_2$ in the γ phase is the most stable form in Li_2CO_3–K_2CO_3 electrolyte among the three allotropic forms of $LiAlO_2$ (α, β, and γ). The electrolyte-filled matrix structure usually possesses about 40% $LiAlO_2$ and 60% carbonate by weight. This is because the structure would be too rigid if the carbonate content is too low and too soft (fluidic) if the carbonate content is too high. To obtain the pastelike plastic property of the structure, the carbonate content has to be within an proper range that is determined by the distribution of the $LiAlO_2$ particle size.

In fact, the important properties of the electrolyte layer, such as the carbonate content, mechanical properties, and effective conductivities, for the cell performance and integrity during the fabrication and operation depend on the size, shape and distribution of the $LiAlO_2$ particles. It has been found that the elongated rod or fiber shape of the particles with submicron sizes is the optimal combination and it naturally yields the adoption of papermaking techniques as the fabrication methods such as hot pressing and tape casting.

The hot-pressing technique involves the pressing of the $LiAlO_2$-carbonate mixture at high pressures (around 5000 psi or 3.4 MPa) and temperatures (just below the melting point of the carbonates). The operation is simple but the resulting electrolyte structure is too porous (>5%) with poor microstructure, leading to limited sizes (<1 m^2) and crack formation; and is too thick (around 1–2 mm) that yields high ohmic resistance (the specific conductivity is around 0.3 S/cm).

Tape-casting fabrication has been generally accepted as the most suitable technique for molten carbonate electrolyte tile. It involves several steps in the process. First, $LiAlO_2$ particles are dispersed in an organic solvent that contains organic binders, plasticizers, and additives. The resulting mixture is then cast over a smooth substrate, and a knife-edge device is used to control the thickness of the tape formed. Continuous production is easily implemented if the substrate is put on a moving belt conveyor. The cast tape is dried and assembled into the fuel cell structure and the organic binder is burned out (evaporated) by heating to about 250 °C–300 °C. Tape casting is also used to produce electrode structures in a similar manner. Tape casting can produce a much thinner electrolyte structure, hence much smaller ohmic resistance.

Table 8.1 presents the characteristics of state-of-the-art molten carbonate fuel cell components, and Table 8.2 provides a quick overview of the progress in the cell component technology over the past half century or so.

8.4.4   Separator Plate

The separator plate has normally fuel gas-flow channels built on one side and oxidant gas-flow channels on the other side, and hence it is exposed to both reducing and oxidizing environment. In addition, it is under the corrosive attack of the carbonate electrolyte because the electrolyte can reach the plate surface through creepage and through evaporation and then condensation on the colder separator plate surface. Therefore, it is still an ongoing effort to find economic materials for the plate. The economics of the selected material include low cost for the material itself and for the fabrication of the plate (or easy for fabrication).

The technique to determine the suitability of a material for the separator is the dual-atmosphere test under the high-risk (i.e., extreme) exposure conditions. The dual-atmosphere test means exposure of the candidate separator material to fuel gas on one side and oxidant gas on the other side at the MCFC operating temperature and pressure. Similarly, single atmosphere test means the exposure to a single reactant-gas environment. The extreme exposure conditions include high hydrogen fuel and lean oxidant, especially near the fuel gas exit and in the wet seal area, and hence the wet

Table 8.1   Characteristics of State-of-the-Art Cell Components for Molten Carbonate Fuel Cells (Adapted from [2, 3])

Component	Property	Current Status
Anode	Material	Ni with 2%–20% Cr/Ni–Al
	Thickness	0.5–1.5 mm
	Porosity	50%–70%
	Pore size	3–6 μm
	Surface area (BET)	0.1–1 m^2/g
Cathode	Material	Lithiated NiO (NiO with 1%–2 wt.% Li)
	Thickness	0.4–0.75 mm
	Porosity	70%–80%
	Pore size	7–15 μm
	Surface area (BET)	0.15 m^2/g (Ni pretest)
		0.5 m^2/g (post-test)
Electrolyte	Material	Alkali carbonate mixture
	Composition	62% Li$_2$CO$_3$–38% K$_2$CO$_3$ by mole
		50% Li$_2$CO$_3$–50% Na$_2$CO$_3$ by mole
		70% Li$_2$CO$_3$–30% K$_2$CO$_3$ by mole
Electrolyte-support matrix	Material	γ-LiAlO$_2$
	Thickness	1.8 mm (hot pressed)
		0.5 mm (tape cast)
	Pore size	0.5–0.8 μm
	Surface area	0.1–12 m^2/g
Electrolyte-filled matrix	Composition	40%–50% LiAlO$_2$ by weight (34%–42% by volume)
		50%–60% carbonates
Current collector	Anode	Ni or Ni-plated steel (perforated) 1-mm thick
	Cathode	Type 316 (perforated) 1-mm thick

seal area on the anode side, especially around the anode gas exit, represents the most severe condition for the plate corrosion.

A whole array of the commercially available Fe- and Ni-base alloys have been tested in order to find the suitable materials for the separator. According to these single- and dual-atmosphere tests, ss 310S, Avesta 600 (FE-28Cr-4Ni-2Mo), and Inconel 601 (NiCr + Al) exhibit the best corrosion resistance in both atmospheres; and high-Cr ($>$21%) austenitic steels such as ss 310 could have a lifetime of at least 20,000–30,000 h when exposed to the cathode environment. On the other hand, severe corrosion occurs for ss 316L and ss 310 when exposed to the anode environment, unless a dense layer of protective surface coatings such as nickel cladding are used. Aluminum oxide (Al$_2$O$_3$) coating is effective for the wet seal areas, since it is transformed to LiAlO$_2$ in situ. Various surface treatment techniques are being investigated for their effectiveness, and Figure 8.5 illustrates the various effective surface protective layers when the separator plate is exposed to the various conditions. Most of studies so far indicate that only aluminum diffusion coating and perhaps high-aluminum stainless steel can provide sufficient protection against corrosion in

Table 8.2   Evolution of Molten Carbonate Fuel Cell Components[4]

Component	ca. 1965	ca. 1975	Current Status
Anode	Pt, Pd, or Ni	Ni with 10 wt% Cr	Ni-10 wt% Cr/Ni–Al 3–6 μm pore size 50%–70% initial porosity 0.5–1.5-mm thickness 0.1–1 m^2/g
Cathode	Ag$_2$O or lithiated NiO	Lithiated NiO	Lithiated NiO 7–15 μm pore size 70%–80% initial porosity 60%–65% after lithiation and oxidation 0.5–0.75-mm thickness 0.5 m^2/g
Electrolyte support matrix (or tile)	MgO	Mixture of α-, β- and γ-LiAlO$_2$ 10–20 m^2/g	γ-LiAlO$_2$ 0.1–12 m^2/g 0.5-mm thickness
Electrolyte[a]	52 Li-48 Na 43.5 Li-31.5 Na-25 K	62 Li–38 K ~60%–65 wt%	62 Li–38 K 50 Li–50 Na 50 Li–50 K ~50 wt%
	"Paste"(Electrolyte filled matrix or tile)	Hot-press "tile" 1.8-mm thickness	Tape cast 0.5-mm thickness

[a] Figures in this row are mol/% of alkali carbonate salt unless otherwise stated.

the wet seal area, especially on the anode side. However, for aluminum-containing steels the surface protective aluminate layer formed has excessive contact resistance to current flow. The aluminum diffusion coating is achieved through vapor deposition at high temperature under a reducing atmosphere, and hence the process is slow and expensive as well. Consequently, the search for suitable alloy materials for the separator is still actively underway, especially when the surface treatment can be avoided.

8.5   PERFORMANCE OF MCFCS

As for all other types of fuel cell, the performance of a single fuel cell or fuel cell stack is characterized by voltage versus current data, the so-called **polarization curve**, for specified operating conditions. These conditions include the temperature and pressure of operation, and the inlet compositions of the fuel and oxidant gases, and the utilization of the fuel and oxidant gases.

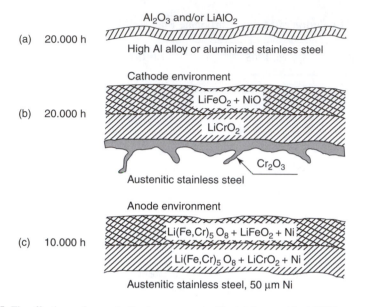

Figure 8.5 The effective surface protective layers on austenitic stainless steel for MCFC separator plates exposed to (a) wet seal; (b) cathode, and (c) anode environment.[3]

The zero-current or open-circuit voltage (OCV) of a fuel cell is identical to the reversible cell potential, provided that the cell is in thermodynamic equilibrium. This is usually a good assumption for high temperature cells except, for example, when corrosion reactions occur at a very high rate, in which case the mixed potential established between the main electrode reaction and corrosion reaction may deviate considerably from the equilibrium value for the main reaction.

The reversible cell potential for a MCFC depends on the gas composition at the fuel electrode (partial pressures of hydrogen, water vapor and carbon dioxide) and at the cathode (partial pressures of oxygen and carbon dioxide) and it can be written for the reaction given in Equations (8.1)–(8.3) as follows

$$E_r(T, P_i) = E_r(T, P) + \frac{RT}{nF} \ln\left\{ \left[\frac{P_{H_2} P_{O_2}^{1/2}}{P_{H_2O}} \right] \left[\frac{P_{CO_{2,c}}}{P_{CO_{2,a}}} \right] \right\} \tag{8.11}$$

where the subscript "c" and "a" represents the partial pressure of CO_2 at the cathode and anode compartment, respectively. The values of the reversible cell potential $E_r(T, P)$ at $P = 1$ atm and $T = 650\,°C$ are given in Table 8.3 for a number of fuel cell reactions involving hydrogen, methane (natural gas) and carbon monoxide. Carbon monoxide is included since it is the product of the methane reforming reaction, and the importance of the water gas shift reaction in MCFCs.

Since the primary fuel for the MCFC is not pure hydrogen gas, rather hydrocarbon fuels, the reformed fuel gas mixture contains various amount of hydrogen, depending on the primary fuel and the reforming process used. The effect of fuel gas

Table 8.3 Reversible and Thermoneutral Cell Potential for a Number of
Fuel Cell Reactions at 1 atm and 650 °C Relevant to MCFCs

Reaction	ΔG (kJ/mol)	E_r (V)	ΔH (kJ/mol)	E_{tn} (V)
$H_2 + \frac{1}{2}O_2 \longrightarrow H_2O$	−196.92	1.020	−247.45	1.282
$CH_4 + 2O_2 \longrightarrow CO_2 + 2H_2O$	−800.89	1.038	−800.64	1.037
$CH_4 + H_2O \longrightarrow CO + 3H_2$	−7.62	0.010	+224.72	−0.291
$CH_4 + CO_2 \longrightarrow 2CO + 2H_2$	−2.04	0.003	+260.62	−0.338
$CH_4 \longrightarrow C + 2H_2$	−16.66	0.043	+89.26	−0.231
$CO + \frac{1}{2}O_2 \longrightarrow CO_2$	−202.51	1.049	−283.01	1.467
$CO + H_2O \longrightarrow CO_2 + H_2$	−5.58	0.029	−35.56	+0.184
$2CO \longrightarrow C + CO_2$	−14.62	0.076	−171.36	+0.888

Note: The corresponding changes in the Gibbs function and enthalpy are also given for the reactions.

composition on the reversible cell potential, as determined from Equation (8.11), is minimal. Between the most H_2-rich (as high as about 80% dry gas basis) and H_2-poor (as low as over 20% dry) compositions, the difference in the reversible cell potential may range only about 70~80 mV, although the difference in the actual cell potential would be much larger due to other loss mechanisms under the actual operating conditions.

The actual cell performance is strongly influenced by the reactant gas-flow configuration, because the local current density across the thin electrode–electrolyte–electrode assembly depends on the local thermodynamic driving force (i.e., the reversible cell voltage E_r), and the local resistance (ohmic and activation polarization). The cell reversible potential E_r changes in the cell along the reactant gas flow direction from the inlet to the outlet due to the variations in the concentrations of fuel and oxidant gas, resulting from the electrochemical consumption in the cell, as described in Chapter 2. The decrease in E_r corresponding to the concentrations at the inlet and outlet may be significant for large cells, and needs to be considered in practice.

The actual cell potential E is equal to the local reversible cell potential E_r minus the losses due to the local ohmic and activation overpotential, or

$$E = E_r - JR_t - J(Z_a + Z_c) \qquad (8.12)$$

where J is the local current density, R_t is the local ohmic resistance, and Z_a and Z_c are the local resistance at the anode and cathode electrode, respectively. These electrode resistance includes contributions from the activation and transport processes, and is normally also dependent on the current density as well. Even though every term on the righthand side of Equation (8.12) varies locally from point to point, the net effect or the actual cell potential E must be a constant, independent of the locations because the cell housing or separator plate is, by design, a good electronic conductor, and hence iso-potential plane.

The local current density may be determined from the change in the reactant flow rate, such as,

$$J = -nF \left(\frac{dN''_{H_2}}{dx} \right) b \tag{8.13}$$

where N''_{H_2} is the molar flux of hydrogen in the fuel gas-flow channel along the flow direction, and b is the width of the fuel gas channel (see Section 4.6.3). The electrode resistance may be determined from modelling the electrode reaction and transport processes, both of them are influenced by the temperature and concentrations. An empirical correlation has been developed by Yuh and Selman[5] from the performance measurements for small laboratory cells with negligible reactant utilization:

$$Z_a = 2.27 \times 10^{-5} \times \left(P_{H_2} \right)^{-0.42} \left(P_{CO_2} \right)^{+0.17} \left(P_{H_2O} \right)^{+1.0} \exp \left(\frac{53,500}{RT} \right) \tag{8.14}$$

$$Z_c = e^{-11.8} \times \left(P_{O_2} \right)^{-0.43} \left(P_{CO_2} \right)^{-0.09} \exp \left(\frac{77,300}{RT} \right) \tag{8.15}$$

where Z is the area specific resistance in $\Omega \cdot cm^2$, and the partial pressures are in atm, the universal gas constant R is in J/(mol $\cdot$ K), and the temperature T is in Kelvin. The local partial pressures of the reactants and products that are needed to determine the reversible cell potential, Equation (8.11), and the area specific resistance Z_a and Z_c, are obtained from a mass balance for each component. Then the cell performance can be determined from this formulation.

Over the past several decades, the performance of single cells has been improved significantly, and the power density increased from about $10\,mW/cm^2$ to more than $150\,mW/cm^2$. During the 1980s both performance and endurance of MCFC stacks showed dramatic improvements. Figure 8.6 illustrates the progress of single cell performance operating at 650 °C and 4.4 atm (65 psia) for low-Btu fuel, i.e., about 17% ($H_2 + CO$).

8.5.1   Effect of Temperature

From Equation (8.12), it is clear that the effect of temperature on the cell potential of MCFCs is through the effect on the reversible cell potential and on the irreversible voltage losses. The influence of temperature on the reversible potential E_r is quite complex for molten carbonate fuel cell, and the complexities arise from the equilibrium composition of the fuel gas due to the shift and reforming reactions taking place at the anode.

In addition to the electrochemical reaction, Equation (8.2), at the anode electrode, several chemical reactions also occur in the anode, including the water–gas shift reaction shown in Equation (8.4), methanation (steam reforming) reaction given in Equation (8.5), carbon deposition (or Boudouard) reaction in Equation (8.6), and

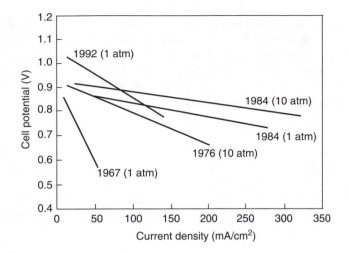

Figure 8.6 Progress in the performance of the MCFC operating on reformate gas and air.[4]

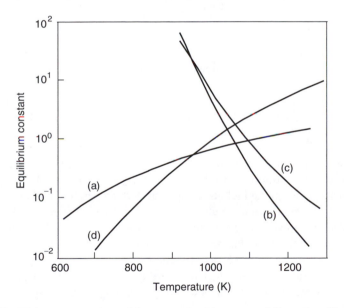

Figure 8.7 Equilibrium constants for partial pressures for: (a) water–gas shift reaction, (b) methane formation (steam reforming) reaction, (c) carbon deposition (Boudouard) reaction, and (d) methane decomposition reaction.[6]

methane decomposition reaction

$$CH_4 \rightleftharpoons C + 2H_2 \qquad (8.16)$$

The temperature dependence of the equilibrium constants for these four equilibrium reactions is shown in Figure 8.7. It is seen that the carbon deposition reaction will suppress the formation of carbon when the temperature is increased, while the methane decomposition reaction will favor the carbon formation. Hence, it becomes clear that for a fixed gas composition of H_2, H_2O, CO, CO_2 and CH_4, there is a critical temperature, T_{c1}, below which the exothermic Boudouard reaction is thermodynamically favored and a temperature, T_{c2}, above which carbon formation by the endothermic decomposition of CH_4 is thermodynamically favored. Therefore, the operating temperature of molten carbon fuel cell falls in the window defined by these two critical temperatures.

To avoid the carbon formation via Boudouard and methane decomposition reaction, sufficient quantity of water is added to anode gas stream in order to promote the competing methane reforming and water–gas shift reaction. As pointed out in Section 8.2, the water–gas shift reaction rapidly reaches equilibrium at the MCFC anode, and consequently CO serves as an indirect source of H_2. The equilibrium constant for partial pressure, defined as

$$K_P(T) = \frac{P_{CO}\, P_{H_2O}}{P_{CO_2}\, P_{H_2}} \tag{8.17}$$

increases with temperature as shown in Figure 8.7, thus the equilibrium composition changes with temperature, utilization and pressure, and as a result, the cell reversible potential is influenced.

The effect of temperature on E_r through the equilibrium composition change as described previously may be best illustrated with the following example: Consider a cell with an oxidant gas mixture of 30% O_2/60% CO_2/10% N_2 and a fuel gas mixture of 80% H_2 and 20% CO_2 on a dry basis. The fuel gas is then saturated with H_2O vapor at the room temperature of 25 °C before being fed to the anode, and its composition now becomes 77.5% H_2, 19.4% CO_2 and 3.1% H_2O. The concentrations of each species in the fuel gas at any elevated temperature and pressure can be calculated by the equilibrium condition established by the water–gas shift reaction. The calculation involves using the equilibrium constant for partial pressures, and the equilibrium concentrations obtained are used in the reversible cell-potential equation, Equation (8.11), to determine E_r as a function of the cell operating temperature and pressure. The results of such calculations are given in Table 8.4 at the pressure of 1 atm, and indicate that E_r decreases with an increase in temperature. This is clear as

$$\left(\frac{\partial E_r}{\partial T}\right)_P = \frac{\Delta S}{nF} \quad \text{and} \quad \Delta S < 0$$

The results shown in Table 8.4 also indicate a weak dependence of the reversible cell potential on temperature. However, the actual cell potential is a much stronger function of temperature due to the significant reduction of various polarizations with temperature. Thus the net effect is that E increases with T. For small cells (8.5 cm^2) operating at 200 mA/cm^2 on constant flow rates (hence variable utilization with current density) of steam-reformed natural gas as the fuel and 30% CO_2/70% air as the

Table 8.4   Effect of Temperature on Equilibrium Composition of Fuel Gas
Mixture Initially Saturated with Water Vapor at 1 atm and 25 °C

Mole Fraction	Temperature (K)			
	298	800	900	1000
H_2	0.775	0.669	0.649	0.643
CO_2	0.194	0.088	0.068	0.053
CO		0.106	0.126	0.141
H_2O	0.031	0.137	0.157	0.172
E_r (V)		1.155	1.143	1.133
K_P		0.2474	0.4538	0.7273

Note: The reversible cell potential is determined with the oxidant gas composition of 30% O_2, 60% CO_2 and 10% N_2. The total pressure is maintained at 1 atm.[4]

oxidant, the temperature effect is correlated as follows.[4]

$$\Delta E_T \ (\text{mV}) = K_T(T - T_1) \tag{8.18}$$

where the temperature coefficient K_T is found to be 2.16 mV/K for 575 °C $\leq T <$ 600 °C; 1.40 mV/K for 600 °C $\leq T <$ 650 °C; and 0.25 mV/K for 650 °C $\leq T <$ 700 °C. Referring to Equation (8.12), and for temperature from 575 to 600 °C, about 1/3 of the total change in E with T is due to change in ohmic losses, and the rest is due to change of activation polarization at the anode and cathode.

Empirical data also suggest that for MCFCs, an operating temperature of 800 °C~900 °C might be optimal for the cell and system performance because such MCFCs match with coal gasifiers nicely. However, materials problems such as the electrolyte evaporative loss and component corrosion become excessive and the performance gain beyond 650 °C is diminishing with higher temperature, as the empirical correlation, Equation (8.18), suggests. Hence, an operating temperature of 650 °C offers an optimization of good performance and long lifetime. A better lifetime could be achieved with lower operating temperature. It might be mentioned that most carbonates do not remain molten below about 520 °C.

8.5.2   Effect of Pressure

The effect of pressure on cell performance can also be divided into two parts: The reversible cell potential and the irreversible potential losses (polarizations). The effect of pressure on the reversible cell potential E_r is determined from the Nernst Equation (8.11). Assuming the total pressure is the same for both anode and cathode gases and all the gas compositions remain the same, the change in E_r is, according to Equation (8.11)

$$\Delta E_{r,P} = E_r(T, P_2) - E_r(T, P_1) = -\frac{(\Delta N)RT}{nF} \ln\left(\frac{P_2}{P_1}\right) \tag{8.19}$$

for a change in pressure from P_1 to P_2. For the overall cell reaction of Equation (8.3): $H_2 + 0.5O_2 \rightleftharpoons H_2O(g)$,

$$\Delta N = 1 - (1 + 0.5) = -0.5 \text{ mol/mole of fuel}$$

Equation (8.19) shows that the change in the reversible cell potential due to the pressure effect is

$$\Delta E_{r,P} = \frac{RT}{2nF} \ln\left(\frac{P_2}{P_1}\right) \tag{8.20}$$

Then at the typical cell operating temperature of 650 °C, the pressure effect on the reversible cell potential becomes

$$\Delta E_{r,P} = 20 \ln\left(\frac{P_2}{P_1}\right) = 46 \log\left(\frac{P_2}{P_1}\right) \quad \text{(mV)} \tag{8.21}$$

Therefore, a 10-fold increase in the cell pressure corresponds to an increase of 46 mV in the reversible cell potential at 650 °C, which indicates a small effect of the cell operating pressure on the reversible cell potential.

For practical MCFCs higher operating pressure leads to enhanced cell potential E due to the increase in the partial pressures of the reactants (i.e., the reactant concentrations), electrode kinetics, gas solubilities and mass transport rates. On the other hand, higher pressure results in undesirable side effects such as carbon deposition due to the Boudouard reaction: $2CO \rightleftharpoons C + CO_2$ and methane formation due to methanation reaction: $CO + 3H_2 \rightleftharpoons CH_4 + H_2O$. In addition, decomposition of CH_4 to carbon and hydrogen is possible $CH_4 \rightleftharpoons C + 2H_2$. According to the Le Chatelier principle, an increase in pressure will favor the reaction to proceed in the direction of volume reduction (hence, the number of mole reduction). Therefore, higher pressure increases carbon deposition and methane formation and inhibits CH_4 decomposition. However, the water–gas shift reaction $CO_2 + H_2 \rightleftharpoons CO + H_2O$ is not affected significantly by an increase in the cell pressure.

Carbon deposition in MCFCs should be avoided because it can plug the gas passages in the porous anode. Methane formation is detrimental to cell performance since each mole of methane formation consumes three moles of hydrogen, which represents a considerable loss of reactant and would reduce the power plant efficiency.

The addition of water and carbon dioxide to the fuel gas modifies the equilibrium gas composition so that the formation of methane is minimized. Carbon deposition can be avoided by increasing the partial pressure of water vapour in the gas stream. Thus, methane formation and carbon deposition at the anode in an MCFC operating on coal-derived gas fuels can be controlled. Carbon deposition boundaries for the C–H–O system can be determined theoretically from the thermodynamic equilibrium calculations. However, experience indicates that carbon deposition can be avoided completely, even for MCFCs operating at the high pressure of 10 atm with a high-CO fuel gas, as long as the gas mixture is sufficiently humidified (e.g., for the high-CO

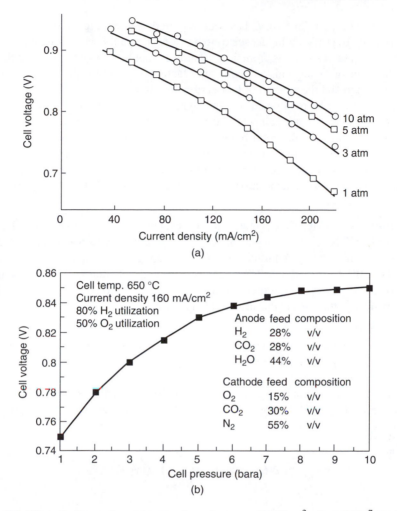

Figure 8.8 Effect of cell operating pressure on the performance of a 300 cm^2 cell at 650 °C[7]. Fuel gas-molar composition: 28% H$_2$, 28% CO$_2$, and 44% N$_2$; oxidant gas-molar composition: 15% O$_2$, 30% CO$_2$, and 55% N$_2$. Fuel utilization: 80%; and oxidant utilization: 50%.

gas at 10 atm, the gas should be saturated with water vapor at 163 °C or higher temperature).

Figure 8.8 shows the pressure effect on the actual cell potential E. At typical operation conditions (160 mA/cm^2 and 650 °C), the potential gain in E due to the reactant gas pressure change may be correlated in the form similar to Equation (8.21)

$$\Delta E_P = k_P \log\left(\frac{P_2}{P_1}\right) \tag{8.22}$$

where $k_P = 104, 84, 76.5$ (mV) has been reported. Some prefers for $k_P = 76.5$ mV, but the value 104 (mV) or higher seems to be more compatible with Equations (8.12)–(8.15) for pressures ranging from 1 to 10 atm. It should also be recognized that the specific value of k_P depends on many other factors such as the operating conditions, the cell design and the method of fabrication, and so on.

It might also be mentioned that the significant pressure effect on the actual as compared to the reversible cell potential results in large part due to the reduction in the cathode resistance and the contribution from the anode resistance reduction is small. This might be because the benefit from the enhanced pressure operation is offset by the adverse effect of reactant consumption via the gas-phase reactions such as the methanation and carbon deposition reactions that are favored at higher pressures.

Example 8.1   A MCFC is operated at 650 °C. The fuel gas contains 28% H_2, 28% CO_2 and 44% N_2 on a dry molar basis, and is saturated with water vapor at 25 °C and 1 atm before being fed to the cell. The oxidant gas has a molar composition of 15% O_2, 30% CO_2, and 55% N_2. Determine the cell potential gain ΔE_P due to the change in the cell operating pressure from 1 to 3, 5 and 10 atm, respectively, if the cell operates at the current density of 160 mA/cm^2.

SOLUTION
We will try to solve this problem by two different approaches: first we will use Equation (8.22) and then Equations (8.12)–(8.15), and the results from the two approaches will be compared as well.

(i) By using the empirical correlation, Equation (8.22), the cell potential gain can be determined easily if the pressure coefficient k_P is fixed. For the moment, we shall choose $k_P = 104$ (mV) so that we have

$$\Delta E_P = 104 \log\left(\frac{P_2}{P_1}\right) \ (\text{mV})$$

The reason for this choice of the pressure coefficient becomes clear when we compare the results obtained from the second approach later on. It should be remembered that this correlation was developed for MCFCs operating at 650 °C and 160 mA/cm^2 and the pressure coefficient would become smaller if the current density is lower.

For the given condition, we have $P_1 = 1$ atm, and $P_2 = 3, 5$, and 10 atm, respectively. Therefore, we obtain correspondingly,

$$\Delta E_P = 49.6, 72.7, \text{ and } 104 \ (\text{mV})$$

for the three P_2 values.

(ii) By using Equations (8.12)–(8.15):
From Equation (8.12), we have

$$E = E_r - JR_t - JZ_c - JZ_a$$

Then the cell potential gain due to the pressure increase can be written as

$$\Delta E_P = \Delta E_{r,P} - \Delta E_{c,P} - \Delta E_{a,P}$$

where we have assumed that the total cell ohmic resistance remains unchanged with the pressure change, a reasonable assumption for MCFCs.

The reversible cell potential change due to the pressure change is given by Equation (8.21), or

$$\Delta E_{r,P} = 46 \log\left(\frac{P_2}{P_1}\right) \quad (mV)$$

For the given condition of $P_1 = 1$ atm, and $P_2 = 3, 5$, and 10 atm, we have, respectively,

$$\Delta E_{r,P} = 21.9, 32.2, \quad \text{and} \quad 46 \quad (mV)$$

for the three P_2 values.

The potential change at the anode and cathode can be obtained from the electrode resistance given by Equations (8.14) and (8.15), respectively. At the given temperature and pressures, the fuel and oxidant gas mixtures can be taken as ideal gas mixture, hence we have

$$P_i = X_i P$$

where P_i and X_i are the partial pressure and mole fraction of the species i in the mixture, and P is the total pressure of the mixture. Then, we have for the cathode

$$\Delta E_{c,P} = J\left(Z_{c,P_2} - Z_{c,P_1}\right)$$
$$= \left[Je^{-11.8}\left(X_{O_2}\right)^{-0.43}\left(X_{CO_2}\right)^{-0.09}\exp\left(\frac{77,300}{RT}\right)\right]\left(P_2^{-0.52} - P_1^{-0.52}\right)$$

Substituting in the parameter values of $J = 160$ mA/cm^2 for the current density, $R = 8.314$ J/(mol $\cdot$ K) for the universal gas constant, $T = 650 + 273 = 923$ K for the temperature, and the given molar composition of $X_{O_2} = 0.15$ and $X_{CO_2} = 0.3$ for the oxidant gas, the previous equation can be simplified to

$$\Delta E_{c,P} = 71.7\left(P_2^{-0.52} - P_1^{-0.52}\right) \quad (mV)$$

where the pressure P_1 and P_2 are in atm. Therefore, the cathode potential loss becomes

$$\Delta E_{r,P} = -31.2, -40.7, \quad \text{and} \quad -50.0 \quad (mV)$$

corresponding to $P_1 = 1$ atm, and $P_2 = 3, 5$, and 10 atm, respectively. The minus sign for the cathode potential loss indicates that a potential gain is made in the cathode when the pressure is increased.

For the anode potential change, we need to calculate the fuel gas composition including the water vapor content. the saturation pressure of water at $25\,°C$ is, from Appendix A1

$$P_{sat} = 3.169 \text{ kPa}$$

hence, the mole fraction for the water vapor in the fuel gas is

$$X_{H_2O} = \frac{3.169}{101.325} = 0.031$$

since the saturation occurs at 1 atm (= 101.325 kPa). Then the mole fraction for H_2 and CO_2 in the fuel gas can be determined from the water vapor content and the dry fuel gas composition as follows

$$X_{H_2} = 0.28 \times (1 - 0.031) = 0.271$$
$$X_{CO_2} = 0.28 \times (1 - 0.031) = 0.271$$

From Equation (8.14), the anode potential change can be determined as

$$\Delta E_{a,P} = J \left(Z_{a,P_2} - Z_{a,P_1} \right)$$
$$= \left[J \times 2.27 \times 10^{-5} \left(X_{H_2} \right)^{-0.42} \left(X_{CO_2} \right)^{+0.17} \left(X_{H_2O} \right)^{+1.0} \exp\left(\frac{53,500}{RT} \right) \right]$$
$$\times \left(P_2^{0.75} - P_1^{0.75} \right)$$

Substituting all the relevant parameter values into the previous expression, we obtain a simplified expression for the pressure dependence:

$$\Delta E_{a,P} = 0.166 \left(P_2^{0.75} - P_1^{0.75} \right) \ (mV)$$

where P_1 and P_2 are also in atm. Then for $P_1 = 1$ atm, and $P_2 = 3$, 5, and 10 atm, respectively, we have for the anode potential change

$$\Delta E_{a,P} = 0.212, 0.389, \quad \text{and} \quad 0.767 \ (mV)$$

Clearly, the electrode resistance at the anode is increased when the operating pressure is increased, so that the resulting anode potential change actually reduces the cell potential. This is opposite to the pressure effect on the cathode potential change.

Summarizing all the potential changes obtained above will give us the cell potential gain arising from the pressure change, due to contribution from the gain in the reversible cell potential, and the change in the cathode and anode resistance. For easy comparison, the various results are summarized in the following table, where the subscripts (i) and (ii) represent the results obtained with the first and second approach, respectively.

P (atm)	$\Delta E_{r,P}$ (mV)	$-\Delta E_{c,P}$ (mV)	$-\Delta E_{a,P}$ (mV)	$\Delta E_{P(ii)}$ (mV)	$\Delta E_{P(i)}$ (mV)
3	21.9	31.2	−0.212	52.9	49.6
5	32.2	40.7	−0.389	72.5	72.7
10	46.0	50.0	−0.767	95.2	104

COMMENTS

1. Notice that Equations (8.14) and (8.15) were obtained for small cells with negligible utilization, therefore, negligibly small utilization is implied in the present calculation, otherwise, reactant consumption and product generation in the cell would have changed the actual composition of the fuel and oxidant gas, which should be counted in reality, although Equations (8.14) and (8.15) probably can be used reasonably with low to moderate utilizations, if we compare the present calculation with the experimental results shown in Figure 8.8.

2. The present calculation indicates that the cell potential gain due to the pressure increase agrees reasonably well (within the typical error ranges) for the results calculated from the two quite different empirical correlations. Although one correlation depends on the pressure logarithmically and the other on the pressure raised to a power, both have origins that can be traced back to Chapter 2 and both can represent the pressure dependence of the cell potential reasonably provided that the pressure change is not excessive.

3. As pointed out earlier, the pressure effect on the anode potential loss is very minimal as compared to the cathode and the total cell-potential change and can be easily neglected without causing any noticeable error.

4. A cathode potential gain occurs as the cell operating pressure is increased; in contrast, a potential loss actually occurs at the anode. This is understandable if one examines the half-cell reaction at the anode and cathode given in Equations (8.1) and (8.2).

8.5.3   Effect of Reactant Gas Composition and Utilization

Utilization of the fuel gas refers to the fraction of the fuel gas that reacts electrochemically in the fuel cell. Because the water gas shift reaction is very rapid in the presence of metals such as Ni, the fuel gas composition at the anode is the equilibrium composition established by the water–gas reaction. Thus, hydrogen that reacts electrochemically at the anode can originate from the hydrogen present in the inlet gas stream and from the hydrogen obtained from CO via the water–gas reaction in the cell. Thus the total hydrogen available for electrochemical reaction is equal to the moles of H_2 and CO that enter the anode gas chamber. As a result, the utilization of the fuel is usually defined as, for MCFCs

$$U_f = \frac{\left(N_{H_2} + N_{CO}\right)_{in} - \left(N_{H_2} + N_{CO}\right)_{out}}{\left(N_{H_2} + N_{CO}\right)_{in}} \tag{8.23}$$

where $N_{H_2,in}$ and $N_{H_2,out}$ stand for the molar flow rate of hydrogen gas at the cell inlet and outlet, respectively; and similar meanings for $N_{CO,in}$ and $N_{CO,out}$.

The utilization of the oxidant is defined similarly. The MCFC cathode has two reactants, O_2 and CO_2, and therefore there are two ways to define U_{ox}. Usually U_{ox} is based on the limiting (deficient) reactant. If O_2 is deficient,

$$U_{ox} = \frac{N_{O_2,in} - N_{O_2,out}}{N_{O_2,in}} \qquad (8.24)$$

In practice, CO_2 is likely to be the limiting reactant in order to minimize the slow NiO cathode dissolution (more on this later), then

$$U_{ox} = \frac{N_{CO_2,in} - N_{CO_2,out}}{N_{CO_2,in}} \qquad (8.25)$$

The effect of gas composition on the reversible cell potential E_r is small, as mentioned earlier. For the H_2-rich and H_2-poor fuel gases, the difference in E_r may only amount to $70 \sim 80\,mV$. On the other hand, the local gas composition in the flow channels varies due to the electrochemical consumption of the reactants in the cell, and the lowest reactant concentration occurs at the cell outlet, which can be determined by the reactant utilization. The cell reversible potential $E_{r,in}$ and $E_{r,out}$, which correspond to the gas composition at the inlet and outlet, respectively, can be determined from Equation (8.11). The difference $E_{r,in} - E_{r,out}$ is sometimes termed the Nernst loss and is an important parameter in the analysis of large cell performance. The Nernst loss for reformed natural gas may be as high as a few hundred mV for the typical fuel utilization of around 80% or higher, if the same utilization is used for the oxidant. In reality, the oxidant utilization is kept much lower than the fuel utilization (approximately half).

Normally, increasing the reactant gas utilization decreases cell performance because of the reduced reactant concentrations near the cell outlet, consequently the electrode kinetics and mass transport rates are lowered.

Oxidant   As Equation (8.1) reveals, the overall electrochemical reaction at the cathode consumes two moles of CO_2 for every mole of O_2 consumed and therefore the cathode gas composition with this 2:1 ratio for the carbon dioxide and oxygen gas concentration yields the optimal cathode performance. On the other hand, CO_2 concentration must be minimized in order to reduce NiO solubility of the cathode and increase cell lifetime (see Section 8.6). However, if CO_2 concentration is too low, such as in the limiting case where no CO_2 is present in the oxidant feed gas at all, the dissociation of carbonate ions

$$CO_3^= \longrightarrow CO_2 + O^=$$

would become significant, and as a result, significant cathode polarization arises due to the electrolyte composition change. Of course, for this situation the cell lifetime is quite limited due to the electrolyte loss.

The effect of the oxidant composition on the actual cell potential is clearly shown in Equations (8.12) and (8.15). Taking into the oxidant utilization into account, the combined effect on the actual cell potential of the oxidant composition at the cell inlet and the utilization in the cell has been correlated empirically and expressed as[4]

$$\Delta E_c = 250 \log \frac{\left(\overline{P}_{CO_2} \overline{P}_{O_2}^{1/2}\right)_2}{\left(\overline{P}_{CO_2} \overline{P}_{O_2}^{1/2}\right)_1} \ (\text{mV}), \quad \text{for } 0.04 \leq \left(\overline{P}_{CO_2} \overline{P}_{O_2}^{1/2}\right) \leq 0.11 \ (8.26)$$

$$\Delta E_c = 99 \log \frac{\left(\overline{P}_{CO_2} \overline{P}_{O_2}^{1/2}\right)_2}{\left(\overline{P}_{CO_2} \overline{P}_{O_2}^{1/2}\right)_1} \ (\text{mV}), \quad \text{for } 0.11 \leq \left(\overline{P}_{CO_2} \overline{P}_{O_2}^{1/2}\right) \leq 0.38 \ \ (8.27)$$

where $\overline{P}_{O_2}$ and $\overline{P}_{CO_2}$ are the average partial pressure of oxygen and carbon dioxide in the cathode, averaged between their values at the cell inlet and the outlet. The cell potential change due to the change in the oxidant gas composition and utilization is denoted as ΔE_c.

Fuel   As mentioned before, the determination of the anode potential is quite elaborate due to a number of other chemical reactions occur simultaneously at the anode. The two most important chemical reactions are the methane steam-reforming that produces the hydrogen and carbon monoxide gas in the first place (assuming natural gas or methane is used as the primary fuel), and the fast water–gas shift reaction whose equilibrium converts the carbon monoxide into hydrogen gas that can participate in the anode electrooxidation process sufficiently fast for current generation. Once the fuel gas composition is determined taking into account of the above reactions, the reversible cell potential can be obtained from Equation (8.11) for a given oxygen gas composition. As discussed in the previous subsection, the maximum cell potential for a given fuel gas composition is obtained when $P_{CO_2}/P_{O_2} = 2$ at the cathode.

It is interesting to note that inert gases added to the oxidant feed gas, for a given P_{CO_2}/P_{O_2} ratio, reduces the reversible cell potential E_r; but adding inert gases to the fuel gas increases E_r, for a given $P_{H_2}/P_{H_2O} P_{CO_2}$ ratio because two moles of products (H_2O and CO_2) are diluted for every mole of reactant (H_2). However, the actual cell potential E is lowered due to the inert gas dilution of the reactant gases.

Again the change in the actual cell potential due to the change in the fuel gas composition and the utilization has been correlated empirically, given as[4]:

$$\Delta E_a = 173 \log \frac{\left[P_{H_2}/(\overline{P}_{CO_2} \overline{P}_{H_2O})\right]_2}{\left[P_{H_2}/(\overline{P}_{CO_2} \overline{P}_{H_2O})\right]_1}, \ (\text{mV}) \tag{8.28}$$

where $\overline{P}_i$ is the average partial pressure of the species i in the fuel gas, averaged between their values at the cell inlet and the outlet; and ΔE_a is the cell potential change due to the change in the fuel gas composition and utilization.

This discussion implies that MCFCs should be operated with low utilizations for both fuel and oxidant, U_f and U_{ox}, in order to achieve better cell performance. On the other hand, low utilization would means inefficient use of fuel and oxidant. As with other types of fuel cell, a compromise must be made to optimize the overall performance. Typically, $U_f = 75\% \sim 85\%$ and $U_{ox} = 50\%$ for practical MCFCs.

8.5.4   Effect of Impurities

Natural gas is considered to be the primary fuel for MCFCs, and in the longer term, gasified coal (or coal gas) is expected to be the major source of fuel gas for MCFCs. Coal contains many contaminants in a wide range of concentrations, of course, depending on the location and quality of the coal mined. Hence, coal-derived fuel also contains a significant amount of contaminants. These contaminants have a strong impact on the corrosion of cell metal components such as electrodes, and as a result, the electrode performance is severely affected. The most significant contaminants include sulfur compounds (such as H_2S in fuel gas and SO_2 in anode exhaust recycled to the cathode along with CO_2) and halides (mainly HCl in the fuel gas). An important issue is to establish the critical concentrations for these contaminants that MCFCs can be operated without significant reduction in both cell performance and lifetime. Table 8.5 shows the typical contaminants from coal-derived fuel gas and

Table 8.5   A List of Typical Contaminants in Coal-Derived Fuel Gas and Possible Negative Effects on MCFCs[3,4]

Class	Contaminants	Possible Effect
Particulates	Coal fines, ash	Plugging of gas passages
Sulfur compounds	H_2S, COS, CS_2, C_4H_4S	Voltage losses
		Reaction with electrolyte via SO_2 (electrolyte sulfation)
		Increased corrosion
Halides	HCl, HF, HBr, $SnCl_2$	Increased corrosion and performance decay (especially cathode)
		Reaction with electrolyte
Nitrogen compounds	NH_3, HCN, N_2	NO_x formation
		Reaction with electrolyte via NO_x
Trace metals	Zn	ZnO precipitation/deposits on electrode
	As, AsH_3	Poisoning Ni catalyst
	Pb, Hg	Surface alloy formation
	Sn (as $SnCl_2$)	Sn precipitation/deposits on electrode
	Se	Poisoning Ni catalyst
	Cd, Te (as H_2Te)	Precipitation/deposits on electrode
		Reaction with electrolyte
Hydrocarbons	C_6H_6, $C_{10}H_8$, $C_{14}H_{10}$	Carbon deposition

their possible negative effects, and the corresponding maximum limits that may be tolerated by MCFCs are provided in Table 8.6.

Sulfur   Sulfur is perhaps the most detrimental contaminant for MCFCs because of significant presence in both the natural gas and the coal-derived gases. Sulfur compounds in low ppm concentrations in the fuel gas degrade the anode performance, however, the adverse effect is reversible and the anode performance can be recovered by operating the cell either at the open-circuit condition or by sulfur-free fuel gas. If the concentrations of sulfur compounds exceed a critical level, the anode Ni electrode can be destroyed by the sulfidation leading to the NiS-Ni eutectic formation and the corresponding anode performance is degraded irreversibly. The critical, hence the tolerance, concentrations of the sulfur compounds are strongly affected by the cell operating conditions, including the cell temperature, pressure, gas concentration, cell components and system operation (e.g., anode exhaust recycle to the cathode, anode exhaust venting, anode gas cleanup, etc.).

Hydrogen sulfide (H_2S) is the principal sulfur compound adversely affecting cell performance. Other compounds like COS and CS_2 are equivalent to H_2S in their effect on MCFCs. At atmospheric pressure and for gas utilization up to about 75%, up to 10-ppm H_2S in the fuel can be tolerated at the anode, and up to 1-ppm SO_2 is acceptable in the oxidant. These tolerant concentration limits can be increased by operating at higher cell temperatures and lower cell pressures.

The mechanisms by which H_2S degrades the cell performance are primarily due to

1. Chemisorption of H_2S on Ni surfaces to block active electrochemical sites, leading to increased activation polarization;

Table 8.6   Maximum Levels of Contaminants that may be Tolerated by MCFCs[3,4,8]

Contaminants	Maximum Limits
Particulates	<0.1 g/l for large particulates (>0.3 μm)
H_2S, COS	<1 ppm
HCl and other halides	<10 ppm
NH_3	<10,000 ppm
Trace metals	
Zn	<20 ppm
As (as AsH_3)	<1 ppm
Pb	<1 ppm
Hg	<35 ppm
Sn	N/A
Cd	<30 ppm
C_2^+	<100 ppm
Tar	<2000 ppm (benzene)

2. Poisoning of catalytic reaction sites for the water–gas reaction, blocking the conversion of CO to more hydrogen for the anode reaction;
3. Oxidation of H_2S to SO_2 in a combustion reaction for the anode exhaust, which are recycled to the cathode along with CO_2, then reacting with carbonate ions in the electrolyte.

Mechanism (1) clearly explains why H_2S does not affect the open-circuit voltage and the cell performance degradation is increased when the current density is increased for the same H_2S concentration in the fuel gas. At MCFC operating condition, catalysts are used to establish rapidly the equilibrium of the water–gas reaction in the anode for additional hydrogen, Mechanism (2) effectively lowers the amount of hydrogen available for the anode reaction, hence the performance deterioration. However, in practical MCFCs, this adverse effect on the shift reaction is minimal because Cr, used in stabilized Ni anodes for resistance to sintering and creepage, also acts as a sulfur tolerant catalyst for the shift reaction.

For practical applications, anode exhaust gas is often recycled, after the residual hydrogen is burned, to the cathode to provide the needed carbon dioxide for cathode reaction. This process also converts sulfur compounds to SO_2 during the burning, and SO_2 is brought to the cathode where it reacts with carbonate ions to produce alkali sulfates. These sulfate ions migrate through the electrolyte to the anode where $SO_4^=$ is reduced to $S^=$, thus increasing the concentration of $S^=$ at the anode. Clearly, Mechanism (3) gradually accumulates sulfur at the anode. For long-term operation (around 40,000 h), sulfur compounds (H_2S, COS and CS_2) have to be ≤ 0.01 ppm in the fuel gas without sulfur removal during the operation, either continuously or periodically. Sulfur tolerance level can be increased to about 0.5 ppm or even higher if sulfur is purged periodically or scrubbed at the burner exit. It then becomes apparent that low cost techniques for sulfur removal are very important for MCFCs and such techniques are still being pursued.

Halides   Severe cathode corrosion occurs due to the attack by halogen containing compounds, degrading the cathode performance significantly. In addition, molten carbonates (Li_2CO_3 and K_2CO_3) can react with HCl and HF to form CO_2, H_2O, and the respective alkali halides, accelerating the electrolyte loss. Both effects reduce not only the cell performance but lifetime as well. Further, the high vapor pressure (high volatility) of LiCl and KCl voids the electrolyte matrix, resulting in reactant crossover. The concentration of Cl^- species in the coal-derived fuel gas is typically in the range of 1~500 ppm. The concentration of HCl should be less than 0.5 ppm in the anode fuel gas for MCFCs, although more work is needed to establish the tolerance level for long term operation. The total concentration of the halogen containing compounds in the fuel gas should be limited as well.

Nitrogen Compounds   Nitrogen compounds like NH_3 and HCN do not have direct impact on MCFCs. However, NO_x may be formed in the burnout of the anode exhaust and recycled to the cathode along with the carbon dioxide. Electrolyte can then react with NO_x to form nitrate salts, causing irreversible loss of the electrolyte. Clearly the tolerance level for nitrogen compounds depends on the amount of NO_x formation during the combustion process and can be increased significantly with the low NO_x combustion technology, including the use of low NO_x burners, catalytic burners, NO_x removal through catalytic converters, and so on. Without these techniques for the reduction of NO_x concentrations, some studies even suggest ammonia (NH_3) concentrations be limited to as low as 0.1–1 ppm.

Solid Particulates   Solid particulates can stick on electrode surface and accumulate there to plug the pores of the electrode, thus blocking the gas passages. They may also cover the catalyst surface to slow down the water–gas shift reaction and the anode electrode reaction. It has been established that the tolerance limit for particulates larger than 3 μm in diameter is 0.1 g/l.

Other Compounds   Other contaminants such as trace metals in the coal and higher hydrocarbons in the anode gas stream can also have an adverse effect on MCFCs, as shown in Table 8.5, and the corresponding tolerance limits for them are given in Table 8.6.

8.5.5   Effect of Current Density

As the cell polarization curve suggests, operating at a higher current density results in a reduction in the cell potential E, because the ohmic, activation, and concentration losses increase with current density. Practical MCFCs are operated in the ohmic polarization dominated region, thus the major loss over the range of current densities of practical application is due to the linear ohmic voltage loss, and it has been found empirically for properly assembled MCFCs that the cell potential change can be correlated with the change in the operating current density as[4]

$$\Delta E_J = -1.21\Delta J \ \text{(mV)} \quad \text{for } 50\,\text{mA/cm}^2 \leq J \leq 150\,\text{mA/cm}^2 \quad (8.29)$$

$$\Delta E_J = -1.76\Delta J \ \text{(mV)} \quad \text{for } 150\,\text{mA/cm}^2 \leq J \leq 200\,\text{mA/cm}^2 \quad (8.30)$$

where J is the current density in mA/cm^2. From the above correlations, the area specific resistance is obtained immediately as 1.21 and 1.76 $\Omega \cdot \text{cm}^2$, respectively, for the low and high current density ranges. As for other types of fuel cell, the majority of the resistance arises from the ion transport through the electrolyte.

8.5.6   Effect of Cell Life

Like other fuel cells, the performance of MCFCs and stacks degrades over time of operation, and the contributing factors affecting this long-term performance degradation described in the following section. For the moment, it suffices to state that the long-term performance and lifetime for MCFCs are still yet to be established with more tests, and the existing data suggest that the cell potential reduction is less than 4~5 mV/1,000 h of operation for small cells in laboratory tests. Certainly longer lifetime can be achieved if the performance decay rate can be reduced.

8.6   LONG-TERM PERFORMANCE AND LIFETIME OF MCFCs

As pointed out earlier, the existing data on long-term performance and cell lifetime are limited, especially for scaled-up cells under the realistic conditions of the practical applications. Existing information suggests that the long-term performance decay that eventually limits the MCFC lifetime is primarily due to (i) electrolyte loss, (ii) contact resistance increase, (iii) electrode creepage and sintering, and (iv) shorting of the electrolyte matrix by nickel particle precipitation.

8.6.1   Electrolyte Loss

A loss of electrolyte, or decrease in the electrolyte filling of the electrolyte matrix, increases the ohmic resistance to the transport of carbonate ions through the electrolyte from the cathode to the anode; and reduces the size of three-phase boundary at each electrode where electrode reactions occur, thereby increasing electrode activation polarization. Significant electrolyte loss results in the reactant gas crossover, which may occur when the loss amounts to 35%–40% of the initial electrolyte loading, although the exact amount depends on many factors such as the electrolyte distribution in the matrix, and the matrix pore sizes and pore size distribution, and so on. Electrolyte loss is the most important and continuously active factor in causing the long-term performance degradation and is primarily a result of electrolyte consumption by the corrosion/dissolution processes of cell components, electric potential-driven electrolyte migration and electrolyte vaporization.

Corrosion/Dissolution of Cell/Stack Components   Due to the strongly corrosive nature of the electrolyte, cell/stack components are corroded in the MCFC working environment. The important phenomena include the dissolution of NiO cathode and corrosion of other components like separator plate. Apparently, electrolyte composition would have a strong influence on the rate of corrosion, and its optimization with the cell performance is essential.

The dissolution of the NiO cathode under the normal MCFC operating conditions follows the reaction

$$NiO + CO_2 \longrightarrow Ni^{2+} + CO_3^{2-} \tag{8.31}$$

Hence, both the solubility and the rate of dissolution of nickel oxide in the carbonate electrolyte depend strongly on the partial pressure of carbon dioxide in the cathode, in addition to the electrolyte composition and cell temperature. In general, the solubility and the dissolution rate increase with the carbon dioxide content and temperature. As described in Section 8.5.3, sufficient content of carbon dioxide is necessary to maintain a good cathode performance. The minimum solubility of NiO in Li–K carbonate eutectic (62% Li_2CO_3 and 38% K_2CO_4) at 650 °C is about 1 molar ppm at the CO_2 partial pressure of about 0.01 atm. Both minimum solubility and the associated carbon-dioxide partial pressure increase with temperature.

However, the effect of cathode dissolution on the cathode performance via the cathode electrode structural changes is minimal, even up to 1/4–1/3 of the electrode mass being dissolved, but the more significant impact is through the electrical shorts created by the nickel particle precipitation in the electrolyte matrix described in Section 8.6.4.

The second route is the corrosion of hardware (i.e., separator plate, current collector, electrodes, etc.), which prevails in the initial stages of cell or stack life. Empirical data suggest that the rate of corrosion may be correlated by the following equation involving the square root of time

$$y = Ct^{1/2} \tag{8.32}$$

where y, in μm, represents the thickness of the material layer that is being corroded, and t is the time in hours. This growth of the corrosion layer with the square root of time may suggest that the corrosion process is dominated by the diffusion mechanism, although in reality the corrosion process may be more complex than this may imply. For stainless steel exposed to the cathode environment with Li–K carbonate eutectic at 650 °C, the empirical constant in Equation (8.32) has been found to be $C = 0.134 \, \mu$m/h$^{1/2}$; and the typical initial corrosion rate is about 8 μm/h in the first 2000 hours, then reduces to an average of 2 μm/kh for the following 10,000 hours of operation.[9]

The corrosion of the anode side of the separator plate is much more severe than the cathode side (approximately 2~5 times higher), and can also be correlated empirically by Equation (8.32). However, the empirical constant C depends on the partial pressure of water vapor. For example, for typical MCFC operations, it was found[3]

$$C(\mu m/h^{1/2}) = \begin{cases} 0.023 & \text{for } P_{H_2O} = 16\% \\ 0.039 & \text{for } P_{H_2O} = 28\% \\ 0.058 & \text{for } P_{H_2O} = 43\% \end{cases} \tag{8.33}$$

Since water is produced in the anode electrode reaction, according to Equation (8.2), water vapor concentration increases strongly in the downstream direction of the anode side, the corrosion rate becomes the highest near the outlet area of the fuel stream.

Figure 8.9 shows the electrolyte loss due to corrosion of Type 316 stainless steel current collector, exposed to both fuel and oxidant gas, in cells operating at 150 mA/cm^2. It is seen that K_2CO_3 loss due to corrosion is almost a constant (about 2 μmol/cm^2), independent of time. Whereas Li_2CO_3 loss is significantly higher, especially as time goes on, and follows the dependence of square root of time described by Equation (8.32), with the empirical constant equal to approximately $C = 0.38 \, \mu$mol/(cm$^2 \cdot$ h$^{1/2}$). Therefore, the total electrolyte loss, dominated by Li_2CO_3 loss, still approximately follows the square-root-of-time dependence, although more complex mechanisms are involved for the corrosion process.

Finally, it might be pointed out that the corrosion of the anode electrode is minimal because anode is operated at the very low potential (unless excessive anode overpotential occurs) in a reducing environment.

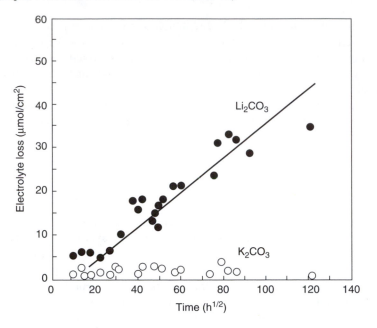

Figure 8.9 Electrolyte loss due to corrosion of Type 316 stainless steel current collector in cells operating at 150 mA/cm^2.[3,10]

Electrolyte Migration   Electrolyte movement under electric potential difference results from different rates of ion migration (or different mobility). For internal manifolding design, this phenomenon causes a net movement of the electrolyte out of the cell, via the wet seal, onto the exterior of hardware enclosing the cell (often called **electrolyte creepage**) because potassium ions move faster than lithium ions under the electric field in the wet seal. Electrolyte creepage is a slow but continuous process, and in severe cases it can create a thin film spreading on the outside surface of the cell hardware, thus establishing ionic contact among the neighboring cells. Then electrolyte movement from one cell to another may occur through the thin surface film, along with the corrosion of the outside surface of the cell hardware.

For external manifolding design, electrolyte motion under the electric field (or migration) becomes significantly important. This is because all the cells in the stack (typically hundreds of cells) are ionically connected via the manifold gasket, which contains a small amount of electrolyte, and thus conducts current from the positive end plate of the stack to the negative endplate. The full stack voltage between the two stack endplates that contain hundreds of electrolyte layers with a total resistance of high value is applied across the manifold gasket, leading to a slow, but continuous one-way movement of the electrolyte. This **ion pumping effect** causes the electrolyte displacement among the cells in the stack, often causing the electrolyte accumulation near the negative endplate (flooding the cells there) and electrolyte depletion for cells near the positive ends. Therefore, electrolyte migration leading to uneven electrolyte distribution among cells in the stack becomes an important issue for externally manifolded stacks.

Electrolyte Vaporization   Electrolyte vaporization (usually Li_2CO_3 and/or K_2CO_3) occurs either directly as carbonate or indirectly as hydroxide. The electrolyte loss due to vaporization increases with increasing temperature, lower pressure and higher reactant gas-flow rate. The rate of evaporation is constant for steady-state operation.

When all these loss mechanisms are considered, the total electrolyte loss is governed by the corrosion of the cell/stack components in the early stage of the operation (up to about 2000 h) and followed by the evaporation of electrolyte. Therefore, the earlier loss of the electrolyte can be described by the square-root-of-time dependence as given in Equation (8.32), and the late stage by the constant rate attributed to electrolyte evaporation. The electrolyte loss due to migration is relatively small because the wet-seal passage is very narrow, resulting in a limited amount of electrolyte creepage. However, if severe corrosion occurs in the wet-seal area, loose contact there would create a large flow passage and then significantly more electrolyte creepage.

8.6.2   Increase in Contact Resistance

The contact resistance usually increases as the cell or stack ages, due to a number of factors. For example, electrode creepage and sintering result in a shrinkage of the thickness of the cell structure deteriorating the surface contact between the adjacent surfaces. The corrosion of cell/stack components, especially the separator plate, leads to the formation of surface oxide layers that have low electrical conductivity; this becomes more significant if the separator plate is made of high-aluminum alloys. The dense surface protection layer formed on the separator plate for corrosion resistance is also poor in electrical conduction. However, the soft-rail design for the separator plate (Section 8.3.2) has a spring-like structure and can accommodate the changes in the cell thickness with good cell-to-cell contact, hence tends to minimize the contact resistance, especially for long-term operation.

8.6.3   Electrode Creepage and Sintering

Electrode creep refers to the deformation of the particulate structure of the electrode under compression and sintering is the coarsening of the particulates. Electrode, especially anode, creepage and sintering, result in increased ohmic resistance and electrode polarization due to the increased contact resistance, reduced electrode void space for reactant gas passage and reduced electrode surface area for the catalysis of electrode reactions. The severe sintering of the nickel anode can be prevented by the addition of about 10% Cr. Other alloys such as Ni–Al and Ni–Cu also have sufficient resistance to sintering. These alloys also provide sufficient resistance to anode electrode creep as well, especially Ni–Al alloy. NiO cathodes have satisfactory sinter and creepage resistance.

8.6.4   Electrolyte Matrix Shorting

As described in Section 8.6.1, the dissolution of NiO cathode creates the dissolved nickel, which diffuses into the electrolyte matrix and is reduced by hydrogen dissolved in the electrolyte and diffused in from the anode. Then nickel particles precipitate in the electrolyte matrix in structures resembling fingering phenomena. When these precipitated nickel particles in fingering forms connect the anode and cathode across the electrolyte matrix, electrical shorting occurs and no useful electrical power output can be obtained, thus the end of useful lifetime is reached. From the NiO-dissolution reaction shown in Equation (8.31), it is clear that the time at which shorting occurs is related to NiO solubility, the CO_2 partial pressure and the cell temperature. From the nickel particle-precipitation process, we know that the shorting phenomenon is also related to the electrolyte matrix structure, such as the porosity, pore size, and more importantly the thickness of the matrix. A thick electrolyte matrix will increase the operation time before the shorting occurs, although it also increases the ohmic resistance of the matrix.

8.7   FUTURE R&D

As the MCFC is in the early stage of precommercial demonstration, its performance including power density and reliability as well as sufficient lifetime has yet to be demonstrated under the practical operating conditions and the operational issues such as the startup procedures and thermal cyclings still need to be developed and perfected. Further research and development are required before the commercialization of the MCFC power systems, and some of the further works should address both fundamental and practical issues. From the fundamental level, the following issues should be tackled:

- **Better performance through improved understanding of the transport and electrochemical kinetic processes in the electrodes and the cells**. Better performance includes higher power density without any sacrifice on efficiency, reliability, and long-term performance and requires the optimization of cell components, such as the electrode structures, and electrolyte composition.
- **Better materials and improved manufacturing processes for low-cost production**. This requires a better anode for creep and sintering resistance, better cathode for minimal dissolution, better separator plate and other cell/stack components (such as current collector) for corrosion resistance. Optimization of electrolyte composition is required to reduce the corrosive nature of the electrolyte, or better electrolyte for better performance and less corrosion.
- **The effect of impurities on the performance and corrosion of cell/stack components, as well as the interaction among the impurities and their combined effect**. This is important because a variety of impurities exist in the coal-derived gas with various possible concentrations, depending on the grade and location of coals mined. Also this will have significant impact on the carbon dioxide recirculation from the anode to the cathode and CO_2 recycling strategies exert considerable impact on the overall MCFC system designs and operations.

From the practical point of view, the optimal cell/stack design should be sought, including the porous structure matching among the electrodes and electrolyte matrix and more importantly, the proper sealing of the cell and stack. An appropriate design for the separator plate is also important to accommodate the structure changes (shrinkage) during the stack operation, having impact on both sealing and long-term performance. Of course, proper procedures for startup and thermal cycling are equally important such as electrolyte melting and wetting of cell components.

8.8   SUMMARY

In this chapter, the molten carbonate fuel cell (MCFC) is described in details. The overall half-cell and whole-cell reactions have been described as well as the other chemical reactions occurring in the anode, and the significance of the water–gas shift reaction in MCFCs has been highlighted. From these, an understanding of the concept of internal reforming for MCFCs (IR-MCFCs), as well as its conceptual advantage and practical limitations should be reached. Fairly detailed coverage has been provided on the components of a molten carbonate fuel cell and stack, their geometrical configurations, and structural characteristics, including the important sealing issues, electrolyte management by capillary action, and the concept of bubble pressure layers, the design of stack manifolds and flow channels, the consideration and various design for the (bipolar) separator plate, as well as stack scaleup issues. In general the reader should grasp the distribution of current density and temperature in MCFC stacks and the pros and cons for the direct and indirect IR-MCFC stacks. Also a general understanding of the typical materials and fabrication techniques used for the various cell/stack components, the issues they face, the possible effects and ways of avoiding the negative effects should be reached. Description of the effect of various operating parameters on the MCFC performance, including the temperature, pressure, reactant concentration and utilization, the effect of various impurities, cell current density, and cell aging was given. You should also become familiar with the definitions for the utilization of the fuel and oxidant in MCFCs and the tolerance limits for various impurities. Long-term performance decay and the factors limiting the MCFC lifetime, including the electrolyte loss via structural corrosion, electrolyte displacement by migration and vaporization, internal resistance increase with cell/stack aging, electrode creepage and sintering, and electrolyte matrix shorting by dissolved nickel precipitation was also outlined. From these coverages, appreciation of the needs existing for various future works in order to advance MCFC technology for commercial application should be reached.

BIBLIOGRAPHY

1. Takashima, S., K. Ohtsuka, N. Kobayashi and H. Fujimura. 1990. *Proc. Second Int. Symp. MCFC Technology* eds. R. R. Selman, H. C. Maru, D. A. Shores and I. Uchida, PV90-16, Pennington, NJ: The Electrochemical Society, 378.
2. Dave, B. B., K. A. Murugesamoorthi, A. Parthasarathy and A.J. Appleby. 1993. Overview of fuel cell Technology. *Fuel Cell Systems*, eds. L.J.M.J. Blomen and M.N. Mugerwa, New York: Plenum Press.
3. Selman, J. R. 1993. Research, development, and demonstration of molten carbonate fuel cell systems. *Fuel Cell Systems*, eds. L. J. M. J. Blomen and M. N. Mugerwa, New York: Plenum Press.

476 Chapter 8: Molten Carbonate Fuel Cells (MCFCs)

4. Hirschenhofer, J. H., D. B. Stauffer and R. R. Engleman. 1994. *Fuel Cells: A Handbook* (Revision 3), U.S. Department of Energy.
5. Yuh, C. Y. and J. R. Selman. 1990. *J. Electrochem. Soc.*, 138: 3542.
6. Rostrup-Nielsen, J. R. 1994. *Catalysis Science and Technology*, eds. J. R. Anderson and M. Boudart, Berlin: Springer-Verlag, 1.
7. Baker, B. S. 1984. *Proc. Symp. MCFC Technology*, eds. J. R. Selman and T. D. Claar, PV84-13, NJ: Pennington, The Electrochemical Society, 15.
8. Mugerwa, M. N. and L. J. M. J. Blomen. 1993. System design and optimization. *Fuel Cell Systems*, eds. L. J. M. J. Blomen and M.N. Mugerwa, New York: Plenum Press.
9. Appleby, A. J. and F. R. Foulkes. 1988. *Fuel Cell Handbook*, New York: Van Nostrand Reinhold.
10. Urushibata H. and T. Murahashi. 1992. *Proc. Intl. fuel Cell Conf.*, Makuhari, Japan. 223.

PROBLEMS

8.1   Describe briefly the advantages and disadvantages as well as the areas of applications for molten carbonate fuel cells.

8.2   Describe the operation principle, such as half-cell and whole-cell reaction, and the primary fuels expected to be used for molten carbonate fuel cells.

8.3   Discuss briefly the typical (or target) operating conditions such as cell voltage, cell current density, temperature, pressure, fuel and oxidant utilization, and chemical-to-electrical energy conversion efficiency.

8.4   Describe the effect of operating conditions on the cell performance.

8.5   Describe the geometrical configuration of cell and stack, typical materials used for, and thickness of, as well as other dimensions of, cell components.

8.6   Describe the concept of internal-reforming molten carbonate fuel cell (IR-MCFC) and its two implementations (i.e., the direct and indirect IR-MCFCs).

8.7   Describe what is unique for the electrolyte management in molten carbonate fuel cell as compared with other liquid electrolyte fuel cells.

8.8   Describe the factors affecting the short- and long-term performance of molten carbonate fuel cells.

8.9   Describe the critical technical barriers to overcome for the commercialization of molten carbonate fuel cells, the possible solutions, and their pros and cons.

8.10 Derive the Nernst equation for the molten carbonate fuel cell given in Equation (8.11) from the general Nernst equation given in Chapter 2.

8.11 Determine the reversible as well as the thermoneutral cell potential for the following reactions at the typical MCFC operating condition of 1 atm and 650 °C:
   (i) $H_2 + \frac{1}{2}O_2 \longrightarrow H_2O$
   (ii) $CH_4 + 2O_2 \longrightarrow CO_2 + 2H_2O$
   (iii) $CH_4 + H_2O \longrightarrow CO + 3H_2$
   (iv) $CO + H_2O \longrightarrow CO_2 + H_2$
   (v) $2CO \longrightarrow C + CO_2$

SOLID OXIDE FUEL CELLS (SOFCs)

9.1 INTRODUCTION

The solid oxide fuel cell (SOFC) has the highest operating temperature (about 1000 °C) among practically all the important fuel cells under development. High temperature is a requirement since it was developed with the intention of operating on coal as the primary fuel, or as a "coal gas cell," meaning that coal-derived fuel gases would be used as the primary fuels. Today, both coal-derived gases and natural gas are being considered as the primary fuel. Solid oxide fuel cell had been sometimes referred to as "the third-generation fuel cell technology," because it was expected that SOFCs would reach the market place after the commercialization of PAFCs (the so-called first generation) and MCFCs (the second generation).

The first solid-state oxygen ion conductor, $(ZrO_2)_{0.85}(Y_2O_3)_{0.15}$, was accidently discovered by Nernst in 1899[1] hence, it is often referred to as the Nernst mass. The first solid oxide fuel cell was proposed by Schottky,[2] one of Nernst's many students to use the Nernst mass as the solid electrolyte and was constructed by Baur and Preis in 1937.[3] However, most of the industrial effort in developing the solid oxide fuel cell for practical application was undertaken since 1958 by Westinghouse Electric Corp., with tubular cell design,[4] the most developed technology among all SOFCs. Innovative cell stacking (connection) was developed with advanced fabrication techniques via plasma spraying. In the early 1980s, monolithic cell structure and planar cell structure were developed and in the later 1980s they were demonstrated for higher power densities than the tubular design, but both are still in the early stage of development. In summary, the solid oxide fuel cell is made of all solid components, including the electrolyte, which is an oxide ion conductor. High operating temperature ($\sim$1000 °C) is adopted to ensure adequate ionic and electronic conductivity for the cell components. The electrolyte of the present SOFCs is oxygen ion conducting. However, it is possible to make the electrolyte proton conducting as well. Since the electrolyte is mainly made of ceramic material, this type of fuel cell is also sometimes referred to as the ceramic fuel cell.[5,6]

The SOFC has a number of advantages when compared with other types of fuel cells. First, since it is made of all solid components, it has a simpler concept, design and construction, the reaction zone at the electrode–electrolyte interface becomes a two-phase, (gas–solid) contact, instead of a three-phase zone for liquid electrolyte fuel cells such as PAFCs and MCFCs. As a result, complex electrolyte management is not needed and the problems of electrolyte depletion affecting cell performance and lifetime and the severe corrosion of the cell/stack components attacked by the corrosive liquid electrolyte are avoided completely. Second, because SOFCs are operated at very high-temperature conditions, the relevant electrochemical kinetics at the electrodes proceed sufficiently fast without the need of Noble metals as catalysts and with low activation polarization; hence the expensive and often specially prepared catalysts are totally avoided. Further, such high operating temperature also makes it possible for the internal reforming of methane and other hydrocarbons to produce hydrogen gas and carbon monoxide and the internal-reforming process is more effective than the corresponding process in the MCFCs. Furthermore, the SOFC has a better ability to tolerate the presence of impurities in the reactant gas streams because of high-temperature operation. Consequently the fuel processing process is much simpler without the need for the extensive steam-reforming and shift conversions that are required by low-temperature fuel cells. Finally, the high-temperature SOFC operation provides a better system match with other components and processes (or balance of plant), such as the coal gasification processes for primary fuel processing; it provides high-quality waste heat for co-generation applications and bottoming cycles utilizing conventional steam or gas turbines for additional electric power generation. Such an integrated SOFC-based power system offers potentially very high-energy conversion efficiency when coals or coal-derived gases are used as the primary fuel; the efficiency of as high as over 70% has been reported based on theoretical analysis.

On the other hand, the solid oxide fuel cell has a number of disadvantages as well. The high operating temperature results in low values of reversible cell potential ($E_r \sim 0.9\,\text{V}$) and very few appropriate materials are available for cell components and the search for them is still under intensive R&D. The current trend is to lower the operating temperature to 550 °C–650 °C (the so-called low-temperature SOFC) or 650 °C–850 °C (the so-called intermediate temperature SOFC) in order to lower the strict requirement for the materials to withstand the SOFC operating conditions. At present, all these low-, intermediate- and high-temperature (about 1000 °C) SOFCs are being pursued and developed. However, the high-temperature SOFC technology is the most developed one, and hence it is described in this chapter, and low- and intermediate-temperature SOFCs are given in Section 9.6 as a part of future R&D efforts. Overall, the entire SOFC technology is still not well developed when compared with the other types of the fuel cells described in earlier chapters.

Because SOFCs have low activation losses and better system match, the overall energy conversion efficiency of greater than 50% is expected for chemical to electrical energy conversion, and the efficiency as high as 65% is possible, corresponding to cell performance of 0.75 V and modest current densities, even without considering the integration with the conventional bottoming turbine systems. The solid oxide fuel cell can be used primarily for electric utility applications, as baseload electricity generation, dispersed, or distributed power generation, and even for onsite co-generation. Although it is limited in mobile applications, it is being considered as power systems for trains and large surface ships, as well as powering the auxilliary equipment for automobiles.

In this chapter, the operating principles for the solid oxide fuel cell are described first, followed by the cell components and innovative geometric configurations. Then, we will present the typical materials for the solid oxide fuel cell along with the typical manufacturing techniques employed. The performance and factors affecting the long-term performance and the lifetime are given next, and a brief overview of the future R&D issues follows.

9.2   BASIC PRINCIPLES AND OPERATIONS

The basic operational principle for the solid oxide fuel cell is illustrated in Figure 9.1. The oxygen gas at the cathode reacts with electrons from the external circuit to form oxide ions. The oxide ions transport through the electrolyte and reach the anode. At the anode hydrogen gas reacts with oxide ions to form water and release electrons, which migrate through the external circuit to reach the cathode, thus completing the electron transfer cycle. The electrons, while going through the external circuit, do work to the electric load before reaching the cathode, thus forming the electric power output. Therefore, the overall electrochemical reaction occurring at each electrode can be written as

$$\frac{1}{2}O_2 + 2e^- \longrightarrow O^{2=} \tag{9.1}$$

at the cathode and

$$H_2 + O^{2=} \longrightarrow H_2O + 2e^- \tag{9.2}$$

at the anode, so that the overall cell reaction is given by

$$\frac{1}{2}O_2 + H_2 \longrightarrow H_2O + \text{Waste Heat} + \text{Electric Energy} \tag{9.3}$$

If carbon monoxide is supplied at the anode instead of the hydrogen gas, then the overall anode reaction becomes

$$CO + O^{2=} \longrightarrow CO_2 + 2e^- \tag{9.4}$$

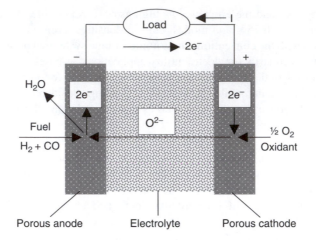

Figure 9.1 Schematic of a solid oxide fuel cell (SOFC) illustrating its operational principle.

while the cathode electrode reaction remains the same as given in Equation (9.1). Then the overall cell reaction becomes

$$CO + \frac{1}{2} \longrightarrow CO_2 + \text{Waste Heat} + \text{Electric Energy} \tag{9.5}$$

Clearly in the SOFC, carbon monoxide is utilized as a fuel directly for the electro-chemical reaction, whereas it is indirectly used as fuel in the molten carbonate fuel cell via the water–gas shift reaction and it is contaminant and has poisoning effect for the low-temperature fuel cells like AFCs, PAFCs, and PEMFCs.

Typically either coal-derived gases or reformed natural gases are used as the fuel, which contains both hydrogen gas and carbon monoxide. If the fuel stream consists of a moles of hydrogen and b moles of carbon monoxide, then the combined anode reaction becomes

$$aH_2 + bCO + (a+b)O^{2=} \longrightarrow aH_2O + bCO_2 + 2(a+b)e^- \tag{9.6}$$

and the combined cathode reaction becomes

$$\frac{1}{2}(a+b)O_2 + 2(a+b)e^- \longrightarrow (a+b)O^{2=} \tag{9.7}$$

Such that the overall cell reaction can be represented by

$$\frac{1}{2}(a+b)O_2 + aH_2 + bCO \longrightarrow aH_2O + bCO_2 + \text{Waste Heat} + \text{Electric Energy} \tag{9.8}$$

The direct utilization of carbon monoxide in the anode electrochemical reaction is especially advantageous for the solid oxide fuel cell. This is because at the lower temperature for the molten carbonate fuel cell, hydrocarbon reforming, such as steam reforming of the natural gas, requires the presence of catalysts to accelerate the reforming kinetics. Although higher temperature reforming can help the kinetics, but

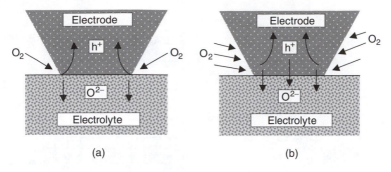

Figure 9.2 Schematic of charge transfer process in two types of electrode: (a) electronically conducting and (b) mixed conducting cathodes.

at the same time more carbon monoxide is formed in the reforming gas, as shown in Figure 6.1. In order to avoid carbon formation and help the water–gas shift reaction for more hydrogen production, sufficient water content is maintained in the reformed fuel gas stream. For the solid oxide fuel cell, catalysts are not required for the steam reforming of natural gas, ideal for internal-reforming arrangement and no further processing is necessary for the reformate gas before being fed to the anode reaction.

However, in reality water vapor is required to be present in the reformed gas stream to avoid carbon formation via the mechanisms of Boudouard reaction and methane decomposition reaction described in Chapter 8. With the presence of water vapor, the water–gas shift reaction proceeds very rapidly. Thus most of the carbon monoxide in the fuel stream is converted to the hydrogen gas at the anode where hydrogen gas is consumed and more water vapor is produced that further helps the shift reaction. On the other hand, the direct electrochemical oxidation of carbon monoxide at the SOFC anode is relatively slow, such that most of the carbon monoxide is still expected to be consumed via the shift reaction.

In solid oxide fuel cells, the cathode and anode electrode reactions occur in a solid–gas two-phase zone and expensive electrocatalysts are avoided due to the high-temperature operation. The actual reaction can occur at the electrode–electrolyte interface if the electrode is made to conduct only electronically; however, the reaction can also occur throughout the entire porous electrode if the electrode is made to conduct both electronically and ionically (oxide ion) — such an electrode is called the **mixed conducting electrode**. A schematic of the charge transfer process at the two types of the electrode is illustrated in Figure 9.2. Current literature should be consulted for more recent development on the materials and electrochemical properties of mixed ionic–electronic conductors as electrodes, such as [7, 8].

9.3   CELL COMPONENTS AND CONFIGURATIONS

9.3.1   Single Cell

The basic structure for a solid oxide fuel cell is the same as for other types of fuel cells described previously, that is, it consists of a solid electrolyte and two electrodes

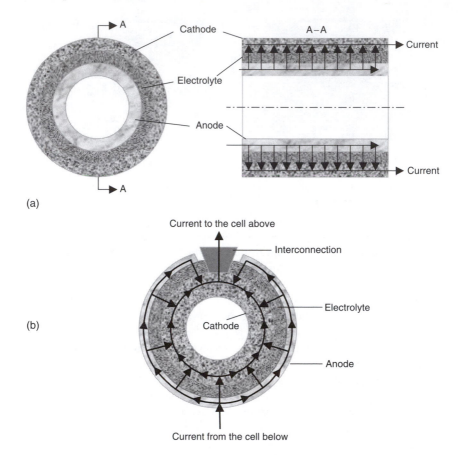

(a)

(b)

Figure 9.3 Schematic of tubular solid oxide fuel cells: (a) Current collection at the cylindrical base and (b) Current collection along the circumferential cylindrical surface.

called a anode and a cathode. However, because all the cell components are solid, including the electrolyte, the solid oxide fuel cell can be made into rich geometrical configurations as compared to the other types of fuel cell. At present, three different configurations are being pursued: tubular, monolithic, and planar. The tubular design is the most advanced, while the other two designs are still in early stages of development. These three designs are described in the following section.

Tubular SOFCs   Solid oxide fuel cells operate at high temperature and severe thermal cycling conditions require that each layer of cell components (anode, electrolyte, and cathode) must have similar thermal expansion coefficient to avoid thermal cracking and delamination. Cylindrical shape of the cell can be made easily with better resistance to the thermal stress damage to the composite structure. Figure 9.3 shows the possible tubular structure for solid oxide fuel cells; the corresponding current collection can be made either at the base of the tube or circumferentially along the curved cylindrical surface. For base current collection, the tube can be made with large diameters, but the tube length is limited (typically to about 1 cm) in order to avoid excessive ohmic polarization; therefore, low power density results. It has been reported that for this design with small cells, a maximum power density of 0.3 W/cm^2 was achieved when operating on hydrogen and oxygen and 0.2 W/cm^2 when operating on the mixture of hydrogen and carbon monoxide as the fuel and air as the oxidant. For circumferential current-collection design as shown in Figure 9.3b, the current follows the path of the cylindrical curved surface, therefore, the tube diameter has to be small, typically on the order of 1.5 cm and the electrode layers cannot be too thin, in order to keep ohmic loss within the acceptable limits, while the tube can be very long, in excess of over 1 m or more. This design can have higher volumetric power density than the base current-collection design, although the power density in terms of the active electrode surface area may not be increased significantly.

For tubular SOFCs, the electrolyte layer forms the middle tube, while either the anode or the cathode tube can be located on the inner side of the electrolyte tube. Also a relatively thick support tube on the inner side is often employed in order to provide the structural support for the thin cell-composite layers. Although earlier cells used the anode on the inner side, more recent cells have the cathode on the inner side and anode on the outside for easy stacking consideration.

The thick support tube is very heavy, significantly affects the specific power density in terms of the weight or volume. It is also expensive, accounting for 50% of the materials cost. Therefore, in recent tubular SOFCs the support tube is discarded. Instead, a thicker cathode inside provides the mechanical support for the entire cell structure.

Monolithic SOFCs   For tubular SOFCs as shown in Figure 9.3, irrespective of the manner in which the current is collected, the current collection belongs to the category of edge collection as described in Section 4.9. Therefore, the performance of the tubular SOFCs is limited. Better performance can be achieved when bipolar arrangement is employed, following Section 4.9. Figure 9.4 provides an illustration of the monolithic solid oxide fuel cell design with co-flow arrangement for both reactant streams. Also Figure 9.4 shows the flow path for the oxide ions and electrons. Clearly, electrons may flow a significant length along the electrode (characteristic of edge current collection), instead of across the electrode thickness, as in a strict bipolar arrangement. In this design, no support material or structure is needed, the ceramic cell components support each other mutually. Consequently, cell performance on a unit active surface-area basis may not be improved significantly compared to the tubular design. However, the power density on a volumetric basis would be very high, in fact, the highest among the three SOFC designs, because the active surface area density per-unit volume is the highest for the three designs. It has been estimated that the power density for monolithic SOFCs may reach as high as 8.08 kW/kg and 4000 kW/m^3, in contrast with 0.1 kW/kg and 140 kW/m^3 for the tubular SOFC design. However, significant effort is needed to fully achieve the potential of this design.

It might be emphasized that the tubular design would inevitably require the parallel flow arrangement for the reactant streams for easy reactant preheat and heat

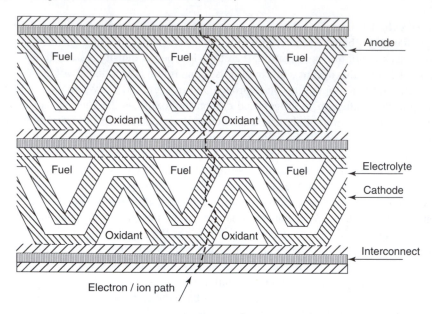

Figure 9.4 Schematic of co-flow type monolithic SOFC.[9]

exchanger arrangement, although cross flow outside the tube is also possible. For the monolithic design, the reactant flow can be easily arranged in co-flow or crossflow.

Planar SOFCs   A truely bipolar arrangement can be achieved if the cell structure is flat, or the so-called **planar design**. Geometrically, the planar SOFCs would look identical to the other types of fuel cells described in earlier chapters. The bipolar current collection provides the least ohmic resistance to the transport of oxide ions and electrons, consequently, best cell performance per active cell-surface area. The maximum power density for this design was reported to be as high as $0.48 \, \text{W/cm}^2$ for small cells when operating on hydrogen and air and $0.9 \, \text{W/cm}^2$ when operating on hydrogen and oxygen — much better than the tubular and monolithic designs. However, on a volumetric basis the performance of planar design is between the tubular and monolithic designs because the volumetric active cell surface area density is between the other two designs. However, planar design is attractive since it is easy for manufacturing process and it uses less materials, hence potentially lower cost.

Table 9.1 provides an overview of the pros and cons for the three different SOFC cell designs described in this section and Table 9.2 provides a summary of the typical dimensions of the cell components for the three different SOFC cell configurations. Recent trend of development for planar SOFCs, as given in Table 9.3, is to use electrolyte-supported structure for high-temperature SOFCs and either anode or cathode electrode support for low- and intermediate-temperature SOFCs. The thick electrolyte layer in the electrolyte-supported cell structure results in a large ohmic overpotential due to the high resistance to the oxide ion transport through the electrolyte. Similarly, electrode-supported cell structures have high resistance to mass transport through the porous electrode, leading to a large concentration overpotential.

Table 9.1 A Comparison for the Three Different SOFC Cell
Configurations

	Advantage	**Disadvantage**
Tubular	Ease of manufacturing No need for gas-tight cell sealing Less thermal cracking due to thermal expansion mismatch (longer life)	Edge current collection Low-power density High materials cost
Planar	Lower fabrication cost Ease in flow arrangement Higher power density	High-temperature gas-tight sealing High assembly effort and cost Stricter requirement on thermal expansion match
Monolithic	Highest power density potential Relative easy assembly Relative easy flow arrangement	Difficult fabrication High-temperature gas-tight sealing Strict requirement on thermal expansion match

Table 9.2 A summary of Typical Dimensions of Cell Components for
Three Different Configurations of Solid Oxide Fuel Cells

Cell	Anode	Cathode	Electrolyte	Interconnection
Tubular[a]	$100\,\mu m$	$1.4\,mm$	$40\,\mu m$	$40\,\mu m$
Monolithic	$50–150\,\mu m$	$50–150\,\mu m$	$50–150\,\mu m$	$50–150\,\mu m$
Planar	$25–100\,\mu m$	$25–100\,\mu m$	$50–250\,\mu m$	$200\,\mu m–1\,mm$ $(+2–6\,mm$ rib height)

[a] Support tube is used for structural integrity (see Table 9.4 for further information). More
recent tubular SOFC design has no support tube, rather a thicker cathode tube (around 2 mm)
is extruded to provide the structural support.

9.3.2 Stack

Although a SOFC stack is also made up of many individual cells like any other fuel
cell stack, the cell-to-cell electrical connection is quite different, depending on the cell
structure described earlier. So is the manner in which the reactant gases are supplied
to each cell. Therefore, the stack structure and configuration corresponding to the
three different cell designs are outlined in the following sections.

Tubular SOFC Stack   The most advanced tubular SOFCs have used circumferential
current collection as illustrated in Figure 9.3b and a typical tubular SOFC stack is
shown in Figure 9.5a. The cell-to-cell electrical connection is accomplished by using
the interconnection contact, which in a way serves the purpose of a bipolar plate in

Table 9.3   A Comparison of Planar SOFC Cell Configurations for High- and Lower-Temperature SOFCs

	High temperature SOFCs (1000 °C)	Lower temperature SOFCs (700 - 800 °C)	
Cell structure	Electrolyte-supported	Anode-supported	Cathode-supported
	Cathode / Electrolyte / Anode	Cathode / Electrolyte / Anode	Cathode / Electrolyte / Anode
Thickness Cathode Electrolyte Anode	50 μm ≥ 150 μm 50 μm	50 μm < 20 μm 300–1500 μm	300–1500 μm < 150 μm 50 μm
Main problem	Large ohmic overpotential	Large concentration overpotential	

other type of fuel cell stack. All the cells in the same rows are connected in parallel between the anodes by using a nickel felt in anode environment, while the cells in different rows are connected in series by using a nickel felt between the anode of one cell in the upper row and the interconnection of another cell in the lower row. This parallel-series connection increases the stack reliability, preventing the stack total failure due to the failure of individual cells in the stack. Figure 9.5b illustrates the side view of one individual tubular cell in the stack, showing the flow organization for both fuel and oxidant stream. Clearly, the fuel and oxidant stream are in a co-flow situation. The oxidant stream is supplied through a smaller tube inserted inside the tubular cell, which is sealed off at one end. This cell design avoids the difficult task of gas-tight sealing at such high temperature. The oxidant preheat from the room temperature to the cell inlet temperature is achieved by burning off the exhaust fuel stream with the exhaust oxidant stream in a combustion chamber located just before the cell inlet for the oxidant.

Figure 9.6 shows the entire tubular SOFC stack with the manifolding arrangement. External manifolding is employed for the fuel supply, while the oxidant supply is provided through internal manifolding to each individual cell.

Monolithic and Planar SOFC Stack   For both monolithic and planar SOFCs, the reactant streams can be arranged in either co-flow or crossflow configurations. Co-flow arrangement is relatively easy for internally manifolded stacks, but it would be more elaborate for external manifolding. The entire stack configuration would be very similar to other types of fuel cell described in the previous chapters. Figure 9.7 shows a short stack of planar SOFCs.

It might be mentioned that the all-solid SOFC has a ceramic support structure, rather than metal as in MCFC, hence it is vulnerable to thermoelastic stresses, especially in its planar or monolithic configuration which requires exact matching

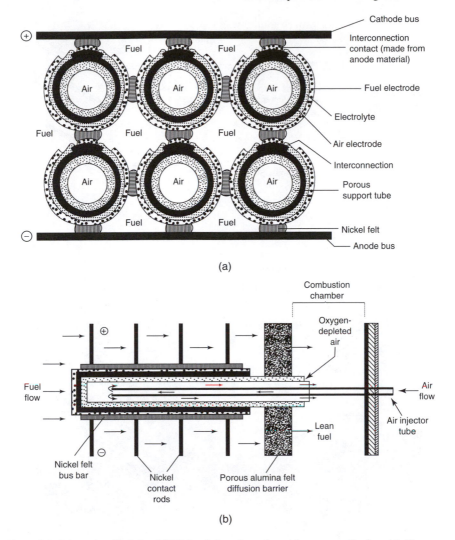

(a)

(b)

Figure 9.5 Schematic of tubular SOFC involving circumferential current collection: (a) The cross-sectional view of the stack and (b) side view of a tubular cell in the stack.[9]

of thermal expansion between the electrode and electrolyte layers in the ultra-thin cell package. These types of SOFC therefore are limited to stack lengths of around 20~25 cm.

9.3.3 System

A SOFC power system also consists of a number of subsystems (called **balance of plants**), just like other types of fuel cell systems described in the previous chapters. It includes fuel processing, oxidant conditioning, thermal management, and power conditioning unit. As pointed out earlier, fuel processing is much simplified compared to

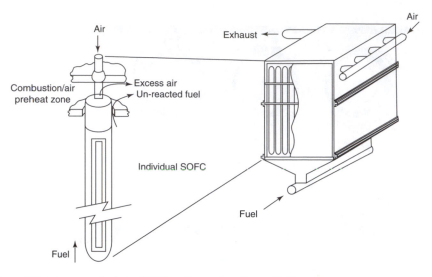

Figure 9.6 Schematic of tubular SOFC stack showing the manifolding arrangement for the fuel and oxidant stream.[10]

Figure 9.7 Planar SOFC stack (Photo Courtesy of Versa Power Systems).

the lower-temperature fuel cells and internal reforming of hydrocarbon fuels is readily achieved to the equilibrium composition with high tolerance to various impurities. Oxidant conditioning mainly involves the preheating of the air from the ambient to the fuel cell inlet temperature, a good thermal integration would provide better system performance. Thermal management is primarily based on the process air flow, which provides the effective cooling of individual cells. The waste heat in the exhaust air-stream, together with the heat from the combustion with the anode exhaust gas, is used for the incoming air-stream preheating. Thus, a high-temperature heat exchanger is integrated with the combustor and the fuel cell stack. In summary, the system complexities are much reduced for the solid oxide fuel cell, as compared to the lower-temperature fuel cells, and the key for good system performance lies in the optimal integration of various system components with effective utilization of waste heat.

The integration of the solid oxide fuel cell system with the conventional steam or gas turbine power generation system deserves a special attention due to extremely high energy efficiency for the combined system which is very much under active investigation. Therefore, readers are referred to the current literature on this important developing area.

9.4   MATERIALS AND MANUFACTURING

As mentioned earlier, the principal challenge for SOFCs is to develop appropriate ceramic materials for the cell components such that each layer is matched with their thermal expansion coefficients and sinterabilities. The materials and fabrication process for each cell/stack component is described in this section.

9.4.1   Cathode

The cathode material of choice is strontium-doped lanthanum manganite ($La_{1-x}Sr_x$ MnO_3, $x = 0.10$–0.15), a p-type semiconductor. It has a high electronic conductivity, essential for low ohmic polarization, especially when the cathode is made thick to provide the structure support (without the support tube in the case of tubular design). The bulk electronic conductivity for $La_{0.5}Sr_{0.5}MnO_3$, for example, is about 294 S/cm at 1000 °C, although the effective conductivity is much lower due to the porous nature of the electrode structure. At the typical SOFC operating condition of 1000 °C, the effective electronic conductivity for the Sr-doped lanthanum manganite cathode is about 100 S/cm. This material also has suitable catalytic properties and dimensional stability during the fabrication process, especially during the electrochemical vapor deposition of the electrolyte and interconnection for the tubular configuration. The thermal expansion coefficients of Sr-doped lanthanum manganite materials are on the order of 1.2×10^{-5} cm/(cm · °C), larger than that of the corresponding electrolyte material. Efforts are being made to reduce this difference in the thermal expansion coefficients in order to reduce the thermal stress and thermal cracking during the thermal cycling.

Other materials have also been considered for the SOFC cathode. For example, metals may be a choice. However, only Noble metals may be used as metal cathode electrodes due to the highly oxidizing environment of the cathode. But because of either melting and sintering (i.e., gold and silver) or high vapor pressure (i.e., palladium) at SOFC operating temperature, only platinum remains a suitable metallic material and this has serious cost implications. Another alternative is to embed metal wires in a matrix of an oxide material with the metal wires acting as the current collector and the porous oxide material provides the path for oxygen transport and acts as the electrocatalyst. However, metal wires are invariably made of Noble metals, such as a platinum wire embedded in porous zirconia. Again cost consideration rules out this approach.

Most attractive cathode electrode materials may be the oxides with mixed conduction (i.e., they can conduct both oxygen oxides and electrons). For such electrodes, the entire electrode structure becomes the electrocatalystic sites for the electrode-charge transfer reaction, increasing significantly the active surface area available for the reaction. As a result, the electrode polarization may be reduced significantly. In this sense, the cathode material of choice, Sr-doped lanthanum manganite, is the best available material for SOFC cathode since it conducts electrons and can also conduct oxygen oxides when partially reduced to create oxygen vacancies at high cathodic-polarization conditions. Because of thermal expansion mismatch with the electrolyte material, ideal cathode material is still to be developed.

9.4.2   Anode

The anode material of choice is a nickel-zirconia cermet, or a mixture of nickel and yttria-stabilized zirconia, $ZrO_2-Y_2O_3$. Such an anode shows a high tolerance to sulfur impurity in the fuel stream. The anode structure has about 20%–40% porosity for the mass transport of reactant and product gases. Nickel serves as the electrocatalyst for anode reaction and as the electronic conductor for the electrons produced at the anode. On the other hand, the doped zirconia acts as a porous support and sintering inhibitor for the nickel metal. But more importantly, the yttria-stabilized zirconia conducts oxygen oxides, thus making the entire anode electrode with mixed conduction and effectively increasing significantly the active anode surface area, consequently minimizing the anode electrode polarization. Nickel in the mixture must be at least 30% by volume in order to maintain adequate and consistent electronic conductivity, but nickel has a high thermal expansion coefficient, as much as 50% higher, when compared to the materials for the other SOFC components. In practice, the optimal nickel content is about 35 vol% in order to balance on the requirement for the electronic conductivity and the thermal expansion match with other SOFC components. At the SOFC operating temperature of 1000 °C, the anode electronic conductivity is about 1000 S/cm for the nickel-zirconia cermet. Better anode materials with mixed conduction are desirable to optimize the conductivity for both electrons and oxygen oxides as well as the thermal expansion match. Although other materials such as Noble metals (platinum) can be used for SOFC anode, they will not be practical from economic considerations.

9.4.3   Electrolyte

Yttria-stabilized zirconia (ZrO_2 doped with 8–10 mol% Y_2O_3) is the material suitable as the SOFC electrolyte because it has conductivity for oxygen oxides over a wide range of O_2 partial pressures ($1-10^{-20}$ atm) without electronic conductivity; in another word, the transport number for oxygen oxide is very close to unity (for 12% yttria-stabilized zirconia), while the transport number for electronic conduction is almost close to zero. It also resists the interdiffusion of cations, Mn and La, into the electrolyte structure. Yttria-stabilized zirconia is very dense with low gas permeability to prevent reactant gas crossover; for example, ZrO_2–Y_2O_3 layer with 92%–93% theoretical bulk density or higher has a hydrogen permeability of less than 10^{-8} cm^2/s. Further, ZrO_2–Y_2O_3 is highly stable in both reducing and oxidizing environment — an important characteristic since electrolyte is exposed to both anode and cathode environment.

However, the yttria-stabilized zirconia electrolyte has low ionic conductivity, about 0.02 S/cm at 800 °C and 0.1 S/cm at 1000 °C. Clearly, the electrolyte has the lowest electric conductivity among all the SOFC components and it is at least an order of magnitude smaller than the typical conductivity of aqueous electrolytes suitable for fuel cell applications shown in Table 4.2. Therefore, the SOFC electrolyte must be made as thin as possible to keep the ohmic loss at acceptable levels, when comparing with the electrolyte ohmic loss for other types of fuel cell. The SOFC electrolyte structure can be made about 40-μm thick by electrochemical vapor deposition (EVD), tape casting as well as other ceramic processing techniques.

The low ionic conductivity of the electrolyte material is the main reason that the solid oxide fuel cell is operated at such a high temperature. A lower-operating temperature (600 °C–800 °C) can still provide fast electrode kinetics and the potential for internal reforming of hydrocarbon fuels, while allow the use of cheaper materials and easy manufacturing processes for other SOFC components. Therefore, the lower-temperature solid oxide fuel cell represents the direction for current and future development and is described in Section 9.6.

9.4.4 Interconnection

In other types of fuel cell described in the previous chapters, a bipolar plate is used for the bipolar current collection of cells in the fuel cell stack. However, for tubular SOFCs, bipolar current collection is not possible and the edge current collection, or cell-to-cell electrical connection, is accomplished by the so-called **interconnection**, shown in Figure 9.3b. Although it is the bipolar plate that provides the electrical connection between the adjacent cells for monolithic and planar SOFCs, the word "interconnection" is carried forward and still being used for these two cases.

The material for the cell interconnection is Mg-doped lanthanum chromite, $LaCr_{1-x}Mg_xO_3$ ($x = 0.02$–0.10). For lanthanum chromite the perovskite phase is stable at a high temperature of around 1000 °C and for a wide range of oxygen concentrations (for oxygen partial pressure P_{O_2} in the range of 1–10^{-29} bars[11]). Pure $LaCrO_3$ is a p-type conductor and has a low electrical conductivity of around 0.6 S/cm at 1000 °C in air. The Mg doping in small proportion increases significantly the electronic conductivity. However, the electronic conductivity of the interconnection at the SOFC operating condition of 1000 °C is only about 2 S/cm, one of the least conductive among all the cell components, only second to the solid oxide electrolyte. Therefore, materials with higher electronic conductivity would be desirable. Although metallic materials have high electronic conductivity, noble metals are too expensive to be practically feasible, and other metals are limited to some nickel alloys which have thermal expansion coefficients much larger than other cell components.

Table 9.4   Evolution of Typical Material for Tubular SOFC components

Component	ca. 1965	ca. 1975	Current Status
Support tube[a]	Yttria-stabilized ZrO_2	Yttria-stabilized ZrO_2	Calcia-stabilized ZrO_2 (15 mol% CaO)
Cathode	Porous Pt	Stabilized ZrO_2 impregnated with praseodymium oxide and covered with SnO-doped Im_2O_3	Sr-doped lanthanum manganite (10 mol% Sr)
Electrolyte	Yttria-stabilized ZrO_2 (0.5-mm thickness)	Yttria stabilized ZrO_2	Yttria stabilized ZrO_2 (8 mol% Y)
Interconnection	Pt	Mn-doped cobalt chromite	Mg-doped lanthanum chromite (10 mol% Mg)
Anode	Porous Pt	Ni/ZrO_2 cermet	Ni/Y_2O_3 stabilized ZrO_2 cermet (35 vol% Ni)

[a] For more recent tubular SOFC design, the support tube is eliminated and a thicker cathode tube (around 2 mm) is extruded to provide the structural support in place of the support tube.

Table 9.4 illustrates the evolution of the typical material for the tubular SOFC components. Most of these materials are also used in the other two SOFC configurations. As described before, the materials are the biggest issue in the development of SOFCs because of its high-temperature operation. A lower temperature eases considerably the restrictions placed on the materials and such has formed the active research for the solid oxide fuel cell. Table 9.5 shows a comparison of the typical materials used for the high- and lower-temperature SOFCs.

9.4.5   Fabrication Techniques

The fabrication process for the three different SOFC configurations all involves making thin layers for the electrodes, electrolyte, and the cell interconnection; depositing one thin layer on top of another and some form of sintering to make the final cell structure. However, the fabrication techniques employed are different for each of the three configurations and vary from developers as well. For the tubular SOFC configuration, the support tube is first made of calcia-stabilized zirconia by extrusion into a cylindrical shape and one end of the tube is plugged to seal the end, followed by sintering. On this support tube a slurry of Sr-doped $LaMnO_3$ powder is deposited by using slurry coating. The resulting composite tube is sintered to produce a thin layer of cathode electrode with a desired thickness. The cell interconnection is then deposited on the support tube-cathode structure by electrochemical vapor deposition (EVD). Next, the electrolyte layer is also made through EVD by providing the zirconium- and yttrium-chloride vapor in a desired ratio to the outer surface of the cathode layer. On the outer

Table 9.5   Typical Materials Used for High- and Lower-temperature SOFCs

	Electrolyte	Anode	Cathode
High T SOFCs	Zirconia doped with 8%–10% yttria (YSZ) • Highly stable in both reducing and oxidizing environments • Ionic conductivity of YSZ 0.02 S/cm at 800 °C 0.1 S/cm at 1000 °C	Cermet made of Ni/YSZ • Good catalytic activity • Thermal expansion coefficient comparable with electrolyte	Strontium-doped (Sr-doped) Lanthanum Manganite
Lower T SOFCs	LaSrGaMgO (LSGM) • Excellent ionic conductivity ≥0.1 S/cm at 800 °C	Samaria-doped Ceria NiO Ni/CeO$_2$ cermet Copper/ceria cermet	Lanthanum strontium cobalite (LSCo)

surface of the dense and uniform electrolyte layer nickel powder slurry is deposited by slurry coating, then yttria-stabilized zirconia is impregnated by EVD to form the nickel-zirconia cermet anode layer. It is standard practice that the resulting cell structure is checked to make sure no gas leakage occurs before any fuel cell operation. Table 9.6 provides a summary of components, structural characteristics, materials used and fabrication processes employed for the tubular solid oxide fuel cell, and Figure 9.8 shows the flowchart for the tubular solid oxide fuel cell fabrication processes.

For the monolithic SOFC configuration, the layers of the anode, cathode, and electrolyte are made into one piece to support each other without a separate support layer. The fabrication process involves either tape casting or tape-calendaring processes. For the tape-casting process, a slurry for each cell component is made by mixing the ceramic powder with organic binders, plasticizers, and solvent(s). A thin layer of each cell component is cast one layer on top of another and the layer thickness is controlled by using the knife edge of a doctor blade. In the tape-calendaring process, a thin, flat tape for each cell component is produced by rolling through a two-roll mill and the resulting three tapes for the anode, electrolyte and cathode are laminated by rolling through a second two-roll mill. The composite tape is then corrugated through molding and stacked into the desired fuel cell stack configuration like that in Figure 9.4. Sintering the stack in one piece at a desired temperature completes the stack fabrication process.

On the other hand, for the planar SOFC configuration each cell component is tape cast and sintered separately at the respective optimal temperature. Then the cell components are assembled into the cell or stack unit as desired. Therefore, the fabrication process for the planar SOFCs is much easier, without the risk posed by the co-firing of all the cell components together. However, the assembly process becomes more time-consuming and complex. Figure 9.9 illustrates the fabrication processes for the planar solid oxide fuel cell. Further details on the fabrication process for the three different SOFC configurations can be found in more advanced texts such as [6].

Table 9.6   Summary of Components, Structural Characteristics, Materials, and Fabrication Processes for the Tubular SOFC

Component	Property	Current Status
Support tube[a]	Materials	Calcia-stabilized zirconia (15 mol% CaO)
	Thermal expansion coefficient	$\sim 10^{-5}$ cm/(cm $\cdot$ °C)
	Thickness	1–1.5 mm
	Diameter	12.8 mm inner diameter
	Porosity	34%–35%
	Fabrication process	Extrusion/sintering
Cathode	Materials	Sr-doped lanthanum manganite (10 mol% Sr)
	Thermal expansion coefficient	1.2×10^{-5} cm/(cm $\cdot$ °C)
	Thickness	0.7–1 mm
	Porosity	20%–40%
	Fabrication process	Slurry coating/sintering
Electrolyte	Materials	Yttria-stabilized ZrO_2 (8 mol% Y)
	Thermal expansion coefficient	1.05×10^{-5} cm/(cm $\cdot$ °C)
	Thickness	0.03–0.04 mm
	Porosity	Close to zero (dense layer)
	Fabrication process	Electrochemical vapor deposition
Interconnection	Materials	Mg-doped lanthanum chromite (10 mol% Mg)
	Thermal expansion coefficient	$\sim 10^{-5}$ cm/(cm $\cdot$ °C)
	Thickness	0.03–0.04 mm
	Porosity	Close to zero (dense layer)
	Fabrication process	Electrochemical vapor deposition
Anode	Materials	Ni/Y_2O_3-stabilized ZrO_2 cermet (35 vol% Ni)
	Thermal expansion coefficient	1.25×10^{-5} cm/(cm $\cdot$ °C)
	Thickness	0.1 mm
	Porosity	20%–40%
	Fabrication process	Slurry coating/EVD

[a] For more recent tubular SOFC design, the support tube is eliminated and a thicker cathode tube (around 2 mm) is extruded to provide the structural support in place of the support tube.

9.5   PERFORMANCE OF SOFCs

Before the performance of SOFC is discussed, we must emphasize that the performance may differ significantly for the three different SOFC configurations, namely, the tubular, monolithic and planar for the cells with different fabrication processes;

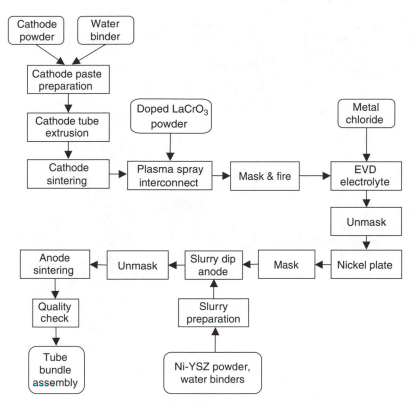

Figure 9.8 Flowchart for the fabrication process of the tubular SOFC.

for the cells developed by different developers, and for the cells with different active areas. Detailed experimental data with parametric studies are fairly limited, and hence our coverage in this section.

SOFC performance may also be written as

$$E = E_r - \eta_{\text{act}} - \eta_{\text{ohm}} - \eta_{\text{conc}} \tag{9.9}$$

where the reversible cell potential is given by the Nernst equation as, corresponding to the cell reaction given in Equation (9.3)

$$E_r(T, P_i) = E_r(T, P) + \frac{RT}{nF} \ln \left(\frac{P_{H_2} P_{O_2}^{1/2}}{P_{H_2O}} \right) \tag{9.10}$$

where $n = 2$ represents the number of electrons transferred during the cell reaction. However, the solid oxide fuel cell is expected to eventually operate on hydrocarbon fuels internally reformed. Hence, the anode stream would contain carbon monoxide which could be oxidized electrochemically as well at the anode, as shown in Equation (9.4); and the reversible cell potential would be the mixed electrode potential due to the electrooxidation of hydrogen and carbon monoxide at the anode, corresponding to the

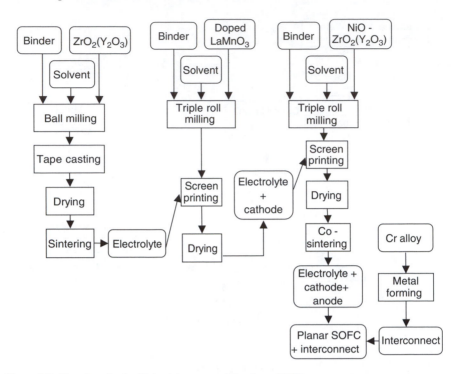

Figure 9.9 Flowchart for the fabrication process of the planar SOFC.

reaction given in Equation (9.8). However, as pointed out earlier, the electrooxidation of carbon monoxide is fairly slow, and the oxidation of carbon monoxide is mainly via the water–gas shift reaction to convert into hydrogen gas. Therefore, the reversible cell potential can still be estimated from Equation (9.10) given, although the determination of the partial pressures for the hydrogen gas and water vapor in the anode compartment becomes complex.

At the SOFC operating temperature of 1000 °C, the polarization due to the electrode activation and mass transport (η_{act} and η_{conc}) is generally considered small due to the fast electrode kinetics and high mass diffusion coefficients at the typical SOFC operating current density of 160 mA/cm^2 or higher, and the main voltage loss is attributed to the ohmic polarization η_{ohm}. This is because the polarization curve for the solid oxide fuel cell operating near the temperature of 1000 °C is almost linear for the cell voltage-current density relation, especially when the current density is reasonably large (for example, for $J > 100$ mA/cm^2 or so). However, the electrochemical reaction at both the anode and cathode can be represented by the Butler–Volmer equation given in Equation (3.70) with a symmetry factor of about 0.5. Then the activation polarization for the anodic and cathodic reaction can be written, respectively, as

$$\eta_{act} = \frac{2RT}{nF} \sinh^{-1}\left(\frac{J}{J_0}\right) \qquad (9.11)$$

Table 9.7   Relative Contribution of Cell Component to Total Ohmic Polarization at 1000 °C for tubular SOFCs[12,14]

	Resistivity ρ ($\Omega \cdot$ cm)	Thickness δ (cm)	Relative Contribution r
Cathode	0.013	0.07	0.65
Anode	0.001	0.01	0.25
Electrolyte	10	0.004	0.09
Interconnection	0.5	0.004	0.01

The total area specific resistance is estimated to be 0.33 $\Omega \cdot$ cm^2.

The exchange current density, including the electrode roughness factor discussed in Section 3.4.4, is estimated to be[13]

$$J_0 = \begin{cases} 0.53 \text{ A/cm}^2 & \text{For anode} \\ 0.20 \text{ A/cm}^2 & \text{For cathode} \end{cases} \tag{9.12}$$

when pure hydrogen is used as the fuel and air as the oxidant. According to Equation (9.12), the activation polarization is sizeable when compared to the ohmic polarization. However, the exchange current density is usually a strong function of temperature which is not reflected in the equation. Therefore, the ability to estimate the activation overpotential as a function of temperature is limited, in fact, with the constancy of the exchange current density given in Equation (9.12), Equation (9.11) would imply that the activation overpotential would increase with the cell operating temperature, which is obviously in contrast with the experimental reality.

The ohmic polarization is contributed by the resistance of the cathode, anode, electrolyte and interconnection. For a tubular cell operating at 1000 °C, the relative contribution to the ohmic polarization has been reported as shown in Table 9.7. The large contribution (65%) from the cathode arises from the long current path along the curved tubular surface (about 1.1 cm), as discussed earlier. Similarly, the second largest contribution (25%) from the anode is also due to the long current path in the anode (about 0.8 cm). The current density through the electrolyte and the interconnection is assumed to be uniform and along their respective thickness, thus their relatively small contribution despite their large resistivity values.

We can certainly expect that the relative contribution from the cell components will be quite different if planar SOFC configuration is considered. Since for planar geometry, the current density can be taken as uniform and along the thickness of each layer, and the relative contribution to the total ohmic polarization by each component can be easily calculated and as given in Table 9.8 if the same thickness of each component is maintained for easy comparison. In this case, clearly the large resistivity of the electrolyte dominates the ohmic polarization. However, the total area specific resistance has reduced significantly from 0.33 $\Omega \cdot$ cm^2 for the tubular configuration to 0.043 $\Omega \cdot$ cm^2 for the planar configuration, a nearly one-order-of-magnitude

Table 9.8 Relative Contribution of Cell Component to Total Ohmic Polarization at 1000 °C for planar SOFCs

	Resistivity ρ ($\Omega \cdot$ cm)	Thickness δ (cm)	Relative Contribution r
Cathode	0.013	0.07	0.021
Anode	0.001	0.01	0.0002
Electrolyte	10	0.004	0.932
Interconnection	0.5	0.004	0.047

Component thickness maintained the same as in Table 9.7. The total area specific resistance is estimated to be 0.043 $\Omega \cdot$ cm^2.

reduction, primarily due to the shortening of the path length for the current. Therefore, we start to see the advantage of the planar structure in reducing the ohmic polarization or enhancing the cell performance.

9.5.1   Effect of Temperature

As discussed before for other types of fuel cell, the effect of temperature is felt through the change in the reversible cell potential and in the various polarization terms, corresponding to Equation (9.9). The reversible cell potential can be easily determined as a function of temperature, based on the materials presented in Chapter 2, and is given in Table 9.9. Clearly, the reversible cell potential decreases almost linearly with cell operating temperature. Therefore for the SOFC, cell potential is reduced at small current densities when the cell operating temperature is increased; while the cell potential at large current densities of the practical importance is actually increased due to significant reduction in the various polarization terms. As a result, there will be a current density at which the polarization curves cross and at this point no effect of temperature will be observed. The current density at which a practical cell is operated is usually much larger than this cross current-density value, hence the effect of temperature at the practical operating current density is dominated by the changes in the various polarizations.

From the previous discussion and Equation (9.9), we have the change in the cell potential as a function of temperature

$$\Delta E_T = -\Delta \eta_{\text{act},T} - \Delta \eta_{\text{ohm},T} \tag{9.13}$$

The change in the activation polarization can be derived based on Equation (9.11)

$$\Delta \eta_{\text{act},T} = \frac{\partial \eta_{\text{act}}}{\partial T} \Delta T = \frac{2R}{nF} \sinh^{-1}\left(\frac{J}{J_0}\right) \Delta T \tag{9.14}$$

where $\Delta T = T - T_1$. Since in practice, a cell is usually operated at a nominal design current density J_1 with a small deviation that might be designated as $\Delta J = J - J_1$,

Table 9.9   Effect of Temperature on the Reversible Cell
Potential for the SOFC Operating at 1 atm

Operating Temperature T (K)	Reversible Cell Potential E_r (V)
1000	0.9983
1100	0.9696
1200	0.9405
1300	0.9113

then we can linearize the current density dependence in Equation (9.14) as

$$\sinh^{-1}\left(\frac{J}{J_0}\right) = \sinh^{-1}\left(\frac{J_1}{J_0}\right) + \left[\sinh^{-1}\left(\frac{J_1}{J_0}\right)\right]' \frac{\Delta J}{J_0} \qquad (9.15)$$

The exchange current density J_0 usually changes with temperature, however, it is neglected in the above analysis because its functional dependence on temperature for the solid oxide fuel cell is not available. Since $[\sinh^{-1}(x)]' = (x^2 + 1)^{-1/2}$, substituting Equation (9.15) into Equation (9.14) and considering Equation (9.12), for $J_1 = 0.16\,\text{A/cm}^2$ we obtain the anodic and cathodic activation polarization change with temperature as

$$\Delta\eta_{\text{act,a}} = -(0.0001556J + 0.7327 \times 10^{-6}) \approx -0.0001556J(T - T_1) \quad (9.16)$$

$$\Delta\eta_{\text{act,c}} = -(0.0003364J + 0.9306 \times 10^{-5}) \approx -0.0003364J(T - T_1) \quad (9.17)$$

where the current density is in A/cm^2, temperature in K and overpotential change in V. In view of the discussion following Equation (9.12), we have considered that the activation overpotential is reduced when temperature is increased, hence the negative sign in the above two equations. As it might be expected, the change in the cathodic activation overpotential is much larger than, in fact over twice as large as, the anodic activation overpotential. The total activation overpotential then becomes

$$\Delta\eta_{\text{act,}T} = -0.0004920J(T - T_1) \qquad (9.18)$$

The change in the ohmic overpotential is due to the resistance in the cell component: cathode, anode, electrolyte, and interconnection (designated by the subscript "c," "a," "e," and "i," respectively); hence it can be expressed as

$$\Delta\eta_{\text{ohm,}T} = \Delta\eta_{\text{ohm,c}} + \Delta\eta_{\text{ohm,a}} + \Delta\eta_{\text{ohm,e}} + \Delta\eta_{\text{ohm,i}}$$

$$= \left(\frac{\partial\eta_{\text{ohm,c}}}{\partial T}\frac{r_c}{\eta_{\text{ohm,c}}} + \frac{\partial\eta_{\text{ohm,a}}}{\partial T}\frac{r_a}{\eta_{\text{ohm,a}}}\right.$$

$$\left. + \frac{\partial\eta_{\text{ohm,e}}}{\partial T}\frac{r_e}{\eta_{\text{ohm,e}}} + \frac{\partial\eta_{\text{ohm,i}}}{\partial T}\frac{r_i}{\eta_{\text{ohm,i}}}\right)\eta_{\text{ohm}} \cdot \Delta T$$

$$= \left(\frac{\partial\rho_c}{\partial T}\frac{r_c}{\rho_c} + \frac{\partial\rho_a}{\partial T}\frac{r_a}{\rho_a} + \frac{\partial\rho_e}{\partial T}\frac{r_e}{\rho_e} + \frac{\partial\rho_i}{\partial T}\frac{r_i}{\rho_i}\right)\eta_{\text{ohm}} \cdot \Delta T \qquad (9.19)$$

Table 9.10 Constants a and b in correlation of Resistivity as Function of Temperature for SOFC cell components

Cell Component	a $(\Omega \cdot cm)$	b (K)
Anode	0.00298	−1392
Cathode	0.00811	600
Electrolyte	0.00294	10350
Interconnection	0.01256	4690

From Equation (9.20) and [15].

where the ohmic loss in the cell component is related to the resistivity as $\eta_{ohm} = \rho \delta J$ and is utilized in the last equality. The relative contribution r of the cell component to the total ohmic overpotential is given in Tables 9.7 and 9.8, respectively, for the tubular and planar configurations.

The resistivity as a function of temperature has been correlated for each cell component as[15]

$$\rho = a \exp\left(\frac{b}{T}\right) \tag{9.20}$$

where the constants a and b are given in Table 9.10 for each cell component. Substituting Equation (9.20) into Equation (9.19) yields

$$\Delta \eta_{ohm,T} = -(b_c r_c + b_a r_a + b_e r_e + b_i r_i) \frac{R'' J}{T^2} \Delta T \tag{9.21}$$

where R'' is the area specific resistance. Using the values of r given in Tables 9.7 and 9.8, respectively, along with the values of b given in Table 9.10, we estimate that the ohmic overpotential change is

$$\Delta \eta_{ohm,T} = -0.00021 J \cdot (T - T_1) \quad \text{For tubular cells} \tag{9.22}$$

$$\Delta \eta_{ohm,T} = -0.00026 J \cdot (T - T_1) \quad \text{For planar cells} \tag{9.23}$$

where the current density J is in A/cm^2, T in K, and $\Delta \eta_{ohm,T}$ in V. It seems that the ohmic overpotential change is less than half of the activation overpotential change.

Now substituting the above results into Equation (9.13), we have the cell potential change as a function of temperature

$$\Delta E_T = (0.00070 \sim 0.00075) J \cdot (T - T_1) \tag{9.24}$$

where the current density J is in A/cm^2, T in K, and ΔE_T in V. This temperature dependence seems to be very weak, compared to the empirical correlations. The coefficient 0.00070~0.00075 $\Omega \cdot cm^2$/K seems to be an order of magnitude smaller than the empirically estimated value of $0.008 \, \Omega \cdot cm^2$ given in [12, 14]. However, empirical evidences exist that the temperature dependence coefficient would change significantly with the composition of the fuel used and the coefficient of $0.008 \, \Omega \cdot cm^2$ was estimated from the experimental data when operating on a fuel consisting of 67% H_2/22% CO/11% H_2O. We may conclude then that the water–gas shift equilibrium reaction and the hydrogen concentration at the reaction sites (which changes the exchange current density) both have a significant contribution to the temperature dependence of the actual cell potential; the lower the hydrogen content in the fuel gas, the stronger the temperature effect on the cell potential.

9.5.2   Effect of Pressure

The pressure effect on the actual cell potential is often expressed in terms of the logarithmic dependence, as for other types of fuel cells

$$\Delta E_P = E(T, P) - E(T, P_1) = k_P \log\left(\frac{P}{P_1}\right) \tag{9.25}$$

For the overall cell reaction given in Equation (9.3), the number of mole changes in the gaseous species for the reaction is $\Delta N = -1/2$. Then the coefficient for the pressure effect on the reversible cell potential becomes, from the Nernst equation shown in Equation (9.10),

$$k_P = \frac{\Delta N R T}{n F} = 0.063 (V) \tag{9.26}$$

at the typical SOFC operating temperature of 1000 °C. This value is fairly consistent with the limited empirical data, which suggest that the pressure coefficient k_P for the actual cell potential would be around 0.059 V.[12,14]

Although higher pressure operation is beneficial for better SOFC performance, the cost and effort involved in making stronger SOFC components (including the connecting pipes) and gas-tight sealing as well as the operational cost involved in the compression of the reactant gases are significant, minimizing the benefit of higher pressure operation. However, for combined SOFC-gas turbine power systems higher pressure operation is particularly worthwhile. However, thermodynamic optimization would be required to couple the particular SOFC and gas turbine bottoming cycle in order to determine the optimal operating pressure for the particular system.

9.5.3   Effect of Reactant Gas Composition and Utilization

Like other types of fuel cells, SOFC performance also changes with reactant concentrations in the cathode and anode compartments. Since the reactant utilization will affect the local reactant concentrations in the anode and cathode compartments for a given set of inlet conditions, it will influence the cell performance as well.

Oxidant   From the Nernst equation for the SOFC shown in Equation (9.10), the effect of the oxidant inlet concentration and utilization is also commonly correlated in terms of the logarithmic dependence on the partial pressure of oxygen in the cathode compartment averaged between the inlet and exit values:

$$\Delta E_c = k_c \log \frac{(\bar{P}_{O_2})}{(\bar{P}_{O_2})_1} \tag{9.27}$$

where the coefficient $k_P = RT/(2nF) = 0.063\,\text{V}$ at $1000\,^\circ\text{C}$ for the reversible cell potential; this value might be larger for the actual cell potential at the practical operating current densities, and a value of $k_P = 0.092\,\text{V}$ has been suggested.[12,14]

Fuel   Similarly, the effect of the hydrogen concentration and utilization on the cell potential can be expressed as, following the Nernst equation given in Equation (9.10)

$$\Delta E_a = k_a \log \frac{(\bar{P}_{H_2}/\bar{P}_{H_2O})}{(\bar{P}_{H_2}/\bar{P}_{H_2O})_1} \tag{9.28}$$

where the coefficient $k_a = RT/(nF) = 0.126\,\text{V}$ at $1000\,^\circ\text{C}$ for the reversible cell potential. Limited experimental data implies a larger value of $k_P = 0.172$ for the actual cell potential at the practical operating current densities when the cell operates on reformed fuel.[12,14]

9.5.4   Effect of Impurities

The effect of impurities on the performance and lifetime of the solid oxide fuel cell is largely unexplored, if compared with the counterpart for the molten carbonate fuel cell described in Section 8.5.4. The possible effect and the maximum tolerance levels of impurities similar to Tables 8.5 and 8.6 for the molten carbonate fuel cell are not available for the solid oxide fuel cell. However, some limited studies indicate that solid oxide fuel cell may be able to tolerate NH_3 upto 5000 ppm, HCl upto 1 ppm, and H_2S upto 0.1 ppm; 1 ppm of H_2S would cause an immediate drop in the cell potential, followed by a gradual decrease with a linear dependence over the time. The adverse effect of H_2S on the cell performance could be reversed by providing clean fuel, free of H_2S, to the anode. Further information may be found in current literature, including [14].

9.5.5   Effect of Current Density

For solid oxide fuel cells operating at $1000\,^\circ\text{C}$, it is generally believed that the ohmic loss plays the dominant role, while the activation and concentration overpotentials are relatively small at the typical operating current density of $160\,\text{mA/cm}^2$ or higher. Under that condition the cell potential will change linearly with the current density such that we have

$$\Delta E_J = -k_J \Delta J \tag{9.29}$$

where the proportionality constant should equal the area specific resistance for the entire cell assembly. For the sample tubular and planar SOFCs given in Tables 9.7

and 9.8, the area specific resistance, excluding the contact resistances and the connecting wires for the current from the cell, is dominated by the cathode resistance for the tubular configuration and by the electrolyte for the planar configuration; and was estimated as 0.33 and 0.043 $\Omega \cdot cm^2$, respectively, for the tubular and planar configurations. However, these values seem to be smaller than the empirical value of $k_J = 0.73 \, \Omega \cdot cm^2$ given in [12, 14]. One should caution for this direct comparison since the cells used to generate the empirical data were not identical to the cells considered in Tables 9.7 and 9.8 in terms of the component sizes and thickness, as well as the materials, etc.

9.6 FUTURE R&D

For commercial applications, the durability and reliability of the solid oxide fuel cell system needs to be demonstrated under the practical conditions along with the cost reduction. The high operating temperature (>1000 °C) of the conventional solid oxide fuel cell (sometimes referred to as the high-temperature SOFC) provides sufficiently high oxygen ion conductivity of the zirconia-based electrolyte, but reduces the reliability and durability (lifetime) of the cells through materials-related issues. Therefore, development of the advanced materials becomes the key issue for the high performance and low cost of the solid oxide fuel cell systems. Other fundamental issues include the kinetics at the anode and cathode electrodes, especially the kinetics for the CO electrochemical oxidation and its competition with the water–gas shift reaction. Proper structural design, including stress analysis of thin composite layers, is important to achieve the cell lifetime without thermal cracking and delamination of the composite cell structure.

A major thrust in the current SOFC development is the lower-temperature solid oxide fuel cell, operating in the range of 600 °C to 800 °C or even lower to about 500 °C, often referred to as the intermediate-temperature solid oxide fuel cell (or IT-SOFC). This lower operating temperature allows the use of cheaper materials and easy fabrication process for the cell components and interconnection (i.e., lower cost) and improved reliability and cell lifetime. However, the conductivity of the yttria-stabilized zirconia electrolyte reduces significantly, as the temperature is lowered, from 0.1 S/cm at 1000 °C to about 0.02 S/cm at 800 °C and about 0.008 S/cm at 700 °C. Therefore, a number of alternative electrolytes have been investigated, and among them ceria-based electrolyte is one of the promising alternatives to the zirconia-based electrolyte at lower operating temperatures; this is because of their high oxygen oxide conductivity at 700 °C[16]. The ceria-based electrolyte, however, shows considerable electronic conductivity, which is detrimental to long-term fuel cell performance. In addition, it is unstable under the hydrogen environment of the anode.[17,18] Oxygen ion conductors, based on a modification of the perovskite, lanthanum gallate, have also been shown to be potential candidate electrolytes for lower temperature SOFC.[19,20]

In addition to the oxide conductor-based lower temperature SOFCs, cation conductor-based lower temperature SOFCs have also been explored.[23] $SrCeO_3$-based oxides were the first class of perovskite materials shown to exhibit proton conductivity at high temperatures in a hydrogen-containing atmosphere.[21] Subsequently, $BaCeO_3$, $SrZrO_3$ and $BaZrO_3$ mixed oxides have also been shown to exhibit high protonic conductivity at lower temperatures than pure oxygen-oxide-conducting electrolytes, such as lanthanum gallate, making them good candidates as the electrolyte for the intermediate-temperature solid oxide fuel cell.[22] However, these variates of the lower temperature SOFCs are still at the very earlier stage of the development; and since the electrolyte is a new material and the electrolyte is the defining component of any fuel cell, new materials for other cell components need to be developed and matched and a whole new type of fuel cell is emerging. A review of materials for oxygen oxide conductor-based IT-SOFCs is available in literature.[24] Almost all IT-SOFCs have not been optimized in the conventional sense and often poor performance is exhibited. Therefore, significant work lies ahead before the technology is matured.

As described in Section 9.3.1, a planar cell structure is an attractive geometry for integration into a stack configuration. However, gas-tight sealing remains a challenge for both high and intermediate temperature SOFCs, just like for other types of fuel cells with planar design. Innovative cell and stack configuration is needed for sealing purpose, in addition to advanced sealing materials.

9.7   SUMMARY

This chapter presents the various aspects of solid oxide fuel cell. Overall half-cell and whole-cell reactions have been described as well as the other chemical reactions occurring in the anode, and an appreciation for the significance of the water–gas shift reaction in SOFCs has been developed. The advantage associated with the all-solid structure for the SOFCs, and the pros and cons associated with the high-temperature operation were also described. Section 9.3 provided the three different SOFC configurations: tubular, monolithic, and planar and the corresponding components of such a cell and stack, as well as system issues. An understanding in general of the different distribution of current density in the cells and stacks corresponding to the three different SOFC configurations should be realized. Section 9.4 developed a general understanding of the typical materials and fabrication techniques used for the various cell/stack components, the issues they face, the possible effects and ways of avoiding the negative effects. The importance of matching the thermal expansion coefficient for the various cell/stack components for good cell performance and cell lifetime was emphasized. The effect of the various operating parameters on the SOFC performance, including the temperature, pressure, reactant concentration and utilization, the effect of impurities and cell current density was outlined. Finally the needs for various future works in order to advance the SOFC technology for commercial application were discussed.

BIBLIOGRAPHY

1.   Nernst, W. Z. 1900. *Electrochemistry*, 6: 141.
2.   Schottky, W., Wiss. Veroff. Siemens Werke. 1935, 14(2): 1–19, *Chem. Abstr.*, 1935, 20: p. 5358.

3. Baur, E. and H. Z. Preis Z. Elecktrochem. 1937, 43: 727–732, 1938, 44: 695–698.
4. Rohr, F. J. 1977. In Proc. Workshop on High Temperature Solic Oxide Fuel Cells (H. S. Isaacs, S. Srinivasan, and I. L. Harry, eds) p. 122.
5. Minh, N. Q. 1993. Ceramic fuel cells. *J. Am. Ceram. Soc.*, 76(3): 563–588.
6. Minh, N. Q. and T. Takahashi. 1995. Science and Technology of Ceramic Fuel Cells, Elsevier, New York.
7. Matsuzaki, Y. and I. Yasuda. 2002. *Solid State Ionics*, 152–153: 463–468.
8. Riess, I. 2003. Solid State Ionics, 157: 1–17.
9. Appleby, A. J. and F. R. Foulkes. 1988. *Fuel Cell Handbook*, Van Nostrand Reinhold, New York.
10. Murugesamoorthi, K. A., S. Srinivasan, and A. J. Appleby. 1993. Research, Development, and Demonstration of Solid Oxide Fuel Cell Systems. In Fuel Cell Systems, ed. by L.J.M.J. Blomen and M.N. Mugerwa, Plenum Press, New York.
11. Nakamura, T., G. Petzow, and L. J. Gauckler. 1979. *Mater. Res. Bull.*, 14: 649.
12. Hirschenhofer, J. H., D. B. Stauffer, and R. R. Engleman. 1994. Fuel Cells: A Handbook (Revision 3), U. S. Department of Energy.
13. Chan, S. H., K. A. Khor, and Z. T. Xia. 2001. *J. Pow. Sour.*, 93: 130–140.
14. U.S. Department of Energy. 2000. *Fuel Cells: A Handbook* (Revision 5).
15. Chan, S. H., C. F. Low, and O. L. Ding. 2002. *J. Pow. Sour.*, 103: 188–200.
16. Steele, B. C. H. 1989. In High Conductivity Solid Ionic Conductors, T. Takahashi, ed. World Scientific, Singapore.
17. Tuller, H. and A. S. Nowick. 1975. *J. Electrochem. Soc.*, 122: 255.
18. Maffei, N. and A. K. Kuriakose. 1998. *Solid State Ionics*, 107: 67.
19. Ishihara, T., H. Matsuda and Y. Takita. 1994. *J. Am. Chem. Soc.*, 116: 3801.
20. Ishihara, T., H. Matsuda and Y. Takita. 1994. *Proc. Electrochem. Soc.*, 94(12): 85.
21. Iwahara, H., T. Esaka, H. Uchida, and N. Maeda. 1981. *Solid State Ionics*, 3–4: 359.
22. Katahira, K., Y. Kohchi, T. Shimura and H. Iwahara. 2000. *Solid State Ionics*, 138: 91.
23. Maffei, N., L. Pelletier and A. McFarlan, Performance characteristics of Gd-doped barium cerate-based fuel cells. *J. Pow Sour.*, to appear.
24. Ralph, J. M., A. C. Schoeler and M. Kumpelt. 2001. Materials for lower temperature SOFC. *J. of Mat. Sci.*, 36: 1161–1172.

PROBLEMS

9.1 Describe briefly the advantages and disadvantages as well as the areas of applications for solid oxide fuel cells.

9.2 Describe the operation principle such as half-cell and whole-cell reaction, the primary fuels expected to be used for solid oxide fuel cells.

9.3 Discuss briefly the typical (or target) operating conditions such as cell voltage, cell current density, temperature, pressure, fuel and oxidant utilization, and chemical-to-electrical energy conversion efficiency.

9.4 Describe the effect of operating conditions on cell performance.

9.5 Describe the geometrical configuration of cell and stack, typical materials used for, and the thickness of, as well as other dimensions of the cell components.

9.6 Describe how the concept of internal reforming could be implemented for the solid oxide fuel cell.

9.7 Discuss the pros and cons for the tubular, monolithic, and planar solid oxide fuel cells.

9.8 Describe the major factors affecting the lifetime of solid oxide fuel cells.

9.9   Describe the critical technical barriers to be overcome for the commercialization of solid oxide fuel cells, the possible solutions, and their pros and cons.

9.10   Derive the Nernst equation for the solid oxide fuel cell given in Equation (9.10) from the general Nernst equation given in Chapter 2.

9.11   Determine the reversible as well as the thermoneutral cell potential for the following reactions at the typical SOFC operating condition of 1 atm and 1000 °C:

   (i) $H_2 + \frac{1}{2}O_2 \longrightarrow H_2O$

   (ii) $CH_4 + 2O_2 \longrightarrow CO_2 + 2H_2O$

   (iii) $CH_4 + H_2O \longrightarrow CO + 3H_2$

   (iv) $CO + H_2O \longrightarrow CO_2 + H_2$

   (v) $2CO \longrightarrow C + CO_2$

DIRECT METHANOL FUEL CELLS (DMFCs)

10.1 INTRODUCTION

For all the fuel cells described in the previous chapters, hydrogen is the direct fuel used in the anode–electrode reaction for the current generation. Since hydrogen gas is not widely available on Earth in pure form, hydrogen is typically regarded as a fuel (energy carrier) produced from some primary fuel such as petroleum oils and natural gases. Hydrogen production is often achieved through steam-reforming, autothermal-reforming, or partial oxidation-reforming,[1,2] and the major components in the reformed gas mixture are hydrogen and carbon dioxide with a few percent of carbon monoxide. As pointed out in previous chapters, carbon dioxide must be removed for alkaline fuel cells, while carbon monoxide must be removed for acid-electrolyte fuel cells including both PAFCs and PEMFCs. The fuel stream processing and cleaning process is not trivial, often contributes significantly to the overall system cost, size and performance reduction.

On the other hand, hydrogen gas as a fuel lacks the distribution infrastructure and has the lowest energy density on a per-unit volume basis. This results in difficulty in onboard hydrogen storage for vehicular applications — which remains an area of intensive current research. At least three viable technologies exist for storing hydrogen onboard of vehicles: compressed hydrogen gas (CH_2), metal hydride adsorption (MH), and cryogenic liquid hydrogen (LH_2).

For a general-purpose vehicle, a typical driving range between successive refuelings requires at least about 5 kg of H_2 to be stored onboard.[3] This amount of hydrogen is equivalent to 19 L or 5 gal of gasoline in terms of the total energy available in the fuel, and it can provide about 320 km (or 200 mi) range for a conventional car (at 17 km/ℓ or 40 mpg). However for a hybrid or fuel cell vehicle 5 kg of hydrogen can provide 640 km (or 400 mi) range (at 34 km/ℓ or 80 mpg — an objective set for future vehicles).

However, to store 5-kg hydrogen at 24.8 MPa (3600 psi) and ambient temperature, the external volume for the pressure vessel would be about 320 ℓ (or 85 gal) as
CH_2, significantly larger than the tank needed for gasoline; such bulky fuel storage
is certainly undesirable for vehicular applications. On the other hand, if the same
amount of hydrogen is stored in metal hydrides, then the resulting metal hydride
storage would weigh about 300 kg for just 5-kg hydrogen; such a heavy fuel storage
device is certainly not desirable for mobile applications. Light and compact hydrogen
storage is possible if hydrogen gas can be cooled to liquid state at cryogenic temperatures (about 22 K or lower) and low pressures (about 0.5 MPa). However, hydrogen
liquefaction consumes a substantial amount of electricity, as much as about 30% of
the LHV of H_2. If this amount of energy is not recovered through complex system
integration and optimization, overall lifecycle efficiency would be lowered substantially. Furthermore, during fueling the low pressure LH_2 tank, a portion of the LH_2
is vaporized in cooling down the tank and connecting hoses, this evaporation loss
amounts to about 8% of the total H_2 pumped into the tank. Another type of evaporation loss occurs during long periods of inactivity due to heat transfer from the
ambient. Hydrogen evaporation loss is not only a safety hazard, it also reduces the
fuel economy and the vehicle range as well. Although liquid hydrogen boiling-off loss
could be eliminated if the cryogenic tank is made of insulated high-pressure vessel;
such a storage tank would be very expensive as well.

Therefore for portable and mobile applications, it is highly desirable to have
a suitable liquid fuel at atmospheric pressure and temperature for easy distribution
and onboard storage. Methanol is such a possibility, and in fact, the most promising fuel that is elaborated on Section 10.2. Section 10.3 is devoted to the difference
between methanol used as a direct or indirect fuel in fuel cells. Section 10.4 presents
the various types of direct methanol fuel cells, and the polymer electrolyte membrane (PEM) DMFCs are considered in Section 10.5. A description and comparison of vapor-feed versus liquid-feed PEM DMFCs is given in Section 10.6. At this
point, we point out that DMFCs commonly referred to today are often the liquid-
feed PEM DMFCs with almost identical materials and configurations for the cells
and stacks as for those PEMFCs described in Chapter 7. As a result, the design,
construction, and materials used are basically the same as for PEMFCs and will not
be described further in this chapter. However, the unique features of PEM DMFC
operation and performance are described in Section 10.7, including anode and cathode reactions as well as methanol crossover through the electrolyte membrane. Similar
to previous chapters, this chapter ends with a brief description on future R&D and a
summary.

10.2   METHANOL AS A FUEL

With the existing social, economic, and technical structure, it is often desirable to
use hydrocarbon fuels for minimal social and economic disruption and smooth pro-
gression. Furthermore, hydrocarbon fuels are easily transported and converted from
the liquid state and methanol is the favored fuel. Methanol is a liquid with a boiling

temperature of 64.7 °C at atmospheric pressure and it provides a potential infrastructure capability for distribution with the existing gasoline supply infrastructure (with little modifications). In fact, most gasoline sold at gas stations contains some methanol, such as M85 which has 15% methanol. Further, methanol has a high volumetric energy density, about $5\,kW \cdot h/\ell$ as compared to about $2.6\,kW \cdot h/\ell$ for LH_2. As a result, methanol as fuel provides easy handling, easy refueling, and easy onboard storage for mobile applications.

Currently, over a billion gallons of methanol is produced annually in the United States alone. It is most economically produced from natural gas (about 78%). However, it is also produced from other sources, such as residual oil (about 7%), coal (about 1%), and biomass (about 14%). Production of methanol from biomass is renewable and can be made carbon-dioxide neutral from the entire lifecycle point of view. Future production methods include using electricity from nuclear energy and carbon dioxide captured from other fossil fuel power systems, thus it might be considered as one of the methods for carbon dioxide sequestration.

Methanol is mainly used at present as antifreeze agents, chemical solvents, and chemical feedstock for other industrial processes. Over the last several decades, the average price of methanol has been cheaper than gasoline. However, current methanol production is insufficient to meet the consumption requirement if it is used as a fuel and the potential cost advantage might disappear without a substantial increase in the production. Although many might consider methanol as a safer fuel as compared to hydrogen, methanol is mildly poisonous and the associated safety issue would be comparable with hydrogen.[2]

10.3 INDIRECT VERSUS DIRECT METHANOL FUEL CELLS

Methanol as fuel can be supplied directly to a fuel cell for direct electrooxidation of methanol for electric current generation; this is often called a **direct methanol fuel cell (DMFC)**. On the other hand, methanol can be first converted into hydrogen-rich gas through an onboard reformer and the hydrogen-rich reformate gas mixture is then fed into a fuel cell for the electrooxidation of hydrogen in the anode, similar to the fuel cells described in Chapters 5–9. This latter approach is often referred to as an **indirect methanol fuel cell**, since it is the hydrogen that is the direct fuel for the fuel cell, while methanol is merely used as a source for hydrogen. Figure 10.1 illustrates a comparison of system complexities between the indirect and direct methanol fuel cell.

Methanol reforming process has a high-energy requirement, and occurs at a temperature of 250 °C to 300 °C, much higher than the usual PEM fuel cell operating temperature. Carbon monoxide removal often requires at least two stages of shift reaction as described in Chapter 6, followed by selective partial oxidation process to reduce further the concentration of carbon monoxide to below about 100 ppm. Even this low concentration of carbon monoxide is still too much for PEM fuel cells, as shown in Chapter 7. In addition, fuel processing including the reforming and cleaning stages adds the size and weight to the fuel cell power system, increases the system complexity for integration, monitoring and control, slows down the system

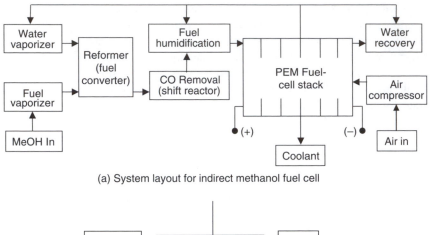

(a) System layout for indirect methanol fuel cell

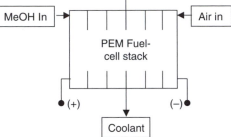

(b) System layout for direct methanol fuel cell

Figure 10.1 A schematic of indirect and direct methanol fuel cells utilizing polymer electrolyte membrane fuel cells.

response time, and adds additional maintenance requirement. Roughly speaking, fuel processing often accounts for approximately one-third of the system size (weight) and cost, with another third due to the fuel cell stack, and the remaining third attributed to the balance of plants. Clearly, onboard fuel processing is undesirable.

On the other hand, from a performance point of view, a fuel cell operating on methanol directly has a reversible cell potential of 1.214 V at the standard temperature and pressure as compared to 1.229 V for a fuel cell operating on hydrogen which might be derived from methanol in the first place. Readers are encouraged to verify these voltage values by following the analysis shown in Chapter 2. Although the best cell performance based on thermodynamic analysis is comparable for both indirect and direct methanol fuel cells, the actual direct methanol fuel cell has only about a third of the fuel cell performance when hydrogen is used as the direct fuel. For example, for the same level of energy conversion efficiency required for practical application, the power density for state of the art PEMFCs reaches as high as 0.6–$0.7\,\mathrm{W/cm^2}$, while a same cell operating on methanol directly can have power density in the range of 0.18–$0.3\,\mathrm{W/cm^2}$. This lower performance of a DMFC arises from a number of causes to be described in Section 10.7 and is clearly a disadvantage from the stack performance point of view. However from the system level, a direct methanol fuel cell does not need the heavy and bulky fuel processing unit that is about a third of the indirect methanol fuel cell as shown in Figure 10.1. As a balance, the performance and cost of a direct methanol fuel cell power system is comparable to an indirect methanol fuel cell power system. Therefore, significant interest in direct methanol fuel cell has accrued over the years.

10.4   VARIOUS TYPES OF DIRECT METHANOL FUEL CELLS

The quest for direct hydrocarbon fuel cells accompanied the initial invention of fuel cells and had since been a great dream of many scientists and engineers. Direct coal fuel cells, for example, were once the aim for large-scale electricity generation even before the development of combustion engine-driven electric generators. However, various efforts led to the conclusion around 1920s–1930s that direct methanol fuel cell would be the best bet if any direct hydrocarbon fuel cell would have the kind of performance expected for practical applications because of the sluggish electrooxidation kinetics of hydrocarbon fuels as compared to hydrogen. That is, electrooxidation of methanol at the fuel cell anode has the fastest kinetics among all hydrocarbon fuels investigated, yielding the highest power density of all direct hydrocarbon fuel cells. Initial development of direct methanol fuel cell technology started in the 1960s. A slow but steady progress had been made over the years with various sizes of DMFC demonstration units developed, and culminated in a large DMFC power unit for a 30-foot long bus in late 1990s; various kinds of electrolyte have also been selected for DMFC including alkaline and acid electrolytes. We briefly describe these various types of direct methanol fuel cells in the following sections, including direct methanol operation in alkaline fuel cells, in molten carbonate fuel cells, in aqueous carbonate fuel cells, in sulphuric acid electrolyte fuel cells, in phosphoric acid fuel cells, and in PEM fuel cells.

10.4.1   Direct Methanol in Alkaline Fuel Cells

In principle, methanol as a fuel can be supplied to any fuel cell with any electrolyte of the choice. However, the electrolyte of choice will determine the specific anode and cathode reactions. For direct methanol operation in alkaline fuel cells (alkaline DMFC), the overall anode reaction is

$$\text{Anode reaction}: CH_3OH + 6OH^- \longrightarrow CO_2 + 5H_2O + 6e^- \qquad (10.1)$$

Since the electrode oxidation reaction of methanol occurs at the electrode–electrolyte interface where carbon dioxide is formed according to Equation (10.1), carbon dioxide reacts with the hydroxyl ions in the alkaline electrolyte to form carbonate, leading to the well-known carbon dioxide poisoning of alkaline fuel cells as described in Chapter 5. Because of this reason, direct methanol (or any other hydrocarbon fuels) operation in alkaline fuel cells is not practical and was abandoned even though it was initially considered in 1960s and 1970s.

Due to this need for the tolerance of carbon dioxide produced during the electrooxidation of methanol, acid electrolyte has to be used, although the acid electrolyte causes problems of corrosion and, more fundamentally, is responsible for the slow electrode kinetics of the reduction of oxygen at the air cathode.

10.4.2   Direct Methanol in Molten Carbonate Fuel Cells

Because of the high operating temperature of molten carbonate fuel cells (around 650 °C), methanol can be easily reformed internally into hydrogen and carbon monoxide mixture, which is then the fuel feeding the anode electrode reaction in MCFCs. The methanol steam-reforming reaction in the MCFC anode compartment is

$$CH_3OH + H_2O \longrightarrow 3H_2 + CO_2 \quad (\Delta H = 49.7\,kJ/mol) \qquad (10.2)$$

In reality, carbon monoxide formation is significant at such high temperature, however, carbon monoxide is not a problem for MCFCs and it can be further converted into hydrogen through the water–gas shift reaction.

The water needed for methanol-reforming reaction is easily available in the MCFC anode since the anode reaction

$$H_2 + CO_3^{2-} \longrightarrow H_2O + CO_2 + 2e^- \qquad (10.3)$$

produces the amount of water needed. Furthermore, methanol steam-reforming is only mildly endothermic; only a small amount of heat is required which can be easily used for waste heat dissipation of the fuel cell without creating significant local temperature decrease in the anode.

However, as pointed out in Chapter 8, MCFCs are not practical for transportation applications because of their high operation temperature that results in a long startup time. But for stationary applications it is more preferable to use natural gas in MCFCs described in Chapter 8, since methanol production and distribution would add cost and lower the lifecycle efficiency when compared to natural gas.

10.4.3   Direct Methanol in Aqueous Carbonate Fuel Cells

Direct methanol can be used effectively in aqueous carbonate fuel cells, and reasonably good performance has been achieved. For example, 0.55 to 0.6 V of cell potential at $150\,mA/cm^2$ current density or higher has been obtained when the cell is operated at 120 psig and 165 °C. However, a significant amount of Noble-metal catalyst loading is required to achieve a good performance, aqueous electrolyte management and cell component corrosion need to be addressed. A credible demonstration of the cell lifetime and reliability is needed for practical applications.

10.4.4   Direct Methanol in Sulphuric Acid Electrolyte Fuel Cells

As pointed out in Chapter 6, sulphuric acid electrolytes would provide good fuel cell performance and are limited in practical use due to its unstable nature in fuel cell operation and high vapor pressure. This limits the operation temperature of sulphuric acid electrolyte fuel cells to about 60 °C or lower. Consequently, cell performances are limited to low current-density operations; hence the output power density is rather low. This makes them unsuitable for practical applications, especially mobile applications.

10.4.5   Direct Methanol in Phosphoric Acid Fuel Cells

At the typical operating temperature of phosphoric acid fuel cell, anode kinetics for the electrooxidation of methanol is slow, resulting in low cell performance. Further, significant methanol crossover with resultant cathode poisoning is another major problem. As a result, the low power density makes it infeasible for mobile applications.

10.4.6   Direct Methanol in PEM Fuel Cells

Direct methanol operation in PEMFCs is currently receiving the most attention for portable and mobile applications as a DMFC. Although early cells employed a combination of PEM and sulfuric acid (H_2SO_4) as electrolyte for better performance, it has been established that a PEM alone can function effectively as an electrolyte for DMFC. Today's DMFCs are essentially PEMFCs with methanol as fuel for direct electrooxidation for electric power generation (PEM DMFCs).

10.5   PEM DMFCs

Since PEM is essentially an acid electrolyte, the overall half-cell reaction at the anode is

$$CH_3OH + H_2O \longrightarrow 6H^+ + 6e^- + CO_2(g) \tag{10.4}$$

The proton is the current-carrying ions transported through the electrolyte to reach the cathode, and the electrons are transported via the external circuit to arrive at the cathode. The overall half-cell reaction at the cathode is then

$$\frac{3}{2}O_2 + 6H^+ + 6e^- \longrightarrow 3H_2O \tag{10.5}$$

Therefore, the overall net-cell reaction becomes

$$CH_3OH + \frac{3}{2}O_2 \longrightarrow CO_2(g) + 2H_2O + We + \text{Waste Heat} \tag{10.6}$$

This cell operation is schematically shown in Figure 10.2.

From the above half-cell reactions it is important to notice that water is needed for the electrooxidation of methanol at the anode and more water is produced at the cathode at the same time. It seems that the water needed at the anode might be provided by the water produced at the cathode. In reality, water is also transported from the anode to the cathode via the PEM due to the electroosmotic drag effect, described in Chapter 7. As a result, methanol–water mixture is supplied to the anode. This approach also provides ample hydration for the PEM.

However, problems arise from this approach. Methanol, just like other alcohol, mixes well with water. Since water is available in the PEM structure for hydration,

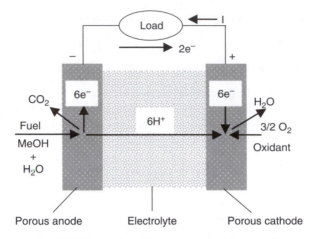

Figure 10.2 A schematic of DMFC operation with a polymer electrolyte membrane as the electrolyte.

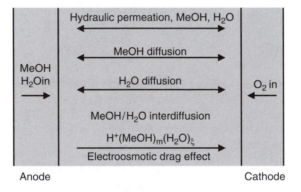

Figure 10.3 A schematic of methanol crossover in DMFCs with a polymer electrolyte membrane as the electrolyte.

methanol spreads into the water in the PEM as well and thus also reaches the water in the cathode. This phenomenon is called methanol crossover from the anode to the cathode and has severe negative consequences on the cell performance. First, methanol crossover to the cathode is not utilized for current generation, hence a drain on the cell energy efficiency. Second, methanol as a fuel at the cathode is electrooxidated, as well where oxygen is reduced. This creates a mixed potential at the cathode, effectively lowers the cathode potential, hence the cell potential difference as well, because the fuel electrooxidation occurs at a much lower electrode potential than the oxygen reduction reaction. Further, the heat of reaction due to the methanol oxidation and oxygen reduction reaction occurring simultaneously at the cathode will also cause the cathode catalyst aging. Methanol crossover can be reduced in DMFCs, but it is difficult to eliminate completely. This issue is taken up again in Section 10.7. Figure 10.3 illustrates the mechanism of methanol crossover through the membrane electrolyte.

10.6   VAPOR-FEED VERSUS LIQUID-FEED PEM DMFCs

For DMFCs with polymer electrolyte membrane as electrolyte, two variations of the cell operation are being investigated. One is methanol supplied to the anode at elevated temperatures (typically 110 °C–130 °C) in gaseous form, called **vapor-feed DMFCs**, another is methanol supplied to the anode at low temperatures (typically around 80 °C) in liquid form, or the so-called **liquid-feed DMFCs**. Both types of DMFCs have their own unique advantages and problems.

For vapor-feed DMFCs, methanol electrooxidation proceeds faster with the higher cell-operation temperature and faster kinetics are beneficial for better cell performance. On the other hand, maintaining adequate membrane hydration is difficult, especially at high current densities, just like in the PEMFCs described in Chapter 7; this will always affect the membrane lifetime which is limited already at higher temperature operation. As a result, both fuel and oxidant streams need humidification before being fed to the cell, another complexity in the design and operation. Another issue is the separation of carbon dioxide in the anode exhaust stream since the residual methanol vapor and carbon dioxide are in gas mixture, and the residual methanol vapor needs to be recirculated back to the anode inlet. Furthermore, methanol is in liquid state at the atmospheric condition, a methanol vaporizer is necessary with significant heat requirement while the stack will require cooling which complicates the thermal management of the entire system.

In comparison, for a liquid-feed DMFC there is no need for a fuel vaporizer and its associated heat source and controls, and no need for complex water and thermal management to provide the membrane hydration. Also for liquid feed, often the liquid methanol–water mixture is used as a coolant for efficient stack thermal control, this also acts as a preheat for the mixture to reach the stack temperature before being fed into the stack for electricity generation. This dual purpose use of the fuel stream eliminates the separate coolant use and simplifies the system design. Further, carbon dioxide removal at the anode exhaust becomes very easy since the carbon dioxide in the gas form can simply bubble off from the liquid methanol-water mixture. All these factors combined together tend to yield significantly lower system size and weight with much easy water and thermal management and longer lifetime. Although lower temperature operation for the liquid-feed DMFC gives lower cell performance, it is not necessarily for the system performance to be lower. Consequently, liquid-feed DMFC is much favored. From now on, by direct methanol fuel cells, we mean exclusively the liquid-feed direct methanol fuel cells using a polymer membrane as the electrolyte.

10.7   LIQUID-FEED PEM DMFCs

From the preceding description, it is clear that liquid-feed PEM DMFCs of current interest are similar, in fact identical, to PEM fuel cells in terms of materials and constructions for the cell and stack components. The only difference is in the type of fuel used in the cell for electricity generation: DMFCs use methanol directly, while

PEMFCs use hydrogen (which could be derived from methanol indirectly through reforming process), for anode reaction. As shown in Figure 10.1, a liquid-feed PEM DMFC has minimal system components, simplified thermal and water management, easy carbon dioxide removal for residual methanol recovery from the anode exhaust, no corrosive electrolyte, and low-temperature operation (70 °C–90 °C) that allows quick startup. Further, methanol infrastructure is considered already in place in comparison with hydrogen. Therefore PEM DMFCs are being developed for small portable devices such as cellular phones and laptop computers, transportation applications, and stationary power supplies.

On the other hand, despite these attractive features, a liquid-feed PEM DMFC has only one-third of the performance of a PEMFC operating on hydrogen, as pointed out earlier. This lower performance is mainly caused by the sluggish kinetics at the anode and cathode as well as significant methanol crossover, and we turn our attention on these issues in the following sections.

10.7.1   Anode Reaction

The overall anodic reaction has been given in Equation (10.4), which indicates that six moles of electrons and protons are produced per-mole of methanol consumed and the corresponding reversible anode potential with respect to the reference hydrogen electrode is about 0.046 V, a small potential indeed. Although this might imply a good production of current per-mole fuel consumed, the actual electrooxidation mechanism is far from the simplistic picture that the overall half-cell reaction might suggest. In fact, the methanol oxidation mechanism is still under active investigation. But the following mechanism is commonly considered for the electrooxidation of methanol in DMFCs

$$Pt + CH_3OH \longrightarrow Pt - (CH_3OH)_{ads} \tag{10.7}$$

$$Pt - (CH_3OH)_{ads} \longrightarrow Pt - (CH_2OH)_{ads} + H^+ + e^- \tag{10.8}$$

$$Pt - (CH_2OH)_{ads} \longrightarrow Pt - (CHOH)_{ads} + H^+ + e^- \tag{10.9}$$

$$Pt - (CHOH)_{ads} \longrightarrow Pt - (COH)_{ads} + H^+ + e^- \tag{10.10}$$

$$Pt - (COH)_{ads} \longrightarrow Pt - (CO)_{ads} + H^+ + e^- \tag{10.11}$$

$$M + H_2O \longrightarrow M - (H_2O)_{ads} \tag{10.12}$$

$$M - (H_2O)_{ads} \longrightarrow M - (OH)_{ads} + H^+ + e^- \tag{10.13}$$

$$Pt - CO_{ads} + M - (H_2O)_{ads} \longrightarrow Pt + M + CO_2 + 2H^+ + 2e^- \tag{10.14}$$

$$Pt - CO_{ads} + M - (OH)_{ads} \longrightarrow Pt + M + CO_2 + H^+ + e^- \tag{10.15}$$

where in the preceding equations, the enclosures within the parenthesis imply a lack of certainty about the exact connectivity of atoms involved. This mechanism might be summarized into four key steps for the oxidation of mechanism:

1.   The first step is the physisorption of methanol at the anode catalyst, which is considered as platinum for the moment, because the membrane electrolyte used

is essentially acid and precious metals like platinum are essential for the oxidation of methanol at the low operating temperatures involved. This step is represented by Equation (10.7).

2. The second step, as shown in Equations (10.8)–(10.11), is the oxidative adsorption of methanol, where the charge transfer reaction occurs sequentially. At the end of this sequential oxidation process, carbon monoxide is formed right at the platinum catalyst surface, as shown in Equation (10.11). Therefore, platinum would be poisoned as shown in Chapter 7. In fact, impurities and more seriously reaction intermediaries like CO, −COH and −CHO can all poison platinum. As a remedy, platinum alloys such as Pt-Ru are used as the anode electrocatalyst in DMFCs for CO tolerance.

3. The third step is the activation of water, typically on the other alloying metals like ruthenium in Pt-Ru catalyst, represented by Equations (10.12) and (10.13).

4. The fourth and last step is the surface oxidation of the adsorbed carbon monoxide on the platinum catalyst surface.

The second step is sensitive to ligand or electronic effects of alloying. The second and third steps are classically known as the bifunctional mechanism, that is methanol oxidative adsorption and water activation for methanol oxidation are described in the second and third steps.

At platinum surface, methanol oxidation occurs at very high anode potential, approximately commences at about 0.5 V (with respect to reference hydrogen electrode) and becomes significant at about 0.6 V. This is in striking contrast with the reversible anode potential of 0.046 V, and about 0.1 V or less for typical operating hydrogen PEM fuel cells. Water activation is always the rate-limiting step on platinum surface. The significant anode overpotential in DMFCs indicates that platinum is not sufficiently active catalyst for the direct oxidation of methanol. The search for a better and more active catalyst for methanol oxidation with CO tolerance is still going on in the hope of reducing the anode overpotential significantly.

The bulk of past research suggests that platinum alloys are the only promising catalyst for methanol oxidation and among the binary, tertiary and quaternary platinum alloys investigated, platinum-ruthenium (Pt-Ru) alloy is the best catalyst (meaning the most active), better than platinum. At Pt-Ru surface, methanol oxidation occurs at about 0.25 V and becomes significant about 0.3 V, much lower than on platinum surface. The rate-limiting process on Pt-Ru surface is the C-H activation since the alloyed catalyst successfully addresses the water activation problem. However, compared to the anode potential in hydrogen PEM fuel cells, Pt-Ru catalyst is still not sufficiently active for methanol oxidation. Consequently, the catalyst loading for DMFCs (about $2 \, mg/cm^2$) is much higher than the corresponding loading in hydrogen PEM fuel cells (about $0.2 \, mg/cm^2$). This almost 10 times higher catalyst loading is also beneficial for more complete fuel consumption in the anode, thus lowering the methanol fuel available for crossover to the cathode side.

In summary, anode overpotential is significant in DMFCs, representing a significant energy loss mechanism despite the high catalyst loadings employed.

10.7.2   Methanol Crossover

Methanol crossover arises due to a number of reasons as shown in Figure 10.3, and is perhaps the most severe problem in limiting the performance and lifetime of DMFCs. First, methanol crossover lowers the fuel utilization (and hence the fuel efficiency) and second, the crossover methanol at the cathode reacts with the oxygen according to

$$CH_3OH + \frac{3}{2}O_2 \longrightarrow CO_2 + 2H_2O \tag{10.16}$$

This will lower oxygen concentration available at the cathode for the electroreduction reaction (hence, increases the oxygen demand) and the heat of reaction also causes the local heating of the membrane and the aging of the cathode catalyst. Due to the high cathode potential, a portion of the crossover methanol is electrooxidized at the catalyst surface, thus creating a mixed electrode potential, effectively lowering the cathode potential.

The effect of the methanol loss due to the crossover is often measured in two different approaches. The first and the easiest measure of the crossover is to express the amount of the methanol crossover in terms of the ratio of the amount of the methanol crossover to the total methanol supplied to the cell, U_{cr}. This ratio represents the fraction of the methanol supplied to the cell that is lost due to the crossover phenomena. Therefore the effective utilization of the methanol is reduced to

$$U_{t,\text{eff}} = 1 - U_{cr} \tag{10.17}$$

This effective utilization is then used to calculate the overall fuel efficiency from the cell or stack's overall free energy conversion efficiency given in Chapter 2. This approach is most often used by practicing engineers. For typical DMFCs, $U_{cr} = 20\%$, hence the effective methanol utilization amounts to 80%.

An alternative approach is to convert the amount of methanol crossover in terms of an equivalent loss in the electrical current, or simply called crossover current, I_{cr}. This current I_{cr} is the current equivalent to what would be produced with the consumption of the crossover methanol if it were consumed properly at the anode. Considering the definition in Chapter 2, we can write the current efficiency for DMFCs as

$$\eta_I = \frac{I}{I + I_{cr}} \tag{10.18}$$

where I is the actual current produced in the cell. This approach is most often used by electrochemists. A typical value for DMFCs is $\eta_I = 80\%$.

For a given DMFC, the rate of methanol crossover can be expressed as a function of the cell current drawn, I, and the methanol concentration, C_0, supplied to the cell. Consider a DMFC operating with the same pressure on both the anode and cathode sides so that the dominant mechanism of methanol crossover is the diffusion and electroosmotic drag of methanol under steady-state condition, as shown in Figure 10.4. The concentration of the methanol decreases linearly in the anode electrode which has a thickness of δ_e and in the membrane of thickness δ_m. However, the difference

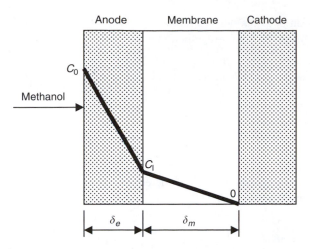

Figure 10.4 Schematic of methanol concentration distribution in the anode and the membrane electrolyte in an operating DMFC.

in the slope represents the fact that more methanol diffusion through the electrode than through the membrane because of methanol reaction at the electrode–membrane interface. In another words, we may state that the rate of methanol flow through the electrode is equal to the rate of methanol reaction at the electrode–membrane interface for current production and the rate of the methanol crossover, or

$$nF\dot{N} = I + I_{cr} \tag{10.19}$$

where n is the number of mole electrons produced per mole methanol reacted at the anode, which is equal to 6 as shown in Equation (10.4), and F is the Faraday constant.

Since diffusion due to concentration gradient is the dominant mechanism in the electrode, we have

$$\dot{N} = AD_e \frac{C_0 - C_I}{\delta_e} \tag{10.20}$$

where A is the electrode active area, D_e is the effective diffusion coefficient in the anode electrode structure and the concentration of methanol at the anode electrode–membrane interface is denoted as C_I. On the other hand, methanol concentration at the cathode electrode–membrane interface can be taken as zero due to the fast oxidation reaction of methanol at the cathode side. In the membrane, methanol crossover arises from the diffusion and electroosmotic drag effect, we can write

$$I_{cr} = nFAD_m \frac{C_I}{\delta_m} + I\xi \frac{C_I}{C_T - C_I} \tag{10.21}$$

where D_m is the effective diffusion coefficient in the membrane, ξ is the electroosmotic coefficient for water, and C_T is the total concentration of methanol and water.

Substitution of Equations (10.20) and (10.21) into Equation (10.19), and combining the terms involving C_I together, yield

$$nFAD_e \frac{C_0}{\delta_e} = I + nFAD_m \frac{C_I}{\delta_m}\left(1 + \frac{D_e}{D_m}\frac{\delta_m}{\delta_e}\right) + I\xi \frac{C_I}{C_T - C_I} \tag{10.22}$$

Now denoting

$$k = \frac{D_e}{D_m}\frac{\delta_m}{\delta_e}$$

and dividing $(1 + k)$ for both sides of the equation, we have the crossover current, after considering Equation (10.21)

$$I_{\text{cr}} = nFAD_e \frac{C_0}{\delta_e}\left(\frac{1}{1+k}\right) - I\left(\frac{1}{1+k}\right)\left(1 - k\xi \frac{C_I}{C_T - C_I}\right) \tag{10.23}$$

Equation (10.23) reveals that for a given operating condition, if the value of k is small, implying strong diffusion mechanism for methanol crossover, the rate of methanol crossover decreases with the cell current (or current density) drawn. This is because as current drawn is increased the concentration of methanol at the anode electrode–membrane interface C_I decreases, resulting in a smaller concentration difference across the membrane that drives the crossover. On the other hand, if the value of k is large, signifying the dominance of electroosmotic drag on the methanol crossover, then the crossover rate increases with the cell current drawn.

For a given DMFC design, the value of k is fixed. Now if the methanol concentration supplied to the cell C_0 is low, correspondingly C_I will be small as well, such that the coefficient of I in Equation (10.23) is negative, indicating that the rate of methanol crossover will decrease as the cell current is increased. On the other hand, if the methanol concentration C_0 is high, C_I will be large as well. Then the coefficient of I in Equation (10.23) becomes positive such that methanol crossover now increases with the cell current. This dependence of the methanol crossover rate on the cell current density has been measured experimentally, as shown in Figure 10.5. It has been established that for Nafion 117 membrane as the electrolyte, the methanol feed concentration C_0 should be less than 1 M, corresponding to 3% by weight methanol in water solution, in order to minimize the effect of methanol crossover. Even at this low concentration, methanol crossover is still appreciable, about 20%.

Therefore, one of the consequences of the methanol crossover is to limit methanol concentration in the feed stream. This exacerbates the anode overpotential since the activation overpotential is substantial for DMFC anode as described in the previous section, and a low methanol feed concentration increases the anode overpotential due to limited rate of mass transfer.

10.7.3   Cathode Reaction

The cathode reaction in the PEM DMFC has been already shown in Equation (10.5), the electroreduction of oxygen is the same as for hydrogen PEM fuel cells. Therefore,

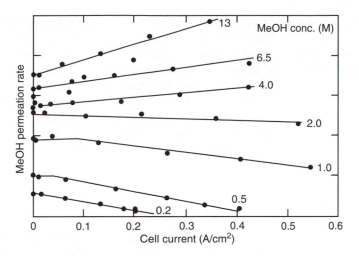

Figure 10.5 Dependence of methanol crossover rate on the cell current density and the methanol concentration in the anode feed stream for a PEM DMFC using Nafion membrane.[4]

the same platinum catalyst is used as in PEM fuel cells or other acid electrolyte fuel cells; oxygen reduction kinetics are very slow. As a result, the cathodic overpotential is substantial, in fact it is significantly worse than in PEM fuel cells because of the crossover methanol as explained earlier. A platinum catalyst in its present form of supported or unsupported Pt is not sufficiently active and efforts are under way to develop or find more active catalysts for the cathode, or catalysts that do not oxidize the crossover methanol to avoid the mixed potential.

From this explanation, the cathode arrangement for DMFCs is basically identical to that of PEM fuel cells. The only exception might be that the cathode stream does not require humidification since the anode feed is a methanol-liquid water mixture, that provides sufficient water for the hydration of membrane.

10.8   FUTURE R&D

The interest and research activities in direct methanol fuel cells are intensifying with significant progresses made.[5-13] The main focuses for the direct methanol fuel cells are, from a fundamental point of view:

1. Improvements to the membrane electrolyte in order to reduce methanol crossover without worsening other performance parameters (e.g., membrane electrical conductivity);
2. Fast anode kinetics in order to reduce anode overpotential. This is often directed at developing/finding better (more active) anode catalysts for fast methanol reaction and CO tolerance;
3. Fast cathode kinetics in order to reduce cathode overpotential. This is also directed at developing/finding better (more active) cathode catalysts for fast oxygen reduction reaction and methanol tolerance,

From the practical system point of view, carbon dioxide removal from the anode reaction sites and from the anode exhaust stream is important. Carbon dioxide gas bubbling off from the reaction sites can block the methanol feed to the reaction sites, especially at high current-density operations and can also take methanol vapor with it from the anode exhaust stream, that would represent a waste of fuel and also might pose a safety concern. Another important engineering issue is again the bipolar plate, including the materials selection (e.g., switch away from graphite), manufacturing process and flow channel designs, and so on. This issue would be in many ways similar to those of PEM fuel cells, although DMFCs do have their own unique features and requirements.

10.9   SUMMARY

This chapter presented the advantages of a methanol fuel cell for portable and mobile applications because of the high volumetric energy density of methanol as a fuel. Indirect and direct methanol fuel cells, their pros and cons and their comparison were considered. A description of the advantages of a direct methanol fuel cell, the various types of direct methanol fuel cells that have been tried or tested in the past, and its present connotation (PEM DMFC) was given. The differences and similarities of vapor-feed and liquid-feed PEM DMFCs were highlighted. Attention was focused on liquid-feed PEM DMFC and its major performance limitations: slow anode and cathode kinetics and methanol crossover. These issues are the major directions of the current and future R&D. It might be good to recognize that a direct methanol fuel cell is still early in its development and significantly more technical progress and engineering effort will be needed to bring it to the commercial stage.

BIBLIOGRAPHY

1. Chan, S. H., O. L. Ding and D. L. Hoang. 2004. A thermodynamic view of partial oxidation, steam reforming and autothermal reforming on methane. *Int. J. Green Energy*, 1: 265–278.
2. Larminie, J. and A. Dicks. 2003. *Fuel Cell Systems Explained*, 2nd ed., New York: Wiley.
3. Aceves, S. M. and G. D. Berry. 1998. Thermodynamics of insulated pressure vessels for vehicular hydrogen storage. *J. Energy Res. Tech.*, 120: 137–142.
4. Ren, X., T. A. Zawodzinski, F. Uribe, H. Dai and S. Gottesfeld. 1995. *In Proton Conducting Membrane Fuel Cells* I, eds. S. Gottesfeld, G. Halpert and A. Landgrebe, The Electrochemical Society, Pennington, NJ: 284.
5. Scott, K., W. M. Taama and P. Argyropolous. 1998. Engineering aspects of the direct methanol fuel cell. *J. Pow. Sour.*, 79: 43–59.
6. Scott, K., W. M. Taama and P. Argyropoulos. 1998. Material aspects of the liquid feed direct methanol fuel cell. *J. App. Electrochem.*, 28: 1389–1397.
7. Sun, Q., et al. 1998. Iron (III) tetramethoxyphenylporphyrin (FeTMPP) as methanol tolerant electrocatalyst for oxygen reduction in direct methanol fuel cells. *J. App. Electrochem.*, 28: 1087–1093.
8. Heinzel, A. and B. M. Barragan. 1999. A review of the state of the art of methanol crossover. *J. Pow. Sour.*, 84: 70–74.
9. McNicol, B. D., D. A. J. Rand and K. R. Willaims. 1999. Direct Methanol-Air Fuel Cells for Road Applications. *J. Pow. Sour.*, 83: 15–31.

10. Argyropoulos, P., K. Scott and W. M. Taama. 2000. The effect of operating conditions on the dynamic response of the direct methanol fuel cell. *Electrochimica Acta*, 45: 1983–1998.

11. Carretta, N., Tricoli, V. and F. Picchioni. 2000. Ionermeric membranes based on partially sulfonated poly(styrene): Synthesis, proton conduction and methanol permeation. *J. Membrane Sci.*, 166: 189–197.

12. Andrian, S. V. and J. Meusinger. 2000. Process analysis of liquid-feed direct methanol fuel cell. *J. Pow. Sour.*, 91: 193–201.

13. Ren, X., P. Zelanay, S. Thomas, J. Davey and S. Gottesfeld. 2000. Recent advances in direct methanol fuel cells at Los Alamos National Laboratory. *J. Pow. Sour.*, 86: 111–116.

PROBLEMS

10.1 Describe briefly the advantages and disadvantages as well as the areas of applications for methanol fuel cells as compared to the hydrogen PEM fuel cells.

10.2 Describe the advantages and disadvantages of using methanol as a fuel for fuel cells as compared to hydrogen.

10.3 Describe the direct and indirect methanol fuel cells and compare their advantage and disadvantage.

10.4 Describe the various types of direct methanol fuel cells, as well as their advantage and disadvantage.

10.5 Describe the operational principle such as half-cell and whole-cell reaction for direct methanol fuel cells using a polymer membrane electrolyte.

10.6 Describe the advantage and disadvantage of PEM direct methanol fuel cells with methanol feed in the vapor or liquid form.

10.7 Identify the major performance limiting factors for liquid-feed PEM direct methanol fuel cells.

10.8 Describe the electrooxidation kinetics of methanol in PEM DMFCs, the type of anode catalyst required, and the cause of a large anode overpotential.

10.9 Identify the mechanism of methanol crossover from the anode to the cathode side in a PEM direct methanol fuel cell.

10.10 Describe the effect of the design and operating conditions on the rate of methanol crossover and why the methanol-feed concentration is limited to a low value in liquid-feed PEM direct methanol fuel cells.

10.11 Identify and describe the major problems associated with the methanol crossover.

10.12 Describe the two approaches to the representation of the effect of methanol crossover.

10.13 Describe the cathode reaction, the type of cathode catalyst used, and the cause of a large cathode overpotential.

10.14 Summarize the major directions for the current and future development of PEM direct methanol fuel cells.

APPENDIX A

THE PROPERTIES OF THE SATURATED WATER

Table A.1.1 The Properties of Saturated Water at Selected Temperatures

Temperature, T (°C)	Pressure, P (kPa)	Latent Enthalpy, h_{fg} (kJ/kg)
5	0.8721	2489.5
10	1.228	2477.7
15	1.705	2465.9
20	2.338	2454.2
25	3.169	2442.3
30	4.246	2430.4
35	5.628	2418.6
40	7.383	2406.8
45	9.593	2394.8
50	12.35	2382.8
55	15.76	2370.7
60	19.94	2358.5
65	25.03	2346.2
70	31.39	2333.8
75	38.58	2321.4
80	47.39	2308.8
85	57.83	2296.0
90	70.13	2283.2
95	84.55	2270.2
100	101.3	2257.0
110	143.3	2230.2
120	198.5	2202.6
130	270.1	2174.2

The following correlation for the determination of the water saturation pressure is taken from the International Association for the Properties of Water and Steam (IAPWS)'s *Industrial Formulation 1997 for the Thermodynamic Properties of Water and Steam*:

$$\frac{P_{\text{sat}}}{P^*} = \left[\frac{2C}{-B + (B^2 - 4AC)^{1/2}} \right]^4 \tag{A.1}$$

where

$$A = \theta^2 + n_1\theta + n_2$$
$$B = n_3\theta^2 + n_4\theta + n_5$$
$$C = n_6\theta^2 + n_7\theta + n_8$$
$$\theta = \frac{T_{\text{sat}}}{T^*} + \frac{n_9}{(T_{\text{sat}}/T^*) - n_{10}}$$

$P^* = 1\,\text{MPa}$ and $T^* = 1\,\text{K}$. The saturation temperature T_{sat} is in Kelvin, and the coefficients n_1 to n_{10} are given in Table A.1.2. The validity range is $273.15\,\text{K} \le T \le 647.096\,\text{K}$, and the accuracy is well within $\pm 0.025\%$. The accuracy improves even further for temperatures less than $100\,°C$. If the saturation pressure is known, the saturation temperature can be determined as

$$\frac{T_{\text{sat}}}{T^*} = \frac{n_{10} + D - [(n_{10} + D)^2 - 4(n_9 + n_{10}D)]^{1/2}}{2} \tag{A.2}$$

where

$$D = \frac{2G}{-F - (F^2 - 4EG)^{1/2}}$$
$$E = \beta^2 + n_3\beta + n_6$$
$$F = n_1\beta^2 + n_4\beta + n_7$$
$$G = n_2\beta^2 + n_5\beta + n_8$$

and $\beta = (P_{\text{sat}}/P^*)^{1/4}$

The coefficients n_i are given in Table A.1.2. The validity range for Equation (A.2) is $611.213\,\text{Pa} \le P \le 22.064\,\text{MPa}$, and the accuracy of the correlation is the same

Table A.1.2   The Coefficients in Equations (A.1) and (A.2)

i	n_i	i	n_i
1	$0.116\,705\,214\,527\,67 \times 10^4$	6	$0.149\,151\,086\,135\,30 \times 10^2$
2	$-0.724\,213\,167\,032\,06 \times 10^6$	7	$-0.482\,326\,573\,615\,91 \times 10^4$
3	$-0.170\,738\,469\,400\,92 \times 10^2$	8	$0.405\,113\,405\,420\,57 \times 10^6$
4	$0.120\,208\,247\,024\,70 \times 10^5$	9	$-0.238\,555\,575\,678\,49$
5	$-0.323\,255\,503\,223\,33 \times 10^7$	10	$0.650\,175\,348\,447\,98 \times 10^3$

as for Equation (A.1). Therefore the correlations for the saturation pressure, and temperature are applicable for the entire ranges of the liquid water saturation curve, from the normal freezing point (at 1 atm) to the critical point.

Other simpler correlations are also available. For example, the following equation correlates well the saturation vapor pressure of water as a function of temperature, taken from J. M. Prausnitz, R. C. Reid, and T. K. Sherwood, 1997. *The Properties of Gases and Liquids — 3*, New York: McGraw-Hill.

$$\ln P_{sat} = -34.625 + 0.258T - 4.8419 \times 10^{-4}T^2 + 3.3282 \times 10^{-7}T^3 \qquad (A.3)$$

where T is in Kelvin and P_{sat} is in Pa.

ENTHALPY OF FORMATION, GIBBS FUNCTION OF FORMATION, AND ABSOLUTE ENTROPY FOR SELECTED SUBSTANCES AT 25 °C AND 1 ATM

Substance	Formula	h_f J/mol	g_f J/mol	s J/(mol · K)
Carbon	C(s)	0	0	5.74
Hydrogen	H_2(g)	0	0	130.68
Nitrogen	N_2(g)	0	0	191.61
Oxygen	O_2(g)	0	0	205.14
Carbon Monoxide	CO(g)	−110,530	−137,150	197.65
Carbon Dioxide	CO_2(g)	−393,522	−394,360	213.80
Water	H_2O(g)	−241,826	−228,590	188.83
Water	H_2O(ℓ)	−285,826	−237,180	69.92
Hydrogen Peroxide	H_2O_2(g)	−136,310	−105,600	232.63
Ammonia	NH_3(g)	−46,190	−16,590	192.33
Methane	CH_4(g)	−74,850	−50,790	186.16
Acetylene	C_2H_2(g)	+226,730	+209,170	200.85
Ethylene	C_2H_4(g)	+52,280	+68,120	219.83
Ethane	C_2H_6(g)	−84,680	−32,890	229.49
Propylene	C_3H_6(g)	+20,410	+62,720	266.94
Propane	C_3H_8(g)	−130,850	−23,490	269.91
n-Butane	C_4H_{10}(g)	−126,150	−15,710	310.12
n-Octane	C_8H_{18}(g)	−208,450	+16,530	466.73
n-Octane	C_8H_{18}(ℓ)	−249,950	+6,610	360.79
n-Dodecane	$C_{12}H_{26}$(g)	−291,010	+50,150	622.83
Benzene	C_6H_6(g)	+82,930	+129,660	269.20
Methyl Alcohol	CH_3OH(g)	−200,670	−162,000	239.70
Methyl Alcohol	CH_3OH(ℓ)	−238,660	−166,360	126.80
Ethyl Alcohol	C_2H_5OH(g)	−235,310	−168,570	282.59
Ethyl Alcohol	C_2H_5OH(ℓ)	−277,690	−174,890	160.70
Oxygen	O(g)	+249,170	+231,770	161.06
Hydrogen	H(g)	+217,999	+203,290	114.72
Nitrogen	N(g)	+472,680	+455,510	153.30
Hydroxyl	OH(g)	+38,987	+34,280	183.70

Adapted from W. Z. Black and J. G. Hartley, 1991. *Thermodynamics — 2*, New York: HarperCollins.

IDEAL GAS SPECIFIC HEAT AT CONSTANT PRESSURE FOR SELECTED SUBSTANCES AT 300 K ALONG WITH MOLECULAR WEIGHT AND GAS CONSTANT

Gas	Formula	Molecular Weight W kg/kmol	Gas Constant kJ/(kg · K)	Specific Heat C_p kJ/(kg · K)
Air	—	28.97	0.2870	1.005
Argon	Ar	39.948	0.2081	0.5203
Butane	C_4H_{10}	58.124	0.1433	1.7164
Carbon Dioxide	CO_2	44.01	0.1889	0.846
Carbon Monoxide	CO	28.011	0.2968	1.040
Ethane	C_2H_6	30.070	0.2765	1.7662
Ethylene	C_2H_4	28.054	0.2964	1.5482
Helium	He	4.003	2.0769	5.1926
Hydrogen	H_2	2.016	4.1240	14.307
Methane	CH_4	16.043	0.5182	2.2537
Neon	Ne	20.183	0.4119	1.0299
Nitrogen	N_2	28.013	0.2968	1.039
Octane	C_8H_{18}	114.230	0.0729	1.7113
Oxygen	O_2	31.999	0.2598	0.918
Propane	C_3H_8	44.097	0.1885	1.6794
Steam	H_2O	18.015	0.4615	1.8723

Adapted from Y. A. Cengel, 1997. *Introduction to Thermodynamics and Heat Transfer*, New York: McGraw-Hill.

IDEAL GAS SPECIFIC HEAT AT CONSTANT
PRESSURE AT VARIOUS TEMPERATURES

T	Air	CO_2	CO	H_2	N_2	O_2	$H_2O(g)^a$
(K)	kJ/(kg · K)	kJ/(kg · K)	kJ/(kg · K)	kJ/(kg · K)	kJ/(kg · K)	kJ/(kg · K)	kJ(kmol · K)
250	1.003	0.791	1.039	14.051	1.039	0.913	33.324
300	1.005	0.846	1.040	14.307	1.039	0.918	33.669
350	1.008	0.895	1.043	14.427	1.041	0.928	34.051
400	1.013	0.939	1.047	14.476	1.044	0.941	34.467
450	1.020	0.978	1.054	14.501	1.049	0.956	34.914
500	1.029	1.014	1.063	14.513	1.056	0.972	35.390
550	1.040	1.046	1.075	14.530	1.065	0.988	35.891
600	1.051	1.075	1.087	14.546	1.075	1.003	36.415
650	1.063	1.102	1.100	14.571	1.086	1.017	36.960
700	1.075	1.126	1.113	14.604	1.098	1.031	37.523
750	1.087	1.148	1.126	14.645	1.110	1.043	38.100
800	1.099	1.169	1.139	14.695	1.121	1.054	38.690
900	1.121	1.204	1.163	14.822	1.145	1.074	39.895
1000	1.142	1.234	1.185	14.983	1.167	1.090	41.118

Adapted from Y. A. Cengel, 1997. *Introduction to Thermodynamics and Heat Transfer*, New York: McGraw-Hill.
[a]Calculated from the third-order polynomial fit given in Appendix A.3C.

IDEAL GAS SPECIFIC HEAT AT CONSTANT PRESSURE AS A FUNCTION OF TEMPERATURE WITH THIRD-ORDER POLYNOMIAL FIT

$$c_P = a + bT + cT^2 + dT^3$$

where T in K, and c_P in kJ/(kmol · K).

Substance	Formula	a	b $(\times 10^{-2})$	c $(\times 10^{-5})$	d $(\times 10^{-9})$	Temperature Range K	% Error Max.	% Error Avg.
Nitrogen	N_2	28.90	−0.1571	0.8081	−2.873	273–1800	0.59	0.34
Oxygen	O_2	25.48	1.520	−0.7155	1.312	273–1800	1.19	0.28
Air		28.11	0.1967	0.4802	−1.966	273–1800	0.72	0.33
Hydrogen	H_2	29.11	−0.1916	0.4003	−0.8704	273–1800	1.01	0.26
Carbon monoxide	CO	28.16	0.1675	0.5372	−2.222	273–1800	0.89	0.37
Carbon dioxide	CO_2	22.26	5.981	−3.501	7.469	273–1800	0.67	0.22
Water vapor	$H_2O(g)$	32.24	0.1923	1.055	−3.595	273–1800	0.53	0.24

Adapted from Y. A. Cengel, 1997. *Introduction to Thermodynamics and Heat Transfer*, New York: McGraw-Hill, p. 845, Table A-2 (c).

SPECIFIC HEAT AT CONSTANT PRESSURE FOR
SATURATED LIQUID WATER $H_2O(\ell)$

Temperature (°C)	Specific Heat, c_p kJ/(kg · K)
0	4.2178
20	4.1818
40	4.1784
60	4.1843
80	4.1964
100	4.2161
120	4.250
140	4.283
160	4.342
180	4.417

Adapted from W. C. Reynolds and H. C. Perkins, 1977.
Engineering Thermodynamics, New York: McGraw-Hill,
p. 661, Table C.3.

IDEAL GAS PROPERTIES FOR
CO, CO_2, H_2, $H_2O(g)$, N_2, O_2:
SPECIFIC HEAT AT CONSTANT PRESSURE, c_p;
SENSIBLE ENTHALPY, $h(T) - h_f(T_{ref} = 298 \text{ K})$;
ENTHALPY OF FORMATION, $h_f(T)$;
ABSOLUTE ENTROPY, $s(T)$; AND
GIBBS FUNCTION OF FORMATION $g_f(T)$

Ideal-gas properties are evaluated at 1 atm. Enthalpy of formation and Gibbs function of formation for compounds are calculated from the elemental substances as

$$h_{f,i}(T) = h_i(T) - \sum_{j \text{ elements}} v'_j h_j(T)$$

$$g_{f,i}(T) = g_i(T) - \sum_{j \text{ elements}} v'_j g_j(T)$$

$$= h_{f,i}(T) - Ts_i(T) - \sum_{j \text{ elements}} v'_j [-Ts_j(T)]$$

The following six tables were originally generated from curve-fit coefficients given in R. J. Kee, F. M. Rupley and J. A. Miller, 1991. "The Chemkin Thermodynamic Data Base," Sandia Report, SAND87-8215B, March, and are adapted from S. R. Turns, 2000. *An Introduction to Combustion, Concepts and Applications — 2.* New York: McGraw-Hill.

Note the properties that are tabulated in Table A.5.1–A.5.6 are

T (K); c_P in (kJ/kmol · K); $[h(T) - h_f(298)]$ in (kJ/kmol); $h_f(T)$ in (kJ/kmol); $s(T)$ in (kJ/kmol · K); and $g_f(T)$ in (kJ/kmol).

Table A.5.1   Carbon Monoxide (CO), $MW = 28.010$, Enthalpy of Formation at 298 λK (kJ/kmol) $= -110,541$

T (K)	c_p (kJ/kmol · K)	$[h(T) - h_f(298)]$ (kJ/kmol)	$h_f(T)$ (kJ/kmol)	$s(T)$ (kJ/kmol · K)	$g_f(T)$ (kJ/kmol)
200	28.687	−2,835	−111,308	186.018	−128,532
298	29.072	0	−110,541	197.548	−137,163
300	29.078	54	−110,530	197.728	−137,328
400	29.433	2,979	−110,121	206.141	−146,332
500	29.857	5,943	−110,017	212.752	−155,403
600	30.407	8,955	−110,156	218.242	−164,470
700	31.089	12,029	−110,477	222.979	−173,499
800	31.860	15,176	−110,924	227.180	−182,473
900	32.629	18,401	−111,450	230.978	−191,386
1000	33.255	21,697	−112,022	234.450	−200,238
1100	33.725	25,046	−112,619	237.642	−209,030
1200	34.148	28,440	−113,240	240.595	−217,768
1300	34.530	31,874	−113,881	243.344	−226,453
1400	34.872	35,345	−114,543	245.915	−235,087
1500	35.178	38,847	−115,225	248.332	−243,674
1600	35.451	42,379	−115,925	250.611	−252,214
1700	35.694	45,937	−116,644	252.768	−260,711
1800	35.910	49,517	−117,380	254.814	−269,164
1900	36.101	53,118	−118,132	256.761	−277,576
2000	36.271	56,737	−118,902	258.617	−285,948
2100	36.421	60,371	−119,687	260.391	−294,281
2200	36.553	64,020	−120,488	262.088	−302,576
2300	36.670	67,682	−121,305	263.715	−310,835
2400	36.774	71,354	−122,137	265.278	−319,057
2500	36.867	75,036	−122,984	266.781	−327,245
2600	36.950	78,727	−123,847	268.229	−335,399
2700	37.025	82,426	−124,724	269.625	−343,519
2800	37.093	86,132	−125,616	270.973	−351,606
2900	37.155	89,844	−126,523	272.275	−359,661
3000	37.213	93,562	−127,446	273.536	−367,684
3100	37.268	97,287	−128,383	274.757	−375,677
3200	37.321	101,016	−129,335	275.941	−383,639
3300	37.372	104,751	−130,303	277.090	−391,571
3400	37.422	108,490	−131,285	278.207	−399,474
3500	37.471	112,235	−132,283	279.292	−407,347
3600	37.521	115,985	−133,295	280.349	−415,192
3700	37.570	119,739	−134,323	281.377	−423,008
3800	37.619	123,499	−135,366	282.380	−430,796
3900	37.667	127,263	−136,424	283.358	−438,557
4000	37.716	131,032	−137,497	284.312	−446,291
4100	37.764	134,806	−138,585	285.244	−453,997
4200	37.810	138,585	−139,687	286.154	−461,677
4300	37.855	142,368	−140,804	287.045	−469,330
4400	37.897	146,156	−141,935	287.915	−476,957
4500	37.936	149,948	−143,079	288.768	−484,558
4600	37.970	153,743	−144,236	289.602	−492,134
4700	37.998	157,541	−145,407	290.419	−499,684

Table A.5.2   Carbon Dioxide (CO$_2$), $MW = 44.011$, Enthalpy of Formation
at 298 λK (kJ/kmol) $= -393,546$

T (K)	c_p (kJ/kmol · K)	$[h(T) - h_f(298)]$ (kJ/kmol)	$h_f(T)$ (kJ/kmol)	$s(T)$ (kJ/kmol · K)	$g_f(T)$ (kJ/kmol)
200	32.387	−3,423	−393,483	199.876	−394,126
298	37.198	0	−393,546	213.736	−394,428
300	37.280	69	−393,547	213.966	−394,433
400	41.276	4,003	−393,617	225.257	−394,718
500	44.569	8,301	−393,712	234.833	−394,983
600	47.313	12,899	−393,844	243.209	−395,226
700	49.617	17,749	−394,013	250.680	−395,443
800	51.550	22,810	−394,213	257.436	−395,635
900	53.136	28,047	−394,433	263.603	−395,799
1000	54.360	33,425	−394,659	269.268	−395,939
1100	55.333	38,911	−394,875	274.495	−396,056
1200	56.205	44,488	−395,083	279.348	−396,155
1300	56.984	50,149	−395,287	283.878	−396,236
1400	57.677	55,882	−395,488	288.127	−396,301
1500	58.292	61,681	−395,691	292.128	−396,352
1600	58.836	67,538	−395,897	295.908	−396,389
1700	59.316	73,446	−396,110	299.489	−396,414
1800	59.738	79,399	−396,332	302.892	−396,425
1900	60.108	85,392	−396,564	306.132	−396,424
2000	60.433	91,420	−396,808	309.223	−396,410
2100	60.717	97,477	−397,065	312.179	−396,384
2200	60.966	103,562	−397,338	315.009	−396,346
2300	61.185	109,670	−397,626	317.724	−396,294
2400	61.378	115,798	−397,931	320.333	−396,230
2500	61.548	121,944	−398,253	322.842	−396,152
2600	61.701	128,107	−398,594	325.259	−396,061
2700	61.839	134,284	−398,952	327.590	−395,957
2800	61.965	140,474	−399,329	329.841	−395,840
2900	62.083	146,677	−399,725	332.018	−395,708
3000	62.194	152,891	−400,140	334.124	−395,562
3100	62.301	159,116	−400,573	336.165	−395,403
3200	62.406	165,351	−401,025	338.145	−395,229
3300	62.510	171,597	−401,495	340.067	−395,041
3400	62.614	177,853	−401,983	341.935	−394,838
3500	62.718	184,120	−402,489	343.751	−394,620
3600	62.825	190,397	−403,013	345.519	−394,388
3700	62.932	196,685	−403,553	347.242	−394,141
3800	63.041	202,983	−404,110	348.922	−393,879
3900	63.151	209,293	−404,684	350.561	−393,602
4000	63.261	215,613	−405,273	353.161	−393,311
4100	63.369	221,945	−405,878	353.725	−393,004
4200	63.474	228,287	−406,499	355.253	−392,683
4300	63.575	234,640	−407,135	356.748	−392,346
4400	63.669	241,002	−407,785	358.210	−391,995
4500	63.753	247,373	−408,451	359.642	−391,629
4600	63.825	253,752	−409,132	361.044	−391,247
4700	63.881	260,138	−409,828	362.417	−390,851

Table A.5.3 Hydrogen (H_2), $MW = 2.016$, Enthalpy of Formation at $298\,\lambda\mathrm{K}$ (kJ/kmol) $= 0$

T (K)	c_p (kJ/kmol · K)	$[h(T) - h_f(298)]$ (kJ/kmol)	$h_f(T)$ (kJ/kmol)	$s(T)$ (kJ/kmol · K)	$g_f(T)$ (kJ/kmol)
200	28.522	−2,818	0	119.137	0
298	28.871	0	0	130.595	0
300	28.877	53	0	130.773	0
400	29.120	2,954	0	139.116	0
500	29.275	5,874	0	145.632	0
600	29.375	8,807	0	150.979	0
700	29.461	11,749	0	155.514	0
800	29.581	14,701	0	159.455	0
900	29.792	17,668	0	162.950	0
1000	30.160	20,664	0	166.106	0
1100	30.625	23,704	0	169.003	0
1200	31.077	26,789	0	171.687	0
1300	31.516	29,919	0	174.192	0
1400	31.943	33,092	0	176.543	0
1500	32.356	36,307	0	178.761	0
1600	32.758	39,562	0	180.862	0
1700	33.146	42,858	0	182.860	0
1800	33.522	46,191	0	184.765	0
1900	33.885	49,562	0	186.587	0
2000	34.236	52,968	0	188.334	0
2100	34.575	56,408	0	190.013	0
2200	34.901	59,882	0	191.629	0
2300	35.216	63,388	0	193.187	0
2400	35.519	66,925	0	194.692	0
2500	35.811	70,492	0	196.148	0
2600	36.091	74,087	0	197.558	0
2700	36.361	77,710	0	198.926	0
2800	36.621	81,359	0	200.253	0
2900	36.871	85,033	0	201.542	0
3000	37.112	88,733	0	202.796	0
3100	37.343	92,455	0	204.017	0
3200	37.566	96,201	0	205.206	0
3300	37.781	99,968	0	206.365	0
3400	37.989	103,757	0	207.496	0
3500	38.190	107,566	0	208.600	0
3600	38.385	111,395	0	209.679	0
3700	38.574	115,243	0	210.733	0
3800	38.759	119,109	0	211.764	0
3900	38.939	122,994	0	212.774	0
4000	39.116	126,897	0	213.762	0
4100	39.291	130,817	0	214.730	0
4200	39.464	134,755	0	215.679	0
4300	39.636	138,710	0	216.609	0
4400	39.808	142,682	0	217.522	0
4500	39.981	146,672	0	218.419	0
4600	40.156	150,679	0	219.300	0
4700	40.334	154,703	0	220.165	0

Table A.5.4   Water (H$_2$O), $MW = 18.016$, Enthalpy of Formation at 298 λK (kJ/kmol) $= -241,845$, Enthalpy of Vaporization (kJ/kmol) $= 44,010$

T (K)	c_p (kJ/kmol · K)	$[h(T) - h_f(298)]$ (kJ/kmol)	$h_f(T)$ (kJ/kmol)	$s(T)$ (kJ/kmol · K)	$g_f(T)$ (kJ/kmol)
200	32.255	−3,227	−240,838	175.602	−232,779
298	33.448	0	−241,845	188.715	−228,608
300	33.468	62	−241,865	188.922	−228,526
400	34.437	3,458	−242,858	198.686	−223,929
500	35.337	6,947	−243,822	206.467	−219,085
600	36.288	10,528	−244,753	212.992	−214,049
700	37.364	14,209	−245,638	218.665	−208,861
800	38.587	18,005	−246,461	223.733	−203,550
900	39.930	21,930	−247,209	228.354	−198,141
1000	41.315	25,993	−247,879	232.633	−192,652
1100	42.638	30,191	−248,475	236.634	−187,100
1200	43.874	34,518	−249,005	240.397	−181,497
1300	45.027	38,963	−249,477	243.955	−175,852
1400	46.102	43,520	−249,895	247.332	−170,172
1500	47.103	48,181	−250,267	250.547	−164,464
1600	48.035	52,939	−250,597	253.617	−158,733
1700	48.901	57,786	−250,890	256.556	−152,983
1800	49.705	62,717	−251,151	259.374	−147,216
1900	50.451	67,725	−251,384	262.081	−141,435
2000	51.143	72,805	−251,594	264.687	−135,643
2100	51.784	77,952	−251,783	267.198	−129,841
2200	52.378	83,160	−251,955	269.621	−124,030
2300	52.927	88,426	−252,113	271.961	−118,211
2400	53.435	93,744	−252,261	274.225	−112,386
2500	53.905	99,112	−252,399	276.416	−106,555
2600	54.340	104,524	−252,532	278.539	−100,719
2700	54.742	109,979	−252,659	280.597	−94,878
2800	55.115	115,472	−252,785	282.595	−89,031
2900	55.459	121,001	−252,909	284.535	−83,181
3000	55.779	126,563	−253,034	286.420	−77,326
3100	56.076	132,156	−253,161	288.254	−71,467
3200	56.353	137,777	−253,290	290.039	−65,604
3300	56.610	143,426	−253,423	291.777	−59,737
3400	56.851	149,099	−253,561	293.471	−53,865
3500	57.076	154,795	−253,704	295.122	−47,990
3600	57.288	160,514	−253,852	296.733	−42,110
3700	57.488	166,252	−254,007	298.305	−36,226
3800	57.676	172,011	−254,169	299.841	−30,338
3900	57.856	177,787	−254,338	301.341	−24,446
4000	58.026	183,582	−254,515	302.808	−18,549
4100	58.190	189,392	−254,699	304.243	−12,648
4200	58.346	195,219	−254,892	305.647	−6,742
4300	58.496	201,061	−255,093	307.022	−831
4400	58.641	206,918	−255,303	308.368	5,085
4500	58.781	212,790	−255,522	309.688	11,005
4600	58.916	218,674	−255,751	310.981	16,930
4700	59.047	224,573	−255,990	312.250	22,861

Table A.5.5   Nitrogen (N$_2$), $MW = 28.013$, Enthalpy of Formation at
298 λK (kJ/kmol) $= 0$

T (K)	c_p (kJ/kmol · K)	$[h(T) - h_f(298)]$ (kJ/kmol)	$h_f(T)$ (kJ/kmol)	$s(T)$ (kJ/kmol · K)	$g_f(T)$ (kJ/kmol)
200	28.793	−2,841	0	179.959	0
298	29.071	0	0	191.511	0
300	29.075	54	0	191.691	0
400	29.319	2,973	0	200.088	0
500	29.636	5,920	0	206.662	0
600	30.086	8,905	0	212.103	0
700	30.684	11,942	0	216.784	0
800	31.394	15,046	0	220.927	0
900	32.131	18,222	0	224.667	0
1000	32.762	21,468	0	228.087	0
1100	33.258	24,770	0	231.233	0
1200	33.707	28,118	0	234.146	0
1300	34.113	31,510	0	236.861	0
1400	34.477	34,939	0	239.402	0
1500	34.805	38,404	0	241.792	0
1600	35.099	41,899	0	244.048	0
1700	35.361	45,423	0	246.184	0
1800	35.595	48,971	0	248.212	0
1900	35.803	52,541	0	250.142	0
2000	35.988	56,130	0	251.983	0
2100	36.152	59,738	0	253.743	0
2200	36.298	63,360	0	255.429	0
2300	36.428	66,997	0	257.045	0
2400	36.543	70,645	0	258.598	0
2500	36.645	74,305	0	260.092	0
2600	36.737	77,974	0	261.531	0
2700	36.820	81,652	0	262.919	0
2800	36.895	85,338	0	264.259	0
2900	36.964	89,031	0	265.555	0
3000	37.028	92,730	0	266.810	0
3100	37.088	96,436	0	268.025	0
3200	37.144	100,148	0	269.203	0
3300	37.198	103,865	0	270.347	0
3400	37.251	107,587	0	271.458	0
3500	37.302	111,315	0	272.539	0
3600	37.352	115,048	0	2733.590	0
3700	37.402	118,786	0	274.614	0
3800	37.452	122,528	0	275.612	0
3900	37.501	126,276	0	276.586	0
4000	37.549	130,028	0	277.536	0
4100	36.597	133,786	0	278.464	0
4200	37.643	137,548	0	279.370	0
4300	37.688	141,314	0	280.257	0
4400	37.730	145,085	0	281.123	0
4500	37.768	148,860	0	281.972	0
4600	37.803	152,639	0	282.802	0
4700	37.832	156,420	0	283.616	0

Table A.5.6   Oxygen (O$_2$), $MW = 31.999$, Enthalpy of Formation at 298 λK (kJ/kmol) $= 0$

T (K)	c_p (kJ/kmol · K)	$[h(T) - h_f(298)]$ (kJ/kmol)	$h_f(T)$ (kJ/kmol)	$s(T)$ (kJ/kmol · K)	$g_f(T)$ (kJ/kmol)
200	28.473	−2,836	0	193.518	0
298	29.315	0	0	205.043	0
300	29.331	54	0	205.224	0
400	30.210	3,031	0	213.782	0
500	31.114	6,097	0	220.620	0
600	32.030	9,254	0	226.374	0
700	32.927	12,503	0	231.379	0
800	33.757	15,838	0	235.831	0
900	34.454	19,250	0	239.849	0
1000	34.936	22,721	0	243.507	0
1100	35.270	26,232	0	246.852	0
1200	35.593	29,775	0	249.935	0
1300	35.903	33,350	0	252.796	0
1400	36.202	36,955	0	255.468	0
1500	36.490	40,590	0	257.976	0
1600	36.768	44,253	0	260.339	0
1700	37.036	47,943	0	262.577	0
1800	37.296	51,660	0	264.701	0
1900	37.546	55,402	0	266.724	0
2000	37.788	59,169	0	268.656	0
2100	38.023	62,959	0	270.506	0
2200	38.250	66,773	0	272.280	0
2300	38.470	70,609	0	273.985	0
2400	38.684	74,467	0	275.627	0
2500	38.891	78,346	0	277.210	0
2600	39.093	82,245	0	278.739	0
2700	39.289	86,164	0	280.218	0
2800	39.480	90,103	0	281.651	0
2900	39.665	94,060	0	283.039	0
3000	39.846	98,036	0	284.387	0
3100	40.023	102,029	0	285.697	0
3200	40.195	106,040	0	286.970	0
3300	40.362	110,068	0	288.209	0
3400	40.526	114,112	0	289.417	0
3500	40.686	118,173	0	290.594	0
3600	40.842	122,249	0	291.742	0
3700	40.994	126,341	0	292.863	0
3800	41.143	130,448	0	293.959	0
3900	41.287	134,570	0	295.029	0
4000	41.429	138,705	0	296.076	0
4100	41.566	142,855	0	297.101	0
4200	41.700	147,019	0	298.104	0
4300	41.830	151,195	0	299.087	0
4400	41.957	155,384	0	300.050	0
4500	42.079	159,586	0	300.994	0
4600	42.197	163,800	0	301.921	0
4700	42.312	168,026	0	302.829	0

Table A.5.7 Curvefit Coefficients for Thermodynamic Properties

$$c_p/R = a_1 + a_2 T + a_3 T^2 + a_4 T^3 + a_5 T^4$$

$$h/RT = a_1 + \frac{a_2}{2}T + \frac{a_3}{3}T^2 + \frac{a_4}{4}T^3 + \frac{a_5}{5}T^4 + \frac{a_6}{T}$$

$$s/R = a_1 \ln T + a_2 T + \frac{a_3}{2}T^2 + \frac{a_4}{3}T^3 + \frac{a_5}{4}T^4 + a_7$$

Species	T(K)	a_1	a_2	a_3	a_4	a_5	a_6	a_7
CO	1,000–5,000	0.03025078E+02	0.14426885E−02	−0.05630827E−05	0.10185813E−09	−0.06910951E−13	−0.14268350E+05	0.0610217E+02
	300–1,000	0.03262451E+02	0.15119409E−02	−0.03881755E−04	0.05581944E−07	0.02474951E−10	−0.14310539E+05	0.04848897E+02
CO₂	1,000–5,000	0.0445623E+02	0.03140168E−01	−0.1278410E−05	0.02393996E−08	−0.1669033E−13	−0.04896696E+06	−0.0955359E+01
	300–1,000	0.02275724E+02	0.09922072E−01	−0.10409113E−04	0.06866686E−07	0.02117280E−10	−0.04837314E+06	0.1018848E+02
H₂	1,000–5,000	0.02991423E+02	0.07000644E−02	−0.05633828E−06	−0.09231578E−10	0.15827519E−14	−0.08350340E+04	−0.13551101E+01
	300–1,000	0.03298124E+02	0.08249441E−02	−0.08143015E−05	−0.09475434E−09	0.04134872E−11	−0.10125209E+04	−0.03294094E+02
H₂O	1,000–5,000	0.02672145E+02	0.03056293E−01	−0.08730260E−05	0.12009964E−09	−0.06391618E−13	−0.02989921E+06	0.06862817E+02
	300–1,000	0.03386842E+02	0.03474982E−01	−0.06354696E−04	0.06968581E−07	0.02506588E−10	−0.03020811E+06	0.02590232E+02
N₂	1,000–5,000	0.02926640E+02	0.14879768E−02	−0.05684760E−05	0.10097038E−09	−0.06753351E−13	−0.09227977E+04	0.05980528E+02
	300–1,000	0.03298677E+02	0.14082404E−02	−0.03963222E−04	0.05641515E−07	0.02444854E−10	−0.10208999E+04	0.03950372E+02
O₂	1,000–5,000	0.03697578E+02	0.06135197E−02	−0.12588420E−06	0.01775281E−09	−0.11364354E−14	−0.12339301E+04	0.03189165E+02
	300–1,000	0.03212936E+02	0.11274864E−02	−0.05756150E−05	0.13138773E−08	−0.08768554E−11	−0.10052490E+04	0.0603473E+02

Source: Kee R. J., F. M. Rupley, and J. A. Miller, 1991. "The Chemkin Thermodynamic Data Base," Sandia Report, SAND87-8215B.

APPENDIX B

THERMOPHYSICAL PROPERTIES OF GASES AT ATMOSPHERIC PRESSURE

T (K)	ρ (kg/m^3)	c_p (kJ/kg · K)	$\mu \cdot 10^7$ (N · s/m^2)	$\nu \cdot 10^6$ (m^2/s)	$k \cdot 10^6$ (W/m · K)	$\alpha \cdot 10^6$ (m^2/s)	Pr
Air							
100	3.5562	1.032	71.1	2.00	9.34	2.54	0.786
150	2.3364	1.012	103.4	4.426	13.8	5.84	0.758
200	1.7458	1.007	132.5	7.590	18.1	10.3	0.737
250	1.3947	1.006	159.6	11.44	22.3	15.9	0.720
300	1.1614	1.007	184.6	15.89	26.3	22.5	0.707
350	0.9950	1.009	208.2	20.92	30.0	29.9	0.700
400	0.8711	1.014	230.1	26.41	33.8	38.3	0.690
450	0.7740	1.021	250.7	32.39	37.3	47.2	0.686
500	0.6964	1.030	270.1	38.79	40.7	56.7	0.684
550	0.6329	1.040	288.4	45.57	43.9	66.7	0.683
600	0.5804	1.051	305.8	52.69	46.9	76.9	0.685
650	0.5356	1.063	322.5	60.21	49.7	87.3	0.690
700	0.4975	1.075	338.8	68.10	52.4	98.0	0.695
750	0.4643	1.087	354.6	76.37	54.9	109	0.702
800	0.4354	1.099	369.8	84.93	57.3	120	0.709
850	0.4097	1.110	384.3	93.80	59.6	131	0.716
900	0.3868	1.121	398.1	102.9	62.0	143	0.720
950	0.3666	1.131	411.3	112.2	64.3	155	0.723
1000	0.3482	1.141	424.4	121.9	66.7	168	0.726
1100	0.3166	1.159	449.0	141.8	71.5	195	0.728
1200	0.2902	1.175	473.0	162.9	76.3	224	0.728
1300	0.2679	1.189	496.0	185.1	82	238	0.719
1400	0.2488	1.207	530	213	91	303	0.703
1500	0.2322	1.230	557	240	100	350	0.685
1600	0.2177	1.248	584	268	106	390	0.688

(*Continued*)

T (K)	ρ (kg/m^3)	c_p (kJ/kg $\cdot$ K)	$\mu \cdot 10^7$ (N $\cdot$ s/m^2)	$\nu \cdot 10^6$ (m^2/s)	$k \cdot 10^6$ (W/m $\cdot$ K)	$\alpha \cdot 10^6$ (m^2/s)	Pr
1700	0.2049	1.267	611	298	113	435	0.685
1800	0.1935	1.286	637	329	120	482	0.683
1900	0.1833	1.307	663	362	128	534	0.677
2000	0.1741	1.337	689	396	137	589	0.672
2100	0.1658	1.372	715	431	147	646	0.667
2200	0.1582	1.417	740	468	160	714	0.655
2300	0.1513	1.478	766	506	175	783	0.647
2400	0.1448	1.558	792	547	196	869	0.630
2500	0.1389	1.665	818	589	222	960	0.613
3000	0.1135	2.726	955	841	486	1570	0.536

Ammonia (NH$_3$)

300	0.6894	2.158	101.5	14.7	24.7	16.6	0.887
320	0.6448	2.170	109	16.9	27.2	19.4	0.870
340	0.6059	2.192	116.5	19.2	29.3	22.1	0.872
360	0.5716	2.221	124	21.7	31.6	24.9	0.872
380	0.5410	2.254	131	24.2	34.0	27.9	0.869
400	0.5136	2.287	138	26.9	37.0	31.5	0.853
420	0.4888	2.322	145	29.7	40.4	35.6	0.833
440	0.4664	2.357	152.5	32.7	43.5	39.6	0.826
460	0.4460	2.393	159	35.7	46.3	43.4	0.822
480	0.4273	2.430	166.5	39.0	49.2	47.4	0.822
500	0.4101	2.467	173	42.2	52.5	51.9	0.813
520	0.3942	2.504	180	45.7	54.5	55.2	0.827
540	0.3795	2.540	186.5	49.1	57.5	59.7	0.824
560	0.3708	2.577	193	52.0	60.6	63.4	0.827
580	0.3533	2.613	199.5	56.5	63.8	69.1	0.817

Carbon Dioxide (CO$_2$)

280	1.9022	0.830	140	7.36	15.20	9.63	0.765
300	1.7730	0.851	149	8.40	16.55	11.0	0.766
320	1.6609	0.872	156	9.39	18.05	12.5	0.754
340	1.5618	0.891	165	10.6	19.70	14.2	0.746
360	1.4743	0.908	173	11.7	21.2	15.8	0.741
380	1.3961	0.926	181	13.0	22.75	17.6	0.737
400	1.3257	0.942	190	14.3	24.3	19.5	0.737
450	1.1782	0.981	210	17.8	28.3	24.5	0.728
500	1.0594	1.02	231	21.8	32.5	30.1	0.725
550	0.9625	1.05	251	26.1	36.6	36.2	0.721
600	0.8826	1.08	270	30.6	40.7	42.7	0.717
650	0.8143	1.10	288	35.4	44.5	49.7	0.712

(Continued)

T (K)	ρ (kg/m^3)	c_p (kJ/kg·K)	$\mu \cdot 10^7$ (N·s/m^2)	$\nu \cdot 10^6$ (m^2/s)	$k \cdot 10^6$ (W/m·K)	$\alpha \cdot 10^6$ (m^2/s)	Pr
700	0.7564	1.13	305	40.3	48.1	56.3	0.717
750	0.7057	1.15	321	45.5	51.7	63.7	0.714
800	0.6614	1.17	337	51.0	55.1	71.2	0.716
Carbon Monoxide (CO)							
200	1.6888	1.045	127	7.52	17.0	9.63	0.781
220	1.5341	1.044	137	8.93	19.0	11.9	0.753
240	1.4055	1.043	147	10.5	20.6	14.1	0.744
260	1.2967	1.043	157	12.1	22.1	16.3	0.741
280	1.2038	1.042	166	13.8	23.6	18.8	0.733
300	1.1233	1.043	175	15.6	25.0	21.3	0.730
320	1.0529	1.043	184	175	26.3	23.9	0.730
340	0.9909	1.044	193	19.5	27.8	26.9	0.725
360	0.9357	1.045	202	21.6	29.1	29.8	0.725
380	0.8864	1.047	210	23.7	30.5	32.9	0.729
400	0.8421	1.049	218	25.9	31.8	36.0	0.719
450	0.7483	1.055	237	31.7	35.0	44.3	0.714
500	0.67352	1.065	254	37.7	38.1	53.1	0.710
550	0.61226	1.076	271	44.3	41.1	62.4	0.710
600	0.56126	1.088	286	51.0	44.0	72.1	0.707
650	0.51806	1.101	301	58.1	47.0	82.4	0.705
700	0.48102	1.114	315	65.5	50.0	93.3	0.702
750	0.44899	1.127	329	73.3	52.8	104	0.702
800	0.42095	1.140	343	81.5	55.5	116	0.705
Hydrogen (H$_2$)							
100	0.24255	11.23	42.1	17.4	67.0	24.6	0.707
150	0.16156	12.60	56.0	34.7	101	49.6	0.699
200	0.12115	13.54	68.1	56.2	131	79.9	0.704
250	0.09693	14.06	78.9	81.4	157	115	0.707
300	0.08078	14.31	89.6	111	183	158	0.701
350	0.06924	14.43	98.8	143	204	204	0.700
400	0.06059	14.48	108.2	179	226	258	0.695
450	0.05386	14.50	117.2	218	247	316	0.689
500	0.04848	14.52	126.4	261	266	378	0.691
550	0.04407	14.53	134.3	305	285	445	0.685
600	0.04040	14.55	142.4	352	305	519	0.678
700	0.03463	14.61	157.8	456	342	676	0 675
800	0.03030	14.70	172.4	569	378	849	0.670
900	0.02694	14.83	186.5	692	412	1030	0.671
1000	0.02424	14.99	201.3	830	448	1230	0.673

(*Continued*)

T (K)	ρ (kg/m^3)	c_p (kJ/kg · K)	$\mu \cdot 10^7$ (N · s/m^2)	$v \cdot 10^6$ (m^2/s)	$k \cdot 10^6$ (W/m · K)	$\alpha \cdot 10^6$ (m^2/s)	Pr
1100	0.02204	15.17	213.0	966	488	1460	0.662
1200	0.02020	15.37	226.2	1120	528	1700	0.659
1300	0.01865	15.59	238.5	1279	568	1955	0.655
1400	0.01732	15.81	250.7	1447	610	2230	0.650
1500	0.01616	16.02	262.7	1626	655	2530	0.643
1600	0.0152	16.28	273.7	1801	697	2815	0.639
1700	0.0143	16.58	284.9	1992	742	3130	0.637
1800	0.0135	16.96	296.1	2193	786	3435	0.639
1900	0.0128	17.49	307.2	2400	835	3730	0.643
2000	0.0121	18.25	318.2	2630	878	3975	0.661
Nitrogen (N$_2$)							
100	3.4388	1.070	68.8	2.00	9.58	2.60	0.768
150	2.2594	1.050	100.6	4.45	13.9	5.86	0.759
200	1.6883	1.043	129.2	7.65	18.3	10.4	0.736
250	1.3488	1.042	154.9	11.48	22.2	15.8	0.727
300	1.1233	1.041	178.2	15.86	25.9	22.1	0.716
350	0.9625	1.042	200.0	20.78	29.3	29.2	0.711
400	0.8425	1.045	220.4	26.16	32.7	37.1	0.704
450	0.7485	1.050	239.6	32.01	35.8	45.6	0.703
500	0.6739	1.056	257.7	38.24	38.9	54.7	0.700
550	0.6124	1.065	274.7	44.86	41.7	63.9	0.702
600	0.5615	1.075	290.8	51.79	44.6	73.9	0.701
700	0.4812	1.098	321.0	66.71	49.9	94.4	0.706
800	0.4211	1.22	349.1	82.90	54.8	116	0.715
900	0.3743	1.146	375.3	100.3	59.7	139	0.721
1000	0.3368	1.167	399.9	118.7	64.7	165	0.721
1100	0.3062	1.187	423.2	138.2	70.0	193	0.718
1200	0.2807	1.204	445.3	158.6	75.8	224	0.707
1300	0.2591	1.219	466.2	179.9	81.0	256	0.701
Oxygen (O$_2$)							
100	3.945	0.962	76.4	1.94	9.25	2.44	0.796
150	2.585	0.921	114.8	4.44	13.8	5.80	0.766
200	1.930	0.915	147.5	7.64	18.3	10.4	0.737
250	1.542	0.915	178.6	11.58	22.6	16.0	0.723
300	1.284	0.920	207.2	16.14	26.8	22.7	0.711
350	1.100	0.929	233.5	21.23	29.6	29.0	0.733
400	0.9620	0.942	258.2	26.84	33.0	36.4	0.737
450	0.8554	0.956	281.4	32.90	36.3	44.4	0.741
500	0.7698	0.972	303.3	39.40	41.2	55.1	0.716
550	0.6998	0.988	324.0	46.30	44.1	63.8	0.726

(Continued)

T (K)	ρ (kg/m^3)	c_p (kJ/kg $\cdot$ K)	$\mu \cdot 10^7$ (N $\cdot$ s/m^2)	$v \cdot 10^6$ (m^2/s)	$k \cdot 10^6$ (W/m $\cdot$ K)	$\alpha \cdot 10^6$ (m^2/s)	Pr
600	0.6414	1.003	343.7	53.59	47.3	73.5	0.729
700	0.5498	1.031	380.8	69.26	52.8	93.1	0.744
800	0.4810	1.054	415.2	86.32	58.9	116	0.743
900	0.4275	1.074	447.2	104.6	64.9	141	0.740
1000	0.3848	1.090	477.0	124.0	71.0	169	0.733
1100	0.3498	1.103	505.5	144.5	75.8	196	0.736
1300	0.2960	1.125	588.4	188.6	87.1	262	0.721
1200	0.3206	1.115	532.5	166.1	81.9	229	0.725
Water Vapor (Steam)							
380	0.5863	2.060	127.1	21.68	24.6	20.4	1.06
400	0.5542	2.014	134.4	24.25	26.1	23.4	1.04
450	0.4902	1.980	152.5	31.11	29.9	30.8	1.01
500	0.4405	1.985	170.4	38.68	33.9	38.8	0.998
550	0.4005	1.997	188.4	47.04	37.9	47.4	0.993
600	0.3652	2.026	206.7	56.60	42.2	57.0	0.993
650	0.3380	2.056	224.7	66.48	46.4	66.8	0.996
700	0.3140	2.085	242.6	77.26	50.5	77.1	1.00
750	0.2931	2.119	260.4	88.84	54.9	88.4	1.00
800	0.2739	2.152	278.6	101.7	59.2	100	1.01
850	0.2579	2.186	296.9	115.1	63.7	113	1.02

THERMOPHYSICAL PROPERTIES OF
SATURATED WATER

Temp., T(K)	Pressure, P (baMs)	Specific Volume (m³/kg) $v_f \cdot 10^3$	v_g	Heat of Vaporization, $h_{fg} \cdot$ (kJ/kg)	Specific Heat (kJ/kg·K) $c_{p,f}$	$c_{p,g}$	Viscosity (N·s/m²) $\mu_f \cdot 10^6$	$\mu_g \cdot 10^6$	Thermal Conductivity (W/m K) $k_f \cdot 10^3$	$k_g \cdot 10^3$	Prandtl Number Pr_f	Pr_g	Surface Tension, $\sigma_f \cdot 10^3$ (N/m)	Expansion Coefficient, $\beta_f \cdot 10^6$ (K⁻¹)	Temp., T(K)
273.15	0.00611	1.000	206.3	2502	4.217	1.854	1750	8.02	569	18.2	12.99	0.815	75.5	-68.05	273.15
275	0.00697	1.000	181.7	2497	4.211	1.855	1652	8.09	574	18.3	12.22	0.817	75.3	-32.74	275
280	0.00990	1.000	130.4	2485	4.198	1.858	1422	8.29	582	18.6	10.26	0.825	74.8	46.04	280
285	0.01387	1.000	99.4	2473	4.189	1.861	1225	8.49	590	18.9	8.81	0.833	74.3	114.1	285
290	0.01917	1.001	69.7	2461	4.184	1.864	1080	8.69	598	19.3	7.56	0.841	73.7	174.0	290
295	0.02617	1.002	51.94	2449	4.181	1.868	959	8.89	606	19.5	6.62	0.849	72.7	227.5	295
300	0.03531	1.003	39.13	2438	4.179	1.872	855	9.09	613	19.6	5.83	0.857	71.7	276.1	300
305	0.04712	1.005	29.74	2426	4.178	1.877	769	9.29	620	20.1	5.20	0.865	70.9	320.6	305
310	0.06221	1.007	22.93	2414	4.178	1.882	695	9.49	628	20.4	4.62	0.873	70.0	361.9	310
315	0.08132	1.009	17.82	2402	4.179	1.888	631	9.69	634	20.7	4.16	0.883	69.2	400.4	315
320	0.1053	1.011	13.98	2390	4.180	1.895	577	9.89	640	21.0	3.77	0.894	68.3	436.7	320
325	0.1351	1.013	11.06	2378	4.182	1.903	528	10.09	645	21.3	3.42	0.901	67.5	471.2	325
330	0.1719	1.016	8.82	2366	4.184	1.911	489	10.29	650	21.7	3.15	0.908	66.6	504.0	330
335	0.2167	1.018	7.09	2354	4.186	1.920	453	10.49	656	22.0	2.88	0.916	65.8	535.5	335
340	0.2713	1.021	5.74	2342	4.188	1.930	420	10.69	660	22.3	2.66	0.925	64.9	566.0	340
345	0.3372	1.024	4.683	2329	4.191	1.941	389	10.89	668	22.6	2.45	0.933	64.1	595.4	345
350	0.4163	1.027	3.846	2317	4.195	1.954	365	11.09	668	23.0	2.29	0.942	63.2	624.2	350
355	0.5100	1.030	3.180	2304	4.199	1.968	343	11.29	671	23.3	2.14	0.951	62.3	652.3	355
360	0.6209	1.034	2.645	2291	4.203	1.983	324	11.49	674	23.7	2.02	0.960	61.4	697.9	360
365	0.7514	1.038	2.212	2278	4.209	1.999	306	11.69	677	24.1	1.91	0.969	60.5	707.1	365
370	0.9040	1.041	1.861	2265	4.214	2.017	289	11.89	679	24.5	1.80	0.978	59.5	728.7	370
373.15	1.0133	1.044	1.679	2257	4.217	2.029	279	12.02	680	24.8	1.76	0.984	58.9	750.1	373.15
375	1.0815	1.045	1.574	2252	4.220	2.036	274	12.09	681	24.9	1.70	0.987	58.6	761	375
380	1.2869	1.049	1.337	2239	4.226	2.057	260	12.29	683	25.4	1.61	0.999	57.6	788	380
385	1.5233	1.053	1.142	2225	4.232	2.080	248	12.49	685	25.8	1.53	1.004	56.6	814	385
390	1.794	1.058	0.980	2212	4.239	2.104	237	12.69	686	26.3	1.47	1.013	55.6	841	390
400	2.455	1.067	0.731	2183	4.256	2.158	217	13.05	688	27.2	1.34	1.033	53.6	896	400
410	3.302	1.077	0.553	2153	4.278	2.221	200	13.42	688	28.2	1.24	1.054	51.5	952	410
420	4.370	1.088	0.425	2123	4.302	2.291	185	13.79	688	29.8	1.16	1.075	49.4	1010	420
430	5.699	1.099	0.331	2091	4.331	2.369	173	14.14	685	30.4	1.09	1.10	47.2	—	430

440	7.333	1.110	0.261	2059	4.36	2.46	162	14.50	682	31.7	1.04	1.12	45.1	—	440
450	9.319	1.123	0.208	2024	4.40	2.56	152	14.85	678	33.1	0.99	1.14	42.9	—	450
460	11.71	1.137	0.167	1989	4.44	2.68	143	15.19	673	34.6	0.95	1.17	40.7	—	460
470	14.55	1.152	0.136	1951	4.48	2.79	136	15.54	667	36.3	0.92	1.20	38.5	—	470
480	17.90	1.167	0.111	1912	4.53	2.94	129	15.88	660	38.1	0.89	1.23	36.2	—	480
490	21.83	1.184	0.0922	1870	4.59	3.10	124	16.23	651	40.1	0.87	1.25	33.9	—	490
500	26.40	1.203	0.0766	1825	4.66	3.27	118	16.59	642	42.3	0.86	1.28	31.6	—	500
510	31.66	1.222	0.0631	1779	4.74	3.47	113	16.95	631	44.7	0.85	1.31	29.3	—	510
520	37.70	1.244	0.0525	1730	4.84	3.70	108	17.33	621	47.5	0.84	1.35	26.9	—	520
530	44.58	1.268	0.0445	1679	4.95	3.96	104	17.72	608	50.6	0.85	1.39	24.5	—	530
540	52.38	1.294	0.0375	1622	5.08	4.27	101	18.1	594	54.0	0.86	1.43	22.1	—	540
550	61.19	1.323	0.0317	1564	5.24	4.64	97	18.6	580	58.3	0.87	1.47	19.7	—	550
560	71.08	1.355	0.0269	1499	5.43	5.09	94	19.1	563	63.7	0.90	1.52	17.3	—	560
570	82.16	1.392	0.0228	1429	5.68	5.67	91	19.7	548	76.7	0.94	1.59	15.0	—	570
580	94.51	1.433	0.0193	1353	6.00	6.40	88	20.4	528	76.7	0.99	1.68	12.8	—	580
590	108.3	1.482	0.0163	1274	6.41	7.35	84	21.5	513	84.1	1.05	1.84	10.5	—	590
600	123.5	1.541	0.0137	1176	7.00	8.75	81	22.7	497	92.9	1.14	2.15	8.4	—	600
610	137.3	1.612	0.0115	1068	7.85	11.1	77	24.1	467	103	1.30	2.60	6.3	—	610
620	159.1	1.705	0.0094	941	9.35	15.4	72	25.9	444	114	1.52	3.46	4.5	—	620
625	169.1	1.778	0.0085	858	10.6	18.3	70	27.0	430	121	1.65	4.20	3.5	—	625
630	179.7	1.856	0.0075	781	12.6	22.1	67	28.0	412	130	2.0	4.8	2.6	—	630
635	190.9	1.935	0.0066	683	16.4	27.6	64	30.0	392	141	2.7	6.0	1.5	—	635
640	202.7	2.075	0.0057	560	26	42	59	32.0	367	155	4.2	9.6	0.8	—	640
645	215.2	2.351	0.0045	361	90	—	54	37.0	331	178	12	26	0.1	—	645
647.3[a]	221.2	3.170	0.0032	0	∞	∞	45	45.0	238	238	∞	∞	0.0	—	647.3[c]

[a]Critical temperature

BINARY DIFFUSION COEFFICIENTS AT ONE ATMOSPHERE[a]

Substance A	Substance B	$T(K)$	$D_{AB}(m^2/s)$
Gases			
NH_3	Air	298	0.28×10^{-4}
H_2O	Air	298	0.26×10^{-4}
CO_2	Air	298	0.16×10^{-4}
H_2	Air	298	0.41×10^{-4}
O_2	Air	298	0.21×10^{-4}
Acetone	Air	273	0.11×10^{-4}
Benzene	Air	298	0.88×10^{-5}
Naphthalene	Air	300	0.62×10^{-5}
Ar	N_2	293	0.19×10^{-4}
H_2	O_2	273	0.70×10^{-4}
H_2	N_2	273	0.68×10^{-4}
H_2	CO_2	273	0.55×10^{-4}
CO_2	N_2	293	0.16×10^{-4}
CO_2	O_2	273	0.14×10^{-4}
O_2	N_2	273	0.18×10^{-4}
Dilute Solutions			
Caffeine	H_2O	298	0.63×10^{-9}
Ethanol	H_2O	298	0.12×10^{-8}
Glucose	H_2O	298	0.69×10^{-9}
Glycerol	H_2O	298	0.94×10^{-9}
Acetone	H_2O	298	0.13×10^{-8}
CO_2	H_2O	298	0.20×10^{-8}
O_2	H_2O	298	0.24×10^{-8}
H_2	H_2O	298	0.63×10^{-8}
N_2	H_2O	298	0.26×10^{-8}

[a]Assuming ideal-gas behavior, the pressure and temperature dependence of the diffusion coefficient for a binary mixture of gases may be estimated from the relation

$$D_{AB} \propto p^{-1} T^{3/2}$$

Index